# Cancer Prevention with Molecular Target Therapies 3.0

# Cancer Prevention with Molecular Target Therapies 3.0

Editor

**Laura Paleari**

MDPI • Basel • Beijing • Wuhan • Barcelona • Belgrade • Manchester • Tokyo • Cluj • Tianjin

*Editor*
Laura Paleari
Research, Innovation and HTA Unit
Liguria Health Authority A.Li.Sa.
Genova
Italy

*Editorial Office*
MDPI
St. Alban-Anlage 66
4052 Basel, Switzerland

This is a reprint of articles from the Special Issue published online in the open access journal *International Journal of Molecular Sciences* (ISSN 1422-0067) (available at: www.mdpi.com/journal/ijms/special_issues/Cancer_Prevention_3).

For citation purposes, cite each article independently as indicated on the article page online and as indicated below:

LastName, A.A.; LastName, B.B.; LastName, C.C. Article Title. *Journal Name* **Year**, *Volume Number*, Page Range.

**ISBN 978-3-0365-7909-2 (Hbk)**
**ISBN 978-3-0365-7908-5 (PDF)**

# Contents

# About the Editor

**Laura Paleari**

I have more than twenty year experience in translational, pre-clinical and clinical cancer research and scientific/medical writing with robust skills in molecular/cellular biology, in vivo testing (including the generation ok knockout, conditional knockout and pdx murine models).

I am a member of the regional HTA Network and a consultant for the National Health Agency (Agenas). I have gained skills in pharmaco-economics and in the management of health systems.

I have served as a member of peer review panels for the major international and national research funding calls: CCM (National Center for Prevention and Disease Control); translational co-fund call of ERACoSysMed, ERANET on System Medicine under Horizon2020; MINDED programme (7th Framework programme, Marie Curie Actions). I am the Author of 70+ scientific papers, being the first/last author/corresponding of the majority.

# Preface to "Cancer Prevention with Molecular Target Therapies 3.0"

Today, the oncologist is like a detective of the human body who, instead of a magnifying glass, uses the new tools of molecular pathology to search not only for genes or molecular targets, to be targeted with innovative anticancer therapies, but also molecular alterations that allow the identification of population groups at risk of developing tumors for preventive purposes. This is the precision oncology thanks to which it is possible today to aim at treatments but also at personalized cancer prevention, based on precision medicine models, through the identification of specific genomic determinants linked to an increased risk of developing cancer. This area includes a series of interventions to identify cancer at an early stage or to avoid the onset of the disease. Molecular pathology is a cornerstone of precision oncology, and today, it is necessary to study not only the single alterations but also the overall modifications of the cellular signal transduction pathways. In this way, the molecular pathologist can provide the clinician with crucial information to drive the therapeutic choice. The so-called histological model, which has long-governed clinical research in oncology and clinical practice, is now flanked by the molecular model. In this approach, the starting point is represented by the organ from which the neoplasia originates, followed by the histological examination, the identification of any molecular alterations and the choice of the drug, through a path of selecting patients who are more likely to respond to treatment. Targeted anticancer therapies are currently at the heart of the development of many anticancer drugs including hormone therapies, signal transduction inhibitors, gene expression modulators, apoptosis inducers, angiogenesis inhibitors, immunotherapies and toxin-releasing compounds. It is important to note that targeted cancer therapies have some limits concerning the possibility of acquired resistance to the treatment. Additionally, it is possible to use molecular targeted therapy in combination with one or more traditional chemotherapy drug. Another limitation of targeted therapy is that drugs for some identified targets are difficult to develop due to the structure of the target and/or the way its function is regulated by the cell. The purpose of this Special Issue was to highlight the importance of research aiming at discovering new and more informative methods of diagnostic molecular testing, targeted therapies mainly in early-stage disease and future directions for precision oncology approaches to understand the tumor evolution and the possible related therapeutic resistance.

**Laura Paleari**
*Editor*

International Journal of *Molecular Sciences*

MDPI

*Article*

# A DSC Test for the Early Detection of Neoplastic Gastric Lesions in a Medium-Risk Gastric Cancer Area

**Valli De Re [1,*], Stefano Realdon [2], Roberto Vettori [1], Alice Zaramella [3,4], Stefania Maiero [2], Ombretta Repetto [1], Vincenzo Canzonieri [5,6], Agostino Steffan [1] and Renato Cannizzaro [2,6]**

1 Immunopathology and Cancer Biomarkers, Centro di Riferimento Oncologico di Aviano, CRO Aviano, National Cancer Institute, IRCCS, 33081 Aviano, Italy
2 Oncological Gastroenterology, Centro di Riferimento Oncologico di Aviano (CRO) IRCCS, 33081 Aviano, Italy
3 Gastroenterology Unit, Veneto Institute of Oncology IOV-IRCCS, Via Gattamelata 64, 35128 Padua, Italy
4 Department of Surgery, Oncology, and Gastroenterology (DiSCOG), University of Padua, Via Giustiniani 2, 35128 Padua, Italy
5 Pathology Unit, Centro di Riferimento Oncologico di Aviano (CRO) IRCCS, 33081 Aviano, Italy
6 Department of Medical, Surgical and Health Sciences, University of Trieste, 34127 Trieste, Italy
* Correspondence: vdere@cro.it; Tel.: +39-0434-659672

**Abstract:** In this study, we aimed to assess the accuracy of the proposed novel, noninvasive serum DSC test in predicting the risk of gastric cancer before the use of upper endoscopy. To validate the DSC test, we enrolled two series of individuals living in Veneto and Friuli-Venezia Giulia, Italy (n = 53 and n = 113, respectively), who were referred for an endoscopy. The classification used for the DSC test to predict gastric cancer risk combines the coefficient of the patient's age and sex and serum pepsinogen I and II, gastrin 17, and anti-*Helicobacter pylori* immunoglobulin G concentrations in two equations: Y1 and Y2. The coefficient of variables and the Y1 and Y2 cutoff points (>0.385 and >0.294, respectively) were extrapolated using regression analysis and an ROC curve analysis of two retrospective datasets (300 cases for the Y1 equation and 200 cases for the Y2 equation). The first dataset included individuals with autoimmune atrophic gastritis and first-degree relatives with gastric cancer; the second dataset included blood donors. Demographic data were collected; serum pepsinogen, gastrin G17, and anti-*Helicobacter pylori* IgG concentrations were assayed using an automatic Maglumi system. Gastroscopies were performed by gastroenterologists using an Olympus video endoscope with detailed photographic documentation during examinations. Biopsies were taken at five standardized mucosa sites and were assessed by a pathologist for diagnosis. The accuracy of the DSC test in predicting neoplastic gastric lesions was estimated to be 74.657% (65% CI; 67.333% to 81.079%). The DSC test was found to be a useful, noninvasive, and simple approach to predicting gastric cancer risk in a population with a medium risk of developing gastric cancer.

**Keywords:** gastric cancer; pepsinogen; gastrin G17; *Helicobacter pylori*; screening

**Citation:** De Re, V.; Realdon, S.; Vettori, R.; Zaramella, A.; Maiero, S.; Repetto, O.; Canzonieri, V.; Steffan, A.; Cannizzaro, R. A DSC Test for the Early Detection of Neoplastic Gastric Lesions in a Medium-Risk Gastric Cancer Area. *Int. J. Mol. Sci.* **2023**, *24*, 3290. https://doi.org/10.3390/ijms24043290

Academic Editor: Monica Currò

Received: 27 September 2022
Revised: 30 January 2023
Accepted: 3 February 2023
Published: 7 February 2023



## 1. Introduction

Gastroscopy is the standard procedure for gastric cancer (GC) diagnosis with a false-negative rate of about 19% [1]. This procedure is invasive, time-consuming, and uncomfortable.

The 5-year GC survival rate is poor, reaching approximately 30% in Europe [2]. Studies in countries such as Japan and Korea have shown a high incidence of GC, with a 30% reduction in GC mortality due to gastroscopy screening programs [3].

Overall, a GC diagnosis at an early/asymptomatic stage not only improves clinical outcomes [4] but is associated with endoscopic resections in most cases, resulting in the now-standard treatment for early gastrointestinal cancers without regional lymph node metastasis [5].

A population with an age-standardized incidence rate (ASIR) of 10 to 20 per 100,000 is considered to have an intermediate risk of developing GC. According to published data, in northern Italy, the ASIR was calculated as 12 to <14 in 2017, on average twofold higher in males (33.9) than in females (17.0) [6,7]; thus, Italy is a geographic area whose population has an intermediate risk of developing GC.

At present, the screening for GC is only performed in countries with an elevated risk of GC, such as Japan and Korea [8]. In a country with an intermediate risk of GC, using gastroscopy as first-line testing alone is not considered feasible due to its invasiveness and expensive cost [9]. On the other hand, the inappropriateness of upper endoscopies leads to decreased diagnostic yield [10,11]. Thus, it is necessary to consider a less invasive and more cost-effective solution to find subjects at risk of developing GC. Moreover, it is necessary to take changes in GC epidemiology that have occurred over the last few decades into consideration. GC epidemiology has changed concomitantly with a reduction in *Helicobacter pylori* (*H. pylori*) infections, but also with an increased incidence of cardia GC and an increase in GC diagnoses and mortality in younger adults [12].

The available pepsinogen (PG) test (PG test) is based on a combination of the serum PG-I level and the PG-I/PG-II ratio, which is used as a marker for chronic atrophic gastritis (CAG), and it has been widely proposed for the selection of patients at risk for GC [13,14].

However, while the PG test is accurate for CAG diagnosis, it suffers from an unsatisfactory specificity in predicting GC due to using the widely standardized cutoff points of PGI $< 70$ ng/mL and PGI/PGII $< 3.0$ [15–17]. Therefore, the ability to predict GC risk needs to be improved.

Recently, the addition of serum gastrin G17 (G17) and anti-*H. Pylori* immunoglobulin (IgG) to the PG test was proposed to improve the diagnosis of atrophic gastritis and GC [18]. PGI is only secreted by the fundic glands; PGII is secreted by the fundic glands, the pylorus, and the Brunner glands; and gastrin G-17 is only secreted by gastric antral G cells. Accordingly, serum PG and G17 levels can be used to localize morphology and detect the extension of gastric lesions [19]. Gastrin G17 might also sustain the proliferation and migration of epithelial gastric cells [20]. *H. pylori* is pathogenetic for GC and a subtype of autoimmune gastritis [21]. Age $\geq 60$ and male sex have been reported to be independent risk factors for GC in several studies [22,23].

Based on our earlier study [24] and two new datasets in this work, first (the discovery part), we developed a model—herein called the DeRe–Steffan–Cannizzaro (DSC) test—to discriminate patients at risk of developing GC. Briefly, the model is based on regression analyses to calculate the coefficient of the patient's age and sex and their PG, anti-*H. Pylori* immunoglobulin (IgG), and serum gastrin G17 levels.

Second (the validation part), we validated the DSC test. Two monocenter studies were conducted involving 53 retrospectively selected (based on their diagnosis) individuals who presented at the Gastroenterology Unit of the Veneto Institute of Oncology IOV-IRCCS, Padua, Italy (validation cohort 1), and 113 consecutive individuals who were referred by physicians to the oncological gastroenterology unit of the Centro di Riferimento Oncologico of Aviano, Italy (validation cohort 2). Physician indications included dyspepsia, preneoplastic lesions, and/or suspected GC. Dyspepsia was defined as upper gastrointestinal symptoms, such as gastric pain and/or burning, without a typical disease and with no clear cause [10,11]. The preneoplastic lesions were diagnosed via gastroscopy and histologically examining the biopsies [25–29]. The GC diagnoses were confirmed via histopathological examinations [30,31]. Data on each patient's age and sex were collected, and expression profiles of serum PG, gastrin G17, and *H. Pylori* IgG were performed in order to group patients based on GC risk classes into negative, neutral, and positive results, respectively. The accuracy of the model was then demonstrated with gastroendoscopy and a histological diagnosis.

This is the first study on GC risk prediction adopting the DSC model. The model showed an overall 74.66% accuracy rate for GC diagnosis, notably improving the detection sensitivity from 15.00% to 70.00% and retaining a good specificity (74.66%) compared with the standard cutoff for PG tests. This study could help physicians make better decisions and allocate proper resources to improve GC prevention, for example, by removing high-grade preneoplastic lesions and selecting patients for endoscopic surveillance. In addition, by diagnosing opportunistic GC at an early stage, the GC survival rate can be improved [2,3].

## 2. Results

### 2.1. Study Design

Based on our previous observations [24] and the two new cohorts in this work (Figure 1, discovery cohort 1 and discovery cohort 2), we developed a model based on the coefficients of the patient's age, sex, and serum PG, *H. Pylori* IgG, and gastrin G17 levels to discriminate patients at an elevated risk of GC. The results of the Y1 and Y2 equations obtained from discovery cohort 1 and discovery cohort 2 are reported in Figure 1. Data from all the study's participants are reported in Table 1. The median G17 level in the female category in discovery cohort 1 was higher than in the other groups due to the prevalence of females affected by autoimmune atrophic gastritis [24]. In discovery and validation cohorts, *H. pylori* IgG and PG II levels were higher in individuals >65 years old, in accordance with the decreased incidence of *H. pylori* infection in the global population [12] and increases in serum PG II levels often found in individuals infected with *H. pylori* [24].

Next, we recruited two validation cohorts, the first consisting of a retrospective series of selected participants from the Veneto region (validation cohort 1, n = 53) and the second consisting of a prospective series of consecutive participants who were referred to gastroenterologists for gastroscopy from the Friuli geographic area (validation cohort 2, n = 113). The study participants' data are reported in Table 1.

### 2.2. DSC Classifications for GC Risk

Two validation cohorts were enrolled in the study. The first cohort consisted of a retrospective series of 53 participants (validation cohort 1); the second cohort consisted of a prospective series of 113 participants consecutively enrolled since May 2020 (validation cohort 2).

The DSC test measures the serum biomarkers and demographic data of the participants. The patients' data are reported in Table 1. Based on the results of both the DSC Y1 and Y2 equations, Y1 < 0.385 and Y2 < 0.294 (as reported in Figure 1 and described in the methods section), we classified participants receiving the serological test as having negative, neutral, or positive DSC results.

The results of the DSC test are reported in Table 2 according to the calculated Y1 and Y2 equations for each participant. Based on these criteria, 19 (35.8%) and 81 (71.7%) participants were classified as negative; 5 (9.4%) and 10 (8.8%) were classified as neutral; and 29 (54.7%) and 22 (19.5%) were classified as positive for GC risk in validation cohort 1 and validation cohort 2, respectively.

### 2.3. Gastroscopy and Histopathological Diagnosis

The same participants were classified after using gastroscopy and histological examinations to diagnose them with GC, dysplasia, severe atrophy (OLGA stages III–IV), or no/moderate-grade atrophy (OLGA stages 0–II). The diagnosis results are summarized in Table 3.

In the first cohort, consisting of a retrospective series of participants (validation cohort 1), the diagnosis was GC in nine patients, dysplasia in five, and severe atrophy in 14. The remaining participants (n = 25) showed no atrophy or mild–moderate atrophy (OLGA stages 0–II).

Of the 113 consecutive participants in the second cohort (validation cohort 2), a neoplastic lesion was diagnosed in two patients (i.e., one advanced, differentiated, diffuse-type GC with middle tumor infiltration; one early antrum GC), dysplasia was diagnosed in three patients (one adenoma with high-grade dysplasia in the corpus; one low-grade dysplasia with intestinal metaplasia and neuroendocrine hyperplasia in the antrum; one dysplasia with intestinal metaplasia in the corpus), a preneoplastic adenoma was diagnosed in one case, and severe atrophy was diagnosed in 15 cases; the remaining participants (n = 92) showed no atrophy or mild–moderate atrophy (OLGA stages 0–II).

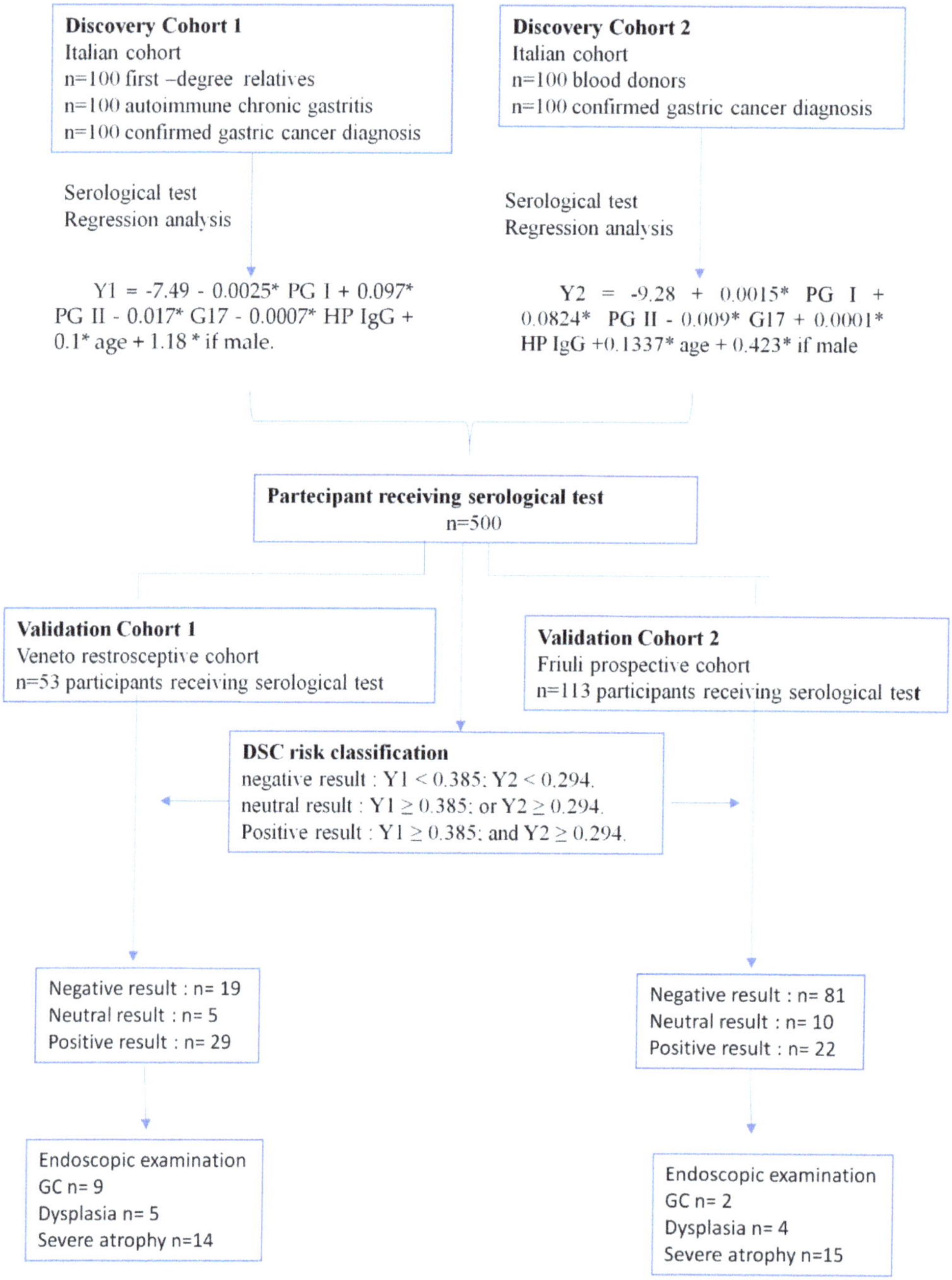


**Figure 1.** Schematic of the study design and main results.

**Table 1.** Distribution of demographic and serological data in the discovery and validation cohorts according to sex and age categories.

| | | Discovery Cohort 1 | | Discovery Cohort 2 | | Validation Cohort 1 | | Validation Cohort 2 n = 113 | |
|---|---|---|---|---|---|---|---|---|---|
| | | Median (IQ: 25–75%) | Median (IQ: 25–75%) | Median (IQ: 25–75%) | Median (IQ: 25–75%) | Median (IQ: 25–75%) | Median (IQ: 25–75%) | Median (IQ: 25–75%) | Median (IQ: 25–75%) |
| **Biomarkers** | **Sex** | **Age ≤ 65** | **Age > 65** | **Age ≤ 65** | **Age > 65** | **Age ≤ 65** | **Age > 65** | **Age ≤ 65** | **Age > 65** |
| G17 | F | 13.9 (3.5–50.3) | 19.5 (3.4–61.8) | 4.0 (1.9–16.3) | 17.6 (2.7–27.9) | 7.3 (3.9–58.6) | 5.3 (3.4–12.4) | 5.1 (2.2–52.2) | 11.1 (2.7–53.1) |
| | M | 9.5 (3.3–32.9) | 15.2 (5.9–38.2) | 2.9 (0.5–8.4) | 17.1 (7.1–38.3) | 8.0 (4.8–36.1) | 16.3 (7.9–32.7) | 5.0 (2.7–13.9) | 31.8 (10.5–119.0) |
| PGI | F | 72 (25.8–119.3) | 87.5 (23.7–141.4) | 63.6 (49.4–173.0) | 127.7 (87.5–210.3) | 78.6 (74.5–139.5) | 102.9 (56.0–251.2) | 60.3 (39.5–105.0) | 78.7 (30.8–144.4) |
| | M | 103.2 (70.4–182.9) | 76.4 (38.0–150.8) | 82.3 (50.5–131.2) | 83.7 (38.0–212.8) | 119.5 (101.4–198.1) | 137.2 (53.2–304.5) | 69.7 (35.0–108.7) | 73.5 (16.9–263.7) |
| PGII | F | 10.1 (7.1–16.9) | 14.0 (8.0–18.1) | 9.0 (4.8–19.8) | 19.3 (16.0–27.1) | 9.9 (7.1–11.2) | 13.0 (9.2–20.5) | 10.9 (6.2–13.7) | 12.7 (10.0–14.8) |
| | M | 13.9 (8.7–21.7) | 12.6 (7.5–16.8) | 9.0 (6.2–17.0) | 13.0 (9.2–20.4) | 11.5 (9.6–14.5) | 12.9 (9.7–19.7) | 9.6 (7.8–12.2) | 11.7 (7.9–19.6) |
| PGI/PGII | F | 6.7 (2.3–11.6) | 6.6 (3.0–9.9) | 8.3 (6.4–10.7) | 6.0 (4.9–7.8) | 10.1 (7.9–11.9) | 9.2 (7.1–10.5) | 7.2 (4.3–9.9) | 6.8 (3.0–8.4) |
| | M | 8.43 (5.5–12.2) | 6.3 (3.0–12.0) | 9.2 (6.3–11.6) | 6.3 (3.0–12.0) | 10.7 (10.3–13.4) | 9.1 (7.4–12.5) | 7.6 (3.4–8.7) | 6.7 (1.8–11.1) |
| H. pylori IgG | F | 14.4 (5.8–72.9) | 38.1 (11.5–60.8) | 27.0 (4.4–76.4) | 47.0 (16.8–71.9) | 5.4 (3.6–21.6) | 14.5 (3.4–41.2) | 5.9 (2.6–16.8) | 10.8 (3.7–32.1) |
| | M | 20.6 (6.0–85.6) | 61.3 (11.5–111.2) | 15.0 (3.85–80.38) | 79.4 (21.7–114.1) | 3.8 (2.6–12.1) | 3.1 (2.6–10.6) | 6.4 (3.7–9.2) | 15.6 (6.3–35.4) |

**Table 2.** DSC GC risk classifications for participants in the validation cohorts.

| DSC | Cohorts | | | | | |
|---|---|---|---|---|---|---|
| GC Risk Classification | Validation 1 | | Validation 2 | | Overall Validation Process | |
| | n. | % | n. | % | n. | % |
| Negative | 19 | 35.8 | 81 | 71.7 | 100 | 60.2 |
| Neutral | 5 | 9.4 | 10 | 8.8 | 15 | 9.0 |
| Positive | 29 | 54.7 | 22 | 19.5 | 51 | 30.7 |
| Total | 53 | | 113 | | 166 | |

**Table 3.** Aggregation of participants according to histopathological diagnosis.

| Diagnosis | Cohorts | | | |
|---|---|---|---|---|
| | Validation Cohort 1 N = 53 | | Validation Cohort 2 N = 113 | |
| | n. Patients | % | n. Patients | % |
| Atrophy (OLGA stages 0–II) | 25 | 47.2 | 92 | 81.4 |
| Severe atrophy (OLGA stages III–IV) | 14 | 26.4 | 15 | 13.3 |
| Dysplasia/preneoplastic lesion | 5 | 9.4 | 4 | 3.5 |
| Gastric cancer | 9 | 17.0 | 2 | 1.8 |

In the patient with an early GC and the four patients with dysplasia, mini-invasive gastric resections were conducted via endoscopy.

### 2.4. DSC Classification Accuracy

A comparison between the DSC classification results and a histological classification was used to calculate the accuracy of the DSC test. A subgroup analysis was performed separately for validation cohort 1 and validation cohort 2, and then for all cases.

In both validation cohorts, we observed a substantially larger risk of GC in individuals classified as DSC-positive (validation cohort 1: 7/9 (77.78%); validation cohort 2: 2/2 (100%)) than those classified as DSC-neutral (validation cohort 1: 1/5 (20.0%); validation cohort 2: 0/10) or DSC-negative (validation cohort 1: 1/19 (5.3%); validation cohort 2: 0/81). The relative risk of GC diagnosis in individuals classified with a positive DSC score was RR 2.90 (95% CI; 0.67–12.67) and RR 20 (95% CI; 0.99–402.45) in validation cohort 1 and validation cohort 2, respectively.

Figure 2 shows the distribution of DSC-positive, -neutral, and -negative results in the validation datasets according to the individual diagnosis of each participant.

Based on the DSC test, 60.2% of the overall participants in the validation cohorts were classified as negative, 9.0% as neutral, and 30.7% as positive for GC risk (Table 2).

The predictive value of the DSC test for the risk of GC was calculated using a diagnostic test; the AUC value was 0.723 considering all participants in the validation datasets (0.614 in validation cohort 1 and 0.749 in validation cohort 2). The overall accuracy was 74.657%, with a calculated disease prevalence of about 0.01% in the general Italian population [6] (see details in Table 4).

### 2.5. Reproducibility of the DSC Method

In 26 participants (17 negative, three neutral, and six positives in the DSC test), the test was repeated after a median interval of 15.5 months (IQR, 10 to 19 months) (Table 5). We found a change in DSC classifications from negative to neutral in two participants with a diagnosis of mild–moderate atrophy (OLGA 0-II category). In the remaining 24 cases, both the classification and the diagnosis remained the same (Table 5).

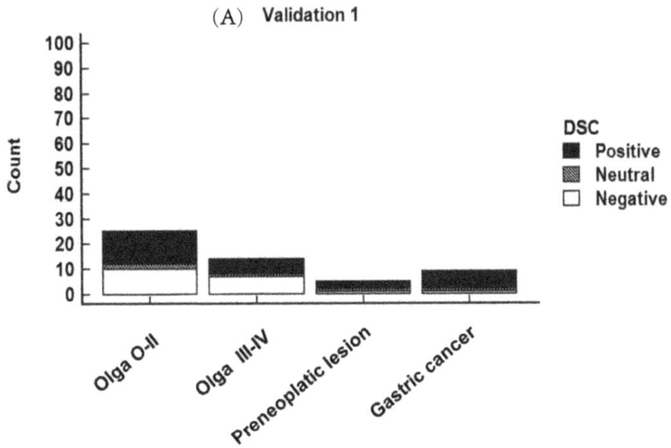


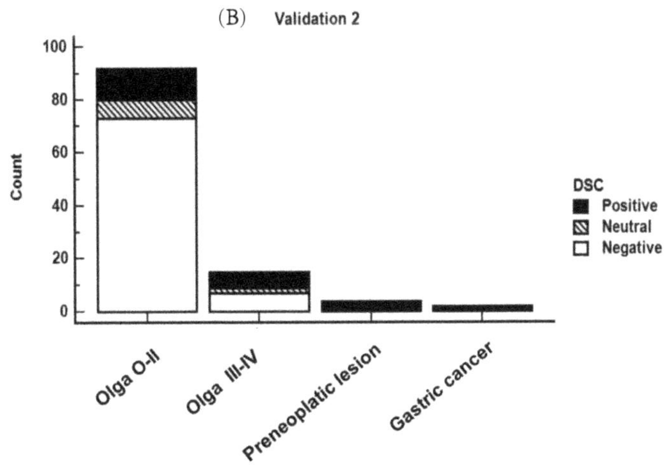


**Figure 2.** Distribution of DSC results taken from the validation 1 (**A**) and validation 2 (**B**) datasets according to the diagnostic categories.

**Table 4.** Prediction value of DSC test for GC risk.

| | Validation 1 53 Selected Retrospective Cases | | Validation 2 113 Consecutive Prospective Cases | | Overall Validation Process 166 Cases | |
|---|---|---|---|---|---|---|
| | **Value** | **95% CI** | **Value** | **95% CI** | **Value** | **95% CI** |
| Sensitivity | 71.429% | 41.896% to 91.611% | 66.667% | 22.278% to 95.673% | 70.000% | 45.721% to 88.107% |
| Specificity | 51.282% | 34.780% to 67.582% | 83.178% | 74.723% to 89.714% | 74.658% | 66.800% to 81.486% |
| AUC | 0.614 | 0.470 to 0.744 | 0.749 | 0.659 to 0.826 | 0.723 | 0.649 to 0.790 |
| Positive LR | 1.466 | 0.924 to 2.327 | 3.963 | 1.957 to 8.024 | 2.762 | 1.852 to 4.120 |
| Negative LR | 0.557 | 0.230 to 1.347 | 0.401 | 0.129 to 1.247 | 0.402 | 0.204 to 0.790 |
| Disease prevalence | | | | | 0.010% | |
| Positive PV | | | | | 0.028% | 0.019% to 0.041% |
| Negative PV | | | | | 99.996% | 99.992% to 99.998% |
| Accuracy | | | | | 74.657% | 67.333% to 81.079% |

LR, likelihood ratio; PV, predictive value.

**Table 5.** Reproducibility of the DSC test.

| DSC Classification | FIRST TEST | At Median Follow-Up (15.5 Months) | Diagnosis | First Test | At Median Follow-Up (15.5 Months) |
|---|---|---|---|---|---|
| | n. (%) | n. (%) | | n. (%) | n. (%) |
| negative | 17 (65.4%) | 15 (57.7%) | OLGA 0–II | 22 (22.3%) | 22 (22.3%) |
| neutral | 3 (11.5%) | 5 (19.2%) | OLGA III–IV | 3 (11.5%) | 3 (11.5%) |
| positive | 6 (23.1%) | 6 (23.1%) | GC | 1 (3.8%) | 1 (3.8%) |

### 2.6. Comparison of the Overall Validation Process ($n = 166$ Cases) Using the DSC Test and the Standardized Pepsinogen Test

The standard cutoff for the PG test is PG I < 70 ng/mL and a PG I/PG II ratio of ≤3.0, resulting in an area under the curve (AUC) of 0.470 (Table 6). For the DSC test, the AUC was 0.723, a better AUC, mostly due to the increase in the sensitivity value from 15.00% to 70.00%. Because the predictive value depends on the prevalence of the disease (0.01% in the general population of Italy), the predictive value for positive cases was 0.03% when using the DSC test and 0.01% when using the PG test, while the overall accuracy remained similar between tests.

**Table 6.** Comparison between the predictive values of the DSC test and pepsinogen tests for GC risk.

| | DSC Test | | PG Test | |
|---|---|---|---|---|
| | Value | 95% CI | Value | 95% CI |
| Sensitivity | 70.00% | 45.721% to 88.107% | 15.00% | 3.207% to 37.893% |
| Specificity | 74.66% | 66.800% to 81.486% | 78.77% | 71.236% to 85.094% |
| AUC | 0.723 | 0.649 to 0.790 | 0.470 | 0.391 to 0.548 |
| Positive LR | 2.762 | 1.852 to 4.120 | 0.71 | 0.238 to 2.099 |
| Negative LR | 0.402 | 0.204 to 0.790 | 1.08 | 0.881 to 1.321 |
| Disease prevalence | 0.01% | | 0.01% | |
| Positive PV | 0.03% | 0.019% to 0.041% | 0.01% | 0.002% to 0.021% |
| Negative PV | 99.996% | 99.992% to 99.998% | 99.989% | 99.987% to 99.991% |
| Accuracy | 74.66% | 67.333% to 81.079% | 78.76% | 71.748% to 84.717% |

LR, likelihood ratio; PV, predictive value.

## 3. Discussion

In this work, we proposed a screening strategy to select individuals at risk for GC based on our newly developed DSC test. The DSC test showed good accuracy (74.66%) with an increase in sensitivity when compared with the PG test, meaning that it could be helpful in identifying gastroenterological patients for opportunistic GC screening in medium-risk areas, e.g., Italy. The DSC test is noninvasive, reproducible, and has high specificity (i.e., a true negative rate). In individuals with a positive DSC classification, gastroscopy and surveillance should be recommended to detect GC at an early stage and improve prevention rates, for example, by using mini-invasive cures.

Earlier studies have confirmed the high risk of GC in patients with atrophic and/or metaplastic gastritis. In particular, two pathological classifications developed from the Sydney System to grade long-standing gastritis and metaplasia (the OLGA and OLGIM classifications) have been shown to be informative in determining severe atrophy/metaplasia (stage III-IV) and an increased risk of GC development [32–34]. Based on these results, gastroenterology guidelines recommend an endoscopic follow-up every 3 years in individuals with a diagnosis of severe atrophy/metaplasia [35]. Performing the DSC test on these patients could be a useful approach for clinical practices better detect individuals at higher risk, who would then be examined using a stricter endoscopic approach in the follow-up.

The results obtained from using the DSC test on the prospective, non-selected, individual cohort (validation cohort 2) are comparable to those achieved using a similar approach in an area of high GC incidence (GC prevalence, 2.84%; ROC curve predicting GC, 0.79).

We propose using this test as an opportunistic screening method in individuals attending the gastroenterology attention to improve the detection rate of GC at an early stage [36].

To the best of our knowledge, there has only been one prospective study that combined the pepsinogen test in a population with a medium GC risk [37], but the study was focused on surveilling patients with precancerous lesions (atrophic gastritis, intestinal metaplasia, dysplasia). The authors showed that only patients with extended precancerous lesions with a low PGI/II $\leq$ 3 ratio and/or OGLIM stage (III-IV) developed high-grade dysplasia or neoplasia at follow-ups after about 57 months. However, it is noteworthy that atrophy is usually associated with *H. pylori* infection and intestinal-type GC. The proposed DSC model introduces the possibility of selecting individuals at high risk of opportunistic neoplastic lesions for further endoscopic examinations. In our study, at a median follow-up of 15.5 months, two individuals out of 17 showed an increase in DSC classification from the negative category to the neutral category, however the histological diagnosis remained at moderate atrophy (OLGA stage 0-II).

GC is a disease of old age, with about one-half of patients with GC being over age 65. A primary characteristic of aging is a progressive loss of physiological gastric tissue integrity that, in turn, leads to impaired function and increased alteration of the PG-I/PG-II ratio, which leads to a decline in gastric acidity [37] and a low circulating concentration of vitamin B12 [38,39]. This could justify a potential decrease in the accuracy of the DSC classification in older subjects (>75 years old); as such, it is necessary to take this aspect into consideration regarding the idea of screening the general population. Therefore, it may be helpful to add other biomarkers to increase the DSC accuracy by reducing the number of false positives in particular in individuals who are >75 years old.

Several studies have reported the serum indicators of GC. Some of them have been proposed in combination with pepsinogen to improve the accurate diagnosis of GC, e.g., sugar carbohydrate antigen 72-4 (CA72-4) [40], CEA, CA12-5, and CA19-9 [41]; metabolites such as hydroxylated sphingomyelins (SM(OH)) and acylcarnitine derivatives (C2, C16, and C18:1) [42]; alcohol dehydrogenase (ADH) activity [43]; interleukin-6 (IL-6); human epididymal protein 4 (HE-4); adiponectin; ferritin and Krebs von den Lungen (KL-6) [44]; soluble T cell immunoglobulin; and mucin domain molecule 3 (sTim-3) [45]. Overall, these results are interesting, but they are all preliminary and need to be evaluated in large prospective studies in combination with the DSC test.

In the last few years, new, increasingly complex technological strategies have emerged, such as the single-cell RNA sequencing (scRNA) transcriptome, which can characterize cellular and molecular networks present in a tissue at the same time [46]. scRNA offers the possibility of finding new GC diagnostic markers by reducing the complexity of gene RNA expression patterns in distinct cell populations. This approach is also useful in deciphering GC pathogenesis and detecting rare, tumor-specific cells at the onset of early GC stages. Currently, only one study has reported potential diagnostic markers for GC using this technology [46], but candidates have also been found in other tissues. Moreover, it is necessary to screen candidates such as secretory proteins, which are abundantly expressed in GC cells and could be evaluated in the serum alone or in addition to the DSC test for GC prediction. Furthermore, their sensitivity and specificity in comparison to the DSC test should be assessed.

Our study had limitations due to the limited number of cases tested and its application in a unique laboratory. Its use in other medium-risk GC populations is necessary to support its validity.

## 4. Materials and Methods

### 4.1. Study Cohorts

This study was based on four cohorts of participants: two discovery cohorts (discovery cohort 1 and discovery cohort 2) and two validation cohorts (validation cohort 1 and validation cohort 2). Demographic and serological data were collected from all participants. Subjects who initially declined consent were excluded from the study.

Discovery cohort 1 included 300 retrospectively selected subjects at risk of GC, i.e., subjects with a family history of GC, autoimmune chronic atrophy gastritis, severe precancerous lesions, or a diagnosis of GC.

Discovery cohort 2 included 200 subjects: blood donors and patients with a GC diagnosis.

Validation cohort 1 consisted of 53 selected individuals from a retrospective series of cases with no atrophy, mild–moderate atrophy, dysplasia, or GC enrolled in the Veneto region, Italy.

Validation cohort 2 consisted of prospective participants recruited from May 2020 to September 2022 (n = 113). Cases were consecutive participants who were referred to a gastroenterologist by a physician for gastroscopy.

Participants from the four cohorts received serological tests to determine their serum PG, G17, and *H. pylori* IgG values. Participant data according to the cohorts are detailed in Table 1.

The study was conducted following the Declaration of Helsinki and approved by the Unique Regional Ethics Committee for Friuli-Venezia Giulia (approval number CRO-2019-46). Written informed consent was obtained from all participants.

### 4.2. Construction of the DSC Model

Data from each participant enrolled in discovery cohort 1 and discovery cohort 2 were used to perform a regression analysis that discriminated patients with GC and determined the coefficients of each variable (i.e., PG, age, sex, *H. pylori* IgG, and gastrin G17)

The resulting Y1 and Y2 equations discriminated patients affected by GC in discovery cohort 1 and discovery cohort 2, respectively.

$$Y1 = -7.49 - 0.0025 \times \text{PG I} + 0.097 \times \text{PG II} - 0.017 \times \text{G17} - 0.0007 \times \text{HP IgG} + 0.1 \times \text{age} + 1.18 \times \text{if male.}$$

$$Y2 = -9.28 + 0.0015 \times \text{PG I} + 0.0824 \times \text{PG II} - 0.009 \times \text{G17} + 0.0001 \times \text{HP IgG} + 0.1337 \times \text{age} + 0.423 \times \text{if male.}$$

To establish the best cutoff for the discrimination, we used an ROC curve analysis.

The combination of the Y1 and Y2 results, based on cutoffs of $Y1 < 0.385$ and $Y2 < 0.294$, determined the DSC classification model.

In detail, subjects positive for $Y1 > 0.385$ and $Y2 > 0.294$ were classified as having high GC risk; subjects positive for only one (Y1 or Y2) were classified as having medium risk; and subjects were deemed low risk when any of the Y1 or Y2 criteria were satisfactory.

### 4.3. DSC Model Assessment in the Validation Cohorts

Participants in the validation cohorts were classified using the DSC model. The DSC classification of each participant was recorded in a database. For each participant, the data were correlated with the diagnosis obtained via gastroscopy and the histological examination of the biopsies.

### 4.4. Serological Data

Blood samples of approximately 5 mL each were obtained from all participants after 10 h of fasting. The tubes were centrifugated for 10 min at ≥10,000 rpm. The serum was stored immediately at −20 °C until an assay was performed. Serologic testing for *H. pylori*-IgG, PGI, PGII, and gastrin G17 was performed using an automated chemiluminescence immunoassay (CLIA) Maglumi analyzer (Medical Systems). Recommended cutoff points, as reported by the manufacturer, were PGI: 70–240 ng/mL, PGII: <13 ng/mL, G17: 2–10 pmol/L, H.p IgG titer: <30 EIU. A combination of PGI < 70 ng/mL and a PGI/PGII ratio of <3 is informative for atrophic gastritis (AG).

### 4.5. Diagnosis

A gastroendoscopy was performed by gastroenterologists on all participants, and biopsies were collected (two biopsies from the antrum, two from the body, and one from the incisure). Gastric mucosae were examined with high-definition (HD) white-light endoscopy and narrow-band imaging (NBI) to improve the visibility of blood vessels and mucosal structures.

Pathological examinations of biopsy samples fixed in buffered formalin (10%) were conducted by an expert pathologist, and the results were reported according to updated OLGA stages to define gastritis [29], and Lauren [30] and Who [31] for GC classifications of gastric tumors and dysplasia. According to the pathological examinations, participants were classified into three groups: without atrophy or metaplasia, with mild–moderate preneoplastic gastric lesions (atrophy or metaplasia), with dysplasia, and with neoplastic lesions.

### 4.6. Statistical Analyses

All analyses were conducted using MedCalc Statistical Software, version 19.0.4 (MedCalc Software bvba, Ostend, Belgium). The results were summarized using age intervals and descriptive statistics. Comparisons between groups were made using a one-way ANOVA test. The predicted DSC scores for each patient were categorized into low-risk, medium-risk, and high-risk groups based on results of the Y1 and Y2 equations as follows: high-risk: Y1 > 0.385 and Y2 => 0.294; medium-risk: Y1 > 0.385 or Y2 => 0.294; low-risk: neither Y1 > 0.385 nor Y2 => 0.294. Considering the histopathology of the gastric lesions, the subjects were categorized into three groups: without lesions, with mild–moderate gastric lesions (chronic and/or autoimmune atrophic gastritis, intestinal metaplasia), and with severe gastric lesions (dysplasia, early GC, and advanced GC). Categorical data were entered into a two-way table by counting the number of observations that fell into each group of variables. A Spearman's rank correlation was used to test the relationship between the DSC and histological categories. To test the accuracy of the high-risk DSC model in predicting the diagnosis of dysplasia and GC, we used a diagnostic test; $p < 0.05$ was considered statistically significant.

## 5. Conclusions

We developed the DSC test model and assessed its accuracy for the classification of people at a high risk of developing GC in two validation datasets. The DSC test achieved a good accuracy of 74.66% and a sensitivity increase of 70.00% compared with 15.00% for the PG test, which supports its potential utility in clinical practice for opportunistic GC identification and the selection of patients at elevated risk for strict follow-ups. It remains to be seen whether the DSC test is effective in follow-ups, since gastric lesions may progress or develop later.

**Author Contributions:** Conceptualization: V.D.R. and R.C.; methodology: V.D.R.; software: R.V. and A.S.; validation: V.D.R., R.C., S.R. and V.C.; formal analysis: V.D.R., S.M., V.C. and A.Z.; resources: R.C. and A.S.; investigation: V.D.R., R.C. and V.C.; data curation: V.D.R.; writing—original draft preparation: V.D.R.; funding acquisition: R.C. and A.S.; writing—review and editing: V.D.R., O.R., R.C. and V.C. All authors have read and agreed to the published version of the manuscript.

**Funding:** We would like to thank the Italian Ministry of Health—Ricerca Corrente for its financial support for this research.

**Institutional Review Board Statement:** The study was conducted according to the guidelines of the Declaration of Helsinki and approved by the Institutional Review Board Unique Regional Ethics Committee for Friuli-Venezia Giulia (approval number CRO-2019-46; date of approval, 14 January 2020). Written informed consent was obtained from all participants.

**Informed Consent Statement:** Informed consent was obtained from all subjects involved in the study.

**Data Availability Statement:** The data presented in this study are available in this article.

**Acknowledgments:** We would like to thank Cinzia Crepaldi and Chiara Corso for their technical support in this research.

**Conflicts of Interest:** The authors declare no conflict of interest.

## References


1. Vradelis, S.; Maynard, N.; Warren, B.F.; Keshav, S.; Travis, S.P.L. Quality Control in Upper Gastrointestinal Endoscopy: Detection Rates of Gastric Cancer in Oxford 2005–2008. *Postgrad. Med. J.* **2011**, *87*, 335–339. [CrossRef] [PubMed]
2. La Vecchia, C.; Bosetti, C.; Lucchini, F.; Bertuccio, P.; Negri, E.; Boyle, P.; Levi, F. Cancer Mortality in Europe, 2000–2004, and an Overview of Trends since 1975. *Ann. Oncol.* **2010**, *21*, 1323–1360. [CrossRef]
3. Jun, J.K.; Choi, K.S.; Lee, H.-Y.; Suh, M.; Park, B.; Song, S.H.; Jung, K.W.; Lee, C.W.; Choi, I.J.; Park, E.-C.; et al. Effectiveness of the Korean National Cancer Screening Program in Reducing Gastric Cancer Mortality. *Gastroenterology* **2017**, *152*, 1319–1328.e7. [CrossRef] [PubMed]
4. Dolcetti, R.; De Re, V.; Canzonieri, V. Immunotherapy for Gastric Cancer: Time for a Personalized Approach? *Int. J. Mol. Sci.* **2018**, *19*, 1602. [CrossRef] [PubMed]
5. Park, C.H.; Yang, D.-H.; Kim, J.W.; Kim, J.-H.; Kim, J.H.; Min, Y.W.; Lee, S.H.; Bae, J.H.; Chung, H.; Choi, K.D.; et al. Clinical Practice Guideline for Endoscopic Resection of Early Gastrointestinal Cancer. *Intest. Res.* **2021**, *19*, 127–157. [CrossRef]
6. GBD 2017 Stomach Cancer Collaborators. The Global, Regional, and National Burden of Stomach Cancer in 195 Countries, 1990-2017: A Systematic Analysis for the Global Burden of Disease Study 2017. *Lancet Gastroenterol. Hepatol.* **2020**, *5*, 42–54. [CrossRef]
7. Registro Tumori Friuli Venezia Giulia 2018. Available online: http://Www.Cro.Sanita.FVG (accessed on 1 April 2020).
8. Sung, H.; Ferlay, J.; Siegel, R.L.; Laversanne, M.; Soerjomataram, I.; Jemal, A.; Bray, F. Global Cancer Statistics 2020: GLOBOCAN Estimates of Incidence and Mortality Worldwide for 36 Cancers in 185 Countries. *CA A Cancer J. Clin.* **2021**, *71*, 209–249. [CrossRef]
9. Areia, M.; Spaander, M.C.; Kuipers, E.J.; Dinis-Ribeiro, M. Endoscopic Screening for Gastric Cancer: A Cost-Utility Analysis for Countries with an Intermediate Gastric Cancer Risk. *United Eur. Gastroenterol. J.* **2018**, *6*, 192–202. [CrossRef] [PubMed]
10. Hassan, C.; Bersani, G.; Buri, L.; Zullo, A.; Anti, M.; Bianco, M.A.; Di Giulio, E.; Ficano, L.; Morini, S.; Di Matteo, G.; et al. Appropriateness of Upper-GI Endoscopy: An Italian Survey on Behalf of the Italian Society of Digestive Endoscopy. *Gastrointest. Endosc.* **2007**, *65*, 767–774. [CrossRef]
11. Buri, L.; Bersani, G.; Hassan, C.; Anti, M.; Bianco, M.A.; Cipolletta, L.; Di Giulio, E.; Di Matteo, G.; Familiari, L.; Ficano, L.; et al. How to Predict a High Rate of Inappropriateness for Upper Endoscopy in an Endoscopic Centre? *Dig. Liver Dis.* **2010**, *42*, 624–628. [CrossRef]
12. Zhang, T.; Chen, H.; Zhang, Y.; Yin, X.; Man, J.; Yang, X.; Lu, M. Global Changing Trends in Incidence and Mortality of Gastric Cancer by Age and Sex, 1990-2019: Findings from Global Burden of Disease Study. *J. Cancer* **2021**, *12*, 6695–6705. [CrossRef] [PubMed]
13. Genta, R.M.; Rugge, M. Assessing Risks for Gastric Cancer: New Tools for Pathologists. *World J. Gastroenterol.* **2006**, *12*, 5622–5627. [CrossRef]
14. Rugge, M.; de Boni, M.; Pennelli, G.; de Bona, M.; Giacomelli, L.; Fassan, M.; Basso, D.; Plebani, M.; Graham, D.Y. Gastritis OLGA-Staging and Gastric Cancer Risk: A Twelve-Year Clinico-Pathological Follow-up Study. *Aliment. Pharmacol. Ther.* **2010**, *31*, 1104–1111. [CrossRef] [PubMed]
15. Terasawa, T.; Nishida, H.; Kato, K.; Miyashiro, I.; Yoshikawa, T.; Takaku, R.; Hamashima, C. Prediction of Gastric Cancer Development by Serum Pepsinogen Test and Helicobacter Pylori Seropositivity in Eastern Asians: A Systematic Review and Meta-Analysis. *PLoS ONE* **2014**, *9*, e109783. [CrossRef] [PubMed]
16. Yanaoka, K.; Oka, M.; Mukoubayashi, C.; Yoshimura, N.; Enomoto, S.; Iguchi, M.; Magari, H.; Utsunomiya, H.; Tamai, H.; Arii, K.; et al. Cancer High-Risk Subjects Identified by Serum Pepsinogen Tests: Outcomes after 10-Year Follow-up in Asymptomatic Middle-Aged Males. *Cancer Epidemiol. Biomark. Prev.* **2008**, *17*, 838–845. [CrossRef] [PubMed]
17. Hamashima, C. Forthcoming Step in Gastric Cancer Prevention: How Can Risk Stratification Be Combined with Endoscopic Screening for Gastric Cancer? *Gut Liver* **2022**, *16*, 811–824. [CrossRef]
18. Malfertheiner, P.; Megraud, F.; O'Morain, C.A.; Gisbert, J.P.; Kuipers, E.J.; Axon, A.T.; Bazzoli, F.; Gasbarrini, A.; Atherton, J.; Graham, D.Y.; et al. Management of Helicobacter Pylori Infection-the Maastricht V/Florence Consensus Report. *Gut* **2017**, *66*, 6–30. [CrossRef]
19. Samloff, I.M. Pepsinogens, Pepsins, and Pepsin Inhibitors. *Gastroenterology* **1971**, *60*, 586–604. [CrossRef]
20. Copps, J.; Murphy, R.F.; Lovas, S. The Production and Role of Gastrin-17 and Gastrin-17-Gly in Gastrointestinal Cancers. *Protein Pept. Lett.* **2009**, *16*, 1504–1518. [CrossRef]
21. De Martel, C.; Ferlay, J.; Franceschi, S.; Vignat, J.; Bray, F.; Forman, D.; Plummer, M. Global Burden of Cancers Attributable to Infections in 2008: A Review and Synthetic Analysis. *Lancet Oncol.* **2012**, *13*, 607–615. [CrossRef]
22. Cai, Q.; Zhu, C.; Yuan, Y.; Feng, Q.; Feng, Y.; Hao, Y.; Li, J.; Zhang, K.; Ye, G.; Ye, L.; et al. Development and Validation of a Prediction Rule for Estimating Gastric Cancer Risk in the Chinese High-Risk Population: A Nationwide Multicentre Study. *Gut* **2019**, *68*, 1576–1587. [CrossRef]
23. Liu, W.; Sun, Y.; Yuan, Y. Analysis of Serum Gastrin-17 and Helicobacter Pylori Antibody in Healthy Chinese Population. *J. Clin. Lab. Anal.* **2020**, *34*, e23518. [CrossRef] [PubMed]
24. De Re, V.; Orzes, E.; Canzonieri, V.; Maiero, S.; Fornasarig, M.; Alessandrini, L.; Cervo, S.; Steffan, A.; Zanette, G.; Mazzon, C.; et al. Pepsinogens to Distinguish Patients With Gastric Intestinal Metaplasia and Helicobacter Pylori Infection Among Populations at Risk for Gastric Cancer. *Clin. Transl. Gastroenterol.* **2016**, *7*, e183. [CrossRef]

25. Miceli, E.; Vanoli, A.; Lenti, M.V.; Klersy, C.; Di Stefano, M.; Luinetti, O.; Dominioni, C.C.; Pisati, M.; Staiani, M.; Gentile, A.; et al. Natural History of Autoimmune Atrophic Gastritis: A Prospective, Single Centre, Long-Term Experience. *Aliment. Pharmacol. Ther.* **2019**, *50*, 1172–1180. [CrossRef] [PubMed]
26. Sotelo, S.; Manterola, C. Morphology and Diagnostic Determinants of Gastric Precancerous Conditions. *Int. J. Morphol.* **2019**, *37*, 917–927. [CrossRef]
27. Rugge, M.; Savarino, E.; Sbaraglia, M.; Bricca, L.; Malfertheiner, P. Gastritis: The Clinico-Pathological Spectrum. *Dig. Liver Dis.* **2021**, *53*, 1237–1246. [CrossRef] [PubMed]
28. Koc, D.O.; Bektas, S. Serum Pepsinogen Levels and OLGA/OLGIM Staging in the Assessment of Atrophic Gastritis Types. *Postgrad. Med. J.* **2022**, *98*, 441–445. [CrossRef]
29. Rugge, M.; Meggio, A.; Pennelli, G.; Piscioli, F.; Giacomelli, L.; De Pretis, G.; Graham, D.Y. Gastritis Staging in Clinical Practice: The OLGA Staging System. *Gut* **2007**, *56*, 631–636. [CrossRef]
30. Lauren, P. The two histological main types of gastric carcinoma: Diffuse and so-called intestinal-type carcinoma. An attempt at a histo-clinical classification. *Acta Pathol. Microbiol. Scand.* **1965**, *64*, 31–49. [CrossRef]
31. Hu, B.; El Hajj, N.; Sittler, S.; Lammert, N.; Barnes, R.; Meloni-Ehrig, A. Gastric Cancer: Classification, Histology and Application of Molecular Pathology. *J. Gastrointest. Oncol.* **2012**, *3*, 251–261. [CrossRef]
32. Capelle, L.G.; de Vries, A.C.; Haringsma, J.; Ter Borg, F.; de Vries, R.A.; Bruno, M.J.; van Dekken, H.; Meijer, J.; van Grieken, N.C.T.; Kuipers, E.J. The Staging of Gastritis with the OLGA System by Using Intestinal Metaplasia as an Accurate Alternative for Atrophic Gastritis. *Gastrointest. Endosc.* **2010**, *71*, 1150–1158. [CrossRef]
33. Rugge, M.; Genta, R.M. Staging and Grading of Chronic Gastritis. *Hum. Pathol.* **2005**, *36*, 228–233. [CrossRef]
34. Yue, H.; Shan, L.; Bin, L. The Significance of OLGA and OLGIM Staging Systems in the Risk Assessment of Gastric Cancer: A Systematic Review and Meta-Analysis. *Gastric. Cancer* **2018**, *21*, 579–587. [CrossRef]
35. Dinis-Ribeiro, M.; Areia, M.; de Vries, A.C.; Marcos-Pinto, R.; Monteiro-Soares, M.; O'Connor, A.; Pereira, C.; Pimentel-Nunes, P.; Correia, R.; Ensari, A.; et al. Management of Precancerous Conditions and Lesions in the Stomach (MAPS): Guideline from the European Society of Gastrointestinal Endoscopy (ESGE), European Helicobacter Study Group (EHSG), European Society of Pathology (ESP), and the Sociedade Portuguesa de Endoscopia Digestiva (SPED). *Endoscopy* **2012**, *44*, 74–94. [CrossRef] [PubMed]
36. Zhou, Q.; Chen, Y.; Pan, J.; Zhou, L.; Lin, J. Application of a Novel Scoring System for Gastric Cancer Opportunistic Screening in Hospital Visits. *BMC Gastroenterol.* **2022**, *22*, 223. [CrossRef] [PubMed]
37. Hurwitz, A.; Brady, D.A.; Schaal, S.E.; Samloff, I.M.; Dedon, J.; Ruhl, C.E. Gastric Acidity in Older Adults. *JAMA* **1997**, *278*, 659–662. [CrossRef]
38. Wright, J.D.; Bialostosky, K.; Gunter, E.W.; Carroll, M.D.; Najjar, M.F.; Bowman, B.A.; Johnson, C.L. Blood Folate and Vitamin B12: United States, 1988-94. In *Vital and Health Statistics. Series 11, Data from the National Health Survey*; National Center for Health Statistics: Washington, DC, USA, 1998; pp. 1–78.
39. Bates, C.J.; Schneede, J.; Mishra, G.; Prentice, A.; Mansoor, M.A. Relationship between Methylmalonic Acid, Homocysteine, Vitamin B12 Intake and Status and Socio-Economic Indices, in a Subset of Participants in the British National Diet and Nutrition Survey of People Aged 65 y and Over. *Eur. J. Clin. Nutr.* **2003**, *57*, 349–357. [CrossRef]
40. Lin, Z.; Bian, H.; Chen, C.; Chen, W.; Li, Q. Application of Serum Pepsinogen and Carbohydrate Antigen 72-4 (CA72-4) Combined with Gastrin-17 (G-17) Detection in the Screening, Diagnosis, and Evaluation of Early Gastric Cancer. *J. Gastrointest. Oncol.* **2021**, *12*, 1042. [CrossRef] [PubMed]
41. Wang, Z.; Mo, T.-M.; Tian, L.; Chen, J.-Q. Gastrin-17 Combined with CEA, CA12-5 and CA19-9 Improves the Sensitivity for the Diagnosis of Gastric Cancer. *Int. J. Gen. Med.* **2021**, *14*, 8087–8095. [CrossRef]
42. Corona, G.; Cannizzaro, R.; Miolo, G.; Caggiari, L.; De Zorzi, M.; Repetto, O.; Steffan, A.; De Re, V. Use of Metabolomics as a Complementary Omic Approach to Implement Risk Criteria for First-Degree Relatives of Gastric Cancer Patients. *Int. J. Mol. Sci.* **2018**, *19*, 750. [CrossRef]
43. Jelski, W.; Chrostek, L.; Laszewicz, W.; Szmitkowski, M. Alcohol Dehydrogenase (ADH) Isoenzyme Activity in the Sera of Patients with Helicobacter Pylori Infection. *Dig. Dis. Sci.* **2007**, *52*, 1513–1516. [CrossRef] [PubMed]
44. Chapelle, N.; Osmola, M.; Martin, J.; Blin, J.; Leroy, M.; Jirka, I.; Moussata, D.; Lamarque, D.; Olivier, R.; Tougeron, D.; et al. Serum Pepsinogens Combined with New Biomarkers Testing Using Chemiluminescent Enzyme Immunoassay for Non-Invasive Diagnosis of Atrophic Gastritis: A Prospective, Multicenter Study. *Diagnostics* **2022**, *12*, 695. [CrossRef] [PubMed]
45. Chen, L.; Hong, J.; Hu, R.; Yu, X.; Chen, X.; Zheng, S.; Qin, Y.; Zhou, X.; Wang, Y.; Zheng, L.; et al. Clinical Value of Combined Detection of Serum STim-3 and Pepsinogen for Gastric Cancer Diagnosis. *Cancer Manag. Res.* **2021**, *13*, 7759–7769. [CrossRef] [PubMed]
46. Zhang, P.; Yang, M.; Zhang, Y.; Xiao, S.; Lai, X.; Tan, A.; Du, S.; Li, S. Dissecting the Single-Cell Transcriptome Network Underlying Gastric Premalignant Lesions and Early Gastric Cancer. *Cell Rep.* **2019**, *27*, 1934–1947.e5. [CrossRef]

International Journal of *Molecular Sciences*

MDPI

*Article*

# Discovery of Small Molecule COX-1 and Akt Inhibitors as Anti-NSCLC Agents Endowed with Anti-Inflammatory Action

**Mehlika Dilek Altıntop [1,*], Gülşen Akalın Çiftçi [2,3], Nalan Yılmaz Savaş [3], İpek Ertorun [4], Betül Can [4], Belgin Sever [1], Halide Edip Temel [2], Özkan Alataş [4] and Ahmet Özdemir [1,*]**

1 Department of Pharmaceutical Chemistry, Faculty of Pharmacy, Anadolu University, 26470 Eskişehir, Turkey
2 Department of Biochemistry, Faculty of Pharmacy, Anadolu University, 26470 Eskişehir, Turkey
3 Graduate School of Health Sciences, Anadolu University, 26470 Eskişehir, Turkey
4 Department of Medical Biochemistry, Faculty of Medicine, Eskisehir Osmangazi University, 26480 Eskişehir, Turkey
* Correspondence: mdaltintop@anadolu.edu.tr (M.D.A.); ahmeto@anadolu.edu.tr (A.Ö.); Tel.: +90-222-335-0580 (ext. 3772) (M.D.A); +90-222-335-0580 (ext. 3780) (A.Ö.)

**Abstract:** Targeted therapies have come into prominence in the ongoing battle against non-small cell lung cancer (NSCLC) because of the shortcomings of traditional chemotherapy. In this context, indole-based small molecules, which were synthesized efficiently, were subjected to an in vitro colorimetric assay to evaluate their cyclooxygenase (COX) inhibitory profiles. Compounds **3b** and **4a** were found to be the most selective COX-1 inhibitors in this series with $IC_{50}$ values of 8.90 μM and 10.00 μM, respectively. In vitro and in vivo assays were performed to evaluate their anti-NSCLC and anti-inflammatory action, respectively. 2-(1*H*-Indol-3-yl)-*N*′-(4-morpholinobenzylidene)acetohydrazide (**3b**) showed selective cytotoxic activity against A549 human lung adenocarcinoma cells through apoptosis induction and Akt inhibition. The in vivo experimental data revealed that compound **3b** decreased the serum myeloperoxidase and nitric oxide levels, pointing out its anti-inflammatory action. Moreover, compound **3b** diminished the serum aminotransferase (particularly aspartate aminotransferase) levels. Based on the in vitro and in vivo experimental data, compound **3b** stands out as a lead anti-NSCLC agent endowed with in vivo anti-inflammatory action, acting as a dual COX-1 and Akt inhibitor.

**Keywords:** Akt; anti-inflammatory action; COX-1; hydrazones; non-small cell lung cancer; thiosemicarbazides

**Citation:** Altıntop, M.D.; Akalın Çiftçi, G.; Yılmaz Savaş, N.; Ertorun, İ.; Can, B.; Sever, B.; Temel, H.E.; Alataş, Ö.; Özdemir, A. Discovery of Small Molecule COX-1 and Akt Inhibitors as Anti-NSCLC Agents Endowed with Anti-Inflammatory Action. *Int. J. Mol. Sci.* **2023**, *24*, 2648. https://doi.org/10.3390/ijms24032648

Academic Editor: Laura Paleari

Received: 5 December 2022
Revised: 29 December 2022
Accepted: 30 December 2022
Published: 31 January 2023



## 1. Introduction

Non-small cell lung cancer (NSCLC), which accounts for the majority (~85%) of lung cancer cases, is by far the primary cause of cancer-related death throughout the world [1]. Despite significant advances in both diagnosis and treatment, the prognosis for patients with NSCLC still remains poor and the 5-year survival rates of the patients are very low [2]. Surgery, chemotherapy, radiotherapy, immunotherapy, and targeted therapy are existing treatment modalities for NSCLC [3]. Clinical outcomes of patients with NSCLC depend on the cancer stage at the time of diagnosis [4]. The early stages of NSCLC carry the maximum potential for therapeutic intervention and, therefore, its early detection is critical for managing the disease and improving the survival rate [4]. However, there are many challenges in the diagnosis of NSCLC, as it is often asymptomatic early in its course [5,6].

The best treatment option for early stage NSCLC continues to be surgical resection. When the disease is diagnosed at an advanced stage, surgical intervention is no longer an option [7]. In this case, radiotherapy and chemotherapy (e.g., platinum-based chemotherapy) become major therapeutic approaches for unresectable NSCLC [1]. Despite their benefits in NSCLC therapy, conventional chemotherapeutic agents destroy normal cells along with cancer cells and, therefore, these drugs cause severe toxicity and adverse effects [8,9].

Two major barriers to NSCLC management are resistance to radio(chemo)therapy and metastasis [1,9], both of which are the main causes of NSCLC-related mortality [10,11].

The above-mentioned drawbacks have shifted the paradigm of cancer therapy from traditional chemotherapy to targeted therapy, a milestone approach that aims to maximize therapeutic benefits with negligible side effects [3].

The lungs are particularly prone to injury and inflammation since the lungs are continuously exposed to the external environment [12]. Mounting evidence has demonstrated the causal link between chronic inflammation and lung cancer. According to epidemiological data, approximately 20% of cancer-related deaths are associated with unabated inflammation [13]. Chronic inflammation plays a multifaceted role in carcinogenesis; conversely, cancer can also lead to inflammation [12]. Inflammation predisposes to the development of lung cancer [14] and can contribute to tumor initiation, promotion, progression, and metastasis [15]. Targeting inflammation stands out as a rational strategy not only for cancer therapy but also for cancer prevention [16]. Nonsteroidal anti-inflammatory drugs (NSAIDs) significantly diminish the risk of developing certain types of cancer (e.g., colon, lung, breast, and prostate cancer) by reducing tumor-related inflammation [13]. Long-term aspirin use has been reported to reduce the incidence and mortality associated with several cancer types. Several possible mechanisms have been suggested to explain the link between NSAID use and cancer prevention. One of those is cyclooxygenase (COX) inhibition, which reduces the production of inflammatory mediators, particularly prostaglandins (PGs) [16].

COX-1 expression has been reported to be up-regulated in tumorigenesis [17] and implicated in multiple aspects of cancer pathophysiology and, therefore, the inhibition of COX-1, by a variety of selective and nonselective inhibitors, is an emerging approach for pharmacologic intervention in cancer. However, there is only one selective COX-1 inhibitor currently prescribed as an NSAID (mofezolac), just in Japan, for the management of pain and inflammation [17–20].

Akt, also known as protein kinase B (PKB), is one of the most frequently hyperactivated protein kinases in a variety of human cancers including NSCLC [21–23]. Akt overactivation affects several downstream effectors and mediates multiple pathways that promote tumorigenesis (e.g., cell survival, growth, and proliferation) [21]. Furthermore, the hyperactivation of Akt intrinsically up-regulates the nuclear factor-κB (NF-κB) pathway, which transcriptionally initiates pro-inflammatory networks to build up the inflammatory tumor microenvironment [24]. Although diverse small molecule Akt inhibitors have been entered in clinical trials, none of them have been approved [25].

Hydrazides-hydrazones are not only versatile intermediates for the synthesis of various heterocyclic compounds but also commonly occurring motifs in drug-like molecules because of their unique features (e.g., serving as both H-bond donors and acceptors) and diverse pharmacological applications for the management of microbial infections, cancer, and inflammation [26–29]. Hydrazones exert pronounced antitumor action through diverse mechanisms including apoptosis induction, cell cycle arrest, angiogenesis inhibition, and inhibition of a plethora of biological targets related to the pathogenesis of cancer, including Akt [29–36]. Moreover, mounting evidence has demonstrated the anti-inflammatory and/or COX inhibitory potential of hydrazones [37–41].

Thiosemicarbazides are sulfur and nitrogen-containing ligands distinguished by their capability to form complexes with transition metals (e.g., iron, zinc, and copper) [42]. Thiosemicarbazides have aroused great interest not only as intermediates for the synthesis of biologically active heterocycles but also privileged motifs in many bioactive pharmaceutical products [42–45]. Thiosemicarbazides/thiosemicarbazones show a wide range of pharmacological activities ranging from anticancer activity to anti-inflammatory potency due to their unique structural features, allowing them to interact with the pivotal residues of biological targets associated with the pathogenesis of many diseases, particularly cancer and inflammation [42–56]. Triapine, a synthetic thiosemicarbazone, is a small molecule antineoplastic agent endowed with ribonucleotide reductase (RNR) inhibitory activity [42–45].

The indole ranks among the top 25 most common nitrogen heterocycles in U.S. Food and Drug Administration (FDA)-approved drugs. It is also a key structural component of an essential amino acid (tryptophan), a monoamine neurotransmitter (serotonin), and countless natural products (e.g., vinca alkaloids) [57]. The diverse applications of the indole core in challenging diseases (e.g., lung cancer, inflammatory diseases) make it one of the most privileged heterocyclic scaffolds for drug discovery [57,58]. Among vinca alkaloids, vinorelbine is the most frequently used antimitotic drug to treat lung cancer and vinblastine, in combination with cisplatin, is used in the management of NSCLC. Nintedanib (in combination with docetaxel), alectinib, osimertinib, anlotinib, and sunitinib are indole-based anti-NSCLC agents (Figure 1) [59]. In general, indole derivatives have been reported to exert marked anti-NSCLC action through diverse mechanisms including the induction of apoptosis, the inhibition of crucial biological targets such as microtubule, topoisomerases, protein kinases (e.g., Akt), and histone deacetylases (HDACs) [58–62]. The indole is also considered to be one of the most eligible scaffolds for anti-inflammatory drug discovery [62–67]. Indomethacin (Figure 2) is one of the most commonly prescribed NSAIDs exerting its action through the inhibition of COXs. Moreover, several experimental studies have revealed that indomethacin shows significant antiproliferative activity against a broad array of cancer (e.g., colorectal, lung) cell lines [68–72].

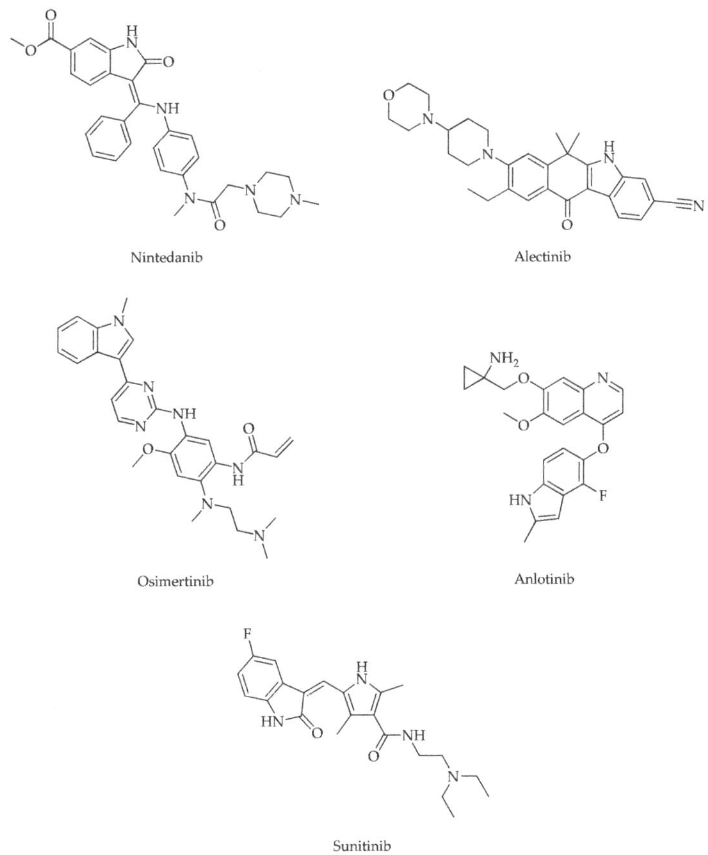


**Figure 1.** Indole-based anti-NSCLC agents.

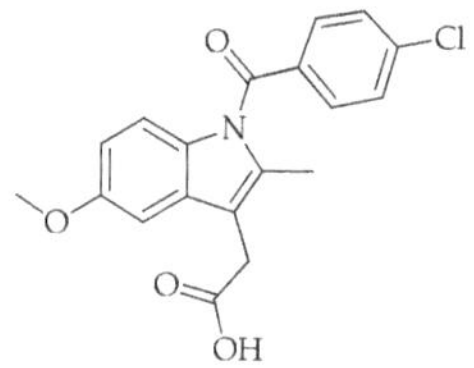

**Figure 2.** Indomethacin.

Taken together, the aforementioned data [26–72] prompted us to design two classes of indole-based small molecules (**3a-j**, **4a-g**) for the targeted therapy of NSCLC. In this context, we performed the synthesis of new hydrazones (**3a-j**) and thiosemicarbazides (**4a-g**) efficiently and conducted in vitro and in vivo assays to assess their potential for the targeted therapy of NSCLC.

## 2. Results

The reaction sequence for the preparation of the hitherto unreported small molecules (**3a-j**, **4a-g**) is depicted in Figure 3, starting from 2-(1*H*-indol-3-yl)acetic acid. The convenient and efficient reaction of compound **2** with aromatic aldehydes or ketones and aryl isothiocyanates yielded new hydrazones (**3a-j**), and thiosemicarbazides (**4a-g**), respectively.

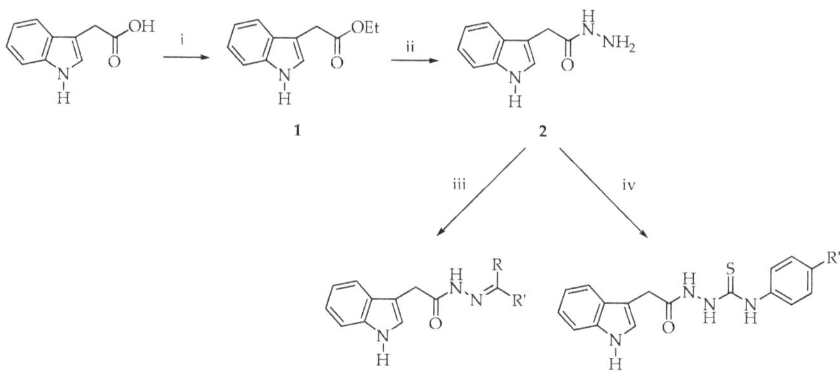

**Figure 3.** The synthetic route for the preparation of compounds **3a-j** and **4a-g**. Reagents and conditions: (i) EtOH, $H_2SO_4$, reflux, 12 h; (ii) $NH_2NH_2.H_2O$, EtOH, reflux, 4 h; (iii) RCHO or RCOR′, EtOH, reflux, 15 h; (iv) $R''C_6H_4NCS$, EtOH, rt, 8 h.

New hydrazones (**3a-j**) and thiosemicarbazides (**4a-g**) were subjected to in vitro assays to determine their COX inhibitory profiles. Among compounds **3a-j**, compound **3a** was found to be a nonselective COX inhibitor with $IC_{50}$ values of 10.35 μM and 12.50 μM for COX-1 and COX-2, respectively (Table 1). On the other hand, compound **3b** was the most selective COX-1 inhibitor ($IC_{50}$ = 8.90 μM) in this series with a selectivity index (SI) value of 0.13.

**Table 1.** COX inhibitory profiles of compounds **3a-j** and positive controls.

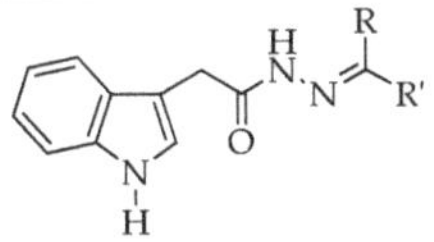

| Compound | R | R′ | $IC_{50}$ (μM) | | SI * |
|---|---|---|---|---|---|
| | | | COX-1 | COX-2 | |
| **3a** | 4-(Pyrrolidin-1-yl)phenyl | H | 10.35 ± 0.35 | 12.50 ± 0.71 | 0.83 |
| **3b** | 4-Morpholinophenyl | H | 8.90 ± 0.14 | 71.00 ± 1.41 | 0.13 |
| **3c** | 4-(Piperidin-1-yl)phenyl | H | >100 | >100 | - |
| **3d** | 4-(4-Methylpiperazin-1-yl)phenyl | H | 78.50 ± 8.50 | >100 | <0.79 |
| **3e** | 4-(Methylsulfonyl)phenyl | H | 83.75 ± 6.25 | 35.00 ± 9.90 | 2.39 |
| **3f** | 4-(Methylsulfonyl)phenyl | $CH_3$ | 93.75 ± 6.25 | 51.00 ± 12.73 | 1.84 |
| **3g** | 4-Morpholinophenyl | $CH_3$ | 51.00 ± 1.00 | 52.50 ± 0.71 | 0.97 |
| **3h** | 4-(2-Morpholinoethoxy)phenyl | H | 38.50 ± 1.5 | 50.50 ± 0.70 | 0.76 |
| **3i** | 1-Methyl-1*H*-indol-3-yl | H | 31.25 ± 1.25 | >100 | <0.31 |
| **3j** | 5-Methoxy-1*H*-indol-3-yl | H | >100 | 44.50 ± 6.36 | >2.25 |
| Indomethacin | - | - | 0.12 ± 0.01 | 0.58 ± 0.08 | 0.21 |
| Celecoxib | - | - | 8.88 ± 0.38 | 2.75 ± 0.05 | 3.23 |

* $IC_{50}$ for COX-1/$IC_{50}$ for COX-2.

Among compounds **4a-g**, compound **4a** was the most selective COX-1 inhibitor ($IC_{50}$ = 10.00 μM) (Table 2). Other compounds did not show any inhibitory potency on COX-1 at the tested concentrations.

**Table 2.** COX inhibitory profiles of compounds **4a-g** and positive controls.

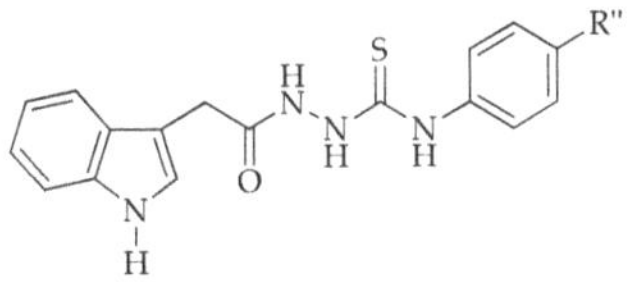

| Compound | R″ | $IC_{50}$ (μM) | | SI * |
|---|---|---|---|---|
| | | COX-1 | COX-2 | |
| **4a** | Br | 10.00 ± 0.13 | 76.50 ± 6.36 | 0.13 |
| **4b** | $CF_3$ | >100 | 61.50 ± 0.71 | >1.63 |
| **4c** | CN | >100 | 56.50 ± 4.49 | >1.77 |
| **4d** | Piperidin-1-ylsulfonyl | >100 | 59.00 ± 8.48 | >1.69 |
| **4e** | 1*H*-Pyrazol-1-yl | >100 | 31.50 ± 2.12 | >3.17 |
| **4f** | 3,4-Methylenedioxy | >100 | 51.50 ± 0.71 | >1.94 |
| **4g** | Benzyloxy | >100 | 59.50 ± 2.12 | >1.68 |
| Indomethacin | - | 0.12 ± 0.01 | 0.58 ± 0.08 | 0.21 |
| Celecoxib | - | 8.88 ± 0.38 | 2.75 ± 0.05 | 3.23 |

* $IC_{50}$ for COX-1/$IC_{50}$ for COX-2.

All compounds were examined for their cytotoxic effects on L929 mouse fibroblast (normal) cells using the MTT test. Based on the in vitro experimental data, compound **3a**, the nonselective COX inhibitor, showed cytotoxicity toward L929 cells with an $IC_{50}$ value of 17.33 μM (Table 3), which is close to its $IC_{50}$ values indicated in Table 1. On the other hand, compounds **3b** and **4a** did not show cytotoxicity against L929 cells at their effective concentrations. As a result, compounds **3b** and **4a** (Figure 4), the selective COX-1 inhibitors in this series, were chosen for further studies.

**Table 3.** $IC_{50}$ values of all compounds for L929 cells.

| Compound | $IC_{50}$ (μM) |
|---|---|
| **3a** | 17.33 ± 2.08 |
| **3b** | 176.67 ± 5.77 |
| **3c** | 21.33 ± 0.58 |
| **3d** | 20.00 ± 1.73 |
| **3e** | 85.00 ± 17.32 |
| **3f** | <3.90 |
| **3g** | <3.90 |
| **3h** | <3.90 |
| **3i** | 42.33 ± 0.58 |
| **3j** | 43.00 ± 1.73 |
| **4a** | 84.00 ± 19.70 |
| **4b** | 22.67 ± 2.08 |
| **4c** | 26.33 ± 3.51 |
| **4d** | 45.67 ± 6.03 |
| **4e** | 62.33 ± 7.51 |
| **4f** | 14.00 ± 5.29 |
| **4g** | 14.00 ± 3.46 |

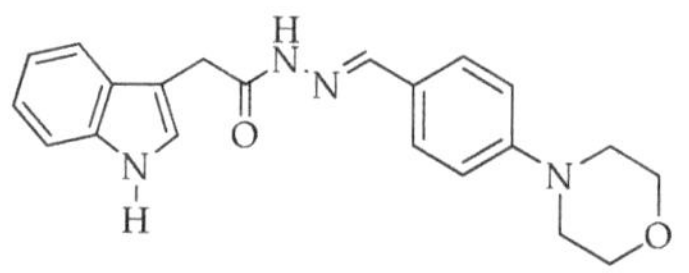

Compound **3b**

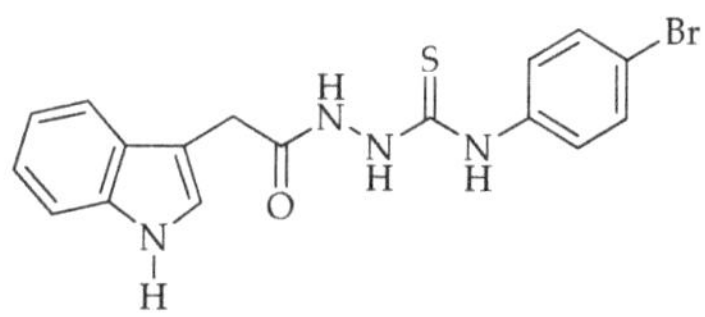

Compound **4a**

**Figure 4.** Selective COX-1 inhibitors in this series.

Compounds **3b** and **4a** were also subjected to the MTT assay to assess their cytotoxicity toward A549 human lung adenocarcinoma cell line. Based on the data presented in Table 4, compound **3b** was found to be the most potent anticancer agent on A549 cells with an $IC_{50}$ value of 89.67 μM compared to cisplatin ($IC_{50}$ = 22.67 μM). On the other hand, compound **4a** showed cytotoxic activity against A549 cells with an $IC_{50}$ value of 179.33 μM.

**Table 4.** $IC_{50}$ values of compounds **3b**, **4a,** and cisplatin for A549 cells.

| Compound | $IC_{50}$ (μM) |
|---|---|
| **3b** | 89.67 ± 10.78 |
| **4a** | 179.33 ± 77.59 |
| Cisplatin | 22.67 ± 4.04 |

After 24 h incubation of A549 cells treated with compounds **3b** and **4a** in this series and cisplatin, flow cytometry-based apoptosis detection assay was performed to identify early and late apoptotic cells using Annexin V-fluorescein isothiocyanate (FITC)/propidium iodide (PI) staining. The percentages of A549 cells undergoing apoptosis (early and late) exposed to compounds **3b** and cisplatin at their $IC_{50}/2$ concentrations were found to be 11.67% and 6.57%, respectively. On the other hand, the percentages of A549 cells undergoing apoptosis (early and late) exposed to compounds **3b** and cisplatin at their $IC_{50}$ concentrations were 12.85% and 4.46%, respectively (Table 5, Figure 5). The percentages of A549 cells undergoing early and late apoptosis exposed to compound **4a** at its $IC_{50}/4$ concentration were 7.34% and 5.19%, respectively. On the other hand, the percentages of A549 cells undergoing early and late apoptosis exposed to compound **4a** at its $IC_{50}/2$ concentration were found to be 7.07% and 6.62%, respectively (Table 5, Figure 5).

**Table 5.** Percentages of typical quadrant analysis of Annexin V FITC/PI flow cytometry of A549 cells treated with compounds **3b**, **4a**, and cisplatin.

| Compound | Early Apoptosis (%) | Late Apoptosis (%) | Necrosis (%) | Viability (%) |
|---|---|---|---|---|
| Control | 4.98 | 1.18 | 0.36 | 93.47 |
| Compound **3b** at $IC_{50}/2$ | 5.89 | 5.78 | 0.36 | 87.97 |
| Compound **3b** at $IC_{50}$ | 6.84 | 6.01 | 0.48 | 86.66 |
| Compound **4a** at $IC_{50}/4$ | 7.34 | 5.19 | 0.37 | 87.10 |
| Compound **4a** at $IC_{50}/2$ | 7.07 | 6.62 | 0.60 | 85.71 |
| Cisplatin at $IC_{50}/2$ | 3.69 | 2.88 | 0.72 | 92.72 |
| Cisplatin at $IC_{50}$ | 2.43 | 2.03 | 2.34 | 93.20 |

A549 cells were cultured for 24 h in medium with compound **4a** (at its $IC_{50}/4$ and $IC_{50}/2$ concentrations), compound **3b,** and cisplatin (at their $IC_{50}/2$ and $IC_{50}$ concentrations). At least 10,000 cells were analyzed per sample, and quadrant analysis was performed.

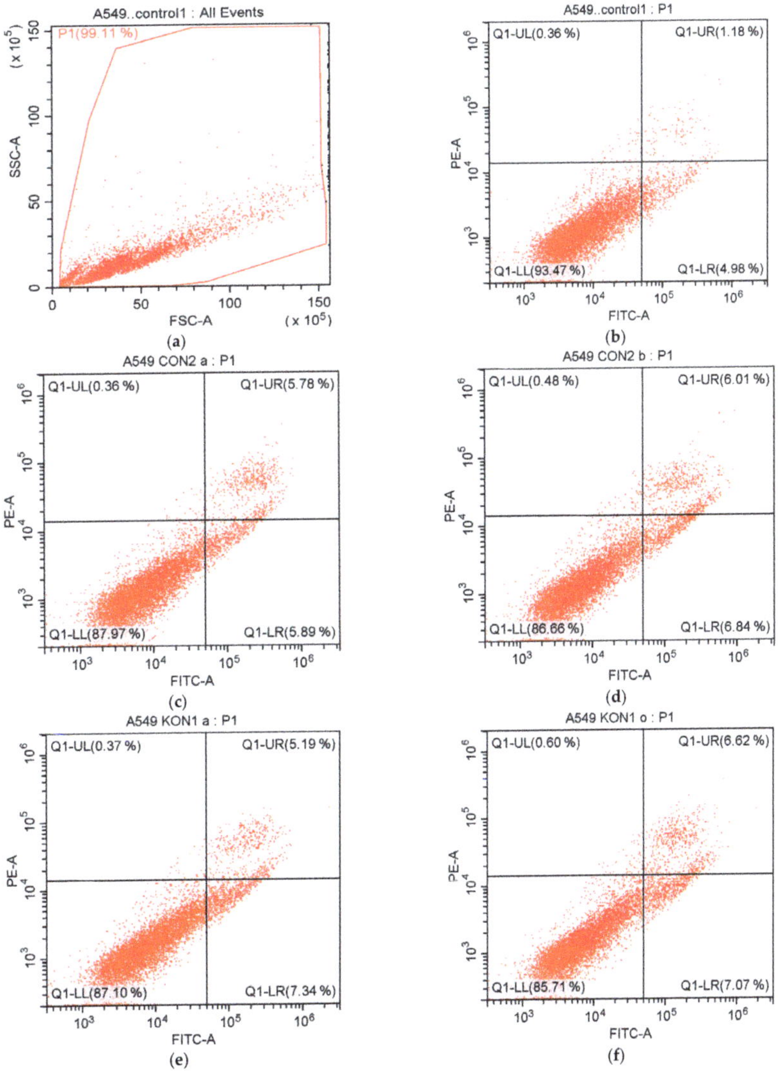


**Figure 5.** *Cont.*

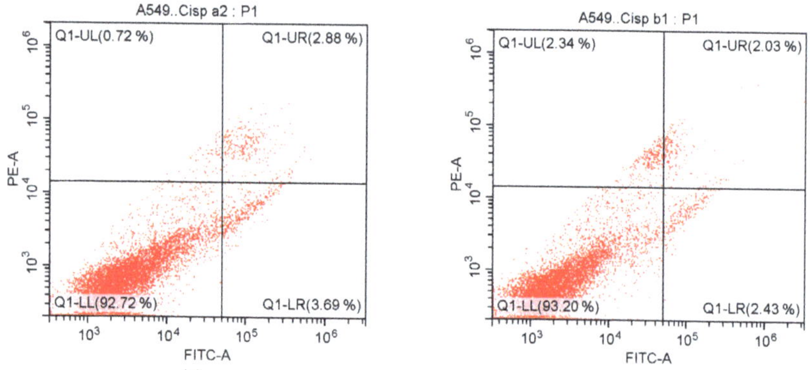


(g) (h)

**Figure 5.** Flow cytometric analysis of A549 cells treated with $IC_{50}/2$ and $IC_{50}$ concentrations of compounds **3b**, **4a**, and cisplatin. At least 10,000 cells were analyzed per sample, and quadrant analysis was performed. Q1-LR, Q1-UR, Q1-LL, and Q1-UL quadrants represent early apoptosis, late apoptosis, viability, and necrosis, respectively. (**a**) Control; (**b**) Control; (**c**) Compound **3b** at $IC_{50}/2$ concentration; (**d**) Compound **3b** at $IC_{50}$ concentration; (**e**) Compound **4a** at $IC_{50}/4$ concentration; (**f**) Compound **4a** at $IC_{50}/2$ concentration; (**g**) Cisplatin at $IC_{50}/2$ concentration; (**h**) Cisplatin at $IC_{50}$ concentration.

Akt inhibition caused by compounds **3b** and **4a** in A549 cells was examined using a colorimetric assay. Compounds **3b** and **4a** caused Akt inhibition in A549 cell line with $IC_{50}$ values of 32.50 and 45.33 μM as compared to GSK690693 ($IC_{50}$= 5.93 μM) (Table 6).

**Table 6.** Akt inhibitory effects of compounds **3b**, **4a**, GSK690693, and cisplatin in A549 cells.

| Compound | $IC_{50}$ (μM) |
|---|---|
| **3b** | 32.50 ± 4.95 |
| **4a** | 45.33 ± 6.51 |
| GSK690693 | 5.93 ± 1.20 |
| Cisplatin | 9.30 ± 2.55 |

The lipopolysaccharide (LPS)-induced sepsis model was used to assess the in vivo anti-inflammatory activities of compounds **3b** and **4a**. According to the data indicated in Table 7, the myeloperoxidase (MPO) activity of the LPS group increased as compared to the control group. However, this increase is not statistically significant. The LPS + compound **3b** group slightly decreased the MPO activity compared to the LPS group, while the LPS + compound **4a** group significantly decreased the MPO activity compared to the LPS group ($p < 0.05$). The decrease in the MPO activity caused by compound **4a** was higher than that caused by the indomethacin therapy.

**Table 7.** Effects of compounds **3b**, **4a**, and indomethacin on MPO levels.

| Groups | MPO (U/L) |
|---|---|
| Control | 1.03 ± 0.51 |
| LPS | 1.79 ± 0.27 |
| LPS + Compound **3b** | 1.50 ± 0.81 |
| LPS + Compound **4a** | 0.84 ± 0.26 # |
| LPS + Indomethacin | 1.06 ± 0.73 |

Values are given as mean ± standard deviation (SD). Significance according to LPS values, #: $p < 0.05$. One-way ANOVA, post-hoc Tukey test $n = 8$.

As presented in Table 8, there was a significant increase in the nitric oxide (NO) level after LPS administration compared to the control ($p < 0.001$). LPS + compound **3b**, LPS + compound **4a**, and LPS + indomethacin caused a significant decrease in the serum NO levels. However, this decrease in LPS + compound **3b** was similar to the control. The NO level was significantly higher in the LPS group than in the control group, while it was markedly lower in the compound **4a** pre-treatment group compared to the LPS group ($p < 0.05$).

**Table 8.** Effects of compounds **3b**, **4a,** and indomethacin on NO levels (μmol/L).

| Groups | 25% Percentile | Median | 75% Percentile |
|---|---|---|---|
| Control | 0 | 0.07 | 0.14 |
| LPS *** | 4.482 | 6.018 | 8.386 |
| LPS + Compound **3b** | 0.316 | 0.667 | 2.772 |
| LPS + Compound **4a** * | 1.369 | 2.509 | 4.0 |
| LPS + Indomethacin * | 0.667 | 1.281 | 3.123 |

Significance relative to control values, *: $p < 0.05$, ***: $p < 0.001$. One-way ANOVA, Kruskal–Wallis test, $n = 8$.

According to the in vivo experiments, the alanine aminotransferase (ALT) level decreased in all groups compared to the LPS group (Table 9). This decrease was greater in the group treated with compound **4a** compared to the group treated with compound **3b**. Likewise, the aspartate aminotransferase (AST) level decreased in all groups compared to the LPS group. This decrease was greater in the group treated with compound **3b** compared to the group treated with compound **4a** and the group treated with indomethacin. However, the decrease caused by the compounds in the ALT and AST levels was not statistically significant compared to the LPS group.

**Table 9.** Effects of compounds **3b**, **4a,** and indomethacin on ALT and AST levels.

| Groups | ALT (U/L) | AST (U/L) |
|---|---|---|
| Control | 45.75 ± 3.81 | 126.60 ± 24.09 |
| LPS | 57.14 ± 22.94 | 139.90 ± 26.34 |
| LPS + Compound **3b** | 50.25 ± 21.86 | 114.90 ± 22.04 |
| LPS + Compound **4a** | 48.29 ± 26.88 | 132.30 ± 37.06 |
| LPS + Indomethacin | 44.00 ± 14.64 | 123.50 ± 32.14 |

Values are given as mean ± SD. One-way ANOVA, post-hoc Tukey test, $n = 8$.

## 3. Discussion

Experimental studies have demonstrated that hydrazones show marked antitumor action through various mechanisms, including the inhibition of Akt [31] or the phosphatidylinositol 3-kinase (PI3K)/Akt signaling pathway [32,36]. *N′*-benzylidene-2-[(4-(4-methoxyphenyl)pyrimidin-2-yl)thio]acetohydrazide was previously reported to exert marked anticancer activity against the 5RP7 H-*ras* oncogene transformed rat embryonic fibroblast cell line via the induction of apoptosis and the inhibition of Akt ($IC_{50}$ = 0.50 μg/mL) [31]. According to western blot data reported by Han et al., (*S*)-2-{[5-[1-(6-methoxynaphtalene-2-yl)ethyl]-4-(4-fluorophenyl)-4*H*-1,2,4-triazole-3-yl]thio}-*N′*-[(5-nitrofuran-2-yl)methylidene]acetohydrazide caused a significant decrease in the epidermal growth factor receptor (EGFR), PI3K, and Akt phosphorylation in PC3 human prostate cancer cells [32]. Bak et al. indicated that 5-hydroxy-7,4′-diacetyloxyflavanone-*N*-phenyl hydrazone (N101-43) induced apoptosis via the up-regulation of Fas/FasL expression, the activation of caspase cascade, and the inhibition of the PI3K/Akt signaling pathway in NSCLC cells [36].

The anti-inflammatory and/or COX inhibitory potential of hydrazones was demonstrated by in vitro and in vivo studies [37–41]. In our previous work, 2-[(1-methyl-1*H*-tetrazol-5-yl)thio]-*N′*-(4-(piperidin-1-yl)benzylidene)acetohydrazide and 2-[(1-methyl-1*H*-tetrazol-5-yl)thio]-*N′*-(4-(morpholin-4-yl)benzylidene)acetohydrazide caused selective COX-1 inhibition [37].

Thiosemicarbazides show pronounced antiproliferative activity toward a variety of tumor cells through diverse mechanisms [42–52]. Our research team reported that 4-(1,3-benzodioxol-5-yl)-1-([1,1′-biphenyl]-4-ylmethylene)thiosemicarbazide showed remarkable anticancer activity against A549 human lung adenocarcinoma and C6 rat glioma cells through apoptosis induction mediated by the disruption of $\Delta\Psi m$ [52].

In vitro and in vivo experimental data revealed that thiosemicarbazides exert marked anti-inflammatory action through several mechanisms including COX inhibition [53–56]. In our recent work [53], 4-[4-(piperidin-1-ylsulfonyl)phenyl]-1-[4-(4-cyanophenoxy)benzylidene]thiosemicarbazide was found to be a selective COX-1 inhibitor with an $IC_{50}$ value of 1.89 μM. On the other hand, 4-[4-(piperidin-1-ylsulfonyl)phenyl]-1-[4-(4-nitrophenoxy)benzylidene]thiosemicarbazide was determined to be a nonselective COX inhibitor (COX-1 $IC_{50}$ = 13.44 μM, COX-2 $IC_{50}$ = 12.60 μM). Based on the LPS-induced sepsis model, these agents diminished the MPO, NO, high-sensitivity C-reactive protein (hsCRP), malondialdehyde (MDA), ALT, and AST levels. Both compounds were identified as potential anti-inflammatory agents [53].

Indole-based small molecules exert a notable anti-NSCLC action through multiple mechanisms such as the induction of apoptosis and the inhibition of crucial biological targets including protein kinases (e.g., Akt) [58–62]. Furthermore, mounting evidence has demonstrated that indole derivatives show marked anti-inflammatory action via COX inhibition [62–67]. In our previous study [65], 3-(5-bromo-1*H*-indol-3-yl)-1-(4-cyanophenyl)prop-2-en-1-one was found to be a nonselective COX inhibitor (COX-1 $IC_{50}$ = 8.10 μg/mL, COX-2 $IC_{50}$ = 9.50 μg/mL), while 3-(5-methoxy-1*H*-indol-3-yl)-1-(4-(methylsulfonyl)phenyl)prop-2-en-1-one inhibited COX-1 ($IC_{50}$ = 8.60 μg/mL). According to the LPS-induced sepsis model, both compounds markedly decreased the MPO, NO, hsCRP, MDA, ALT, and AST levels. Both indole derivatives were identified as potential anti-inflammatory agents [65].

Based on the aforementioned studies [26–72], two classes of indole-based small molecules (**3a-j**, **4a-g**) for the targeted therapy of NSCLC were designed. In this context, we carried out the synthesis of new hydrazones (**3a-j**) and thiosemicarbazides (**4a-g**) efficiently and performed in vitro and in vivo experiments to assess their potential for the targeted therapy of NSCLC.

Among compounds **3a-j**, compound **3a** was determined to be a nonselective COX inhibitor with $IC_{50}$ values of 10.35 μM and 12.50 μM for COX-1 and COX-2, respectively, while compound **3b** was found to be a selective COX-1 inhibitor ($IC_{50}$ = 8.90 μM). Compound **3b** exhibited COX-2 inhibitory activity with an $IC_{50}$ value of 71.00 μM. The SI values of compounds **3a** and **3b** were determined to be 0.83 and 0.13, respectively. In particular, the pyrrolidine ring enhanced the inhibitory effects on both COXs, whereas the morpholine substitution caused selective COX-1 inhibitory potency. The replacement of the morpholine ring (compound **3b**) with the piperidine ring (compound **3c**) or the piperazine ring (compound **3d**) led to a substantial drop in COX-1 inhibitory activity. Compound **3c**, carrying a piperidine ring, showed the lowest COX inhibition (>100 μM) in this series.

According to the in vitro data related to the inhibitory effects of compounds **3b** and **3h** on COXs, it can be concluded that the ethoxy linker between the morpholine and the benzene rings diminishes COX-1 inhibition, while it enhances COX-2 inhibition. Taking into account the inhibitory effects of compounds **3b** and **3g** on COXs, it is important to note that the methyl branching decreases COX-1 inhibition and increases COX-2 inhibition.

The SI values of compounds **3e** and **3f** were found to be 2.39 and 1.84, respectively, indicating that the methylsulfonyl group significantly enhances COX-2 selectivity.

Based on the experimental results related to the inhibitory effects of compounds **3i** and **3j** on COXs, the methyl substituent at the 1$^{st}$ position of the indole scaffold enhances COX-1 selectivity, while the methoxy substituent at the 5$^{th}$ position of the indole core enhances COX-2 selectivity.

Among compounds **4a-g**, the most selective COX-1 inhibitor was found to be compound **4a** ($IC_{50}$ = 10.00 μM, SI = 0.13). It can be concluded that the bromo substituent at the 4th position of the phenyl moiety significantly enhanced the COX-1 inhibitory potency.

Other compounds did not exhibit any inhibitory activity towards COX-1 at the tested concentrations. Thiosemicarbazides tested in this work, except for compound **4a**, were found to have a tendency to inhibit the COX-2 enzyme.

Among the indole-based small molecules (**3a-j**, **4a-g**), compounds **3b** and **4a**, selective COX-1 inhibitors in this series, were chosen for further studies since both compounds did not exert cytotoxicity toward L929 (normal) cells at their effective concentrations reported for their COX-1 inhibitory activity.

To investigate their potential as anti-NSCLC agents, their cytotoxic effects on A549 cells were evaluated by means of the MTT assay protocol. Based on the experimental data, compound **3b** was the most potent anticancer agent on A549 cell line with an $IC_{50}$ value of 89.67 μM. It can be concluded that the anticancer activity of compound **3b** against A549 cells is selective since the $IC_{50}$ value of compound **3b** for L929 cells is 176.67 μM. On the other hand, compound **4a** showed cytotoxic activity against A549 and L929 cells with $IC_{50}$ values of 179.33 μM and 84.00 μM, respectively. The cytotoxic activity of compound **4a** toward A549 cells was found to be nonselective at its $IC_{50}$ value. For this reason, the $IC_{50}/4$ and $IC_{50}/2$ concentrations of compound **4a** were applied in the flow cytometry analyses of apoptosis and the Akt inhibition assay.

In cancer, cells lose their ability to undergo apoptosis, resulting in uncontrolled proliferation [73]. The induction of apoptosis is reported to be an intriguing modality for the management of all types of cancer since apoptosis evasion is a hallmark of cancer and is nonspecific to the cause or the type of the cancer [74]. Based on the flow cytometry-based apoptosis detection assay performed in this work, A549 cells treated with compounds **3b** and **4a** underwent apoptosis, pointing out their apoptosis-inducing efficacy.

Akt participates in the pathogenesis of NSCLC and, therefore, the inhibition of Akt by natural and synthetic agents stands out as a rational strategy for cancer therapy [21–25]. The colorimetric assay conducted in this study revealed that the Akt inhibitory activity of compound **3b** ($IC_{50}$ = 32.50 μM) was more notable than that of compound **4a** ($IC_{50}$ = 45.33 μM) in A549 cells.

Sepsis is described as a life-threatening organ dysfunction provoked by a dysregulated host response to infection [53]. Inflammatory imbalance plays a fundamental role in the pathogenesis of sepsis and occurs throughout the whole process of sepsis [75]. The parameters related to inflammation are crucial for evaluating a sepsis case [76]. In this work, the LPS-induced sepsis model was used to evaluate the in vivo anti-inflammatory activities of compounds **3b** and **4a**.

MPO is linked to several diseases, particularly those in which strong infiltration of polymorphonuclear cells (PMNs) and acute or chronic inflammation are involved. MPO contributes to the pathophysiology of diverse diseases such as rheumatoid arthritis, atherosclerosis, pulmonary fibrosis, renal glomerular injury, multiple sclerosis, Huntington's disease, Alzheimer's disease, Parkinson's disease, liver diseases, diabetes, obesity, and cancer. MPO is reported to promote tumor initiation and progression. MPO participates in the regulation of tumor growth, apoptosis, migration, and metastasis [77]. In this work, compound **3b** caused a slight decrease in the MPO activity compared to the LPS group, whereas compound **4a** significantly diminished the MPO activity compared to the LPS group ($p < 0.05$).

Sepsis is characterized by a robust rise in NO production throughout the body that is driven by inducible NO synthase (iNOS) [78]. Due to the key role of NO in the pathogenesis of inflammation as a signaling molecule [78–80], herein the effects of compounds **3b** and **4a** on the serum NO levels were evaluated. The in vivo experimental data revealed that compounds **3b** and **4a** diminished the serum NO levels.

Aminotransferases, also referred to as transaminases, are commonly used as markers of hepatocellular injury in nonclinical toxicology studies and clinical trials. In general, aminotransferase activity in blood (serum or plasma) is elevated in the hepatocellular damage induced by diseases or drugs such as anti-inflammatory drugs [81–84]. Based on the in vivo experimental data performed in this work, both compounds caused a decrease in the serum aminotransferase levels. In particular, compound **3b** diminished the serum AST level more than indomethacin.

Taking into account the knowledge obtained from the in vitro and in vivo assays, compound **3b** can be considered as a lead compound for the targeted therapy of NSCLC due to its direct cytotoxic effects on A549 cells as well as its possible effects on the tumor microenvironment (e.g., tumor-related inflammation).

## 4. Materials and Methods

### *4.1. Chemistry*

The chemicals were procured from commercial suppliers and were used without further purification. Melting points (M.p.) were determined on the Electrothermal IA9200 digital melting point apparatus (Staffordshire, UK) and were uncorrected. Thin Layer Chromatography (TLC) was performed on TLC Silica gel 60 F254 aluminum sheets (Merck, Darmstadt, Germany) using petroleum ether:ethyl acetate solvent system (1:1). IR spectra were recorded on the IRPrestige-21 Fourier Transform Infrared spectrophotometer (Shimadzu, Tokyo, Japan). $^{1}$H and $^{13}$C NMR spectra were recorded on the Varian Mercury 400 NMR spectrometer (Agilent, Palo Alto, CA, USA). HRMS spectra were recorded on the LC/MS IT-TOF system (Shimadzu, Tokyo, Japan) using the electrospray ionization (ESI) technique.

#### 4.1.1. Preparation of ethyl 2-(1*H*-indol-3-yl)acetate (**1**)

Compound **1** was synthesized starting from 2-(1*H*-indol-3-yl)acetic acid according to a previous work [85].

#### 4.1.2. Preparation of 2-(1*H*-indol-3-yl)acetohydrazide (**2**)

Compound **2** was obtained by the reaction of compound **1** with hydrazine hydrate according to a previous work [85].

#### 4.1.3. General Method for the Preparation of *N′*-benzylidene/(1-arylethylidene)-2-(1*H*-indol-3-yl)acetohydrazide Derivatives (**3a-j**)

A mixture of compound **2** and aromatic aldehyde or ketone in ethanol was heated under reflux for 15 h. At the end of this period, the precipitate was filtered off and dried. The product was crystallized from ethanol.

2-(1*H*-Indol-3-yl)-*N′*-[4-(pyrrolidin-1-yl)benzylidene]acetohydrazide (**3a**)

Yield: 78%. M.p.: 302–303 °C. IR $\nu_{max}$ (cm$^{-1}$): 3196.05, 3074.53, 3043.67, 2966.52, 2914.44, 2873.94, 2848.86, 1668.43, 1595.13, 1546.91, 1521.84, 1487.12, 1460.11, 1431.18, 1386.82, 1350.17, 1323.17, 1292.31, 1249.87, 1224.80, 1174.65, 1163.08, 1118.71, 1047.35, 1001.06, 983.70, 958.62, 929.69, 914.26, 856.39, 804.32, 719.45, 682.80. $^{1}$H NMR (400 MHz, DMSO-$d_6$): 2.02–2.05 (m, 4H), 3.40–3.42 (m, 4H), 3.60 and 4.02 (2s, 2H), 6.92–7.07 (m, 4H), 7.21 (dd, *J* = 2.4 Hz, 12.8 Hz, 1H), 7.31–7.35 (m, 1H), 7.48–7.59 (m, 3H), 7.88 and 8.08 (2s, 1H), 10.85 and 10.88 (2s, 1H), 11.05 and 11.28 (2s, 1H). $^{13}$C NMR (100 MHz, DMSO-$d_6$): 25.41 (2CH$_2$), 32.13 (CH$_2$), 47.70 (2CH$_2$), 108.79 (C), 111.74 (CH), 115.13 (2CH), 118.74 (CH), 119.19 (CH), 121.33 (CH), 124.32 (CH), 124.75 (C), 127.90 (C), 128.27 (2CH), 136.45 (C), 146.90 (CH), 152.26 (C), 172.72 (C). HRMS (ESI) (*m/z*): [M + H]$^+$ calcd. for $C_{21}H_{22}N_4O$: 347.1866, found: 347.1864.

2-(1*H*-Indol-3-yl)-*N′*-(4-morpholinobenzylidene)acetohydrazide (**3b**)

Yield: 85%. M.p.: 306–307 °C. IR $\nu_{max}$ (cm$^{-1}$): 3275.13, 3178.69, 3055.24, 2964.59, 2922.16, 2870.08, 2825.72, 1660.71, 1604.77, 1558.48, 1541.12, 1519.91, 1506.41, 1489.05, 1456.26, 1446.61, 1425.40, 1392.61, 1375.25, 1338.60, 1313.52, 1301.95, 1259.52, 1224.80, 1186.22, 1176.58, 1159.22, 1109.07, 1095.57, 1062.78, 1045.42, 1006.84, 958.62, 921.97, 875.68, 858.32, 846.75, 823.60, 798.53, 786.96, 742.59, 682.80. $^{1}$H NMR (400 MHz, DMSO-$d_6$): 3.19 (t, *J* = 4.41 Hz, 4.62 Hz, 4H), 3.72–3.75 (m, 4H), 3.61 and 4.02 (2s, 2H), 6.92–7.08 (m, 4H), 7.21 (dd, *J* = 2.4 Hz, 12.8 Hz, 1H), 7.31–7.35 (m, 1H), 7.49–7.60 (m, 3H), 7.88 and 8.07 (2s,

1H), 10.85 and 10.89 (2s, 1H), 11.05 and 11.28 (2s, 1H). $^{13}$C NMR (100 MHz, DMSO-$d_6$): 32.13 ($CH_2$), 53.79 (2$CH_2$), 66.40 (2$CH_2$), 108.79 (C), 111.74 (CH), 115.13 (2CH), 118.74 (CH), 119.19 (CH), 121.33 (CH), 124.32 (CH), 124.75 (C), 127.91 (C), 128.27 (2CH), 136.45 (C), 146.90 (CH), 152.29 (C), 172.73 (C). HRMS (ESI) ($m/z$): $[M + H]^+$ calcd. for $C_{21}H_{22}N_4O_2$: 363.1816, found: 363.1824.

2-(1*H*-Indol-3-yl)-*N'*-[4-(piperidin-1-yl)benzylidene]acetohydrazide (**3c**)

Yield: 80%. M.p.: 265–266 °C. IR $\nu_{max}$ (cm$^{-1}$): 3203.76, 3082.25, 3034.03, 2972.31, 2935.66, 2856.58, 2825.72, 1668.43, 1598.99, 1552.70, 1514.12, 1448.54, 1427.32, 1384.89, 1350.17, 1282.66, 1247.94, 1220.94, 1182.36, 1124.50, 1024.20, 962.48, 914.26, 858.32, 804.32, 721.38, 651.94. $^{1}$H NMR (400 MHz, DMSO-$d_6$): 1.58 (brs, 6H), 3.32 (brs, 4H), 3.60 and 4.02 (2s, 2H), 6.92–7.08 (m, 4H), 7.21 (dd, $J$ = 2.4 Hz, 12.8 Hz, 1H), 7.31–7.35 (m, 1H), 7.48–7.59 (m, 3H), 7.88 and 8.08 (2s, 1H), 10.85 and 10.89 (2s, 1H), 11.05 and 11.28 (2s, 1H). $^{13}$C NMR (100 MHz, DMSO-$d_6$): 24.40 ($CH_2$), 25.42 (2$CH_2$), 32.12 ($CH_2$), 48.95 (2$CH_2$), 108.79 (C), 111.74 (CH), 115.13 (2CH), 118.74 (CH), 119.19 (CH), 121.33 (CH), 124.32 (CH), 124.75 (C), 127.91 (C), 128.27 (2CH), 136.45 (C), 146.90 (CH), 152.26 (C), 172.73 (C). HRMS (ESI) ($m/z$): $[M + H]^+$ calcd. for $C_{22}H_{24}N_4O$: 361.2023, found: 361.2031.

2-(1*H*-Indol-3-yl)-*N'*-[4-(4-methylpiperazin-1-yl)benzylidene]acetohydrazide (**3d**)

Yield: 81%. M.p.: 218–220 °C. IR $\nu_{max}$ (cm$^{-1}$): 3398.57, 3205.69, 3165.19, 3111.18, 3043.67, 2939.52, 2883.58, 2831.50, 1649.49, 1602.85, 1517.98, 1446.61, 1427.32, 1409.96, 1377.17, 1340.53, 1286.52, 1232.51, 1184.29, 1159.22, 1141.86, 1124.50, 1105.21, 1080.14, 1001.06, 956.69, 943.19, 921.97, 806.25, 794.67, 742.59, 686.66. $^{1}$H NMR (400 MHz, DMSO-$d_6$): 2.21 (s, 3H), 2.40–2.44 (m, 4H), 3.20–3.22 (m, 4H), 3.59 and 4.02 (2s, 2H), 6.92–7.08 (m, 4H), 7.21 (dd, $J$ = 2.4 Hz, 12.8 Hz, 1H), 7.31–7.35 (m, 1H), 7.48–7.59 (m, 3H), 7.88 and 8.08 (2s, 1H), 10.85 and 10.89 (2s, 1H), 11.05 and 11.28 (2s, 1H). $^{13}$C NMR (100 MHz, DMSO-$d_6$): 32.13 ($CH_2$), 46.21 ($CH_3$), 47.69 (2$CH_2$), 54.89 (2$CH_2$), 108.79 (C), 111.74 (CH), 115.13 (2CH), 118.74 (CH), 119.19 (CH), 121.33 (CH), 124.32 (CH), 124.75 (C), 127.91 (C), 128.27 (2CH), 136.45 (C), 146.90 (CH), 152.28 (C), 172.73 (C). HRMS (ESI) ($m/z$): $[M + H]^+$ calcd. for $C_{22}H_{25}N_5O$: 376.2132, found: 376.2148.

2-(1*H*-Indol-3-yl)-*N'*-(4-methylsulfonylbenzylidene)acetohydrazide (**3e**)

Yield: 86%. M.p.: 264–265 °C. IR $\nu_{max}$ (cm$^{-1}$): 3344.57, 3205.69, 3055.24, 2929.87, 2897.08, 1668.43, 1604.77, 1556.55, 1489.05, 1454.33, 1408.04, 1365.60, 1328.95, 1313.52, 1290.38, 1242.16, 1222.87, 1199.72, 1145.72, 1089.78, 1055.06, 1018.41, 983.70, 972.12, 956.69, 943.19, 869.90, 835.18, 792.74, 769.60, 750.31, 729.09, 686.66, 651.94. $^{1}$H NMR (400 MHz, DMSO-$d_6$): 3.24 (s, 3H), 3.67 and 4.09 (2s, 2H), 6.94–7.09 (m, 2H), 7.25 (dd, $J$ = 2.4 Hz, 9.6 Hz, 1H), 7.34 (t, $J$ = 8.0 Hz, 8.4 Hz, 1H), 7.56–7.60 (m, 1H), 7.90–7.96 (m, 4H), 8.07 and 8.31 (2s, 1H), 10.87 and 10.93 (2s, 1H), 11.52 and 11.76 (2s, 1H). $^{13}$C NMR (100 MHz, DMSO-$d_6$): 31.87 ($CH_2$), 43.64 ($CH_3$), 108.10 (C), 111.51 (CH), 118.53 (CH), 118.87 (CH), 121.12 (CH), 124.18 (CH), 127.31 (C), 127.50 (2CH), 127.70 (2CH), 136.19 (C), 139.35 (C), 141.21 (C), 144.37 (CH), 173.17 (C). HRMS (ESI) ($m/z$): $[M + H]^+$ calcd. for $C_{18}H_{17}N_3O_3S$: 356.1063, found: 356.1071.

2-(1*H*-Indol-3-yl)-*N'*-[1-(4-methylsulfonylphenyl)ethylidene]acetohydrazide (**3f**)

Yield: 81%. M.p.: 204–205 °C. IR $\nu_{max}$ (cm$^{-1}$): 3342.64, 3190.26, 3088.03, 3032.10, 3005.10, 2924.09, 2848.86, 1668.43, 1585.49, 1562.34, 1489.05, 1456.26, 1417.68, 1394.53, 1338.60, 1296.16, 1280.73, 1226.73, 1188.15, 1145.72, 1093.64, 1070.49, 1008.77, 977.91, 964.41, 852.54, 839.03, 788.89, 758.02, 740.67, 717.52, 700.16, 688.59. $^{1}$H NMR (400 MHz, DMSO-$d_6$): 2.30 (s, 3H), 3.26 (s, 3H), 3.81 and 4.12 (2s, 2H), 6.95–7.07 (m, 2H), 7.20–7.36 (m, 2H), 7.54–7.61 (m, 1H), 8.01 (d, $J$ = 8.0 Hz, 2H), 8.15 (d, $J$ = 8.8 Hz, 2H), 10.63 and 10.66 (2s, 1H), 10.84 and 10.89 (2s, 1H). $^{13}$C NMR (100 MHz, DMSO-$d_6$): 14.96 ($CH_3$), 31.87 ($CH_2$), 43.42 ($CH_3$), 108.10 (C), 111.26 (CH), 118.25 (CH), 118.64 (CH), 121.12 (CH), 123.84 (CH), 126.70 (C), 127.11 (2CH), 127.39 (2CH), 141.55 (C), 142.06 (C), 142.97 (C), 156.20 (C), 173.17 (C). HRMS (ESI) ($m/z$): $[M + H]^+$ calcd. for $C_{19}H_{19}N_3O_3S$: 370.1220, found: 370.1202.

2-(1*H*-Indol-3-yl)-*N'*-[1-(4-morpholinophenyl)ethylidene]acetohydrazide (**3g**)

Yield: 79%. M.p.: 198–199 °C. IR $\nu_{max}$ (cm$^{-1}$): 3269.34, 3080.32, 3047.53, 2966.52, 2916.37, 2848.86, 1668.43, 1608.63, 1593.20, 1546.91, 1516.05, 1454.33, 1444.68, 1417.68,

1379.10, 1361.74, 1340.53, 1301.95, 1263.37, 1236.37, 1197.79, 1118.71, 1068.56, 1051.20, 1026.13, 937.40, 923.90, 864.11, 821.68, 798.53, 742.59, 729.09, 648.08. $^1$H NMR (400 MHz, DMSO-$d_6$): 2.31 (s, 3H), 3.19 (t, *J* = 4.41 Hz, 4.62 Hz, 4H), 3.72–3.75 (m, 4H), 3.60 and 4.02 (2s, 2H), 6.92–7.08 (m, 4H), 7.21 (dd, *J* = 2.4 Hz, 12.8 Hz, 1H), 7.31–7.35 (m, 1H), 7.48–7.59 (m, 3H), 10.85 and 10.89 (2s, 1H), 11.05 and 11.27 (2s, 1H). $^{13}$C NMR (100 MHz, DMSO-$d_6$): 14.96 ($CH_3$), 32.13 ($CH_2$), 53.79 ($2CH_2$), 66.40 ($2CH_2$), 108.79 (C), 111.74 (CH), 115.13 (2CH), 118.74 (CH), 119.19 (CH), 121.33 (CH), 124.32 (CH), 124.75 (C), 127.91 (C), 128.27 (2CH), 136.45 (C), 143.30 (C), 156.18 (C), 172.73 (C). HRMS (ESI) (*m/z*): $[M + H]^+$ calcd. for $C_{22}H_{24}N_4O_2$: 377.1972, found: 377.1982.

2-(1*H*-Indol-3-yl)-*N'*-[4-(2-morpholinoethoxy)benzylidene]acetohydrazide (**3h**)

Yield: 83%. M.p.: 132–135 °C. IR $\nu_{max}$ ($cm^{-1}$): 3383.14, 3319.49, 3196.05, 3045.60, 2958.80, 2918.30, 2850.79, 1664.57, 1604.77, 1548.84, 1510.26, 1456.26, 1421.54, 1355.96, 1340.53, 1303.88, 1240.23, 1201.65, 1170.79, 1116.78, 1049.28, 1010.70, 983.70, 952.84, 925.83, 860.25, 831.32, 742.59, 646.15. $^1$H NMR (400 MHz, DMSO-$d_6$): 2.45–2.47 (m, 4H), 2.66–2.70 (m, 2H), 3.55–3.58 (m, 4H), 4.04 (s, 2H), 4.08–4.12 (m, 2H), 6.95–7.07 (m, 4H), 7.23 (dd, *J* = 2.4 Hz, 12.4 Hz, 1H), 7.34 (t, *J* = 8.0 Hz, 1H), 7.57–7.65 (m, 3H), 7.94 and 8.16 (2s, 1H), 10.86 and 10.91 (2s, 1H), 11.15 and 11.39 (2s, 1H). $^{13}$C NMR (100 MHz, DMSO-$d_6$): 31.85 ($CH_2$), 53.79 ($2CH_2$), 57.12 ($CH_2$), 65.64 ($CH_2$), 66.35 ($2CH_2$), 108.44 (C), 111.48 (CH), 115.05 (2CH), 118.47 (CH), 118.91 (CH), 121.07 (CH), 124.07 (CH), 127.18 (C), 127.63 (C), 128.41 (2CH), 136.19 (C), 146.18 (CH), 159.90 (C), 172.64 (C). HRMS (ESI) (*m/z*): $[M + H]^+$ calcd. for $C_{23}H_{26}N_4O_3$: 407.2078, found: 407.2071.

2-(1*H*-Indol-3-yl)-*N'*-[(1-methyl-1*H*-indol-3-yl)methylene]acetohydrazide (**3i**)

Yield: 84%. M.p.: 221–224 °C. IR $\nu_{max}$ ($cm^{-1}$): 3414.00, 3147.83, 3101.54, 3061.03, 2980.02, 2945.30, 2908.65, 2819.93, 1651.07, 1612.49, 1570.06, 1539.20, 1502.55, 1462.04, 1452.40, 1421.54, 1404.18, 1377.17, 1346.31, 1332.81, 1321.24, 1253.73, 1244.09, 1197.79, 1157.29, 1139.93, 1120.64, 1087.85, 1072.42, 1045.42, 1008.77, 948.98, 933.55, 900.76, 856.39, 808.17, 785.03, 744.52, 734.88, 673.16. $^1$H NMR (400 MHz, DMSO-$d_6$): 3.64 and 4.13 (2s, 2H), 3.79 (s, 3H), 6.94–7.18 (m, 3H), 7.22–7.28 (m, 2H), 7.36 (t, *J* = 8.4 Hz, 8.8 Hz, 1H), 7.47 (t, *J* = 8.4 Hz, 9.2 Hz, 1H), 7.64 (t, *J* = 7.6 Hz, 1H), 7.73 (d, *J* = 2.4 Hz, 1H), 8.20–8.38 (m, 2H), 10.86 and 10.92 (2s, 1H), 10.99 and 11.19 (2s, 1H). $^{13}$C NMR (100 MHz, DMSO-$d_6$): 31.94 ($CH_2$), 32.90 ($CH_3$), 108.74 (C), 110.42 (CH), 110.83 (CH), 111.49 (C), 118.50 (CH), 118.93 (CH), 120.89 (CH), 121.10 (CH), 121.86 (CH), 122.80 (CH), 123.96 (CH), 124.70 (C), 127.70 (C), 133.81 (CH), 136.23 (C), 137.76 (C), 142.98 (CH), 172.09 (C). HRMS (ESI) (*m/z*): $[M + H]^+$ calcd. for $C_{20}H_{18}N_4O$: 331.1553, found: 331.1538.

2-(1*H*-Indol-3-yl)-*N'*-[(5-methoxy-1*H*-indol-3-yl)methylene]acetohydrazide (**3j**)

Yield: 82%. M.p.: 231–233 °C. IR $\nu_{max}$ ($cm^{-1}$): 3415.93, 3373.50, 3049.46, 3012.81, 2958.80, 2931.80, 2877.79, 2829.57, 1654.92, 1614.42, 1577.77, 1539.20, 1487.12, 1456.26, 1421.54, 1396.46, 1354.03, 1342.46, 1307.74, 1292.31, 1261.45, 1213.23, 1182.36, 1176.58, 1130.29, 1105.21, 1087.85, 1072.42, 1049.28, 1022.27, 1006.84, 950.91, 923.90, 856.39, 810.10, 744.52, 725.23, 671.23, 651.94. $^1$H NMR (400 MHz, DMSO-$d_6$): 3.59 (s, 3H), 3.74 and 4.16 (2s, 2H), 6.83 (dd, *J* = 2.4 Hz, 8.8 Hz, 1H), 6.93–7.10 (m, 2H), 7.27–7.38 (m, 3H), 7.64 (t, *J* = 8.8 Hz, 9.2 Hz, 1H), 7.72–7.80 (m, 2H), 8.23 and 8.41 (2s, 1H), 10.85 and 10.91 (2s, 1H), 11.02 and 11.18 (2s, 1H), 11.40 (s, 1H). $^{13}$C NMR (100 MHz, DMSO-$d_6$): 31.90 ($CH_2$), 55.01 ($CH_3$), 103.47 (CH), 108.71 (C), 111.50 (C), 111.59 (CH), 112.39 (CH), 112.64 (CH), 118.52 (CH), 118.83 (CH), 121.14 (CH), 123.88 (CH), 124.83 (C), 127.73 (C), 130.56 (C), 132.19 (CH), 136.19 (C), 143.76 (CH), 154.55 (C), 172.04 (C). HRMS (ESI) (*m/z*): $[M + H]^+$ calcd. for $C_{20}H_{18}N_4O_2$: 347.1503, found: 347.1505.

#### 4.1.4. General Method for the Preparation of 4-aryl-1-[2-(1*H*-indol-3-yl)acetyl]thiosemicar bazide Derivatives (**4a-g**)

A mixture of compound **2** and aryl isothiocyanate in ethanol was stirred at room temperature for 8 h. The precipitate was filtered off. The product was crystallized from ethanol.

4-(4-Bromophenyl)-1-[2-(1*H*-indol-3-yl)acetyl]thiosemicarbazide (**4a**)

Yield: 87%. M.p.: 187–189 °C. IR $\nu_{max}$ ($cm^{-1}$): 3390.86, 3311.78, 3286.70, 3207.62, 3143.97, 3057.17, 2997.38, 2927.94, 1680.00, 1647.21, 1620.21, 1589.34, 1544.98, 1506.41, 1485.19, 1452.40, 1419.61, 1352.10, 1309.67, 1282.66, 1247.94, 1207.44, 1138.00, 1087.85, 1074.35, 1049.28, 1004.91, 987.55, 871.82, 823.60, 792.74, 736.81, 715.59, 669.30. $^1$H NMR (400 MHz, DMSO-$d_6$): 3.63 (s, 2H), 6.97 (t, *J* = 6.8 Hz, 1H), 7.07 (t, *J* = 6.8 Hz, 1H), 7.25 (d, *J* = 2.4 Hz, 1H), 7.34 (d, *J* = 8.0 Hz, 1H), 7.42 (d, *J* = 8.0 Hz, 2H), 7.51 (d, *J* = 8.4 Hz, 2H), 7.59 (d, *J* = 7.6 Hz, 1H), 9.59 (brs, 1H), 9.73 (s, 1H), 10.10 (brs, 1H), 10.89 (s, 1H). $^{13}$C NMR (100 MHz, DMSO-$d_6$): 31.18 ($CH_2$), 108.39 (C), 111.75 (CH), 118.80 (CH), 119.26 (CH), 121.46 (CH), 122.70 (C), 124.44 (CH), 127.70 (C), 129.40 (2CH), 131.37 (2CH), 136.51 (C), 139.05 (C), 170.35 (C), 181.10 (C). HRMS (ESI) (*m*/*z*): $[M + H]^+$ calcd. for $C_{17}H_{15}BrN_4OS$: 403.0223, found: 403.0204.

4-(4-Trifluoromethylphenyl)-1-[2-(1*H*-indol-3-yl)acetyl]thiosemicarbazide (**4b**)

Yield: 80%. M.p.: 184–186 °C. IR $\nu_{max}$ ($cm^{-1}$): 3392.79, 3315.63, 3292.49, 3223.05, 3163.26, 3070.68, 2995.45, 2927.94, 1681.93, 1649.14, 1616.35, 1568.13, 1544.98, 1504.48, 1454.33, 1419.61, 1357.89, 1321.24, 1246.02, 1224.80, 1209.37, 1184.29, 1163.08, 1132.21, 1120.64, 1112.93, 1085.92, 1070.49, 1012.63, 985.62, 846.75, 788.89, 736.81, 711.73, 665.44. $^1$H NMR (400 MHz, DMSO-$d_6$): 3.64 (s, 2H), 6.97 (t, *J* = 6.8 Hz, 1H), 7.07 (t, *J* = 6.8 Hz, 1H), 7.26 (d, *J* = 2.0 Hz, 1H), 7.34 (d, *J* = 8.0 Hz, 1H), 7.60 (d, *J* = 7.6 Hz, 1H), 7.67–7.75 (m, 4H), 9.75 (brs, 1H), 9.88 (s, 1H), 10.14 (brs, 1H), 10.89 (s, 1H). $^{13}$C NMR (100 MHz, DMSO-$d_6$): 31.16 ($CH_2$), 108.36 (C), 111.76 (CH), 118.80 (CH), 119.25 (CH), 121.45 (CH), 123.45 (CH), 124.46 (2CH), 125.64 (C), 126.15 (C), 127.69 (2CH), 132.50 (C), 136.51 (C), 143.45 (C), 170.35 (C), 181.10 (C). HRMS (ESI) (*m*/*z*): $[M + H]^+$ calcd. for $C_{18}H_{15}F_3N_4OS$: 393.0991, found: 393.0989.

4-(4-Cyanophenyl)-1-[2-(1*H*-indol-3-yl)acetyl]thiosemicarbazide (**4c**)

Yield: 89%. M.p.: 180–182 °C. IR $\nu_{max}$ ($cm^{-1}$): 3425.58, 3313.71, 3284.77, 3201.83, 3145.90, 3059.10, 2995.45, 2956.87, 2914.44, 2223.92, 1680.00, 1651.07, 1620.21, 1602.85, 1541.12, 1510.26, 1475.54, 1454.33, 1409.96, 1334.74, 1290.38, 1244.09, 1226.73, 1203.58, 1174.65, 1136.07, 1093.64, 1060.85, 1012.63, 975.98, 837.11, 790.81, 769.60, 734.88, 692.44. $^1$H NMR (400 MHz, DMSO-$d_6$): 3.64 (s, 2H), 6.97 (t, *J* = 7.2 Hz, 7.6 Hz, 1H), 7.07 (t, *J* = 7.2 Hz, 7.6 Hz, 1H), 7.26 (s, 1H), 7.35 (d, *J* = 7.6 Hz, 1H), 7.60 (d, *J* = 7.2 Hz, 1H), 7.78 (s, 4H), 9.76 (brs, 1H), 9.97 (s, 1H), 10.16 (brs, 1H), 10.90 (s, 1H). $^{13}$C NMR (100 MHz, DMSO-$d_6$): 31.17 ($CH_2$), 108.04 (C), 109.58 (C), 111.49 (CH), 118.53 (C), 118.97 (CH), 119.16 (CH), 121.20 (CH), 124.20 (CH), 127.40 (C), 129.32 (2CH), 132.55 (2CH), 136.24 (C), 143.87 (C), 170.35 (C), 181.10 (C). HRMS (ESI) (*m*/*z*): $[M + H]^+$ calcd. for $C_{18}H_{15}N_5OS$: 350.1070, found: 350.1063.

4-[4-(Piperidin-1-ylsulfonyl)phenyl]-1-[2-(1*H*-indol-3-yl)acetyl]thiosemicarbazide (**4d**)

Yield: 85%. M.p.: 182–184 °C. IR $\nu_{max}$ ($cm^{-1}$): 3390.86, 3288.63, 3197.98, 3089.96, 2939.52, 2850.79, 1645.28, 1595.13, 1550.77, 1496.76, 1467.83, 1404.18, 1336.67, 1315.45, 1276.88, 1244.09, 1226.73, 1215.15, 1149.57, 1093.64, 1053.13, 1028.06, 1012.63, 983.70, 929.69, 860.25, 839.03, 819.75, 777.31, 752.24, 738.74, 719.45, 698.23, 667.37. $^1$H NMR (400 MHz, DMSO-$d_6$): 1.33–1.36 (m, 2H), 1.50–1.54 (m, 4H), 2.87 (t, *J* = 4.8 Hz, 5.2 Hz, 4H), 3.66 (s, 2H), 6.98 (t, *J* = 7.2 Hz, 1H), 7.08 (t, *J* = 7.2 Hz, 1H), 7.27 (d, *J* = 2.0 Hz, 1H), 7.36 (d, *J* = 7.6 Hz, 1H), 7.61 (d, *J* = 8.0 Hz, 1H), 7.67 (d, *J* = 8.4 Hz, 2H), 7.83 (d, *J* = 8.4 Hz, 2H), 9.75 (brs, 1H), 9.93 (s, 1H), 10.16 (brs, 1H), 10.90 (s, 1H). $^{13}$C NMR (100 MHz, DMSO-$d_6$): 23.37 ($CH_2$), 25.14 ($2CH_2$), 31.18 ($CH_2$), 47.08 ($2CH_2$), 108.35 (C), 111.78 (CH), 118.83 (CH), 119.26 (CH), 121.48 (CH), 124.49 (CH), 125.05 (2CH), 127.68 (C), 128.11 (2CH), 135.47 (C), 136.53 (C), 143.89 (C), 170.34 (C), 181.11 (C). HRMS (ESI) (*m*/*z*): $[M + H]^+$ calcd. for $C_{22}H_{25}N_5O_3S_2$: 472.1472, found: 472.1452.

4-[4-(1*H*-Pyrazol-1-yl)phenyl]-1-[2-(1*H*-indol-3-yl)acetyl]thiosemicarbazide (**4e**)

Yield: 85%. M.p.: 196–198 °C. IR $\nu_{max}$ ($cm^{-1}$): 3305.99, 3223.05, 3167.12, 3134.33, 3095.75, 3061.03, 2999.31, 2933.73, 1680.00, 1647.21, 1622.13, 1573.91, 1546.91, 1523.76, 1454.33, 1421.54, 1396.46, 1359.82, 1332.81, 1317.38, 1305.81, 1249.87, 1222.87, 1199.72, 1159.22, 1136.07, 1124.50, 1089.78, 1043.49, 1033.85, 1008.77, 985.62, 935.48, 840.96, 792.74, 758.02, 744.52, 717.52, 667.37. $^1$H NMR (400 MHz, DMSO-$d_6$): 3.65 (s, 2H), 6.53 (t, *J* = 2.4 Hz,

1H), 6.99 (t, $J$ = 7.2 Hz, 1H), 7.08 (t, $J$ = 7.2 Hz, 1H), 7.27 (d, $J$ = 2.0 Hz, 1H), 7.36 (d, $J$ = 8.0 Hz, 1H), 7.55 (d, $J$ = 8.0 Hz, 1H), 7.62 (d, $J$ = 8.0 Hz, 2H), 7.74 (d, $J$ = 1.6 Hz, 1H), 7.80 (d, $J$ = 8.8 Hz, 2H), 8.46 (d, $J$ = 2.4 Hz, 1H), 9.65 (brs, 1H), 9.71 (s, 1H), 10.13 (s, 1H), 10.90 (s, 1H). $^{13}$C NMR (100 MHz, DMSO-$d_6$): 31.20 ($CH_2$), 108.23 (C), 108.45 (CH), 111.77 (CH), 118.56 (CH), 118.82 (CH), 119.28 (2CH), 121.48 (CH), 124.47 (CH), 126.80 (CH), 127.73 (C), 128.09 (2CH), 136.53 (C), 137.64 (2C), 141.30 (CH), 170.32 (C), 181.10 (C). HRMS (ESI) ($m/z$): $[M + H]^+$ calcd. for $C_{20}H_{18}N_6OS$: 391.1336, found: 391.1334.

4-(1,3-Benzodioxol-5-yl)-1-[2-(1*H*-indol-3-yl)acetyl]thiosemicarbazide (**4f**)

Yield: 81%. M.p.: 168–170 °C. IR $\nu_{max}$ (cm$^{-1}$): 3300.20, 3209.55, 3149.76, 3057.17, 2929.87, 2897.08, 1678.07, 1643.35, 1591.27, 1539.20, 1500.62, 1481.33, 1454.33, 1419.61, 1334.74, 1282.66, 1240.23, 1197.79, 1122.57, 1089.78, 1037.70, 981.77, 923.90, 850.61, 815.89, 808.17, 790.81, 731.02, 698.23. $^1$H NMR (400 MHz, DMSO-$d_6$): 3.62 (s, 2H), 6.02 (s, 2H), 6.72 (d, $J$ = 8.4 Hz, 1H), 6.87 (d, $J$ = 8.0 Hz, 1H), 6.96–7.09 (m, 3H), 7.26 (s, 1H), 7.34 (d, $J$ = 8.4 Hz, 1H), 7.60 (d, $J$ = 8.0 Hz, 1H), 9.47 (brs, 1H), 9.55 (s, 1H), 10.06 (s, 1H), 10.89 (s, 1H). $^{13}$C NMR (100 MHz, DMSO-$d_6$): 30.65 ($CH_2$), 101.21 ($CH_2$), 107.38 (C), 107.97 (CH), 111.25 (CH), 118.29 (2CH), 118.78 (CH), 120.95 (CH), 123.95 (CH), 127.24 (CH), 133.08 (C), 136.02 (2C), 144.59 (C), 146.56 (C), 170.35 (C), 181.10 (C). HRMS (ESI) ($m/z$): $[M + H]^+$ calcd. for $C_{18}H_{16}N_4O_3S$: 369.1016, found: 369.0998.

4-[4-(Benzyloxy)phenyl]-1-[2-(1*H*-indol-3-yl)acetyl]thiosemicarbazide (**4g**)

Yield: 88%. M.p.: 198–200 °C. IR $\nu_{max}$ (cm$^{-1}$): 3394.72, 3290.56, 3213.41, 3155.54, 3059.10, 3032.10, 2939.52, 2873.94, 1681.93, 1649.14, 1618.28, 1564.27, 1546.91, 1504.48, 1456.26, 1417.68, 1381.03, 1359.82, 1294.24, 1244.09, 1219.01, 1170.79, 1138.00, 1089.78, 1051.20, 999.13, 912.33, 879.54, 829.39, 790.81, 734.88, 702.09, 646.15. $^1$H NMR (400 MHz, DMSO-$d_6$): 3.64 (s, 2H), 5.10 (s, 2H), 6.97–7.01 (m, 3H), 7.06–7.10 (m, 1H), 7.27–7.29 (m, 3H), 7.33–7.47 (m, 6H), 7.61 (d, $J$ = 8.0 Hz, 1H), 9.47 (brs, 1H), 9.53 (s, 1H), 10.07 (s, 1H), 10.90 (s, 1H). $^{13}$C NMR (100 MHz, DMSO-$d_6$): 30.68 ($CH_2$), 69.34 ($CH_2$), 107.99 (C), 111.27 (CH), 114.22 (2CH), 118.31 (2CH), 118.79 (CH), 120.96 (CH), 123.95 (CH), 127.25 (C), 127.67 (2CH), 127.80 (2CH), 128.41 (2CH), 132.15 (C), 136.03 (C), 137.08 (C), 155.77 (C), 170.39 (C), 181.17 (C). HRMS (ESI) ($m/z$): $[M + H]^+$ calcd. for $C_{24}H_{22}N_4O_2S$: 431.1536, found: 431.1554.

### 4.2. *Biochemistry*

#### 4.2.1. In Vitro COX Inhibition Assay

COX (ovine) Colorimetric Inhibitor Screening Assay (Cayman, Ann Arbor, MI, USA) was conducted to detect the peroxidase component of COX-1 and COX-2 according to the manufacturer's instructions [53]. The assay was performed in triplicate. Half maximal inhibitory concentration ($IC_{50}$) data (μM) were expressed as mean ± SD.

#### 4.2.2. Cell Culture and Drug Treatment

A549 human lung adenocarcinoma and L929 mouse fibroblast cell lines were obtained from American Type Culture Collection (ATCC) (Manassas, VA, USA). Both cell lines were cultured, and drug treatments were carried out as previously reported [31,86].

#### 4.2.3. MTT Assay

MTT assay was conducted as previously explained in the literature [87] with small modifications [86]. Cisplatin was used as a positive control. The assay was performed in triplicate. $IC_{50}$ data (μM) were expressed as mean ± SD.

#### 4.2.4. Flow Cytometry-Based Apoptosis Detection

FITC Annexin V Apoptosis Detection kit (BD Pharmingen, San Jose, CA, USA) was applied based on the manufacturer's instructions after the incubation of A549 cells with compound **4a** (at its $IC_{50}/4$ and $IC_{50}/2$ concentrations), compound **3b**, and cisplatin (at their $IC_{50}/2$ and $IC_{50}$ concentrations) for 24 h [87].

4.2.5. Determination of Akt Inhibition

After A549 cells were incubated with compounds **3b** (22.42 μM, 44.84 μM, 89.67 μM), **4a** (22.42 μM, 44.84 μM, 89.67 μM), Akt inhibitor GSK690693 (3.61 μM, 7.23 μM, 14.45 μM), and cisplatin (5.67 μM, 11.34 μM, 22.67 μM) for 24 h, Akt Colorimetric In-Cell ELISA Kit (Thermo Fisher Scientific, Waltham, MA, USA) was used according to the manufacturer's instructions [87]. The assay was performed in triplicate. $IC_{50}$ data (μM) were expressed as mean ± SD.

4.2.6. Experimental Animals

Male albino Sprague Dawley rats (~250–300 g) were procured from the Medical and Surgical Experimental Animals Application and Research Center of Eskisehir Osmangazi University (ESOGU). In the animal house, the rats were housed in stainless steel cages under standard atmospheric conditions at 22 ± 1 °C and exposed to 12 h/12 h light/dark cycle [53]. Food and water were given ad libitum. All experiments and protocols reported in this work were approved by ESOGU Animal Experiments Local Ethics Committee (10 December 2018/700).

4.2.7. Chemicals and Drug Administrations

Compounds **3b**, **4a**, and indomethacin (Sigma-Aldrich, St. Louis, MO, USA) were dissolved in 5% dimethyl sulfoxide (DMSO) and then diluted. The final DMSO concentration in the solution was 0.5% (*v*/*v*). The agents were administered by gastric intubation. LPS (Sigma-Aldrich, St. Louis, MO, USA) (1 mg/kg) dissolved in 0.9% sodium chloride solution was intraperitoneally injected only once on the 7th day for the experimentally induced sepsis model [53].

4.2.8. In vivo Experimental Design

Rats were randomly divided into five groups ($n$ = 8) as control group, LPS group, test groups (**3b** and **4a**), and reference group. 0.5% DMSO was used as control solution for LPS group. Indomethacin (5 mg/kg) was used as a reference agent. Control group (Group I) was fed with basal rat chow throughout the experimental period. LPS group (Group II) was fed with basal rat chow for six days (only 0.5% DMSO was administered by gastric intubation) and LPS was injected intraperitoneally in 0.9% sodium chloride solution only once on the 7th day. Groups III, IV, and V were fed with basal rat chow and compound **3b** (10 mg/kg/day), compound **4a** (10 mg/kg/day), and indomethacin were administered, respectively, by gastric intubation for six days. Then, LPS was injected intraperitoneally in 0.9% sodium chloride solution only once on the 7th day for three groups as well. After 24 h of LPS injection, all rats were sacrificed by ketamine (80 mg/kg) ve xylazine (10 mg/kg) anesthesia via intraperitoneal route. Blood samples were collected via cardiac puncture in tubes containing gel for obtaining serum [53].

Serum ALT and AST levels were determined using enzyme-based Roche Diagnostics kit in Roche Modular Systems analyzer by photometric assay [53] based on the manufacturer's instructions. The other serum samples were stored at −80 °C (Thermo Electron, Waltham, MA, USA) for subsequent analyses of MPO and NO levels.

4.2.9. Determination of MPO Levels

Suzuki's assay [88] was performed with slight modifications [53]. The rate of MPO-catalyzed oxidation of 3,3′,5,5′-tetramethylbenzidine (TMB) was followed by recording the absorbance increase at 655 nm for 5 min. Taking into account the linear phase of the reaction, the absorbance change was measured per minute. The enzyme activity was expressed as the amount of the enzyme producing one absorbance change per minute under assay conditions [53].

#### 4.2.10. Determination of NO Levels

Nitrate and nitrite, which represent the best index of the entire NO production, are the stable end products of NO *in vivo*. Nitrate in serum was assayed by a slight modification of the Cd-reduction method as reported by Cortas and Wakid [89].

#### 4.2.11. Statistical Analyses

The data used in statistical analyses were obtained from eight animals for each group and statistically evaluated by means of Statistical Package for the Social Sciences (SPSS) for Windows 17.0. Comparisons were performed by one-way ANOVA (Tukey for post-hoc analyses) test. Differences between groups were considered statistically significant at a level of $p < 0.05$.

## 5. Conclusions

In this paper, two classes of indole-based small molecules (**3a-j**, **4a-g**) were designed and synthesized for the targeted therapy of NSCLC. Based on the data gathered from the COX colorimetric inhibitor screening assay, compounds **3b** and **4a** were found to be the selective COX-1 inhibitors in this series with $IC_{50}$ values of 8.90 and 10.00 µM, respectively. In vitro and in vivo assays were conducted to assess their potential for the targeted therapy of NSCLC. The experimental data demonstrate that compound **3b** exerts selective anticancer activity against A549 cells through apoptosis induction and Akt inhibition. Compound **3b** also caused a substantial drop in the serum MPO and NO levels, pointing out its potential as an anti-inflammatory agent. Moreover, compound **3b** decreased the serum aminotransferase (particularly AST) levels. Taken together, compound **3b** stands out as a lead anti-NSCLC agent endowed with in vivo anti-inflammatory action acting as a dual COX-1 and Akt inhibitor. In the view of this work, a new generation of indole-based small molecules with enhanced antitumor potency could be designed through the molecular modification of compound **3b** for the targeted therapy of NSCLC.

**Supplementary Materials:** The following supporting information can be downloaded at: https://www.mdpi.com/article/10.3390/ijms24032648/s1.

**Author Contributions:** Conceptualization, M.D.A., G.A.Ç. and A.Ö.; methodology, M.D.A., G.A.Ç., N.Y.S., İ.E., B.C., H.E.T., Ö.A. and A.Ö.; software, M.D.A., G.A.Ç., İ.E. and B.C.; validation, M.D.A., G.A.Ç., İ.E. and B.C.; formal analysis, M.D.A., G.A.Ç., İ.E. and B.C.; investigation, M.D.A.; resources, M.D.A., G.A.Ç., Ö.A. and A.Ö.; writing—original draft preparation, M.D.A.; writing—review and editing, M.D.A., G.A.Ç., N.Y.S., İ.E., B.C., B.S., H.E.T., Ö.A. and A.Ö.; visualization, M.D.A.; project administration, M.D.A. and A.Ö.; funding acquisition, M.D.A. and A.Ö. All authors have read and agreed to the published version of the manuscript.

**Funding:** This research was supported by Anadolu University Scientific Research Projects Commission under the grant no: 1902S013. The APC was funded by Anadolu University Scientific Research Projects Commission under the grant no: 2107S205.

**Institutional Review Board Statement:** The animal study protocol was approved by the Animal Experiments Local Ethics Committee of Eskisehir Osmangazi University (protocol no: 700 and date of approval: 10.12.2018).

**Informed Consent Statement:** Not applicable.

**Data Availability Statement:** Data are contained within the article or Supplementary Material.

**Conflicts of Interest:** The authors declare no conflict of interest.

## References

1. Li, L.; Zhu, T.; Gao, Y.-F.; Zheng, W.; Wang, C.-J.; Xiao, L.; Huang, M.-S.; Yin, J.-Y.; Zhou, H.-H.; Liu, Z.-Q. Targeting DNA damage response in the radio(chemo)therapy of non-small cell lung cancer. *Int. J. Mol. Sci.* **2016**, *17*, 839. [CrossRef] [PubMed]
2. Chen, K.; Shang, Z.; Dai, A.-l.; Dai, P.-l. Novel PI3K/Akt/mTOR pathway inhibitors plus radiotherapy: Strategy for non-small cell lung cancer with mutant RAS gene. *Life Sci.* **2020**, *255*, 117816. [CrossRef] [PubMed]

3. Jayan, A.P.; Anandu, K.R.; Madhu, K.; Saiprabha, V.N. A pharmacological exploration of targeted drug therapy in non-small cell lung cancer. *Med. Oncol.* **2022**, *39*, 147. [CrossRef] [PubMed]
4. Mithoowani, H.; Febbraro, M. Non-small-cell lung cancer in 2022: A review for general practitioners in oncology. *Curr. Oncol.* **2022**, *29*, 1828–1839. [CrossRef]
5. Arya, S.K.; Bhansali, S. Lung cancer and its early detection using biomarker-based biosensors. *Chem. Rev.* **2011**, *111*, 6783–6809. [CrossRef]
6. Gyoba, J.; Shan, S.; Roa, W.; Bédard, E.L. Diagnosing lung cancers through examination of micro-RNA biomarkers in blood, plasma, serum and sputum: A review and summary of current literature. *Int. J. Mol. Sci.* **2016**, *17*, 494. [CrossRef]
7. Nascimento, A.V.; Bousbaa, H.; Ferreira, D.; Sarmento, B. Non-small cell lung carcinoma: An overview on targeted therapy. *Curr. Drug Targets.* **2015**, *16*, 1448–1463. [CrossRef]
8. Dilruba, S.; Kalayda, G.V. Platinum-based drugs: Past, present and future. *Cancer Chemother. Pharmacol.* **2016**, *77*, 1103–1124. [CrossRef]
9. Iksen; Pothongsrisit, S.; Pongrakhananon, V. Targeting the PI3K/AKT/mTOR signaling pathway in lung cancer: An update regarding potential drugs and natural products. *Molecules* **2021**, *26*, 4100. [CrossRef]
10. Zhu, T.; Bao, X.; Chen, M.; Lin, R.; Zhuyan, J.; Zhen, T.; Xing, K.; Zhou, W.; Zhu, S. Mechanisms and future of non-small cell lung cancer metastasis. *Front. Oncol.* **2020**, *10*, 585284. [CrossRef]
11. Xue, Y.; Hou, S.; Ji, H.; Han, X. Evolution from genetics to phenotype: Reinterpretation of NSCLC plasticity, heterogeneity, and drug resistance. *Protein Cell* **2017**, *8*, 178–190. [CrossRef]
12. Cho, W.C.; Kwan, C.K.; Yau, S.; So, P.P.; Poon, P.C.; Au, J.S. The role of inflammation in the pathogenesis of lung cancer. *Expert. Opin. Ther. Targets.* **2011**, *15*, 1127–1137.
13. Greene, E.R.; Huang, S.; Serhan, C.N.; Panigrahy, D. Regulation of inflammation in cancer by eicosanoids. *Prostaglandins Other Lipid Mediat.* **2011**, *96*, 27–36. [CrossRef]
14. Greten, F.R.; Grivennikov, S.I. Inflammation and cancer: Triggers, mechanisms, and consequences. *Immunity* **2019**, *51*, 27–41. [CrossRef]
15. Guven Maiorov, E.; Keskin, O.; Gursoy, A.; Nussinov, R. The structural network of inflammation and cancer: Merits and challenges. *Semin. Cancer Biol.* **2013**, *23*, 243–251. [CrossRef]
16. Todoric, J.; Antonucci, L.; Karin, M. Targeting inflammation in cancer prevention and therapy. *Cancer Prev. Res.* **2016**, *9*, 895–905. [CrossRef]
17. Perrone, M.G.; Scilimati, A.; Simone, L.; Vitale, P. Selective COX-1 inhibition: A therapeutic target to be reconsidered. *Curr. Med. Chem.* **2010**, *17*, 3769–3805. [CrossRef]
18. Vitale, P.; Scilimati, A.; Perrone, M.G. Update on SAR Studies toward new COX-1 selective inhibitors. *Curr. Med. Chem.* **2015**, *22*, 4271–4292. [CrossRef]
19. Pannunzio, A.; Coluccia, M. Cyclooxygenase-1 (COX-1) and COX-1 inhibitors in cancer: A review of oncology and medicinal chemistry literature. *Pharmaceuticals* **2018**, *11*, 101. [CrossRef]
20. Vitale, P.; Tacconelli, S.; Perrone, M.G.; Malerba, P.; Simone, L.; Scilimati, A.; Lavecchia, A.; Dovizio, M.; Marcantoni, E.; Bruno, A.; et al. Synthesis, pharmacological characterization, and docking analysis of a novel family of diarylisoxazoles as highly selective cyclooxygenase-1 (COX-1) inhibitors. *J. Med. Chem.* **2013**, *56*, 4277–4299. [CrossRef]
21. Hers, I.; Vincent, E.E.; Tavaré, J.M. Akt signalling in health and disease. *Cell Signal.* **2011**, *23*, 1515–1527. [CrossRef] [PubMed]
22. Nitulescu, G.M.; Margina, D.; Juzenas, P.; Peng, Q.; Olaru, O.T.; Saloustros, E.; Fenga, C.; Spandidos, D.A.; Libra, M.; Tsatsakis, A.M. Akt inhibitors in cancer treatment: The long journey from drug discovery to clinical use. *Int. J. Oncol.* **2016**, *48*, 869–885. [CrossRef] [PubMed]
23. Huang, J.; Chen, L.; Wu, J.; Ai, D.; Zhang, J.-Q.; Chen, T.-G.; Wang, L. Targeting the PI3K/AKT/mTOR signaling pathway in the treatment of human diseases: Current status, trends, and solutions. *J. Med. Chem.* **2022**, *65*, 16033–16061. [CrossRef] [PubMed]
24. Tang, F.; Wang, Y.; Hemmings, B.A.; Rüegg, C.; Xue, G. PKB/Akt-dependent regulation of inflammation in cancer. *Semin. Cancer Biol.* **2018**, *48*, 62–69. [CrossRef]
25. Guo, K.; Tang, W.; Zhuo, H.; Zhao, G. Recent advance of Akt inhibitors in clinical trials. *ChemistrySelect* **2019**, *4*, 9040–9044. [CrossRef]
26. Mali, S.N.; Thorat, B.R.; Gupta, D.R.; Pandey, A. Mini-review of the importance of hydrazides and their derivatives—Synthesis and biological activity. *Eng. Proc.* **2021**, *11*, 21.
27. Popiołek, Ł. Hydrazide-hydrazones as potential antimicrobial agents: Overview of the literature since 2010. *Med. Chem. Res.* **2017**, *26*, 287–301. [CrossRef]
28. Mathew, B.; Suresh, J.; Ahsan, M.J.; Mathew, G.E.; Usman, D.; Subramanyan, P.N.S.; Safna, K.F.; Maddela, S. Hydrazones as a privileged structural linker in antitubercular agents: A review. *Infect. Disord. Drug Targets.* **2015**, *15*, 76–88. [CrossRef]
29. Wahbeh, J.; Milkowski, S. The use of hydrazones for biomedical applications. *SLAS Technol.* **2019**, *24*, 161–168. [CrossRef]
30. Şenkardeş, S.; Han, M.İ.; Gürboğa, M.; Bingöl Özakpinar, Ö.; Küçükgüzel, Ş.G. Synthesis and anticancer activity of novel hydrazone linkage-based aryl sulfonate derivatives as apoptosis inducers. *Med. Chem. Res.* **2022**, *31*, 368–379. [CrossRef]
31. Güngör, E.M.; Altıntop, M.D.; Sever, B.; Akalın Çiftçi, G. Design, synthesis, in vitro and in silico evaluation of new hydrazone-based antitumor agents as potent Akt inhibitors. *Lett. Drug Des. Discov.* **2020**, *17*, 1380–1392. [CrossRef]

32. Han, M.İ.; Bekçi, H.; Uba, A.I.; Yıldırım, Y.; Karasulu, E.; Cumaoğlu, A.; Karasulu, H.Y.; Yelekçi, K.; Yılmaz, Ö.; Küçükgüzel, Ş.G. Synthesis, molecular modeling, in vivo study, and anticancer activity of 1,2,4-triazole containing hydrazide–hydrazones derived from (*S*)-naproxen. *Arch. Pharm. Chem. Life Sci.* **2019**, *352*, e1800365. [CrossRef]
33. Viswanathan, A.; Kute, D.; Musa, A.; Konda Mani, S.; Sipilä, V.; Emmert-Streib, F.; Zubkov, F.I.; Gurbanov, A.V.; Yli-Harja, O.; Kandhavelu, M. 2-(2-(2,4-Dioxopentan-3-ylidene)hydrazineyl)benzonitrile as novel inhibitor of receptor tyrosine kinase and PI3K/AKT/mTOR signaling pathway in glioblastoma. *Eur. J. Med. Chem.* **2019**, *166*, 291–303. [CrossRef]
34. Chen, X.; Li, H.; Luo, H.; Lin, Z.; Luo, W. Synthesis and evaluation of pyridoxal hydrazone and acylhydrazone compounds as potential angiogenesis inhibitors. *Pharmacology* **2019**, *104*, 244–257. [CrossRef]
35. Alam, M.S.; Lee, D.U. Synthesis, biological evaluation, drug-likeness, and in silico screening of novel benzylidene-hydrazone analogues as small molecule anticancer agents. *Arch. Pharm. Res.* **2016**, *39*, 191–201. [CrossRef]
36. Bak, Y.; Kim, H.; Kang, J.W.; Lee, D.H.; Kim, M.S.; Park, Y.S.; Kim, J.H.; Jung, K.Y.; Lim, Y.; Hong, J.; et al. A synthetic naringenin derivative, 5-hydroxy-7,4′-diacetyloxyflavanone-*N*-phenyl hydrazone (N101-43), induces apoptosis through up-regulation of Fas/FasL expression and inhibition of PI3K/Akt signaling pathways in non-small-cell lung cancer cells. *J. Agric. Food Chem.* **2011**, *59*, 10286–10297. [CrossRef]
37. Altıntop, M.D.; Sever, B.; Temel, H.E.; Kaplancıklı, Z.A.; Özdemir, A. Design, synthesis and in vitro COX inhibitory profiles of a new series of tetrazole-based hydrazones. *Eur. J. Life Sci.* **2022**, *1*, 20–27. [CrossRef]
38. Medeiros, M.A.M.B.; Gama e Silva, M.; de Menezes Barbosa, J.; Martins de Lavor, É.; Ribeiro, T.F.; Macedo, C.A.F.; de Souza Duarte-Filho, L.A.M.; Feitosa, T.A.; de Jesus Silva, J.; Fokoue, H.H.; et al. Antinociceptive and anti-inflammatory effects of hydrazone derivatives and their possible mechanism of action in mice. *PLoS ONE* **2021**, *16*, e0258094. [CrossRef]
39. Abdelgawad, M.A.; Labib, M.B.; Abdel-Latif, M. Pyrazole-hydrazone derivatives as anti-inflammatory agents: Design, synthesis, biological evaluation, COX-1,2/5-LOX inhibition and docking study. *Bioorg. Chem.* **2017**, *74*, 212–220. [CrossRef]
40. Gorantla, V.; Gundla, R.; Jadav, S.S.; Anugu, S.R.; Chimakurthy, J.; Nidasanametla, S.K.; Korupolu, R. Molecular hybrid design, synthesis and biological evaluation of *N*-phenyl sulfonamide linked *N*-acyl hydrazone derivatives functioning as COX-2 inhibitors: New anti-inflammatory, anti-oxidant and anti-bacterial agents. *New J. Chem.* **2017**, *41*, 13516–13532. [CrossRef]
41. Kaplancikli, Z.A.; Altintop, M.D.; Ozdemir, A.; Turan-Zitouni, G.; Khan, S.I.; Tabanca, N. Synthesis and biological evaluation of some hydrazone derivatives as anti-inflammatory agents. *Lett. Drug Des. Discov.* **2012**, *9*, 310–315. [CrossRef]
42. Acharya, P.T.; Bhavsar, Z.A.; Jethava, D.J.; Patel, D.B.; Patel, H.D. A review on development of bio-active thiosemicarbazide derivatives: Recent advances. *J. Mol. Struct.* **2021**, *1226*, 129268. [CrossRef]
43. Shakya, B.; Yadav, P.N. Thiosemicarbazones as potent anticancer agents and their modes of action. *Mini-Rev. Med. Chem.* **2020**, *20*, 638–661. [CrossRef] [PubMed]
44. Moorthy, N.S.H.N.; Cerqueira, N.M.F.S.A.; Ramos, M.J.; Fernandes, P.A. Aryl- and heteroaryl-thiosemicarbazone derivatives and their metal complexes: A pharmacological template. *Recent Pat. Anticancer. Drug Discov.* **2013**, *8*, 168–182. [CrossRef] [PubMed]
45. Chapman, T.R.; Kinsella, T.J. Ribonucleotide reductase inhibitors: A new look at an old target for radiosensitization. *Front. Oncol.* **2012**, *1*, 56. [CrossRef]
46. Icharam Narkhede, H.; Shridhar Dhake, A.; Rikhabchand Surana, A. Synthesis and screening of thiosemicarbazide-dithiocarbamate conjugates for antioxidant and anticancer activities. *Bioorg. Chem.* **2022**, *124*, 105832. [CrossRef]
47. Kozyra, P.; Korga-Plewko, A.; Karczmarzyk, Z.; Hawrył, A.; Wysocki, W.; Człapski, M.; Iwan, M.; Ostrowska-Leśko, M.; Fornal, E.; Pitucha, M. Potential Anticancer Agents against Melanoma Cells Based on an As-Synthesized Thiosemicarbazide Derivative. *Biomolecules* **2022**, *12*, 151. [CrossRef]
48. de Oliveira, J.F.; Lima, T.S.; Vendramini-Costa, D.B.; de Lacerda Pedrosa, S.C.B.; Lafayette, E.A.; da Silva, R.M.F.; de Almeida, S.M.V.; de Moura, R.O.; Ruiz, A.L.T.G.; de Carvalho, J.E.; et al. Thiosemicarbazones and 4-thiazolidinones indole-based derivatives: Synthesis, evaluation of antiproliferative activity, cell death mechanisms and topoisomerase inhibition assay. *Eur. J. Med. Chem.* **2017**, *136*, 305–314. [CrossRef]
49. Zaltariov, M.F.; Hammerstad, M.; Arabshahi, H.J.; Jovanović, K.; Richter, K.W.; Cazacu, M.; Shova, S.; Balan, M.; Andersen, N.H.; Radulović, S.; et al. New iminodiacetate-thiosemicarbazone hybrids and their copper(II) complexes are potential ribonucleotide reductase R2 inhibitors with high antiproliferative activity. *Inorg. Chem.* **2017**, *56*, 3532–3549. [CrossRef]
50. Pape, V.F.; Tóth, S.; Füredi, A.; Szebényi, K.; Lovrics, A.; Szabó, P.; Wiese, M.; Szakács, G. Design, synthesis and biological evaluation of thiosemicarbazones, hydrazinobenzothiazoles and arylhydrazones as anticancer agents with a potential to overcome multidrug resistance. *Eur. J. Med. Chem.* **2016**, *117*, 335–354. [CrossRef]
51. Altıntop, M.D.; Atlı, Ö.; Ilgın, S.; Demirel, R.; Özdemir, A.; Kaplancıklı, Z.A. Synthesis and biological evaluation of new naphthalene substituted thiosemicarbazone derivatives as potent antifungal and anticancer agents. *Eur. J. Med. Chem.* **2016**, *108*, 406–414. [CrossRef]
52. Altıntop, M.D.; Temel, H.E.; Sever, B.; Akalın Çiftçi, G.; Kaplancıklı, Z.A. Synthesis and evaluation of new benzodioxole-based thiosemicarbazone derivatives as potential antitumor agents. *Molecules* **2016**, *21*, 1598. [CrossRef]
53. Altıntop, M.D.; Sever, B.; Akalın Çiftçi, G.; Ertorun, İ.; Alataş, Ö.; Özdemir, A. A new series of thiosemicarbazone-based anti-inflammatory agents exerting their action through cyclooxygenase inhibition. *Arch. Pharm. Chem. Life Sci.* **2022**, *355*, e2200136. [CrossRef]

54. Da Fonseca, A.G.; Fernandes Ribeiro Dantas, L.L.S.; Rodrigues, J.P.; Alencar Filho, M.P.D.C.; De Melo Rêgo, M.J.B.; Da Rocha Pitta, M.G.; De Moraes Gomes, P.A.T.; De Melo Silva, V.G.; Lima Leite, A.C.; Furtado, A.A.; et al. PA-Int5: An isatin-thiosemicarbazone derivative that exhibits anti-nociceptive and anti-inflammatory effects in Swiss mice. *Biomed. Rep.* **2021**, *15*, 61. [CrossRef]
55. Jacob, Í.T.T.; Gomes, F.O.S.; de Miranda, M.D.S.; de Almeida, S.M.V.; da Cruz-Filho, I.J.; Peixoto, C.A.; da Silva, T.G.; Moreira, D.R.M.; de Melo, C.M.L.; de Oliveira, J.F.; et al. Anti-inflammatory activity of novel thiosemicarbazone compounds indole-based as COX inhibitors. *Pharmacol. Rep.* **2021**, *73*, 907–925. [CrossRef]
56. Subhashree, G.R.; Haribabu, J.; Saranya, S.; Yuvaraj, P.; Anantha Krishnan, D.; Karvembu, R.; Gayathri, D. In vitro antioxidant, antiinflammatory and in silico molecular docking studies of thiosemicarbazones. *J. Mol. Struct.* **2017**, *1145*, 160–169. [CrossRef]
57. Vitaku, E.; Smith, D.T.; Njardarson, J.T. Analysis of the structural diversity, substitution patterns, and frequency of nitrogen heterocycles among U.S. FDA approved pharmaceuticals. *J. Med. Chem.* **2014**, *57*, 10257–10274. [CrossRef]
58. Chadha, N.; Silakari, O. Indoles as therapeutics of interest in medicinal chemistry: Bird's eye view. *Eur. J. Med. Chem.* **2017**, *134*, 159–184. [CrossRef]
59. Dhuguru, J.; Skouta, R. Role of indole scaffolds as pharmacophores in the development of anti-lung cancer agents. *Molecules* **2020**, *25*, 1615. [CrossRef]
60. Jia, Y.; Wen, X.; Gong, Y.; Wang, X. Current scenario of indole derivatives with potential anti-drug-resistant cancer activity. *Eur. J. Med. Chem.* **2020**, *200*, 112359. [CrossRef]
61. Wan, Y.; Li, Y.; Yan, C.; Yan, M.; Tang, Z. Indole: A privileged scaffold for the design of anti-cancer agents. *Eur. J. Med. Chem.* **2019**, *183*, 111691. [CrossRef] [PubMed]
62. Kumari, A.; Singh, R.K. Medicinal chemistry of indole derivatives: Current to future therapeutic prospectives. *Bioorg. Chem.* **2019**, *89*, 103021. [CrossRef] [PubMed]
63. Nisha; Singh, S.; Sharma, N.; Chandra, R. The indole nucleus as a selective COX-2 inhibitor and anti-inflammatory agent (2011–2022). *Org. Chem. Front.* **2022**, *9*, 3624–3639. [CrossRef]
64. Moraes, A.D.T.d.O.; Miranda, M.D.S.d.; Jacob, Í.T.T.; Amorim, C.A.d.C.; Moura, R.O.d.; Silva, S.Â.S.d.; Soares, M.B.P.; Almeida, S.M.V.d.; Souza, T.R.C.d.L.; Oliveira, J.F.d.; et al. Synthesis, in vitro and in vivo biological evaluation, COX-1/2 inhibition and molecular docking study of indole-*N*-acylhydrazone derivatives. *Bioorg. Med. Chem.* **2018**, *26*, 5388–5396. [CrossRef]
65. Özdemir, A.; Altintop, M.D.; Turan-Zitouni, G.; Çiftçi, G.A.; Ertorun, İ.; Alataş, Ö.; Kaplancikli, Z.A. Synthesis and evaluation of new indole-based chalcones as potential antiinflammatory agents. *Eur. J. Med. Chem.* **2015**, *89*, 304–309. [CrossRef]
66. da Silva Guerra, A.S.H.; do Nascimento Malta, D.J.; Laranjeira, L.P.M.; Maia, M.B.S.; Colaço, N.C.; de Lima, M.d.C.A.; Galdino, S.L.; Pitta, I.d.R.; Gonçalves-Silva, T. Anti-inflammatory and antinociceptive activities of indole-imidazolidine derivatives. *Int. Immunopharmacol.* **2011**, *11*, 1816–1822. [CrossRef]
67. Kalgutkar, A.S.; Marnett, A.B.; Crews, B.C.; Remmel, R.P.; Marnett, L.J. Ester and amide derivatives of the nonsteroidal antiinflammatory drug, indomethacin, as selective cyclooxygenase-2 inhibitors. *J. Med. Chem.* **2000**, *43*, 2860–2870. [CrossRef]
68. Sarvepalli, S.; Parvathaneni, V.; Chauhan, G.; Shukla, S.K.; Gupta, V. Inhaled indomethacin-loaded liposomes as potential therapeutics against Non-Small Cell Lung Cancer (NSCLC). *Pharm. Res.* **2022**, *39*, 2801–2815. [CrossRef]
69. Harras, M.F.; Sabour, R.; Ammar, Y.A.; Mehany, A.B.M.; Farrag, A.M.; Eissa, S.I. Design synthesis and cytotoxicity studies of some novel indomethacin-based heterocycles as anticancer and apoptosis inducing agents. *J. Mol. Struct.* **2021**, *1228*, 129455. [CrossRef]
70. Guo, Y.-C.; Chang, C.-M.; Hsu, W.-L.; Chiu, S.-J.; Tsai, Y.-T.; Chou, Y.-H.; Hou, M.-F.; Wang, J.-Y.; Lee, M.-H.; Tsai, K.-L.; et al. Indomethacin inhibits cancer cell migration via attenuation of cellular calcium mobilization. *Molecules* **2013**, *18*, 6584–6596. [CrossRef]
71. de Groot, D.J.A.; van der Deen, M.; Le, T.K.P.; Regeling, A.; de Jong, S.; de Vries, E.G.E. Indomethacin induces apoptosis via a MRP1-dependent mechanism in doxorubicin-resistant small-cell lung cancer cells overexpressing MRP1. *Br. J. Cancer* **2007**, *97*, 1077–1083. [CrossRef]
72. Hull, M.A.; Gardner, S.H.; Hawcroft, G. Activity of the non-steroidal anti-inflammatory drug indomethacin against colorectal cancer. *Cancer Treat. Rev.* **2003**, *29*, 309–320. [CrossRef]
73. Mohammad, R.M.; Muqbil, I.; Lowe, L.; Yedjou, C.; Hsu, H.Y.; Lin, L.T.; Siegelin, M.D.; Fimognari, C.; Kumar, N.B.; Dou, Q.P.; et al. Broad targeting of resistance to apoptosis in cancer. *Semin. Cancer Biol.* **2015**, *35*, S78–S103. [CrossRef]
74. Pfeffer, C.M.; Singh, A.T.K. Apoptosis: A target for anticancer therapy. *Int. J. Mol. Sci.* **2018**, *19*, 448. [CrossRef]
75. Huang, M.; Cai, S.; Su, J. The pathogenesis of sepsis and potential therapeutic targets. *Int. J. Mol. Sci.* **2019**, *20*, 5376. [CrossRef]
76. Crimi, E.; Sica, V.; Slutsky, A.S.; Zhang, H.; Williams-Ignarro, S.; Ignarro, L.J.; Napoli, C. Role of oxidative stress in experimental sepsis and multisystem organ dysfunction. *Free Radic. Res.* **2006**, *40*, 665–672. [CrossRef]
77. Valadez-Cosmes, P.; Raftopoulou, S.; Mihalic, Z.N.; Marsche, G.; Kargl, J. Myeloperoxidase: Growing importance in cancer pathogenesis and potential drug target. *Pharmacol. Ther.* **2022**, *236*, 108052. [CrossRef]
78. Fortin, C.F.; McDonald, P.P.; Fülöp, T.; Lesur, O. Sepsis, leukocytes, and nitric oxide (NO): An intricate affair. *Shock* **2010**, *33*, 344–352. [CrossRef]
79. Sharma, J.N.; Al-Omran, A.; Parvathy, S.S. Role of nitric oxide in inflammatory diseases. *Inflammopharmacology* **2007**, *15*, 252–259. [CrossRef]
80. Wu, C.C. Nitric oxide and inflammation. *Curr. Med. Chem. Anti-Inflamm. Anti-Allergy Agents* **2004**, *3*, 217–222. [CrossRef]
81. Kobayashi, A.; Suzuki, Y.; Sugai, S. Specificity of transaminase activities in the prediction of drug-induced hepatotoxicity. *J. Toxicol. Sci.* **2020**, *45*, 515–537. [CrossRef] [PubMed]

82. Catarro, M.; Serrano, J.L.; Ramos, S.S.; Silvestre, S.; Almeida, P. Nimesulide analogues: From anti-inflammatory to antitumor agents. *Bioorg. Chem.* **2019**, *88*, 102966. [CrossRef] [PubMed]
83. Kunutsor, S.K.; Apekey, T.A.; Khan, H. Liver enzymes and risk of cardiovascular disease in the general population: A meta-analysis of prospective cohort studies. *Atherosclerosis* **2014**, *236*, 7–17. [CrossRef] [PubMed]
84. Limdi, J.K.; Hyde, G.M. Evaluation of abnormal liver function tests. *Postgrad. Med. J.* **2003**, *79*, 307–312. [CrossRef]
85. Kumar, D.; Kumar, N.M.; Noel, B.; Shah, K. A series of 2-arylamino-5-(indolyl)-1,3,4-thiadiazoles as potent cytotoxic agents. *Eur. J. Med. Chem.* **2012**, *55*, 432–438. [CrossRef]
86. Altıntop, M.D.; Özdemir, A.; Temel, H.E.; Demir Cevizlidere, B.; Sever, B.; Kaplancıklı, Z.A.; Akalın Çiftçi, G. Design, synthesis and biological evaluation of a new series of arylidene indanones as small molecules for targeted therapy of non-small cell lung carcinoma and prostate cancer. *Eur. J. Med. Chem.* **2022**, *244*, 114851. [CrossRef]
87. Mosmann, T. Rapid colorimetric assay for cellular growth and survival: Application to proliferation and cytotoxicity assays. *J. Immunol. Methods* **1983**, *16*, 55–63. [CrossRef]
88. Suzuki, K.; Ota, H.; Sasagawa, S.; Sakatani, T.; Fujikura, T. Assay method for myeloperoxidase in human polymorphonuclear leukocytes. *Anal. Biochem.* **1983**, *132*, 345–352. [CrossRef]
89. Cortas, N.K.; Wakid, N.W. Determination of inorganic nitrate in serum and urine by a kinetic cadmium-reduction method. *Clin. Chem.* **1990**, *36*, 1440–1443. [CrossRef]

International Journal of
*Molecular Sciences*

MDPI

*Review*

# Modulating Glycolysis to Improve Cancer Therapy

**Chaithanya Chelakkot [1,†], Vipin Shankar Chelakkot [2,†], Youngkee Shin [3,4] and Kyoung Song [5,*]**

1 Bio-MAX/N-Bio, Seoul National University, Seoul 08826, Republic of Korea
2 Department of Cancer Biology, Lerner Research Institute, Cleveland Clinic, Cleveland, OH 44195, USA
3 Laboratory of Molecular Pathology and Cancer Genomics, Research Institute of Pharmaceutical Science, Department of Pharmacy, Seoul National University, Seoul 08826, Republic of Korea
4 Department of Molecular Medicine and Biopharmaceutical Sciences, Graduate School of Convergence Science and Technology, Seoul National University, Seoul 08826, Republic of Korea
5 College of Pharmacy, Duksung Women's University, Seoul 01366, Republic of Korea
* Correspondence: songseoul17@duksung.ac.kr
† These authors contributed equally to this work.

**Abstract:** Cancer cells undergo metabolic reprogramming and switch to a 'glycolysis-dominant' metabolic profile to promote their survival and meet their requirements for energy and macromolecules. This phenomenon, also known as the 'Warburg effect,' provides a survival advantage to the cancer cells and make the tumor environment more pro-cancerous. Additionally, the increased glycolytic dependence also promotes chemo/radio resistance. A similar switch to a glycolytic metabolic profile is also shown by the immune cells in the tumor microenvironment, inducing a competition between the cancer cells and the tumor-infiltrating cells over nutrients. Several recent studies have shown that targeting the enhanced glycolysis in cancer cells is a promising strategy to make them more susceptible to treatment with other conventional treatment modalities, including chemotherapy, radiotherapy, hormonal therapy, immunotherapy, and photodynamic therapy. Although several targeting strategies have been developed and several of them are in different stages of pre-clinical and clinical evaluation, there is still a lack of effective strategies to specifically target cancer cell glycolysis to improve treatment efficacy. Herein, we have reviewed our current understanding of the role of metabolic reprogramming in cancer cells and how targeting this phenomenon could be a potential strategy to improve the efficacy of conventional cancer therapy.

**Keywords:** glycolysis; cancer metabolism; combination therapy

**Citation:** Chelakkot, C.; Chelakkot, V.S.; Shin, Y.; Song, K. Modulating Glycolysis to Improve Cancer Therapy. *Int. J. Mol. Sci.* **2023**, *24*, 2606. https://doi.org/10.3390/ijms24032606

Academic Editors: Laura Paleari and Ana Cristina Gonçalves

Received: 15 November 2022
Revised: 18 January 2023
Accepted: 20 January 2023
Published: 30 January 2023



## 1. Introduction

Cancer cells reprogram their metabolism to promote growth, metastasis, and survival. They exhibit an increased glycolytic dependency and show an elevated glucose uptake and fermentation of glucose to lactate to meet the heightened anabolic needs for cancer cell proliferation [1]. Increased glycolysis is not only important for meeting the energy needs of the cells but is also crucial for the generation of metabolic intermediates necessary for macromolecule synthesis in cancer cells [2,3]. This phenomenon, often referred to as the 'Warburg effect,' is observed even in the presence of completely functional mitochondria [4]. The Warburg effect has been studied for over 90 years, and several studies have explored the mechanisms governing the increased glycolytic dependency of cancer cells. Several oncogenic proteins and tumor suppressors, including the hypoxia-inducible factor (HIF-1), Myc, p53, and PI3K/Akt/mTOR pathway, have been implicated in regulating this cancer cell-specific metabolic reprogramming.

The altered metabolism in the cancer cells provides an avenue for developing cancer cell-specific therapeutic targets and anti-cancer agents. Indeed, therapeutic strategies that target glycolysis and cancer cell-specific biosynthetic pathways are a major focus area in cancer research. Although the increased glycolytic dependency of neoplastic cells suggests the potential therapeutic efficacy of glycolytic inhibitors in cancer treatment, glycolytic inhibition alone is not effective in a clinical setting [5]. Targeting metabolism, especially in combination with chemotherapy, is expected to improve therapy responses and may help overcome drug resistance [6]. Elevated glycolysis in cancer cells and the resulting lactic acidosis modulate the tumor stroma to a pro-tumorigenic microenvironment. Targeting glycolytic changes in the tumor microenvironment has been shown to be a safe and effective strategy to enhance therapeutic efficacy [7,8]. In this review, we explore and discuss glycolytic modulation in cancer cells and how it could aid as a therapeutic strategy in combination therapies with chemotherapy and radiotherapy, immunotherapy, hormonal therapy, and photodynamic therapy (PDT).

## 2. Modulating Glycolysis to Improve Chemotherapy and Radiotherapy

Metabolic modulation has been shown to sensitize cancer cells toward chemotherapy and radiotherapy. Increased glycolysis, facilitated by an increased glucose uptake, is the major energy source in cancer cells apart from being the major source of macromolecules for cell proliferation and survival [9]. Recently it was demonstrated that glycolysis-addicted cancer cells show metabolic rewiring via mTORC1 activation [10,11]. Sustained mTORC1 activation bypasses glycolysis by directing the glucose flux toward the pentose phosphate pathway. Metabolic rewiring, including dysregulated glycolysis, elevated ATP production, and cell-death escape mechanisms, are the major culprits for therapeutic resistance in cancer cells. The intracellular ATP level in cancer cells is also associated with metastasis and stemness. Thus, targeting glycolysis or intracellular/extracellular ATP levels [12–14] is a promising strategy to sensitize cancer cells toward chemotherapy. Several studies have reported that glycolytic inhibitors improve the efficacy of cancer treatment and that glycolytic inhibition is a promising strategy when used as a combination therapy with other treatment modalities. In line with this, inhibiting glycolytic enzymes, hexokinase (HK) [15], pyruvate kinase (PK) [16], and lactate dehydrogenase (LDH) [17], have shown sensitizing effects with several chemotherapeutic agents [5].

### 2.1. Targeting Glucose Transporters and Glucose Uptake to Improve Chemotherapy

The inhibition of glucose transporters (GLUTs), a critical rate-limiting step in glucose metabolism, modulates the therapeutic efficacy of several drugs. GLUT expression is elevated in several types of cancers and is associated with a poor prognosis suggesting their key role in cancer cell metabolism [18–22]. Glycolysis inhibition using inhibitors of GLUT1, when combined with routine cancer therapy, has proven to be relevant in potentiating their effects in a synergistic manner in pre-clinical studies for several cancers [19,23–26]. GLUT1 inhibition curbs the self-renewing capacity and tumor-initiating potential of cancer stem cells and has a substantial significance from a therapeutic perspective [27].

A widely studied small molecule inhibitor of GLUT1, WZB117 synergistically inhibits breast cancer cells by inducing DNA damage when treated in combination with an allosteric AKT inhibitor [24,28]. WZB117 also sensitizes breast cancer cells toward treatment with adriamycin [29] and radioresistant breast cancer cells to radiotherapy [27,30]. Previous studies have reported the synergistic effects of GLUT1 inhibitor #43 in melanoma cells and have shown that GLUT1 inhibition induces apoptosis, intracellular reactive oxygen species (ROS) generation, and the loss of mitochondrial membrane potential. Combination therapy with GLUT1 inhibitor #43 enhances the DNA-damaging effects of cisplatin by regulating the AKT/mTOR pathways [24]. Another GLUT1 inhibitor, BAY-876, enhances the cisplatin-mediated inhibition of esophageal squamous cell carcinoma [23]. siRNA-mediated GLUT1 inhibition also showed similar results and improved the efficacy of low-dose cisplatin treatment [23]. In vivo studies in uterine cancer, patient-derived models have shown that glycolytic activation contributes to the stemness of uterine endometrial cancer, and BAY-876-mediated GLUT inhibition synergistically suppressed endometrial cancer cell proliferation when used in combination with paclitaxel [31].

Combination therapy with GLUT modulators can also improve the bioavailability of chemotherapeutic drugs. The co-treatment of paclitaxel with silybin (a GLUT modulator) significantly improved the oral bioavailability of the drug in several in vitro and in vivo studies and overcame the major drawback of the limited oral bioavailability of paclitaxel [32,33]. A nanomedicine-based combination therapy using GLUT1 inhibitor and chemotherapeutic agent, curcumin, deprived cancer cells of glucose and sensitized cancer cells to chemotherapy, induced apoptosis, improved anti-tumor effects, and alleviated side-effects in vitro and in vivo [34]. Thus, combination therapy with GLUT1 inhibitors might be a rational therapeutic strategy and could also allow for low-dose treatment with chemotherapy drugs providing a paradigm for high-efficacy, low-toxicity therapeutic options [35].

Another straightforward and interesting strategy to improve the effectiveness of chemotherapy is to deprive cancer cells of glucose [36]. Icard et al. proposed that the modulation of glucose intake in combination with chemotherapy could improve the efficacy of the drug via the deprivation of ATP to cancer cells. A recent study showed that intermittent fasting throughout chemotherapy was well tolerated in patients and reduced chemotherapy-induced toxicity as measured by hematologic, metabolic, and inflammatory parameters [37]. Contrastingly, an opposite effect was reported in a pre-clinical model of pancreatic ductal adenocarcinoma (PDAC), wherein a relative glucose abundance sensitized PDAC cells to chemotherapy. Hyperglycemic patients with stage IV PDAC showed an enhanced response to chemotherapy, possibly via impaired glutathione biosynthesis [38]. A case report on non-small cell lung cancer (NSCLC) with bone and brain metastasis reported the efficacy of glucose uptake inhibition in combination with chemotherapy as a palliative treatment strategy. Fasting-induced hypoglycemia or insulin-induced hypoglycemia combined with low-dose chemotherapy could benefit cancer patients, particularly those who do not tolerate the conventional dosage of drugs [39].

### 2.2. Targeting Glycolysis Enzymes to Improve Chemotherapy and Radiotherapy

The enhanced glycolysis in the cancer cells correlates with an upregulation and activation of critical glycolytic enzymes. Targeting the key glycolytic enzymes is a promising strategy to rewire the altered tumor metabolism, to sensitize (or re-sensitize, when resistance develops) cancers to chemotherapy.

HK catalyzes the first step in glucose metabolism and converts glucose to glucose-6-phosphate. Four HK isoforms, HK1-4, with different cellular distributions and glucose affinity have been identified. HK1 and HK2 are located on the outer mitochondrial membrane and are associated with AKT-mediated cell survival [40,41]. Further, HK2 is associated with the recurrence and poor prognosis of breast cancer (BC) [42]. HK2 expression is also elevated in lung cancer, and shows significant association with the tumor stage. The deletion of the Hk2 gene in lung cancer cells ameliorated glucose-derived ribonucleotides and glutamine-derived carbon utilization in anaplerosis [43]. Targeting HK2 inhibits cell proliferation and shifts the metabolic profile of cancer cells from glycolytic to oxidative phosphorylation (OXPHOS) [43,44].

2-Deoxy glucose (2-DG), an HK inhibitor, has been shown to sensitize cancer cells to chemotherapy and radiotherapy and is being investigated in clinical trials. 2-DG is a glucose analog that triggers the intracellular accumulation of 2-deoxy-d-glucose-6-phosphate (2-DG6P), inhibiting the function of HK and glucose-6-phosphate isomerase [45]. Glycolysis inhibition with 2-DG can improve the therapeutic efficacy of trastuzumab in treating HER2+ BC [46]. Similarly, the therapeutic efficacy of paclitaxel was enhanced when treated in combination with 2-DG in in vivo studies in NSCLC and osteosarcoma models [47]. A recent study showed that 2-DG could sensitize glioblastoma cells to chloroethyl nitrosourea by regulating glycolysis, intracellular ROS generation, and endoplasmic reticulum stress induction [48]. Another study reported that combining 2-DG with autophagy inhibiting drug hydroxychloroquine enhances apoptosis in BC cells. The inhibition of autophagy combined with 2-DG induced the accumulation of misfolded proteins in the endoplasmic reticulum and resulted in sustained endoplasmic reticulum stress, induced through the pERK-eIF2$\alpha$-ATF4-CHOP axis to enhance apoptosis in BC cells [49]. Although a few clinical trials (NCT05314933, NCT00096707) are investigating the toxicity, tolerability, pharmacokinetics, and recommended dose of 2-DG in advanced tumors [50], additional studies and trials are required to characterize the mechanism of action and treatment benefits of 2-DG in cancer.

Curcumin, another compound with a proven anti-tumor effect, is also known to inhibit HK expression by inhibiting transcriptional repressor SLUG. Combination therapy of curcumin with docetaxel demonstrated a high response rate, low-toxicity, and improved patient tolerance in prostate cancer [51].

3-Bromopyruvate acid (3BrPA) is another classic glycolytic inhibitor, which inhibits several enzymes in the glycolytic pathway, including HK and LDH, and is a potent inhibitor of cancer cell growth [52–55]. Combining 3BrPA with rapamycin enhanced the anti-tumor efficacy through the dual inhibition of mTOR signaling and glycolysis in LC and neuroblastoma [56,57]. In BC cells, 3BrPA enhanced the expression of thioredoxin interacting protein (TXNIP) and inhibited HK2 expression via c-Myc downregulation [58], and enhanced tamoxifen-induced cytotoxicity in vitro [59]. The combination regimen with tamoxifen also enhanced oxidative stress and reduced glutathione levels in cells, and affected tumor angiogenesis and metastasis in animal models [59]. Further, the intra-cranial delivery of 3BrPA with temozolomide showed synergistic effects and increased survival in animal models of glioma [60]. Moreover, enhanced therapeutic efficacy was demonstrated when 3BrPA was combined with sorafenib in murine models of liver cancer [61]. 3BrPA can also enhance the anti-tumor effect of low-dose radiation via the reprogramming of mitochondrial metabolism and hindering of ATP generation [62]. 3BrPA also inhibits monocarboxylate transporter 1 (MCT1) expression, which mediates the bidirectional transport of lactate in cancer cells, and sensitizes cancer cells to ionizing radiation [63].

Elevated glycolysis in cancer cells results in the conversion of pyruvate to lactate, even under aerobic conditions. Lactate is excreted at high levels from tumor cells and acts as a metabolic fuel and oncometabolite with signaling properties. Lactate utilization by tumor cells depends on the expression of monocarboxylic transporters (MCTs), which are upregulated in cancer cells [64,65]. MCTs facilitate the shuttle of lactate from cancer cells to neighboring cells, tumor stroma, and tumor-associated endothelial cells and induce metabolic rewiring. Lactate is involved in several tumorigenic functions of cancer cells, including tumor microenvironment modulation and tumor angiogenesis [66]. Targeting MCTs has been shown to suppress tumor growth in cancer cells [67]. MCT1 and MCT4 inhibitors impair leukemia cell proliferation and enhance their sensitivity toward chemotherapy [68]. Currently, a phase 1 clinical trial is evaluating the toxicity and pharmacokinetic profile of AZD3965, an MCT inhibitor in cancer therapy for B-cell lymphoma [69,70].

LDH is a key glycolytic enzyme that is elevated in aggressive cancers and is essential for tumor maintenance [71–73]. LDH-A is regulated by numerous oncogenic transcription factors, including c-Myc and HIF-1, and is closely associated with malignant phenotypes of cancer cells [74]. LDH-A overexpression upregulated AKT phosphorylation and PI3K, which upregulated cyclin D1 and c-Myc expression in LC cells [75–78]. LDH overexpression is also associated with epithelial–mesenchymal transition (EMT)-related genes, SNAIL and SLUG, and is thus involved in regulating the metastatic progression of cancer cells [79]. LDH-A levels could thus serve as a biomarker for cancer diagnosis and prognosis [80–82]. LDH inhibition induces oxidative stress, impacting cancer stem cells' renewable capacity. The overburden of mitochondrial complex II is speculated to account for the increased ROS production in LDH-A-inhibited cancer stem cells [83,84]. The widely studied LDH-A inhibitors include pyruvate analog, oxamate (OXM), and the NADH competitive inhibitor, gossypol. LDH-A inhibition with OXM triggers a specific tumor reduction in brain tumors by reducing ATP levels, increasing ROS production, and inducing apoptosis [85]. OXM also induces autophagy via the AKT-mTOR signaling pathway in certain cancer cell types [86]. Recent pre-clinical and clinical studies have supported the combined use of LDH inhibitors with concurrent treatments as a promising strategy in cancer therapy [85]. Combination therapy of OXM with other chemotherapeutic drugs, an including mTORC1 inhibitor and phenformin (phenethylbiguanidine; an anti-diabetic agent), have shown synergistic effects suggesting their implications in combination therapies [87,88]. A triple combination therapy of doxorubicin with metformin and OXM induced autophagy and apoptosis in colorectal cancers by downregulating hypoxia-induced HIF-1 expression [85].

Enolase (ENO-1), which converts 2-phosphoglycerate (2PG) to phosphoenol pyruvate (PEP), is emerging as a promising target for cancer therapy, partially owing to its diverse functions apart from being a major enzyme in glycolysis [89–91]. The overexpression of ENO-1 is associated with disease progression, metastasis-free survival, and overall survival in colorectal cancer, BC, gastric cancer, gliomas, head and neck cancer, and leukemia. ENO-1 promotes tumor cell progression via a plethora of mechanisms, including inducing angiogenesis, evading immune suppression and growth suppressors, and resisting cell death [92–94]. Small molecule inhibitors of ENO-1 have been shown to inhibit cancer cell growth [95–97]. POMHEX, a selective enolase inhibitor, has been shown to selectively inhibit the tumor cell progression of glioma cells in vivo by triggering apoptosis, showing a favorable safety profile and tolerance in non-human primates [98]. A previous study identified a potent inhibitor of ENO1, macrosphelide A, which demonstrates anti-cancer effects by simultaneously inactivating ENO1, aldolase, and fumarase [95].

6-phosphofructokinase/fructose-2,6-bisphophatase (PFKFBs) catalyzes the first irreversible step in glycolysis, which is the conversion of fructose-6-phosphate to fructose-1,6-bis-phosphate. As with most other glycolytic enzymes, PFKFB activity and expression are enhanced in many cancers. The selective inhibition of PFKFB3 displays broad anti-tumor activity in syngenic pre-clinical models and early human studies by inducing necroptotic cell death, apoptosis, cell cycle arrest, and inhibiting invasion [99,100]. A phase-1 dose escalation study for PFK-158, a first-in-human, first-in-class, small molecule inhibitor of PFKFB3, showed commendable tolerance and tumor burden reduction in pancreatic cancer, renal cell carcinoma, and adenocystic carcinoma patients [101–103]. The synergistic effect of PFK-158 with other FDA-approved targeted-chemotherapy agents can potentially improve their chemotherapy efficacy and is being validated. In gynecologic cancers, it was shown that PFK-158 improves lipophagy and sensitizes platinum therapy-resistant cells to carboplatin/oxaliplatin therapy [104]. Combining PFKFB3 inhibition with standard chemotherapy can thus be a novel strategy to improve the outcome in gynecologic and endometrial cancer patients who are resistant to therapy or have advanced, recurrent diseases [89].

Besides its glycolytic function, PFKFB3 is a crucial player in regulating endothelial cells, tumor angiogenesis, and tumor vascularization [105,106]. A transient inhibition of PFKFB3 in endothelial cells using 3-(3-pyridinyl)-1-(4-pyridinyl)-2-propen-1-one (3PO) induced tumor vessel normalization, impaired metastasis, and improved chemotherapy [107,108]. The PFKFB3 inhibitor, AZ67, inhibited angiogenesis in vivo, independent of glycolysis regulation [109]. A combination therapy with PFKFB3 inhibitor and VEGF inhibitor, bevacizumab, improved tumor vasculature, alleviated tumor hypoxia, normalized lactate production, and improved the efficacy and delivery of doxorubicin in glioblastoma [110].

Pyruvate kinase M2 (PKM2), which catalyzes the conversion of PEP to pyruvate, is upregulated in numerous cancers and has emerged as a critical regulator of cancer cell metabolism [111,112]. Apart from being a key enzyme in glycolysis, nuclear PKM2 regulates the expression of GLUT1 and LDH-A through positive feedback to further support glycolytic metabolism [113]. PKM2 expression is upregulated under hypoxic conditions and induces tumor angiogenesis and metastasis [114]. The association of PKM2 expression with poor prognosis and overall survival indicates that PKM2 level could serve as a prognostic or diagnostic marker for cancers [115–117]. PKM2 inhibition induces apoptosis and tumor regression in xenograft models of different cancer types [89] and plays a role in maintaining redox homeostasis and glutathione turnover. PKM2 inhibition also increases the efficacy of docetaxel treatment in vitro and xenograft models of LC [118,119]. In NSCLC patients who received platinum therapy in a first-line setting, tumors with low PKM2 expression showed significantly longer progression-free survival and overall survival [120]. In tumor xenograft models of NSCLC, combination therapy with PKM2 siRNA and chemotherapeutic agents increased apoptosis and inhibited tumor growth [121].

Targeting Glyceraldehyde 3-phosphate dehydrogenase (GAPDH) is being explored as an alternative approach for inhibiting glycolysis [122]. GAPDH catalyzes the first step in which energy is derived from NADH in the 'pay-off-phase' of glycolysis. NADH, the first molecule generated in this phase, is critical for regulating intracellular ROS and redox balance. Targeting GAPDH triggers the accumulation of glucotrioses such as glyceraldehyde-3-phosphate and dihydroxy acetone phosphate in the cells, the partial degradation of which results in the formation of cytotoxic methylglyoxal [123]. Thus, inhibiting GAPDH not only depletes ATP but also triggers cytotoxicity through the upregulation of ROS and the accumulation of methylglyoxal [122]. 3-BrPA, discussed above, is a potent inhibitor of GAPDH and was shown to deplete intracellular ATP. Additionally, 3-BrPA showed high specificity and selectivity for GAPDH both in vitro and in vivo [122,124,125].

Although targeting glycolytic enzymes can improve the efficacy of chemotherapy and radiotherapy, the ubiquitous nature of glycolysis and glycolytic enzymes presents the challenge of the systemic toxicity of glycolysis inhibition. The selective targeting of cancer-specific enzymes or enzyme isoforms and the targeted delivery of therapeutic agents could circumvent this challenge.

### 2.3. Modulating Glycolysis to Overcome Drug Resistance

Aberrant glycolysis is a major contributor to drug resistance in cancer [126,127]. The mechanism underlying glucose metabolism reprogramming-induced drug resistance is not clearly understood. Increased glucose uptake induces gemcitabine resistance in pancreatic cancer, doxorubicin resistance in BC, and cisplatin resistance in genitourinary cancers [6,128]. It is thought that glucose metabolism reprogramming in cancer cells induces DNA repair and immune suppression in the tumor microenvironment, contributing to drug resistance. Anabolic alterations could account for the increased nucleotide demand required for the efficient repair of chemotherapy/radiation-induced DNA damage. The DNA repair pathways in reprogrammed cancer cells induces the activation of several pro-tumorigenic signaling pathways, including Wnt, PI3K/AKT, NF-κB, and MAPK, triggering prolonged cancer cell survival and apoptosis resistance [89,129]. Aberrant glycolysis can also promote DNA repair by increasing nucleotide turn over by enhancing the hexosamine biosynthetic pathway (HBP) and pentose phosphate pathway (PPP) [130,131]. By limiting pyruvate flux into OXPHOS, upregulated glycolysis also enables cancer cells to reduce the ROS accumulation in cells, another mechanism by which metabolic reprogramming contributes to resistance to therapy. Increasing evidence also suggests that the activation of Wnt, PI3K/AKT, and Notch pathways activate autophagy which also contributes to cancer cell survival and resistance to therapy, whereas inhibiting autophagy sensitizes cancer cells to therapy [89,132,133]. Autophagy, thus, downregulates cell metabolism leading to cancer cell quiescence and survival, inducing radio-resistance [134].

Metabolic changes in the tumor cells happen hand-in-hand with similar reprogramming of the tumor microenvironment. This metabolic reprogramming induces immunosuppression and immune escape of cancer cells and contributes to the development of resistance to chemotherapy and radiotherapy [135,136]. The upregulation of glycolytic enzyme HK2 suppresses the mTOR-S6K signaling pathway and blocks chemotherapy-induced apoptosis by binding to voltage-dependent anion channels, and suppresses the formation of mitochondrial permeability transition pores, contributing to chemoresistance. Aberrant glycolytic pathways in cancer stem cells also play critical roles in contributing to resistance to therapy via enhancing cancer cell stemness by activating the PI3K/AKT pathway and upregulating the stem cell-like properties [89]. Enhanced exosomal secretion from cancer stem cells also activates neighboring cancer cells toward stemness and promotes chemo/radio-resistance [137,138]. The overexpression of ENO-1 in cancer cells can also contribute to cisplatin resistance in different cancer types [139] and is considered a biomarker to predict prognosis and drug resistance in cancers [85,140].

Glycolytic inhibitors are reported to sensitize cancer cells to chemotherapy and radiotherapy, thereby overcoming resistance to therapy. 3BrPA aids in dissociating HK2 from the mitochondrial complex and improves therapy response to daunorubicin [15]. A combination therapy of curcumin and docetaxel has demonstrated improved drug response and tolerance in a clinical study in prostate cancer patients [141]. 2-DG, a glycolytic inhibitor that modulates several glycolytic enzymes, restores sensitivity to adriamycin in ER+ BC cells. In HER2+ BC, trastuzumab inhibits tumor growth by downregulating heat shock factor 1 (HSF1) and LDH-A, thereby inhibiting glycolysis. A combination therapy of trastuzumab with LDH-A siRNA-mediated glycolysis inhibition synergistically inhibited tumor growth in trastuzumab-resistant breast cancer cells, suggesting their usefulness in overcoming drug resistance. Further, the combination therapy of trastuzumab with glycolytic inhibitor 2-DG and oxamate increased the sensitivity of ErbB2-positive cancer cells to therapy. An allosteric inhibitor of phosphoglycerate mutase (PGAM), a glycolytic enzyme that converts 3-phosphoglycerate to 2-phosphoglycerate, has been shown to overcome erlotinib resistance in NSCLC. PGAM inhibition alters the ERK and AKT signaling pathways and induces oxidative stress and ROS production to overcome erlotinib resistance [142].

ENO-1 overexpression has been associated with chemoresistance in prostate and pancreatic cancer cells [102,143–145]. Cisplatin-resistant gastric cancer cells also exhibit enhanced glycolysis by upregulating ENO-1. ENO1 inhibition in cisplatin-resistant cells increased sensitivity to the therapy by activating apoptotic pathways or inducing autophagy [96]. In ovarian cancer cells, inhibiting ENO-1 expression increased cell senescence and improved cisplatin resistance [146]. Hypoxia-induced resistance to gemcitabine is a critical issue in PDAC treatment. A recent study demonstrated that the shRNA-based downregulation of ENO-1 modulated redox homeostasis, increased intracellular ROS concentration, and sensitized resistant PDAC cells to gemcitabine treatment. In ovarian cancer models, PFKFB3 inhibitors, 3-PO and PFK-158, impaired metabolic reprogramming-induced stemness and chemoresistance, possibly by modulating apoptosis via the NF-κB pathway [147,148].

PKM2 can contribute to chemoresistance against cisplatin and gemcitabine treatment in different cancer types. PKM2 overexpression has been reported to be a biomarker for cancer resistance. PKM2 regulates the DNA repair mechanism in addition to glucose metabolism and induces resistance to genotoxic damage, driving treatment resistance [149,150]. Targeting PKM2 sensitizes cancer cells to treatment. In NSCLC, shRNA-based silencing of PKM2 enhanced radiation-induced autophagy in vitro and in vivo [149] and increased the sensitivity to docetaxel treatment [119]. PKM2 expression also correlated with a resistance to platinum-based therapy in colorectal cancer [151].

A few studies, however, have reported contradicting results, where PKM2 activation was shown to act as a chemosensitizer in some cancer types. In a study by Anastasiou et al., an increase in intracellular ROS concentration in response to therapy was shown to inhibit PKM2, which in turn diverted the glucose flux into PPP generating redox potential for the detoxification of ROS. These regulatory properties of PKM2 confer an additional advantage to cancer cells to tolerate therapy-induced oxidative stress. The endogenous expression of oxidation-resistant-PKM2 mutants increased oxidative stress and impaired tumor progression [152]. PKM2 activation could thus be an attractive strategy in cancer therapy. High levels of PKM2 activate the mTOR-HIF1α pathway and are associated with a positive chemotherapy response in cervical cancer patients treated with cisplatin-neoadjuvant chemotherapy [153,154]. The high expression of PKM2 has been shown to enhance drug response to epirubicin and 5-fluorouracil in BC [155], whereas a decrease in PKM2 levels/activity contributes to cisplatin/oxaliplatin resistance in cervical cancer, colorectal cancer, and gastric cancer cells [153,156,157].

The development of resistance is frequently encountered in cancer treatment, and the link between cancer cell metabolism and the development of resistance is becoming more apparent. Targeting metabolic enzymes is an efficient strategy to re-sensitize the resistant cells to chemotherapy and radiotherapy. However, the clinical application of glycolysis inhibition to overcome drug resistance has remained limited. Future studies should identify the key metabolic shifts that contribute to the development of drug resistance and explore their potential as drug targets to improve the sensitivity of cancers to chemotherapeutic agents and radiotherapy and to overcome the development of resistance. Figure 1 summarizes how targeting glycolysis can be used to modulate cancer therapy (Figure 1).

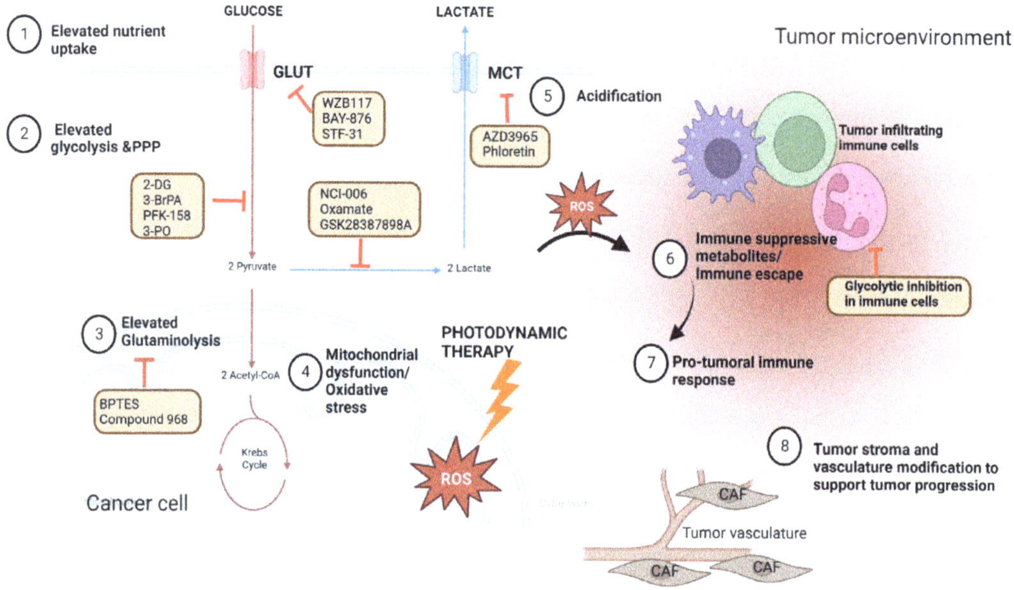


**Figure 1.** Targeting glycolysis to improve cancer therapy. The cancer cells show enhanced dependency on glycolysis that may be targeted to improve the treatment efficacy of conventional cancer therapy modalities, including chemotherapy, radiotherapy, immunotherapy, hormonal therapy, and photodynamic therapy. Glycolysis metabolism can be potentially targeted by limiting glucose uptake (targeting glucose transporters), targeting glycolysis enzymes, targeting glutaminolysis, targeting lactate synthesis, targeting MCT, or targeting mitochondrial complexes. The increased glycolysis in the cancer cells increase the release of lactic acid to the tumor microenvironment, acidifying it and making it pro-cancer and immunosuppressive. The tumor-infiltrating immune cells also show a similar shift toward glucose metabolism increasing the competition for glucose in the tumor microenvironment. Modulating glycolysis in the immune cells can potentially improve immune therapy. This figure was created using the Biorender app.

## 3. Targeting Glycolysis to Enhance Immunotherapy

The advances in our understanding of the remarkable potential of the immune system to fight cancer have garnered tremendous attention on immunotherapy for cancer.

Arguably, immunotherapy is now being considered one of the most promising therapeutic strategies for several types of cancers, and immune checkpoint blockade (ICB)- and adoptive-cell therapy (ACT)-based therapeutic strategies have been approved for several cancers [158–160]. However, several reports have shown that a high percentage of patients initially fail to respond to these interventions or acquire resistance in the long run [161,162], limiting the application of these promising strategies. Several studies have reported that the metabolic reprogramming of the cancer cell that leads to the development of an immunosuppressive tumor microenvironment is one of the main contributors to the reduced efficacy of immunotherapy [163]. Further, it has been suggested that metabolic interventions can significantly enhance the efficacy of immunotherapy [164,165]. Therefore, understanding the challenges the immune cells face in the harsh immunosuppressive tumor microenvironment and identifying strategic interventions to overcome these challenges would contribute to improving immunotherapy.

### *3.1. Glucose Metabolism in the Immune Cells of the Tumor Microenvironment*

It is known that cancer cells undergo a metabolic reprogramming called the 'Warburg effect' or aerobic glycolysis in response to hypoxia and oncogenic signals, such as Myc and PI3K [165]. This preference for glycolysis is also shared by other rapidly proliferating cells in the TME, including effector T-cells and M1-like macrophages, to satisfy their increased energy requirements [166]. In contrast, other cells in the TME, including memory T-cells, regulatory T-cells (Tregs), and M2-like macrophages, rely on fatty acid oxidation (FAO) to satisfy their energy needs.

The naïve T-cells utilize TCA-coupled OXPHOS as their primary energy source [167]. On MHC activation, the T-cell receptor (TCR) and CD28 activate the PI3K-AKT-mTORC1 and Myc signaling pathways and induce metabolic reprogramming [167]. The effector T-cells upregulate aerobic glycolysis and enhance their anabolic metabolism for cancer-killing and clone expansion. In addition, the glycolytic intermediates support effector T-cell activation and cytokine generation. Phosphoenolpyruvate (PEP), a glycolytic metabolite, blocks sarco/endoplasmic reticulum $Ca^{2+}$-ATPase (SERCA)-mediated endoplasmic reticulum calcium uptake and the nuclear factor of activated T-cells (NFAT) signaling, enabling TCR signaling [168]. In addition, phosphoenolpyruvate carboxykinase 1 (PCK1) catalyzes the conversion of oxaloacetate (OAA) into PEP, and the overexpression of PCK1 enhances the cancer-killing functions of adoptive transferred CD4+ and CD8+T-cells [168]. However, although CD8+ T-cells experiencing continuous stimulation or hypoxia differentiated into functional effectors in vitro, it rapidly drove T-cell dysfunction and exhaustion [169].

Similar to the CD8+ T-cells, the functions of the CD4+ T-cells are also affected by specific metabolic reprogramming. Increased glycolysis promotes IL-2, TNFα, and IFN secretion in the CD4+ T-cells, and the inhibition of glycolysis drives the functional and metabolic exhaustion of the CD4+ T-cells [170,171]. In line with this, the inflammatory CD4+ T-cells (Th1 and Th17) show enhanced glycolysis. The Th17 cells exclusively express the pyruvate dehydrogenase (PDH) inhibitor, pyruvate dehydrogenase kinase isozyme 1 (PDHK1), which, when downregulated, leads to the selective reduction of Th17 cells [172]. On the other hand, Tregs show upregulated OXPHOS and FAO [172], and their expression of FOXP3 inhibits Myc and attenuates PI3K-AKT-mTORC1 axis-mediated activation of glycolysis and increases oxidation and catabolic metabolism, rendering a survival advantage in the TME [64,173]. Furthermore, the Tregs utilize lactate metabolism, and culturing them in high-glucose conditions decreases their stability [167]. The mechanisms by which the low levels of oxygen, high levels of lactate, and the high competition for glucose potentially contribute to T-cell dysfunction in the tumor microenvironment is been summarized in Figure 2.

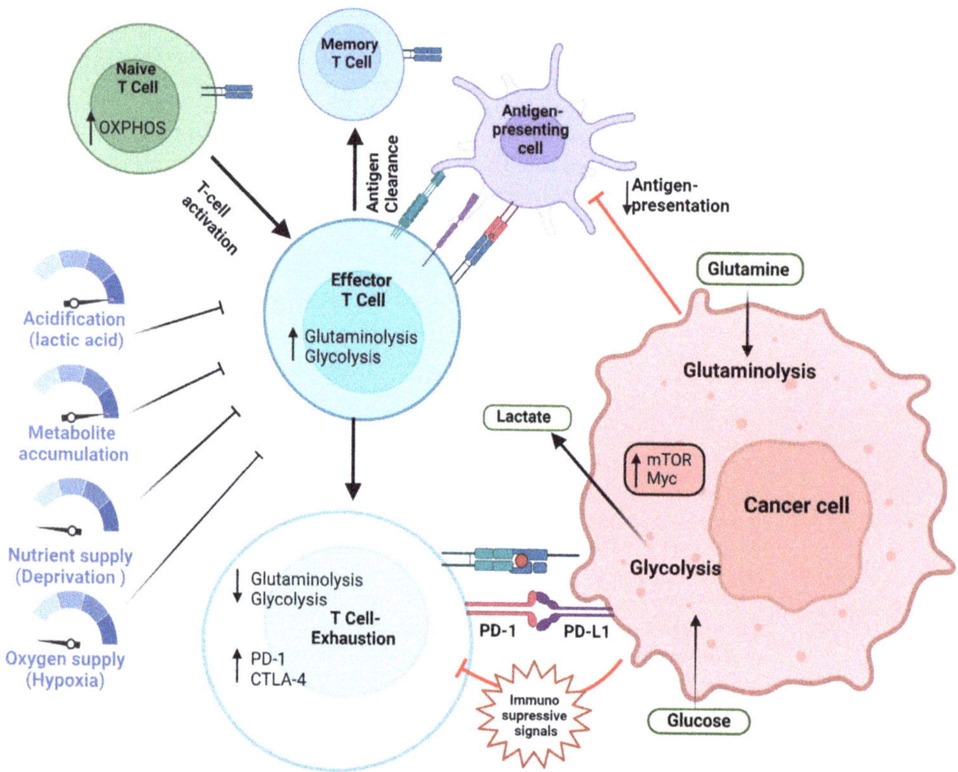


**Figure 2.** Naïve T-cells relay on oxidative metabolism. After activation, the effector T-cells increase glycolysis to support their function. On antigen clearance, the effector T-cells enter a memory state. On antigen persistence, as with long-term tumor elimination, inhibitory receptors such as PD-1 and CTLA4 reprogram the T-cell metabolism leading to metabolic impairments. Exhausted T-cells show reduced glycolysis and glutaminolysis and dependence of fatty acid oxidation. The low levels of oxygen, high levels of lactate, and the high competition for glucose potentially contribute to T-cell dysfunction in the tumor microenvironment. The image was created using Biorender app.

The M1-like macrophages preferentially utilize glycolysis to sustain the inflammatory phenotype [174], while the M2-like macrophages depend on TCA and FAO to maintain an immunosuppressive phenotype [175]. The highly acidic environment of melanoma was found to induce tumor-associated macrophages (TAM) toward a cancer-promoting phenotype [176]. The tumor-associated neutrophils (TAN), on the other hand, exhibit pro- or anti-tumor effects in different cancer microenvironments. In pancreatic ductal adenocarcinoma (PDAC), it was shown that the TANs undergo an LDH-A-mediated glycolytic switch and exhibit a tumor-promoting phenotype [177]. Triple-negative BC (TNBC) cells with an accelerated glycolysis support myeloid-derived suppressor cell (MDSC) development and facilitate CD8+T-cell inhibition and cancer progression [178]. In addition, an increase in lactate generation enhances the tumor-promoting capacity of MDSCs [179].

The TME is characterized by a decrease in nutrients, insufficient vasculature, increased lactate accumulation, and hypoxia, conditions that affect the cancer-killing capacity of the T-cell. The cancer cells and the immune cells compete for glucose utilization, and the activated glycolysis of the cancer cells endows them with an advantage over the immune cells, impairing the function and survival of the effector T-cells. It has been shown that the expression of glycolysis-related genes, such as *ALDOA*, *ALDOC*, *ENO2*,

*GAPDH, GPI,* and *PFKM,* negatively correlates with T-cell infiltration in melanoma and NSCLC [180]. The lower availability of glucose in the TME affects the glycolytic capacity of the T-cells [168], inducing T-cell exhaustion [181]. Furthermore, the lack of glucose also affects mitochondrial functions, promoting terminal CD8+ T-cell exhaustion [182]. In addition to the competition for glucose, the increased production of lactate due to the increased rate of glycolysis contributes to the immunosuppressive character of the TME. Lactate metabolism has been shown to contribute to oncogenesis [183], and LDH has been reported to be a marker for poor prognosis and immunotherapy efficacy [184,185]. The low pH in the TME triggers reduced CD25 and TCR expression in the effector CD8+ T-cells and inactivated STAT5 and ERK signaling, affecting anti-tumor immunity [186]. Lactate also affects TCR signaling [187], and exposure to large amounts of lactate makes the Tregs switch to OXPHOS for regenerating NAD+; however, the Tregs fail to maintain the NAD+/NADH balance through this pathway [64]. Moreover, lactate in the TME promotes the expression of pro-inflammatory cytokines, including IL-23 and IL-17, supporting tumorigenesis and impairing anti-tumor immunity [188]. The poor vasculature and the increased metabolism of the cancer cells contribute to the formation of a hypoxic environment, further affecting anti-tumor immunity. Hypoxia triggers epigenetic reprogramming of effector T-cells, reducing their functional capabilities [189,190]. Although hypoxia-inducible factor-1$\alpha$ (HIF-1$\alpha$), expressed in response to hypoxia, can induce Tregs and bind to the promoter region of the FOXP3 to promote transcription [191] (22988108), it contributes to the development of an immunosuppressive TME by enhancing cancer-promoting immune cell functions [192,193].

### 3.2. Signaling Mechanisms Regulating Glycolysis

The signaling pathways that modulate glycolysis in the immune cells can be potentially targeted to improve their anti-tumor functions, and drugs such as metformin and phenformin have been extensively tested in the clinical setting [194–196]. The liver kinase B1 (LKB1)-AMPK pathway and the PI3K-AKT-mTOR axis are the two primary signaling mechanisms that modulate glycolysis in the cell. The LKB1-dependent kinases regulate metabolic pathways by targeting several effectors, including AMPK [197]. LKB1 loss upregulates GLUT1 and hexokinase 2 (HK2), increasing T-cell glycolytic transcription and flux [198].

On the other hand, the deregulation of the PI3K-AKT-mTOR axis promotes HIF-1 activation and GLUT1 expression [199,200]. PI3K activation increases AKT phosphorylation and glycolytic flux via LDH-A in T-cells [201]. AKT is the primary glycolysis regulator in both cancer and immune cells. AKT activation induces GLUT1 and LDH-A expression [202], activates HK1 by promoting HK2 and PFK2 phosphorylation [203], and inhibits PDH by activating pyruvate dehydrogenase kinase-1 (PDK1) [204], resulting in activated glycolysis. mTOR regulates the expression of HIF-1$\alpha$, the major transcription factor regulating several glycolytic enzymes and GLUT1 [199]. mTOR kinases determine effector and memory CD8+ T-cell fates, and blocking mTOR promotes T-cell effector functions [205,206].

### 3.3. Targeting Glycolysis to Improve Immunotherapy

Adoptive-cell transfer (ACT) and immune checkpoint blockade (ICB) are the two primary strategies for immunotherapy. The interplay between anti-tumor immunity and cancer metabolism suggests that combining immunotherapy with glycolysis-targeted therapy is a promising strategy to improve treatment efficacy. ACT utilizes therapeutic-modified immune cells to directly boost anti-tumor immunity. Ex vivo-expanded T-cells and T-cells engineered to express antigen-specific TCRs or chimeric antigen receptors (CARs) are primarily used for ACT [207]. ACT has shown promising efficacies, particularly with CD19-specific CAR-T cells, in the treatment of B-cell acute lymphoblastic leukemia and B-cell lymphomas [208]. However, responses in other cancers have been poor. It has been suggested that optimizing T-cell metabolism to support robust initial and durable T-cell responses for target cells and the TME would improve and broaden their applicability.

CAR-T cells can be engineered to express specific signaling domains, and endowing them with specific properties can tailor them for effector activity and long-lasting memory. The two FDA-approved CARs carry a CD19-targeting extracellular domain coupled with an intracellular signaling domain from CD3ζ and the co-stimulatory molecules CD28 or 4-1BB [209]. Linking TCR signaling with the co-stimulatory signals enables the CARs to elicit effector functions in the absence of additional inflammatory or co-stimulatory stimuli [210]. CAR-T cells carrying the CD28 domain have increased glycolysis and show enhanced effector responses but are short-lived. On the other hand, CAR-T cells expressing the 4-1BB domain show elevated OXPHOS and FAO and display a memory phenotype [211]. This is in line with the normal physiological functions of the co-stimulatory molecules. CD28 stimulation activates PI3K/AKT/mTOR, promoting glycolysis and effector differentiation, whereas 4-1BB activates AMP-driven FAO and OXPHOS [211]. Based on this understanding, it has been proposed that introducing signaling mutations could improve CAR-T cell effector functions and survival. For example, it was shown that mutations of the YMNM signaling motif of CD28 increase CAR-T cell survival and reduce T-cell exhaustion, enabling enhanced tumor control [212]. Furthermore, additional co-stimulatory molecules modulating T-cell metabolism may be incorporated into CAR constructs to improve T-cell function. For example, ICOS, a CD28 family member, promotes glycolysis and mTORC1 activity in T follicular helper cells [213]; GITR agonists enhance cellular metabolism to support CD8+ T-cell proliferation and effector cytokine production [214]; OX40 is associated with enrichment of glycolysis and lipid metabolism transcripts; and OX40 agonists enhanced lipid uptake in Tregs [215]. Similar to the CAR-T cells, T-cells procured from tumors and expanded ex vivo, or modified to express engineered TCRs, can also be optimized through metabolic manipulations, and T-cells with low glycolysis rates can be generated or selected for longevity while retaining effector functions [216].

The in vitro stimulation and the T-cell engineering phases of the ACT strategy provides the added advantage of the opportunity to modify T-cell metabolism and mitochondria without affecting other cells and tissues and thus prevent any potential boosting of the cancer cell metabolism from the intervention. It was also shown that blocking glutamine metabolism increases T-cell function [217]. Furthermore, supplementing the culture medium with glutamine antagonist 6-Diazo-5-oxo-l-norleucine (DON) enhanced CAR-T cell FAO and reduced glycolysis, making the CAR-T cells remain in a more undifferentiated state [218]. Inhibiting glycolysis using the HK-2 inhibitor, 2-DG, before ACT induced a memory-like phenotype in the T-cells, enabling a more efficient control of tumors and prolonged survival in an animal model [219]. Similarly, CD19-CAR-T cells treated with AKT inhibitors showed reduced glycolysis and a memory-like phenotype and had robust tumor elimination potential [220]. Conversely, modulating mitochondrial OXPHOS and FAO would enable CAR-T cell longevity and the continued expression of a memory-like phenotype. Furthermore, a transient glucose restriction followed by glucose re-exposure could enhance the tumor-clearing efficacy of CD8+ T-cells [221].

Unlike ACT, which directs anti-tumor immunity through pharmaceutical interventions, ICB aims to modify inhibitory signals to activate endogenous anti-tumor-specific T-cells. Furthermore, ICB is primarily targeted against solid tumors, which show a greater influence of the TME in modulating T-cell metabolism. In addition to the initial challenge of T-cells infiltrating the tumors, ICB must overcome several obstacles, including tumor-infiltrating lymphocyte (TIL) exhaustion, the upregulation of inhibitory receptors and epigenetic modifications, metabolic adaptations resulting in nutrient deficits, impaired translocation of GLUT1 to the cell surface, the downregulation of glycolytic enzymes, GAPDH and ENO-1, and dysregulated and fragmented mitochondria with increased ROS generation [222–225]. It was reported that the TIL inflammatory function could be enhanced by rescuing TIL metabolism by expressing PCK to promote gluconeogenesis and replacing the intracellular glycolytic intermediates or by improving mitochondrial metabolism by treating with pyruvate or acetate [225]. The immunosuppressive environment of the TME and chronic antigen stress direct the T-cells to an exhausted state, characterized by the expression of immune checkpoints and decreased cytotoxicity. The checkpoint molecules, including CD28, CD40L, and cytotoxic T lymphocyte-associated protein-4 (CTLA-4), along with TCR signaling, drive TIL exhaustion and contribute to the persistence of the exhausted state. The concurrent metabolic adaptations further enhance the immunosuppressive effects of the checkpoints. Thus, targeting glycolytic regulators and metabolites that support the immune checkpoint-directed T-cell inhibition is a potential strategy to improve the efficacy of ICB.

The most well-known and commonly targeted immune checkpoints in cancer are the programmed cell death protein-1 (PD-1) and CTLA-4. CTLA-4 is expressed primarily on T-cells and plays an immunosuppressive role during the initial phase of T-cell activation and downregulates T-cell activation-triggered glycolysis. PD-1 is activated after TCR activation and impedes glucose uptake and glycolysis while promoting FAO. Thus, these checkpoint signals prevent T-cell activation and inflammation. It was shown that PD-1–deficient T-cells maintain higher metabolic activity in chronic infection [226,227]. Thus, blocking PD-1 and CTLA-4 relieves PI3K/AKT/mTORC1 signaling and allows increased T-cell stimulation and metabolism, inducing an effector-like phenotype. In addition, this metabolic shift induces epigenetic reprogramming inducing effector functions and longevity. Although PD-1 and CTLA-4 are the most extensively targeted ICB candidates, several other co-inhibitory and co-stimulatory molecules modulate T-cell metabolism. Notably, T-cell immunoglobulin mucin receptor 3 (TIM-3) downregulates glycolysis and GLUT1 expression [228], while LAG3 downregulates OXPHOS [229]. Similarly, inhibiting 4-1BB and OX40 enhances T-cell OXPHOS and promotes effector function and longevity [230]. Taken together, immune checkpoint molecules induce metabolic dysfunction and thus affect anti-tumor immunity, suggesting that combining ICB with metabolic regulators is a potential strategy to improve treatment efficacy.

### 3.4. Glycolysis-Targeting Therapies to Improve Immunotherapy Efficacy

mTOR is an oncogenic molecule that contributes to the regulation of metabolism in TILs. The inhibition of the mTOR signaling axis downregulates the malignant phenotype of cancer cells. Owing to their inhibitory effects, several rapamycin analogs have been approved for treating cancers [231,232]. However, it was shown that mTOR inhibition could diminish anti-tumor immunity [194] as these inhibitors directly affect the lineage differentiation-determining glycolytic activity in T-cells. It was shown that rapamycin suppresses Th17 differentiation and promotes Treg differentiation under TGF$\beta$ induction [233,234]. Further, the activation of AKT-mTORC1 signaling was associated with T-cell function restoration and the reduced expression of PD-1 and TIM-3 [235]. Additionally, the over-activation of mTORC1 affects the immunosuppressive function of Tregs, while low mTORC1 levels enhance Treg activity [236]. These suggest that an optimized inhibition of the PI3K-mTORC1 signaling axis is critical for improving the efficacy of immunotherapies.

Metformin has shown promising effects in different cancers [195] and has been shown to regulate metabolism by interacting with AMPK, the PI3K-AKT-mTOR axis, and HIF-1$\alpha$ [237,238]. Additionally, metformin promotes the cancer-killing capacity of CD8+ T-cells by modulating glycolysis [239–241] and downregulates immune checkpoint expression and glycolytic flux through HIF-1$\alpha$ inhibition [242,243]. Furthermore, it was shown that metformin stops the cancer cells from using the lactate and ketone bodies produced by cancer-associated fibroblasts as nutrients and thus suppresses cancer progression [244]. A recent study combined 2-DG, an HK inhibitor [245], BAY-876, a GLUT-1 inhibitor, and chloroquine and developed the nano-drug, D/B/CQ@ZIF-8@CS, which inhibited glycolysis and improved anti-CTLA-4 immunotherapy by reducing Treg metabolic fitness [246]. Tregs pretreated with 2-DG showed enhanced inhibition of T-cell proliferation in ovarian cancer [247]. Furthermore, HK upregulates PD-L1 expression in cancer cells, and combining the HK inhibitor, Lonidamine, with anti-PD-1 therapy improved cancer cell elimination in a mouse model [248].

Knocking out glucose-6-phosphate isomerase (GPI), the enzyme that catalyzes the conversion of glucose-6-phosphate (G6P) to fructose-6-phosphate (F6P), upregulated OXPHOS and sustained the survival of cancer cells [245]. Additionally, GPI inhibition selectively eliminated inflammatory encephalitogenic and colitogenic Th17 cells without affecting the homeostatic microbiota-specific Th17 cells [249]. However, it remains unknown whether GPI-targeted therapies would improve the efficacy of immunotherapy.

The GAPDH inhibitor, dimethyl fumarate (DMF), promotes the oxidative PPP and inhibits glycolysis and OXPHOS in cancer cells. This reduces the competition between cancer cells and T-cells for glucose consumption and promotes the efficacy of ICB and IL-2 therapy [250]. Low-dose osimertinib was shown to inhibit GAPDH and tumor endothelial glycolysis and promote vascularization and immune cell infiltration and thus improve the efficacy of anti-PD-1 therapy [251]. Inhibiting fructose-2,6-bisphosphatase 3 (PFKFB3), which is upregulated in several cancers, repressed glycolysis and upregulated PD-L1 expression [193]. On the contrary, glucose deficiency upregulated PD-L1 through the EGFR/ERK/c-Jun pathway, leading to the upregulation of PFKFB3, and promoted glycolysis [252,253]. This suggests the existence of a positive feedback loop between metabolism and checkpoint molecules [254]. A dual-target drug comprising paclitaxel and the PFKFB3 inhibitor, PFK15, blocked cancer-associated fibroblast-mediated cancer cell growth and reduced the lactate concentration in the TME [193]. PFK15 was also shown to upregulate PD-1 and LAG-3 expression in the context of type 1 diabetes [171]. Pyruvate kinase isoform M2 (PKM2), the final rate-limiting enzyme in glycolysis, promotes PD-L1 expression in macrophages, DCs, and tumor cells and contributes toward accelerated tumor progression [255]. High PKM2 expression was associated with a poor prognosis of pancreatic ductal adenocarcinoma, and the knocking down of PKM2 improved the efficacy of anti-PD-1/PD-L1 therapy [256].

ENO-1, which catalyzes the conversion of 2-phosphoglycerate to PEP and also acts as a plasminogen receptor and a DNA-binding protein, was shown to be overexpressed in several cancers. [64]. A pan-cancer analysis showed that ENO-1 expression correlated with immune cell infiltration, including B cells, CD8+ and CD4+ T-cells, macrophages, neutrophils, and dendritic cells [257]. The presence of autoantibodies against ENO-1 correlated with a better prognosis in PDA, suggesting that ENO-1 was a good molecular candidate for improving immune cell response to cancers [258]. Antibodies against ENO-1 were detected in approximately 60% of patients with PDAC, and ENO-1-specific T-cell responses are observed in patients who have the anti-ENO-1 antibodies. In line with this, an ENO-1 DNA vaccine induced an antibody and cellular response and increased the median survival in mouse models of PDA [259]. ENO-1-targeting DNA vaccines have shown prophylactic and therapeutic potential in PDAC mouse models by inducing complement-dependent cytotoxicity and immune cell response [144,259,260]. In a spontaneous mouse model of PDAC, co-treatment with gemcitabine and ENO-1 DNA vaccine enhanced CD4 anti-tumor activity and impaired tumor progression [261,262]. Additionally, a recent study showed that targeting ENO-1 using specific antibodies targets multiple TME niches involved in prostate cancer (PC) progression and bone metastasis via a plasmin-related mechanism [263]. These show the potential of ENO-1 targeted therapies in improving the efficacy of immunotherapies.

Optimizing T-cell metabolism is a promising strategy for improving cancer immunotherapy. Metabolic modification can potentially increase stemness and long-term memory, enhance effector functions, and reduce T-cell exhaustion. Future studies should identify key metabolic transitions and regulatory steps that are differently regulated in cancer cells and immune cells and develop effective targeting strategies to enhance the synergistic effects of metabolic modulation and cancer immunotherapy [264].

## 4. Targeting Glycolysis to Enhance Hormonal Therapy

Hormonal therapy has shown remarkable advancement as a therapeutic strategy for cancers dependent on hormones, especially in breast, prostate, and other gynecological cancers. Aromatase inhibitors (AI), estrogen receptor (ER) antagonists, ER modulators, anti-estrogens, and GnRH agonists are effective therapeutic drugs and have shown high success rates in patients with hormone-sensitive recurring or metastatic gynecologic malignancies [265]. Hormone therapy interferes with the hormone-dependency of cancer cells by limiting hormone production in the body [266]. While hormonal therapy has improved survival and reduced recurrence in different cancer types [266], de novo or acquired resistance to hormonal therapy is a major clinical problem that requires the development of innovative strategies [264]. Resistance to hormonal therapy invariably occurs in most patients with ER+ metastatic BC and castration-resistance PC (CRPC) [267]. Metabolic reprogramming is an inherent feature of endocrine-resistant cancer cells, implicating that combination therapy with metabolic regulators and conventional hormonal therapy might be beneficial in overcoming resistance [268]. However, it is unclear whether metabolic rewiring is a cause or consequence of endocrine resistance, and several studies are investigating the cross-talk between hormone signaling and cancer cell metabolism [269]. Somatic mutation in estrogen receptors is related to the clinical development of the resistance to hormone therapies [268,270–272]. The Y537S mutation in ER-$\alpha$ enhanced mitochondrial metabolism and glycolysis in BC cells. The Y537S mutation is also associated with poor clinical outcomes, suggesting that enhanced glucose metabolism is a highly conserved mechanism of endocrine resistance [268].

Elevated glucose levels resembling hyperglycemia in BC cells have been attributed to a reduced response to tamoxifen therapy and could act as a marker for responses to hormonal therapy [273,274]. An increased glycolytic rate is a characteristic feature of tamoxifen-resistant cells, and inhibiting glycolysis is expected to restore tamoxifen sensitivity [273,275]. Elevated glycolysis in BC cells is also associated with mitochondrial malfunction and upregulated AKT/mTOR and HIF-1$\alpha$ signaling pathways. Tamoxifen-resistant BC cells escape cell death by increasing autophagy through the inactivation of TOR-S6K via the HK2 pathway [276]. Glycolytic inhibition by the knockdown of HK2 or 3BrPA treatment downregulated AKT/mTOR signaling and could be a therapeutic strategy to overcome tamoxifen resistance in BC [277]. A recent study investigated the potency of a combination therapy employing low-dose tamoxifen (ER$\alpha$ antagonist) and metabolism inhibitors, 2-DG and CB-839 (glutaminolysis inhibitor), in improving the anti-proliferation effect in tamoxifen-resistant ER$\alpha$-positive BC cells. The triple combination showed superior cell growth inhibition by inducing apoptosis and c-Myc downregulation; however, a combination of tamoxifen with 2-DG did not show significantly strong inhibition of cell viability [278,279]. The pharmacological inhibition of glycolysis with PFK-158, a PFKFB3 inhibitor, with tamoxifen or fulvestrant has been explored as a potential therapeutic intervention to overcome endocrine resistance. PFKFB3 upregulation, with an elevated basal expression of PFKFB3 mRNA, is observed in endocrine-therapy-resistant BC cells and is associated with adverse recurrence-free survival in BC patients. The anti-tumor effect of PFK-158 is exacerbated when combined with tamoxifen and fulvestrant treatment [280]. PFKFB3 inhibition activated necroptotic markers receptor-interacting kinase 1 (RIPK1) and mixed lineage kinase domain-like pseudokinase (MLKL), implicating the possible mechanism of PFK-158-induced cell death [281]. In a long-term estrogen deprivation model (LTED) of AI resistance, cancer cells were demonstrated to have increased glycolysis dependency. The inhibition of glycolysis with HK2 inhibitors, along with AI, and letrozole, reduced cell viability [282]. Dietary interventions that target metabolic rewiring have also been shown to improve the efficacy of endocrine therapy in liver metastatic BC patients. Metastatic burden in the liver increases with increasing carbohydrate percentage in the diet. A fasting-mimicking diet increased the efficacy of fulvestrant treatment and reduced the metastatic burden in BC liver metastatic models, providing a proof-of-concept for a more straightforward strategy to circumvent drug resistance, with potential applicability in other cancer types as well [283].

Metabolic reprogramming is emerging as a crucial mechanism contributing to resistance to endocrine therapy in PC [284]. The expression of key glycolytic enzymes, including

LDH-A, MCT-4, and GLUT1, is elevated in mCRPC patients [285]. Glycolysis inhibition by targeting GLUT1 plays an important role in drug response in prostate PC [286]. In PC, elevated androgen levels increase glucose uptake and upregulate the expression of GLUTs, implying a cross-talk between androgen signaling and glycolytic pathway, a mechanism that protects PC cells from glucose-deprivation-induced oxidative stress [287,288]. NF-κB-mediated GLUT1 overexpression and upregulated glucose metabolism are associated with enzalutamide resistance in PC [289,290]. In xenograft models of CRPC and enzalutamide-resistant PC patients, GLUT1 inhibition significantly reduced tumor volume and displayed synergistic effects with androgen receptor (AR)-targeted therapy [286]. Glycolytic inhibitors, gossypol (LDH-A inhibitor), and AZD3965 (MCT-1 inhibitor) are currently in clinical trials as potential glycolysis-targeting agents in mCRPC. Progesterone treatment was reported to have an anti-tumor effect in glioblastoma multiform (GBM) in vitro and in vivo and improved the efficacy of temozolomide [291]. Recently it was shown that high-dose progesterone treatment inhibits GBM growth by inhibiting the key modulators of glycolytic metabolism. This early observation highlighted the potential of progesterone in metabolic reprogramming; however, more direct evidence is essential to validate this, and future studies should determine the synergistic effect of direct glycolytic inhibitors and progesterone in GBM treatment [291].

Endocrine resistance remains a major clinical barrier that requires the development of novel strategies to circumvent the resistance. Several mechanisms that contribute to endocrine resistance have been identified. Metabolic rewiring is frequently observed in most cancer cells that exhibit resistance, and targeting glucose metabolism with well-established glycolytic inhibitors has shown to enhance the sensitivity to endocrine therapy in breast and PC models. The mutual interplay between glucose metabolism and androgen receptor/ER signaling implies that combination approaches of endocrine therapy with metabolic modulators could be a standard-of-care to overcome resistance. Dietary interventions that modulate glucose metabolism have also been demonstrated to be an interesting strategy for evading resistance to therapy. Well-designed clinical trials are urgently needed to elucidate the clinical utility of the strategies mentioned above and to develop metabolic drugs as routine standard-of-care in endocrine-resistant cancer patients in clinical settings.

## 5. Targeting Glycolysis to Improve Photodynamic Therapy

Photodynamic therapy (PDT) is a relatively new, minimally invasive therapeutic procedure that relies on the selective accumulation of a photosensitive compound in the cancer cells, which, on excitation with light of an appropriate wavelength, would generate ROS, predominantly singlet oxygen, within the cancer cells, and eventually kill the cancer cell, with minimal damage to the surrounding tissue [292–294]. Although PDT is widely used to treat several cancers, its efficacy is limited by several factors, including the effective irradiation of deep tissue. Therefore, several studies have attempted to improve the efficacy of PDT by combining it with other chemotherapeutic agents. It has been shown that glycolytic inhibitors disrupt cancer cell metabolism, elevate the cellular ROS level, and disrupt the mitochondria, resulting in cell death [295]. Therefore, when combined with PDT, glycolytic inhibitors could, in theory, enhance the levels of cellular ROS and thus trigger increased cancer cell death.

5-aminolevulinic acid (5-ALA) is one of the most commonly used photosensitizers for photodynamic therapy. 5-ALA is a naturally occurring non-proteinogenic δ-amino acid synthesized in the mitochondria by the condensation of glycine and succinyl-CoA by mitochondrial 5-ALA synthase (ALAS). This is the first committed step toward heme biosynthesis. The final precursor of heme is Protoporphyrin IX (PpIX), which is a highly potent photosensitizer. The exogenous supplementation of 5-ALA overrides the normal feedback inhibition of ALAS and results in the accumulation of PpIX selectively in cancer cells, owing to the differences in the heme biosynthesis pathway enzyme activities between the cancer cells and normal cells. This cancer-specific accumulation of PpIX is exploited for selectively purging cancer cells by PDT and for the visualization of tumor tissue by photodynamic diagnosis (PD). 5-ALA was approved by the U.S FDA in 2017 as an adjunct for the visualization of malignant tissue in grade III and IV glioma (NDA 208630/SN0014) and is currently used in the clinic to guide the resection of malignant glioma and glioblastoma. In addition to its role as a precursor of PpIX, 5-ALA has been reported to enhance aerobic bioenergetics [296], promote mitochondrial protein expression, and stimulate heme-oxygenase-1, triggering heme degradation [297]. Figure 3 describes the heme biosynthetic pathway, and how it is correlated with glycolysis.

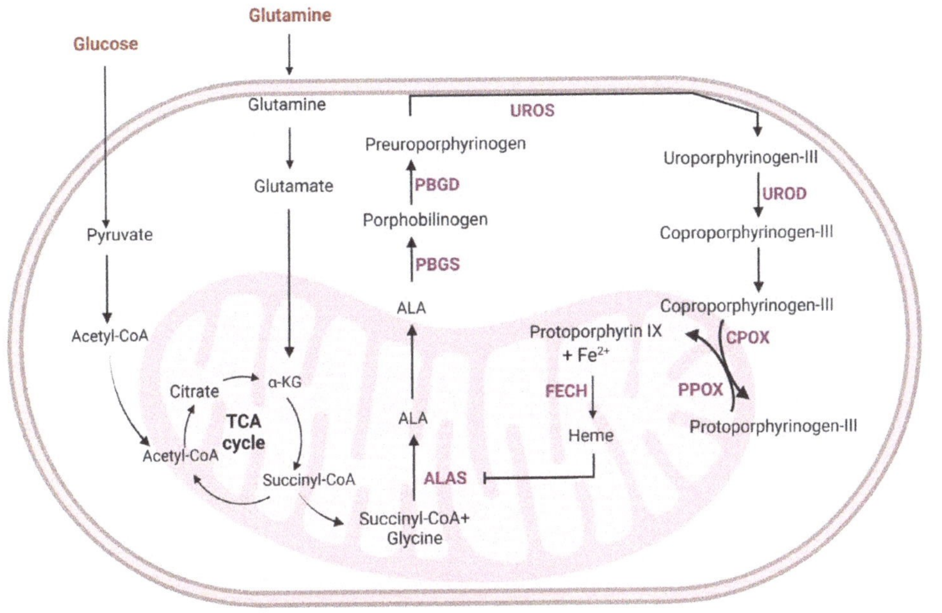


**Figure 3.** The heme biosynthesis is connected with glucose and glutamine metabolism. Enhanced glycolysis and glutaminolysis in the cancer cell might contribute to the elevated rate of heme synthesis and the upregulation of several heme biosynthesis pathway enzymes in the cancers. Addition of exogenous 5-ALA bypasses the feedback inhibition of ALAS and increase the accumulation of protoporphyrin IX (PpIX) in the cancer cells. The elevated accumulation of PpIX is exploited for fluorescence-guided detection (photodynamic diagnosis) of cancers. Further, irradiating PpIX-accumulating cancer cells with light generated singlet oxygen and ROS that kills the cancer cell, with minimal damage to the surrounding tissue (photodynamic therapy; PDT). The interdependency of glucose metabolism and the heme synthesis pathways suggest that targeting glycolysis could enable the modulation of PpIX accumulation in the cancer cells. The image was created using Biorender app.

A recent study showed that 5-ALA is a potent competitive inhibitor of LDH with efficacies comparable to oxamate (OXM) and tartronate (TART) [298]. They showed that treatment with 5-ALA induced glycolysis inhibition and triggered cell death in glioblastoma cell lines. Further, it was shown that up to 25% of the 5-ALA was used for glycolysis inhibition in these cells, leaving a lower amount of 5-ALA for conversion to PpIX and subsequent use as a photosensitizer for PDT and PD. Treating the glioblastoma cells with an LDH inhibitor before 5-ALA treatment enhanced the efficacy of PDT by 15%. Precise delineation of the tumor–normal tissue is of critical importance, especially in brain cancer surgeries, and a 15% increase in PD efficacy is a significant improvement.

Another study showed that exogenous 5-ALA suppressed oxidative metabolism and glycolysis and reduced cell proliferation in ovarian and BC cells [299]. Further, 5-ALA also destabilized Bach1 and inhibited cancer cell migration. The study also showed that

5-ALA-induced suppression of oxidative metabolism and glycolysis was mediated through different mechanisms in BC and ovarian cancer but involved Bach1 destabilization, AMPK activation, and the induction of oxidative stress. Additionally, an inverse relationship between oxidative metabolism and ALA sensitivity was revealed.

It was also shown that the administration of 5-ALA for 6 weeks reduced the plasma glucose levels in rats without affecting their plasm insulin levels and induced HO-1 expression in the liver and white adipose tissue [297]. An increase in HO-1 indicates an increase in heme in the liver, which promotes the formation of nuclear receptor subfamily 1 (Rev-Erbα) with its corepressor nuclear receptor co-repressor 1 (NCOR), which in turn downregulates the enzymes involved in gluconeogenesis such as PEPCK and G6Pase, resulting in reduced glucose production in the liver [300]. On the other hand, 5-ALA administration enhances glucose metabolism in the adipocytes by decreasing the total amount of adipose tissue or decreasing the number of mitochondria in the adipose tissue [301]. It has been shown that the induction of peroxisome proliferator-activated receptor γ coactivator 1-α (PGC-1α), a master transcription coactivator, increases heme synthesis by upregulating ALAS [302]. Glucose intake repressed PGC-1α mediated ALAS in this study, suggesting that nutrient stress could trigger increased heme synthesis. Nutrient stress is a characteristic feature of the tumor microenvironment, and earlier studies have shown that BC cells (MCF7) grown under glucose deprivation produced higher levels of PpIX than those cultured under standard conditions [66,303]. Furthermore, co-treatment with glycolytic inhibitors and 5-ALA reduced intracellular PpIX levels [304]. This has been attributed to the inactivation of ABC transporters induced by ATP depletion, which in turn decreased the flux of precursors into the cell [303,305]. Interestingly, other studies have reported an increase in cellular PpIX accumulation with the inhibition of ABC transporters [306,307]. On the other hand, a combined treatment with 5-ALA-PDT and dichloroacetate, an inhibitor of pyruvate dehydrogenase, showed improved efficacy in BC (MCF 7) cells [308].

In another study, an 18 h glucose deprivation prior to PDT reduced intracellular glutathione levels and increased the cytotoxicity of PDT [309]. A similar increase in PDT efficacy was observed when inhibitors of glutathione synthesis (buthionine-sulfoximine) or its regeneration (1,3-bis-(2-chlorethyl)-1-nitrosourea) were used for co-treatment with PDT [310]. Changes in the availability of glycolytic substrates affect NADPH availability in the cells. NADPH is a critical agent involved in the anti-oxidative defense mechanisms of the cell. As mentioned, PDT relies on increased ROS in the cancer cells to kill them. PDT-based ROS generation under conditions of impaired glucose and glutathione metabolism results in much higher intracellular ROS levels, contributing to increased efficacy. In line with this, ROS scavengers were shown to protect the cells from ALA-PDT-induced damage [311].

A recent study showed that the metabolic reprogramming toward aerobic glycolysis in cancer cells contributes to the development of resistance to 5-ALA-PDT. Further, they showed that treatment with metformin reduced aerobic glycolysis and increased OXPHOS in squamous cell carcinoma cells, and improved the cytotoxic effect of PDT by increasing PpIX production, ROS generation, and AMPK expression, and inhibiting the AKT/mTOR pathway [312]. In another study, combining glycolysis inhibitor 2-DG with 5-ALA induced an enhanced accumulation of PpIX in HepG2 liver cancer cells [313], contributing to improved treatment efficacy. Further, treatment with 2-DG and 3-bromopyruvate (3-BP) synergistically improved the efficacy of PDT in BC cells [314].

Oncogenic transformation has been reported to upregulate glycolytic enzymes and contribute to the increased exogenous ALA-treatment-induced PpIX accumulation in cells [315]. The increased PpIX accumulation in cancer cells is often attributed to their lower ferrochelatase (FECH) levels. FECH is the terminal enzyme in the heme synthesis pathway that catalyzes the chelation of PpIX with the ferrous ion to generate heme. Lower FECH levels have shown a significant association with the cancer grade, the TNM stage, unfavorable prognosis, and impaired immune cell infiltration in clear cell renal cell carcinoma [316]. However, oncogenic Ras/MEK has been shown to increase the conversion of

PpIX to heme by increasing HIF-1α expression and thereby promoting FECH activity [317]. Inhibiting MEK decreased HIF-1α expression and FECH activity and increased 5-ALA-induced PpIX accumulation in cancers [317,318]. Oncogenic Ras/MEK signaling-induced HIF-1α expression has also been implicated in increasing glycolytic flux and driving cancer progression [319,320]. Although 5-ALA-induced PpIX accumulation in cancer cells is a well-documented and well-studied concept, our knowledge of the underlying mechanisms remains largely vague. The inter-dependencies between oncogenic transformation, glycolytic flux, and heme metabolism should be further studied to identify the optimal targeting strategy to enhance 5-ALA-induced PpIX accumulation in cancer cells and to improve the efficacies of PDT and PD.

## 6. Conclusions

The metabolic signatures of tumor cells are different from normal cells, which allows the tumor cells to adapt to the increased energy and metabolite demands [321,322]. Several signaling mechanisms that regulate or hijack the canonical reactions in glucose metabolism have been identified; however, there is no universal mechanism underlying the reprogramming of glucose metabolism in all cancers. Elevated glucose metabolism, hypoxia-induced GLUT, LDH-A, and PFKFB3 overexpression, and the AKT- and c-Myc-mediated transcriptional activation of HK2 are observed in most cancer cells, and these metabolic changes may be exploited for developing effective therapeutic approaches. The tumor microenvironmental regulation and immune suppressive effects of metabolic rewiring are also crucial players in cancer cell progression, survival, and resistance. Apart from glycolytic modulation, mitochondrial dysfunction, elevated ROS production, and dysregulated TCA, cycle enzymes also participate in oncogenic signaling and tumor progression, which were not discussed in this review, and are elegantly reviewed elsewhere [323,324].

Elevated glucose metabolism and nutrient uptake have been exploited for tumor diagnosis in the clinic, and F-18 fluoro-2-deoxyglucose PET is widely used in the clinical staging of cancers [325,326]. However, glycolysis inhibition has not been exploited to its full potential in cancer therapy and it has not translated into the clinic. The inhibition of glycolysis may have undesirable consequences as normal cells also use the same glycolytic enzymes, and it is hence necessary to identify the enzymes or enzyme isoforms that are specifically upregulated or preferentially used by cancer cells. Additionally, several of the glycolysis inhibitors that have been developed for clinical use have been shown to have off-target effects, killing non-cancerous cells [327]. In addition, though the inhibition of glycolysis might inhibit cancer cell proliferation, cancer cells may adapt by upregulating OXPHOS or glutaminolysis, which could result in the development of resistance to therapy, in addition to co-morbidities such as cachexia in patients. This rewiring of metabolic pathways also poses challenges to precision therapies. In fact, previous studies have highlighted the dispensability of the Warburg effect in cancer cell growth and have shown that a complete disruption of glycolysis would require the deletion of both LDHA and LDHB genes. The study evaluated a potent LDH A/B dual inhibitor GNE-140 and demonstrated that the "glycolytic Warburg" phenotype of tumor cells depends on both LDH A and LDH B expression, and is a dispensable phenotype that can be replaced by OXPHOS [328]. This emphasizes the importance of evaluating the simultaneous inhibition of glycolysis and OXPHOS as a therapeutic strategy. In cell lines that are innately resistant to GNE-140 as they predominantly use OXPHOS, the inhibition of OXPHOS sensitized the cells to GNE-140 treatment [329]. Understanding the pathways that contribute to resistance in glycolysis targeting drugs is hence crucial [245,329–331].

Despite early indications, glycolytic inhibitors such as 2-DG have failed in the clinical setting due to their limited effects as a monotherapy agent. In addition, it might be necessary to combine glycolytic inhibitors with other agents that target alternate pathways that are activated in response to glycolytic inhibition. Studies have shown promising effects when metformin (that inhibits mitochondrial complex 1) was used along with glycolytic inhibitors. The simultaneous targeting of MCT1 inhibitor AZD3965 and the mitochondrial respiratory complex 1 inhibitor IACS01759 is a more clinically relevant strategy compared with targeting only one pathway in B-cell lymphomas [332]. Using glycolytic inhibitors as adjuvant therapy with already approved conventional therapies, including chemotherapy, radiotherapy, immunotherapy, and photodynamic therapy, is an attractive strategy, and several early studies have shown promising results. Dietary interventions and lifestyle changes also affect the metabolic landscape of cancer cells, and studies should address this issue in detail to emphasize their clinical implications.

Though metabolic rewiring unquestionably affects cancer cell proliferation, the translation of metabolic reprogramming to the clinic must overcome several hurdles [333]. In-depth analytical and extensive pre-clinical studies should identify targetable metabolic enzymes/enzyme isoforms that are efficacious in different tumor types with minimal toxicity to normal cells. Another major challenge in the clinical development of cancer therapeutics is the need to identify patient groups that would benefit from the therapy. Dependency on metabolic pathways is not universal for all cancer types, which makes it all the more important to identify appropriate patient groups to avoid unwanted adverse effects and toxicity. Technical hurdles in measuring metabolism in vivo also add more complexity. It is essential to integrate genetic and metabolic biomarkers and tissue information to optimize the criteria for patient selection. An interesting study published recently showed the possibility of using glycolytic enzymes as a surrogate for glycolytic activity in cancer cells, using liquid biopsy. Cells expressing high levels of HK2 were identified in both cytokeratin (CK)-positive and CK-negative CTC populations isolated from lung adenocarcinoma patients [334]. The possibility of monitoring the glycolytic metabolic rewiring in real-time using liquid biopsy samples and downstream single-cell level molecular, genomic, and metabolic studies is likely to create a paradigm shift in the clinical utility of glycolytic modulations in cancer therapy. Well-designed clinical trials that unravel the metabolic dependencies of different cancer types and their association with genetic and histopathological information might answer several questions on the complex and sophisticated nature of cancer metabolism. Metabolomic studies, with powerful technological backing, would take the lead in the early identification of cancer risk factors and aid in improving cancer therapy and cancer screening, diagnosis, and therapy monitoring in the coming years.

**Author Contributions:** Conceptualization, C.C. and V.S.C. resources, C.C. and V.S.C.; writing—original draft preparation, C.C., and V.S.C.; writing—review and editing, C.C., V.S.C., and K.S.; supervision, Y.S.; project administration, K.S.; funding acquisition, Y.S. and K.S. All authors have read and agreed to the published version of the manuscript.

**Funding:** This work was supported by the National Research Foundation of Korea (NRF) grant funded by the Korea government (MSIT) (No. 202200070001, No. 202100400001).

**Institutional Review Board Statement:** Not applicable.

**Informed Consent Statement:** Not applicable.

**Data Availability Statement:** Not applicable.

**Conflicts of Interest:** The authors declare no conflict of interest.

## References

1. Vander Heiden, M.G.; Cantley, L.C.; Thompson, C.B. Understanding the Warburg effect: The metabolic requirements of cell proliferation. *Science* **2009**, *324*, 1029–1033. [CrossRef] [PubMed]
2. DeBerardinis, R.J.; Lum, J.J.; Hatzivassiliou, G.; Thompson, C.B. The biology of cancer: Metabolic reprogramming fuels cell growth and proliferation. *Cell Metab.* **2008**, *7*, 11–20. [CrossRef] [PubMed]
3. Pedersen, P.L. Tumor mitochondria and the bioenergetics of cancer cells. *Prog. Exp. Tumor Res.* **1978**, *22*, 190–274. [CrossRef] [PubMed]
4. Liberti, M.V.; Locasale, J.W. The Warburg Effect: How Does it Benefit Cancer Cells? *Trends Biochem. Sci.* **2016**, *41*, 211–218. [CrossRef]
5. Lemberg, K.M.; Gori, S.S.; Tsukamoto, T.; Rais, R.; Slusher, B.S. Clinical development of metabolic inhibitors for oncology. *J. Clin. Investig.* **2022**, *132*, e148550. [CrossRef]

6. Peng, J.; Cui, Y.; Xu, S.; Wu, X.; Huang, Y.; Zhou, W.; Wang, S.; Fu, Z.; Xie, H. Altered glycolysis results in drug-resistant in clinical tumor therapy. *Oncol. Lett.* **2021**, *21*, 369. [CrossRef]
7. Brand, A.; Singer, K.; Koehl, G.E.; Kolitzus, M.; Schoenhammer, G.; Thiel, A.; Matos, C.; Bruss, C.; Klobuch, S.; Peter, K.; et al. LDHA-Associated Lactic Acid Production Blunts Tumor Immunosurveillance by T and NK Cells. *Cell Metab.* **2016**, *24*, 657–671. [CrossRef]
8. Sonveaux, P.; Vegran, F.; Schroeder, T.; Wergin, M.C.; Verrax, J.; Rabbani, Z.N.; De Saedeleer, C.J.; Kennedy, K.M.; Diepart, C.; Jordan, B.F.; et al. Targeting lactate-fueled respiration selectively kills hypoxic tumor cells in mice. *J. Clin. Investig.* **2008**, *118*, 3930–3942. [CrossRef]
9. Gatenby, R.A.; Gillies, R.J. Why do cancers have high aerobic glycolysis? *Nat. Rev. Cancer* **2004**, *4*, 891–899. [CrossRef]
10. Pusapati, R.V.; Daemen, A.; Wilson, C.; Sandoval, W.; Gao, M.; Haley, B.; Baudy, A.R.; Hatzivassiliou, G.; Evangelista, M.; Settleman, J. mTORC1-Dependent Metabolic Reprogramming Underlies Escape from Glycolysis Addiction in Cancer Cells. *Cancer Cell* **2016**, *29*, 548–562. [CrossRef]
11. Fu, V.; Moroishi, T.; Guan, K.L. Glycoholics Anonymous: Cancer Sobers Up with mTORC1. *Cancer Cell* **2016**, *29*, 432–434. [CrossRef]
12. Fiorillo, M.; Ozsvari, B.; Sotgia, F.; Lisanti, M.P. High ATP Production Fuels Cancer Drug Resistance and Metastasis: Implications for Mitochondrial ATP Depletion Therapy. *Front. Oncol.* **2021**, *11*, 740720. [CrossRef]
13. Wang, T.; Ma, F.; Qian, H.L. Defueling the cancer: ATP synthase as an emerging target in cancer therapy. *Mol. Ther. Oncolytics* **2021**, *23*, 82–95. [CrossRef]
14. Vultaggio-Poma, V.; Sarti, A.C.; Di Virgilio, F. Extracellular ATP: A Feasible Target for Cancer Therapy. *Cells* **2020**, *9*, 2496. [CrossRef]
15. Rai, Y.; Yadav, P.; Kumari, N.; Kalra, N.; Bhatt, A.N. Hexokinase II inhibition by 3-bromopyruvate sensitizes myeloid leukemic cells K-562 to anti-leukemic drug, daunorubicin. *Biosci. Rep.* **2019**, *39*, BSR20190880. [CrossRef]
16. Almouhanna, F.; Blagojevic, B.; Can, S.; Ghanem, A.; Wolfl, S. Pharmacological activation of pyruvate kinase M2 reprograms glycolysis leading to TXNIP depletion and AMPK activation in breast cancer cells. *Cancer Metab.* **2021**, *9*, 5. [CrossRef]
17. Zhou, M.; Zhao, Y.; Ding, Y.; Liu, H.; Liu, Z.; Fodstad, O.; Riker, A.I.; Kamarajugadda, S.; Lu, J.; Owen, L.B.; et al. Warburg effect in chemosensitivity: Targeting lactate dehydrogenase-A re-sensitizes taxol-resistant cancer cells to taxol. *Mol. Cancer* **2010**, *9*, 33. [CrossRef]
18. Carvalho, K.C.; Cunha, I.W.; Rocha, R.M.; Ayala, F.R.; Cajaiba, M.M.; Begnami, M.D.; Vilela, R.S.; Paiva, G.R.; Andrade, R.G.; Soares, F.A. GLUT1 expression in malignant tumors and its use as an immunodiagnostic marker. *Clinics* **2011**, *66*, 965–972. [CrossRef]
19. Yu, M.; Yongzhi, H.; Chen, S.; Luo, X.; Lin, Y.; Zhou, Y.; Jin, H.; Hou, B.; Deng, Y.; Tu, L.; et al. The prognostic value of GLUT1 in cancers: A systematic review and meta-analysis. *Oncotarget* **2017**, *8*, 43356–43367. [CrossRef]
20. Xintaropoulou, C.; Ward, C.; Wise, A.; Queckborner, S.; Turnbull, A.; Michie, C.O.; Williams, A.R.W.; Rye, T.; Gourley, C.; Langdon, S.P. Expression of glycolytic enzymes in ovarian cancers and evaluation of the glycolytic pathway as a strategy for ovarian cancer treatment. *BMC Cancer* **2018**, *18*, 636. [CrossRef]
21. Shin, E.; Koo, J.S. Glucose Metabolism and Glucose Transporters in Breast Cancer. *Front. Cell Dev. Biol.* **2021**, *9*, 728759. [CrossRef] [PubMed]
22. Meng, Y.; Xu, X.; Luan, H.; Li, L.; Dai, W.; Li, Z.; Bian, J. The progress and development of GLUT1 inhibitors targeting cancer energy metabolism. *Future Med. Chem.* **2019**, *11*, 2333–2352. [CrossRef] [PubMed]
23. Sawayama, H.; Ogata, Y.; Ishimoto, T.; Mima, K.; Hiyoshi, Y.; Iwatsuki, M.; Baba, Y.; Miyamoto, Y.; Yoshida, N.; Baba, H. Glucose transporter 1 regulates the proliferation and cisplatin sensitivity of esophageal cancer. *Cancer Sci.* **2019**, *110*, 1705–1714. [CrossRef] [PubMed]
24. Weng, H.C.; Sung, C.J.; Hsu, J.L.; Leu, W.J.; Guh, J.H.; Kung, F.L.; Hsu, L.C. The Combination of a Novel GLUT1 Inhibitor and Cisplatin Synergistically Inhibits Breast Cancer Cell Growth By Enhancing the DNA Damaging Effect and Modulating the Akt/mTOR and MAPK Signaling Pathways. *Front. Pharmacol.* **2022**, *13*, 879748. [CrossRef] [PubMed]
25. Pliszka, M.; Szablewski, L. Glucose Transporters as a Target for Anticancer Therapy. *Cancers* **2021**, *13*, 4184. [CrossRef]
26. Wu, Q.; Ba-Alawi, W.; Deblois, G.; Cruickshank, J.; Duan, S.; Lima-Fernandes, E.; Haight, J.; Tonekaboni, S.A.M.; Fortier, A.M.; Kuasne, H.; et al. GLUT1 inhibition blocks growth of RB1-positive triple negative breast cancer. *Nat. Commun.* **2020**, *11*, 4205. [CrossRef]
27. Shibuya, K.; Okada, M.; Suzuki, S.; Seino, M.; Seino, S.; Takeda, H.; Kitanaka, C. Targeting the facilitative glucose transporter GLUT1 inhibits the self-renewal and tumor-initiating capacity of cancer stem cells. *Oncotarget* **2015**, *6*, 651–661. [CrossRef]
28. Li, Y.L.; Weng, H.C.; Hsu, J.L.; Lin, S.W.; Guh, J.H.; Hsu, L.C. The Combination of MK-2206 and WZB117 Exerts a Synergistic Cytotoxic Effect Against Breast Cancer Cells. *Front. Pharmacol.* **2019**, *10*, 1311. [CrossRef]
29. Chen, Q.; Meng, Y.Q.; Xu, X.F.; Gu, J. Blockade of GLUT1 by WZB117 resensitizes breast cancer cells to adriamycin. *Anticancer Drugs* **2017**, *28*, 880–887. [CrossRef]
30. Zhao, F.; Ming, J.; Zhou, Y.; Fan, L. Inhibition of Glut1 by WZB117 sensitizes radioresistant breast cancer cells to irradiation. *Cancer Chemother. Pharmacol.* **2016**, *77*, 963–972. [CrossRef]

31. Mori, Y.; Yamawaki, K.; Ishiguro, T.; Yoshihara, K.; Ueda, H.; Sato, A.; Ohata, H.; Yoshida, Y.; Minamino, T.; Okamoto, K.; et al. ALDH-Dependent Glycolytic Activation Mediates Stemness and Paclitaxel Resistance in Patient-Derived Spheroid Models of Uterine Endometrial Cancer. *Stem Cell Rep.* **2019**, *13*, 730–746. [CrossRef]
32. Lee, C.K.; Choi, J.S. Effects of silibinin, inhibitor of CYP3A4 and P-glycoprotein in vitro, on the pharmacokinetics of paclitaxel after oral and intravenous administration in rats. *Pharmacology* **2010**, *85*, 350–356. [CrossRef]
33. Pashaei-Asl, F.; Pashaei-Asl, R.; Khodadadi, K.; Akbarzadeh, A.; Ebrahimie, E.; Pashaiasl, M. Enhancement of anticancer activity by silibinin and paclitaxel combination on the ovarian cancer. *Artif. Cells Nanomed. Biotechnol.* **2018**, *46*, 1483–1487. [CrossRef]
34. Li, X.; Jiang, C.; Wang, Q.; Yang, S.; Cao, Y.; Hao, J.N.; Niu, D.; Chen, Y.; Han, B.; Jia, X.; et al. A "Valve-Closing" Starvation Strategy for Amplification of Tumor-Specific Chemotherapy. *Adv. Sci.* **2022**, *9*, e2104671. [CrossRef]
35. Tilekar, K.; Upadhyay, N.; Iancu, C.V.; Pokrovsky, V.; Choe, J.Y.; Ramaa, C.S. Power of two: Combination of therapeutic approaches involving glucose transporter (GLUT) inhibitors to combat cancer. *Biochim. Biophys. Acta Rev. Cancer* **2020**, *1874*, 188457. [CrossRef]
36. Icard, P.; Teboul, B.; El Baze, P. A Simple Method to Optimize the Effectiveness of Chemotherapy: Modulation of Glucose Intake During Chemotherapy. *Anticancer Res.* **2017**, *37*, 6199–6202. [CrossRef]
37. Omar, E.M.; Omran, G.A.; Mustafa, M.F.; El-Khodary, N.M. Intermittent fasting during adjuvant chemotherapy may promote differential stress resistance in breast cancer patients. *J. Egypt. Natl. Cancer Inst.* **2022**, *34*, 38. [CrossRef]
38. Vaziri-Gohar, A.; Hue, J.J.; Graor, H.G.; Abbas, A.; Zarei, M.; Hajihassani, O.; Titomihelakis, G.; Feczko, J.; Rathore, M.; Wang, R.; et al. Increased glucose availability sensitizes pancreatic cancer to chemotherapy. *bioRxiv* **2022**. [CrossRef]
39. Liu, Y.; Zhang, Y.; Mao, X.; Qi, Q.; Zhu, M.; Zhang, C.; Pan, X.; Ling, Y. Palliative treatment efficacy of glucose inhibition combined with chemotherapy for non-small cell lung cancer with widespread bone and brain metastases: A case report. *Biomed. Rep.* **2017**, *7*, 553–557. [CrossRef]
40. Gottlob, K.; Majewski, N.; Kennedy, S.; Kandel, E.; Robey, R.B.; Hay, N. Inhibition of early apoptotic events by Akt/PKB is dependent on the first committed step of glycolysis and mitochondrial hexokinase. *Genes Dev.* **2001**, *15*, 1406–1418. [CrossRef]
41. Ahn, K.J.; Hwang, H.S.; Park, J.H.; Bang, S.H.; Kang, W.J.; Yun, M.; Lee, J.D. Evaluation of the role of hexokinase type II in cellular proliferation and apoptosis using human hepatocellular carcinoma cell lines. *J. Nucl. Med.* **2009**, *50*, 1525–1532. [CrossRef] [PubMed]
42. Sato-Tadano, A.; Suzuki, T.; Amari, M.; Takagi, K.; Miki, Y.; Tamaki, K.; Watanabe, M.; Ishida, T.; Sasano, H.; Ohuchi, N. Hexokinase II in breast carcinoma: A potent prognostic factor associated with hypoxia-inducible factor-1alpha and Ki-67. *Cancer Sci.* **2013**, *104*, 1380–1388. [CrossRef] [PubMed]
43. Patra, K.C.; Wang, Q.; Bhaskar, P.T.; Miller, L.; Wang, Z.; Wheaton, W.; Chandel, N.; Laakso, M.; Muller, W.J.; Allen, E.L.; et al. Hexokinase 2 is required for tumor initiation and maintenance and its systemic deletion is therapeutic in mouse models of cancer. *Cancer Cell* **2013**, *24*, 213–228. [CrossRef] [PubMed]
44. Zajicek, G. The normal and the pathological. *Harefuah* **1994**, *127*, 24–25. [PubMed]
45. Zhang, X.D.; Deslandes, E.; Villedieu, M.; Poulain, L.; Duval, M.; Gauduchon, P.; Schwartz, L.; Icard, P. Effect of 2-deoxy-D-glucose on various malignant cell lines in vitro. *Anticancer Res.* **2006**, *26*, 3561–3566.
46. Zhao, Y.; Liu, H.; Liu, Z.; Ding, Y.; Ledoux, S.P.; Wilson, G.L.; Voellmy, R.; Lin, Y.; Lin, W.; Nahta, R.; et al. Overcoming trastuzumab resistance in breast cancer by targeting dysregulated glucose metabolism. *Cancer Res.* **2011**, *71*, 4585–4597. [CrossRef]
47. Maschek, G.; Savaraj, N.; Priebe, W.; Braunschweiger, P.; Hamilton, K.; Tidmarsh, G.F.; De Young, L.R.; Lampidis, T.J. 2-deoxy-D-glucose increases the efficacy of adriamycin and paclitaxel in human osteosarcoma and non-small cell lung cancers in vivo. *Cancer Res.* **2004**, *64*, 31–34. [CrossRef]
48. Sun, X.; Fan, T.; Sun, G.; Zhou, Y.; Huang, Y.; Zhang, N.; Zhao, L.; Zhong, R.; Peng, Y. 2-Deoxy-D-glucose increases the sensitivity of glioblastoma cells to BCNU through the regulation of glycolysis, ROS and ERS pathways: In vitro and in vivo validation. *Biochem. Pharmacol.* **2022**, *199*, 115029. [CrossRef]
49. Zhou, N.; Liu, Q.; Wang, X.; He, L.; Zhang, T.; Zhou, H.; Zhu, X.; Zhou, T.; Deng, G.; Qiu, C. The combination of hydroxychloroquine and 2-deoxyglucose enhances apoptosis in breast cancer cells by blocking protective autophagy and sustaining endoplasmic reticulum stress. *Cell Death Discov.* **2022**, *8*, 286. [CrossRef]
50. Raez, L.E.; Papadopoulos, K.; Ricart, A.D.; Chiorean, E.G.; Dipaola, R.S.; Stein, M.N.; Rocha Lima, C.M.; Schlesselman, J.J.; Tolba, K.; Langmuir, V.K.; et al. A phase I dose-escalation trial of 2-deoxy-D-glucose alone or combined with docetaxel in patients with advanced solid tumors. *Cancer Chemother. Pharmacol.* **2013**, *71*, 523–530. [CrossRef]
51. Geng, C.; Li, J.; Ding, F.; Wu, G.; Yang, Q.; Sun, Y.; Zhang, Z.; Dong, T.; Tian, X. Curcumin suppresses 4-hydroxytamoxifen resistance in breast cancer cells by targeting SLUG/Hexokinase 2 pathway. *Biochem. Biophys. Res. Commun.* **2016**, *473*, 147–153. [CrossRef]
52. Pedersen, P.L. Warburg, me and Hexokinase 2: Multiple discoveries of key molecular events underlying one of cancers' most common phenotypes, the "Warburg Effect", i.e., elevated glycolysis in the presence of oxygen. *J. Bioenerg. Biomembr.* **2007**, *39*, 211–222. [CrossRef]
53. Pedersen, P.L.; Mathupala, S.; Rempel, A.; Geschwind, J.F.; Ko, Y.H. Mitochondrial bound type II hexokinase: A key player in the growth and survival of many cancers and an ideal prospect for therapeutic intervention. *Biochim. Biophys. Acta* **2002**, *1555*, 14–20. [CrossRef]
54. Nelson, K. 3-Bromopyruvate kills cancer cells in animals. *Lancet Oncol.* **2002**, *3*, 524. [CrossRef]

55. Fan, T.; Sun, G.; Sun, X.; Zhao, L.; Zhong, R.; Peng, Y. Tumor Energy Metabolism and Potential of 3-Bromopyruvate as an Inhibitor of Aerobic Glycolysis: Implications in Tumor Treatment. *Cancers* **2019**, *11*, 317. [CrossRef]
56. Zhang, Q.; Pan, J.; Lubet, R.A.; Komas, S.M.; Kalyanaraman, B.; Wang, Y.; You, M. Enhanced antitumor activity of 3-bromopyruvate in combination with rapamycin in vivo and in vitro. *Cancer Prev. Res.* **2015**, *8*, 318–326. [CrossRef]
57. Gan, L.; Ren, Y.; Lu, J.; Ma, J.; Shen, X.; Zhuang, Z. Synergistic Effect of 3-Bromopyruvate in Combination with Rapamycin Impacted Neuroblastoma Metabolism by Inhibiting Autophagy. *Onco Targets Ther.* **2020**, *13*, 11125–11137. [CrossRef]
58. Li, J.; Pan, J.; Liu, Y.; Luo, X.; Yang, C.; Xiao, W.; Li, Q.; Yang, L.; Zhang, X. 3-Bromopyruvic acid regulates glucose metabolism by targeting the c-Myc/TXNIP axis and induces mitochondria-mediated apoptosis in TNBC cells. *Exp. Ther. Med.* **2022**, *24*, 520. [CrossRef]
59. Attia, Y.M.; El-Abhar, H.S.; Al Marzabani, M.M.; Shouman, S.A. Targeting glycolysis by 3-bromopyruvate improves tamoxifen cytotoxicity of breast cancer cell lines. *BMC Cancer* **2015**, *15*, 838. [CrossRef]
60. Wicks, R.T.; Azadi, J.; Mangraviti, A.; Zhang, I.; Hwang, L.; Joshi, A.; Bow, H.; Hutt-Cabezas, M.; Martin, K.L.; Rudek, M.A.; et al. Local delivery of cancer-cell glycolytic inhibitors in high-grade glioma. *Neuro-Oncol.* **2015**, *17*, 70–80. [CrossRef]
61. Kim, Y.; Na, J.; Youn, H.; Yu, J.S.; Kang, K.W.; Chung, J.-K. Enhanced therapeutic efficacy of combined Sorafenib and 3-bromopyruvate in murine orthotopic liver cancer assessed by bioluminescence image. *J. Nucl. Med.* **2016**, *57*, 1331.
62. Roy, S.K.; Dukic, T.; Bhandary, B.; Acharya, A.; Tu, K.J.; Ko, Y.H.; Shukla, H.D. 3-Bromopyruvate in combination with radiation inhibits pancreatic cancer growth by dismantling mitochondria and ATP generation in a preclinical mouse model. *Cancer Res.* **2022**, *82*, 5243–5243. [CrossRef]
63. Skaripa-Koukelli, I.; Hauton, D.; Walsby-Tickle, J.; Thomas, E.; Owen, J.; Lakshminarayanan, A.; Able, S.; McCullagh, J.; Carlisle, R.C.; Vallis, K.A. 3-Bromopyruvate-mediated MCT1-dependent metabolic perturbation sensitizes triple negative breast cancer cells to ionizing radiation. *Cancer Metab.* **2021**, *9*, 37. [CrossRef] [PubMed]
64. Angelin, A.; Gil-de-Gomez, L.; Dahiya, S.; Jiao, J.; Guo, L.; Levine, M.H.; Wang, Z.; Quinn, W.J., 3rd; Kopinski, P.K.; Wang, L.; et al. Foxp3 Reprograms T Cell Metabolism to Function in Low-Glucose, High-Lactate Environments. *Cell Metab.* **2017**, *25*, 1282–1293.e1287. [CrossRef]
65. Brooks, G.A. The Science and Translation of Lactate Shuttle Theory. *Cell Metab.* **2018**, *27*, 757–785. [CrossRef]
66. Involvement of physicians in capital punishment. *N. Y. State J. Med.* **1991**, *91*, 271–272.
67. Morais-Santos, F.; Granja, S.; Miranda-Goncalves, V.; Moreira, A.H.; Queiros, S.; Vilaca, J.L.; Schmitt, F.C.; Longatto-Filho, A.; Paredes, J.; Baltazar, F.; et al. Targeting lactate transport suppresses in vivo breast tumour growth. *Oncotarget* **2015**, *6*, 19177–19189. [CrossRef]
68. Saulle, E.; Spinello, I.; Quaranta, M.T.; Pasquini, L.; Pelosi, E.; Iorio, E.; Castelli, G.; Chirico, M.; Pisanu, M.E.; Ottone, T.; et al. Targeting Lactate Metabolism by Inhibiting MCT1 or MCT4 Impairs Leukemic Cell Proliferation, Induces Two Different Related Death-Pathways and Increases Chemotherapeutic Sensitivity of Acute Myeloid Leukemia Cells. *Front. Oncol.* **2020**, *10*, 621458. [CrossRef]
69. Halford, S.E.; Walter, H.; McKay, P.; Townsend, W.; Linton, K.; Heinzmann, K.; Dragoni, I.; Brotherton, L.; Veal, G.; Siskos, A.; et al. Phase I expansion study of the first-in-class monocarboxylate transporter 1 (MCT1) inhibitor AZD3965 in patients with diffuse large B-cell lymphoma (DLBCL) and Burkitt lymphoma (BL). *J. Clin. Oncol.* **2021**, *39*, 3115. [CrossRef]
70. Halford, S.E.R.; Jones, P.; Wedge, S.; Hirschberg, S.; Katugampola, S.; Veal, G.; Payne, G.; Bacon, C.; Potter, S.; Griffin, M.; et al. A first-in-human first-in-class (FIC) trial of the monocarboxylate transporter 1 (MCT1) inhibitor AZD3965 in patients with advanced solid tumours. *J. Clin. Oncol.* **2017**, *35* (Suppl. 15), 2516. [CrossRef]
71. Claps, G.; Faouzi, S.; Quidville, V.; Chehade, F.; Shen, S.; Vagner, S.; Robert, C. The multiple roles of LDH in cancer. *Nat. Rev. Clin. Oncol.* **2022**, *19*, 749–762. [CrossRef]
72. Flores, A.; Sandoval-Gonzalez, S.; Takahashi, R.; Krall, A.; Sathe, L.; Wei, L.; Radu, C.; Joly, J.H.; Graham, N.A.; Christofk, H.R.; et al. Increased lactate dehydrogenase activity is dispensable in squamous carcinoma cells of origin. *Nat. Commun.* **2019**, *10*, 91. [CrossRef]
73. Fantin, V.R.; St-Pierre, J.; Leder, P. Attenuation of LDH-A expression uncovers a link between glycolysis, mitochondrial physiology, and tumor maintenance. *Cancer Cell* **2006**, *9*, 425–434. [CrossRef]
74. Lu, R.; Jiang, M.; Chen, Z.; Xu, X.; Hu, H.; Zhao, X.; Gao, X.; Guo, L. Lactate dehydrogenase 5 expression in Non-Hodgkin lymphoma is associated with the induced hypoxia regulated protein and poor prognosis. *PLoS ONE* **2013**, *8*, e74853. [CrossRef]
75. Zha, X.; Wang, F.; Wang, Y.; He, S.; Jing, Y.; Wu, X.; Zhang, H. Lactate dehydrogenase B is critical for hyperactive mTOR-mediated tumorigenesis. *Cancer Res.* **2011**, *71*, 13–18. [CrossRef]
76. Shim, H.; Dolde, C.; Lewis, B.C.; Wu, C.S.; Dang, G.; Jungmann, R.A.; Dalla-Favera, R.; Dang, C.V. c-Myc transactivation of LDH-A: Implications for tumor metabolism and growth. *Proc. Natl. Acad. Sci. USA* **1997**, *94*, 6658–6663. [CrossRef]
77. He, T.L.; Zhang, Y.J.; Jiang, H.; Li, X.H.; Zhu, H.; Zheng, K.L. The c-Myc-LDHA axis positively regulates aerobic glycolysis and promotes tumor progression in pancreatic cancer. *Med. Oncol.* **2015**, *32*, 187. [CrossRef]
78. Chen, L.; Wu, Q.; Xu, X.; Yang, C.; You, J.; Chen, F.; Zeng, Y. Cancer/testis antigen LDHC promotes proliferation and metastasis by activating the PI3K/Akt/GSK-3beta-signaling pathway and the in lung adenocarcinoma. *Exp. Cell Res.* **2021**, *398*, 112414. [CrossRef]
79. Jin, L.; Chun, J.; Pan, C.; Alesi, G.N.; Li, D.; Magliocca, K.R.; Kang, Y.; Chen, Z.G.; Shin, D.M.; Khuri, F.R.; et al. Phosphorylation-mediated activation of LDHA promotes cancer cell invasion and tumour metastasis. *Oncogene* **2017**, *36*, 3797–3806. [CrossRef]

80. Ferraris, A.M.; Giuntini, P.; Gaetani, G.F. Serum lactic dehydrogenase as a prognostic tool for non-Hodgkin lymphomas. *Blood* **1979**, *54*, 928–932. [CrossRef]
81. Mohammad, G.H.; Olde Damink, S.W.; Malago, M.; Dhar, D.K.; Pereira, S.P. Pyruvate Kinase M2 and Lactate Dehydrogenase A Are Overexpressed in Pancreatic Cancer and Correlate with Poor Outcome. *PLoS ONE* **2016**, *11*, e0151635. [CrossRef] [PubMed]
82. Petrelli, F.; Cabiddu, M.; Coinu, A.; Borgonovo, K.; Ghilardi, M.; Lonati, V.; Barni, S. Prognostic role of lactate dehydrogenase in solid tumors: A systematic review and meta-analysis of 76 studies. *Acta Oncol.* **2015**, *54*, 961–970. [CrossRef] [PubMed]
83. Xie, H.; Hanai, J.; Ren, J.G.; Kats, L.; Burgess, K.; Bhargava, P.; Signoretti, S.; Billiard, J.; Duffy, K.J.; Grant, A.; et al. Targeting lactate dehydrogenase–a inhibits tumorigenesis and tumor progression in mouse models of lung cancer and impacts tumor-initiating cells. *Cell Metab.* **2014**, *19*, 795–809. [CrossRef] [PubMed]
84. Le, A.; Cooper, C.R.; Gouw, A.M.; Dinavahi, R.; Maitra, A.; Deck, L.M.; Royer, R.E.; Vander Jagt, D.L.; Semenza, G.L.; Dang, C.V. Inhibition of lactate dehydrogenase A induces oxidative stress and inhibits tumor progression. *Proc. Natl. Acad. Sci. USA* **2010**, *107*, 2037–2042. [CrossRef]
85. Altinoz, M.A.; Ozpinar, A. Oxamate targeting aggressive cancers with special emphasis to brain tumors. *Biomed. Pharmacother.* **2022**, *147*, 112686. [CrossRef]
86. Zhao, Z.; Han, F.; Yang, S.; Wu, J.; Zhan, W. Oxamate-mediated inhibition of lactate dehydrogenase induces protective autophagy in gastric cancer cells: Involvement of the Akt-mTOR signaling pathway. *Cancer Lett.* **2015**, *358*, 17–26. [CrossRef]
87. Miskimins, W.K.; Ahn, H.J.; Kim, J.Y.; Ryu, S.; Jung, Y.S.; Choi, J.Y. Synergistic anti-cancer effect of phenformin and oxamate. *PLoS ONE* **2014**, *9*, e85576. [CrossRef]
88. Xing, B.C.; Wang, C.; Ji, F.J.; Zhang, X.B. Synergistically suppressive effects on colorectal cancer cells by combination of mTOR inhibitor and glycolysis inhibitor, Oxamate. *Int. J. Clin. Exp. Pathol.* **2018**, *11*, 4439–4445.
89. Almaguel, F.A.; Sanchez, T.W.; Ortiz-Hernandez, G.L.; Casiano, C.A. Alpha-Enolase: Emerging Tumor-Associated Antigen, Cancer Biomarker, and Oncotherapeutic Target. *Front. Genet.* **2020**, *11*, 614726. [CrossRef]
90. He, P.; Naka, T.; Serada, S.; Fujimoto, M.; Tanaka, T.; Hashimoto, S.; Shima, Y.; Yamadori, T.; Suzuki, H.; Hirashima, T.; et al. Proteomics-based identification of alpha-enolase as a tumor antigen in non-small lung cancer. *Cancer Sci.* **2007**, *98*, 1234–1240. [CrossRef]
91. Hsiao, K.C.; Shih, N.Y.; Chu, P.Y.; Hung, Y.M.; Liao, J.Y.; Chou, S.W.; Yang, Y.Y.; Chang, G.C.; Liu, K.J. Anti-alpha-enolase is a prognostic marker in postoperative lung cancer patients. *Oncotarget* **2015**, *6*, 35073–35086. [CrossRef]
92. Fu, Q.F.; Liu, Y.; Fan, Y.; Hua, S.N.; Qu, H.Y.; Dong, S.W.; Li, R.L.; Zhao, M.Y.; Zhen, Y.; Yu, X.L.; et al. Alpha-enolase promotes cell glycolysis, growth, migration, and invasion in non-small cell lung cancer through FAK-mediated PI3K/AKT pathway. *J. Hematol. Oncol.* **2015**, *8*, 22. [CrossRef]
93. Zhan, P.; Zhao, S.; Yan, H.; Yin, C.; Xiao, Y.; Wang, Y.; Ni, R.; Chen, W.; Wei, G.; Zhang, P. alpha-enolase promotes tumorigenesis and metastasis via regulating AMPK/mTOR pathway in colorectal cancer. *Mol. Carcinog.* **2017**, *56*, 1427–1437. [CrossRef]
94. Capello, M.; Ferri-Borgogno, S.; Riganti, C.; Chattaragada, M.S.; Principe, M.; Roux, C.; Zhou, W.; Petricoin, E.F.; Cappello, P.; Novelli, F. Targeting the Warburg effect in cancer cells through ENO1 knockdown rescues oxidative phosphorylation and induces growth arrest. *Oncotarget* **2016**, *7*, 5598–5612. [CrossRef]
95. Song, K.; Rajasekaran, N.; Chelakkot, C.; Lee, H.S.; Paek, S.M.; Yang, H.; Jia, L.; Park, H.G.; Son, W.S.; Kim, Y.J.; et al. Macrosphelide A Exhibits a Specific Anti-Cancer Effect by Simultaneously Inactivating ENO1, ALDOA, and FH. *Pharmaceuticals* **2021**, *14*, 1060. [CrossRef]
96. Qian, X.; Xu, W.; Xu, J.; Shi, Q.; Li, J.; Weng, Y.; Jiang, Z.; Feng, L.; Wang, X.; Zhou, J.; et al. Enolase 1 stimulates glycolysis to promote chemoresistance in gastric cancer. *Oncotarget* **2017**, *8*, 47691–47708. [CrossRef]
97. Huang, C.K.; Sun, Y.; Lv, L.; Ping, Y. ENO1 and Cancer. *Mol. Ther. Oncolytics* **2022**, *24*, 288–298. [CrossRef]
98. Lin, Y.H.; Satani, N.; Hammoudi, N.; Yan, V.C.; Barekatain, Y.; Khadka, S.; Ackroyd, J.J.; Georgiou, D.K.; Pham, C.D.; Arthur, K.; et al. An enolase inhibitor for the targeted treatment of ENO1-deleted cancers. *Nat. Metab.* **2020**, *2*, 1413–1426. [CrossRef]
99. Jones, B.C.; Pohlmann, P.R.; Clarke, R.; Sengupta, S. Treatment against glucose-dependent cancers through metabolic PFKFB3 targeting of glycolytic flux. *Cancer Metastasis Rev.* **2022**, *41*, 447–458. [CrossRef]
100. Zhu, W.; Ye, L.; Zhang, J.; Yu, P.; Wang, H.; Ye, Z.; Tian, J. PFK15, a Small Molecule Inhibitor of PFKFB3, Induces Cell Cycle Arrest, Apoptosis and Inhibits Invasion in Gastric Cancer. *PLoS ONE* **2016**, *11*, e0163768. [CrossRef]
101. Telang, S.; O'Neal, J.; Tapolsky, G.; Clem, B.; Kerr, A.; Imbert-Ferndandez, Y.; Chesney, J. Discovery of a PFKFB3 inhibitor for phase I trial testing that synergizes with the B-Raf inhibitor vemurafenib. *Cancer Metab.* **2014**, *2*, 1–2. [CrossRef]
102. Phase 1 Safety Study of ACT-PFK-158, 2HCl in Patients with Advanced Solid Malignancies_NCT02044861. Available online: https://clinicaltrials.gov/ct2/show/NCT02044861 (accessed on 30 September 2015).
103. Redman, R.A.; Pohlmann, P.R.; Kurman, M.R.; Tapolsky, G.; Chesney, J. A phase I, dose-escalation, multicenter study of ACT-PFK-158, 2HCl in patients with advanced solid malignancies explores a first-in-human inhibitor of glycolysis. *J. Clin. Oncol.* **2015**, *33*, TPS494. [CrossRef]
104. Yan, S.; Zhou, N.; Zhang, D.; Zhang, K.; Zheng, W.; Bao, Y.; Yang, W. PFKFB3 Inhibition Attenuates Oxaliplatin-Induced Autophagy and Enhances Its Cytotoxicity in Colon Cancer Cells. *Int. J. Mol. Sci.* **2019**, *20*, 5415. [CrossRef] [PubMed]
105. De Bock, K.; Georgiadou, M.; Carmeliet, P. Role of endothelial cell metabolism in vessel sprouting. *Cell Metab.* **2013**, *18*, 634–647. [CrossRef]

106. De Bock, K.; Georgiadou, M.; Schoors, S.; Kuchnio, A.; Wong, B.W.; Cantelmo, A.R.; Quaegebeur, A.; Ghesquiere, B.; Cauwenberghs, S.; Eelen, G.; et al. Role of PFKFB3-driven glycolysis in vessel sprouting. *Cell* **2013**, *154*, 651–663. [CrossRef]
107. Schoors, S.; De Bock, K.; Cantelmo, A.R.; Georgiadou, M.; Ghesquiere, B.; Cauwenberghs, S.; Kuchnio, A.; Wong, B.W.; Quaegebeur, A.; Goveia, J.; et al. Partial and transient reduction of glycolysis by PFKFB3 blockade reduces pathological angiogenesis. *Cell Metab.* **2014**, *19*, 37–48. [CrossRef]
108. Cantelmo, A.R.; Conradi, L.C.; Brajic, A.; Goveia, J.; Kalucka, J.; Pircher, A.; Chaturvedi, P.; Hol, J.; Thienpont, B.; Teuwen, L.A.; et al. Inhibition of the Glycolytic Activator PFKFB3 in Endothelium Induces Tumor Vessel Normalization, Impairs Metastasis, and Improves Chemotherapy. *Cancer Cell* **2016**, *30*, 968–985. [CrossRef]
109. Emini Veseli, B.; Van Wielendaele, P.; Delibegovic, M.; Martinet, W.; De Meyer, G.R.Y. The PFKFB3 Inhibitor AZ67 Inhibits Angiogenesis Independently of Glycolysis Inhibition. *Int. J. Mol. Sci.* **2021**, *22*, 5970. [CrossRef]
110. Zhang, J.; Xue, W.; Xu, K.; Yi, L.; Guo, Y.; Xie, T.; Tong, H.; Zhou, B.; Wang, S.; Li, Q.; et al. Dual inhibition of PFKFB3 and VEGF normalizes tumor vasculature, reduces lactate production, and improves chemotherapy in glioblastoma: Insights from protein expression profiling and MRI. *Theranostics* **2020**, *10*, 7245–7259. [CrossRef]
111. Hsu, M.C.; Hung, W.C. Pyruvate kinase M2 fuels multiple aspects of cancer cells: From cellular metabolism, transcriptional regulation to extracellular signaling. *Mol. Cancer* **2018**, *17*, 35. [CrossRef]
112. Allen, A.E.; Locasale, J.W. Glucose Metabolism in Cancer: The Saga of Pyruvate Kinase Continues. *Cancer Cell* **2018**, *33*, 337–339. [CrossRef]
113. Yang, W.; Zheng, Y.; Xia, Y.; Ji, H.; Chen, X.; Guo, F.; Lyssiotis, C.A.; Aldape, K.; Cantley, L.C.; Lu, Z. ERK1/2-dependent phosphorylation and nuclear translocation of PKM2 promotes the Warburg effect. *Nat. Cell Biol.* **2012**, *14*, 1295–1304. [CrossRef]
114. Azoitei, N.; Becher, A.; Steinestel, K.; Rouhi, A.; Diepold, K.; Genze, F.; Simmet, T.; Seufferlein, T. PKM2 promotes tumor angiogenesis by regulating HIF-1alpha through NF-kappaB activation. *Mol. Cancer* **2016**, *15*, 3. [CrossRef]
115. Wu, J.; Hu, L.; Chen, M.; Cao, W.; Chen, H.; He, T. Pyruvate kinase M2 overexpression and poor prognosis in solid tumors of digestive system: Evidence from 16 cohort studies. *Onco Targets Ther.* **2016**, *9*, 4277–4288. [CrossRef]
116. Zhu, H.; Luo, H.; Zhu, X.; Hu, X.; Zheng, L.; Zhu, X. Pyruvate kinase M2 (PKM2) expression correlates with prognosis in solid cancers: A meta-analysis. *Oncotarget* **2017**, *8*, 1628–1640. [CrossRef]
117. Tennant, D.A.; Duran, R.V.; Gottlieb, E. Targeting metabolic transformation for cancer therapy. *Nat. Rev. Cancer* **2010**, *10*, 267–277. [CrossRef]
118. Shi, H.S.; Li, D.; Zhang, J.; Wang, Y.S.; Yang, L.; Zhang, H.L.; Wang, X.H.; Mu, B.; Wang, W.; Ma, Y.; et al. Silencing of pkm2 increases the efficacy of docetaxel in human lung cancer xenografts in mice. *Cancer Sci.* **2010**, *101*, 1447–1453. [CrossRef]
119. Yuan, S.; Qiao, T.; Zhuang, X.; Chen, W.; Xing, N.; Zhang, Q. Knockdown of the M2 Isoform of Pyruvate Kinase (PKM2) with shRNA Enhances the Effect of Docetaxel in Human NSCLC Cell Lines In Vitro. *Yonsei Med. J.* **2016**, *57*, 1312–1323. [CrossRef]
120. Papadaki, C.; Sfakianaki, M.; Lagoudaki, E.; Giagkas, G.; Ioannidis, G.; Trypaki, M.; Tsakalaki, E.; Voutsina, A.; Koutsopoulos, A.; Mavroudis, D.; et al. PKM2 as a biomarker for chemosensitivity to front-line platinum-based chemotherapy in patients with metastatic non-small-cell lung cancer. *Br. J. Cancer* **2014**, *111*, 1757–1764. [CrossRef]
121. Guo, W.; Zhang, Y.; Chen, T.; Wang, Y.; Xue, J.; Zhang, Y.; Xiao, W.; Mo, X.; Lu, Y. Efficacy of RNAi targeting of pyruvate kinase M2 combined with cisplatin in a lung cancer model. *J. Cancer Res. Clin. Oncol.* **2011**, *137*, 65–72. [CrossRef]
122. Ganapathy-Kanniappan, S.; Kunjithapatham, R.; Geschwind, J.F. Glyceraldehyde-3-phosphate dehydrogenase: A promising target for molecular therapy in hepatocellular carcinoma. *Oncotarget* **2012**, *3*, 940–953. [CrossRef] [PubMed]
123. Thornalley, P.J.; Rabbani, N. Glyoxalase in tumourigenesis and multidrug resistance. *Semin. Cell Dev. Biol.* **2011**, *22*, 318–325. [CrossRef] [PubMed]
124. Ganapathy-Kanniappan, S.; Vali, M.; Kunjithapatham, R.; Buijs, M.; Syed, L.H.; Rao, P.P.; Ota, S.; Kwak, B.K.; Loffroy, R.; Geschwind, J.F. 3-bromopyruvate: A new targeted antiglycolytic agent and a promise for cancer therapy. *Curr. Pharm. Biotechnol.* **2010**, *11*, 510–517. [CrossRef] [PubMed]
125. Ganapathy-Kanniappan, S.; Kunjithapatham, R.; Geschwind, J.F. Anticancer efficacy of the metabolic blocker 3-bromopyruvate: Specific molecular targeting. *Anticancer Res.* **2013**, *33*, 13–20.
126. Gao, Y.; Zhang, T.; Terai, H.; Ficarro, S.B.; Kwiatkowski, N.; Hao, M.F.; Sharma, B.; Christensen, C.L.; Chipumuro, E.; Wong, K.K.; et al. Overcoming Resistance to the THZ Series of Covalent Transcriptional CDK Inhibitors. *Cell Chem. Biol.* **2018**, *25*, 135–142.e5. [CrossRef]
127. Marcucci, F.; Rumio, C. Glycolysis-induced drug resistance in tumors-A response to danger signals? *Neoplasia* **2021**, *23*, 234–245.e5. [CrossRef]
128. Sun, H.; Wang, H.; Wang, X.; Aoki, Y.; Wang, X.; Yang, Y.; Cheng, X.; Wang, Z.; Wang, X. Aurora-A/SOX8/FOXK1 signaling axis promotes chemoresistance via suppression of cell senescence and induction of glucose metabolism in ovarian cancer organoids and cells. *Theranostics* **2020**, *10*, 6928–6945. [CrossRef]
129. Efimova, E.V.; Takahashi, S.; Shamsi, N.A.; Wu, D.; Labay, E.; Ulanovskaya, O.A.; Weichselbaum, R.R.; Kozmin, S.A.; Kron, S.J. Linking Cancer Metabolism to DNA Repair and Accelerated Senescence. *Mol. Cancer Res.* **2016**, *14*, 173–184. [CrossRef]
130. Kohnken, R.; Kodigepalli, K.M.; Wu, L. Regulation of deoxynucleotide metabolism in cancer: Novel mechanisms and therapeutic implications. *Mol. Cancer* **2015**, *14*, 176. [CrossRef]

131. Le, T.M.; Poddar, S.; Capri, J.R.; Abt, E.R.; Kim, W.; Wei, L.; Uong, N.T.; Cheng, C.M.; Braas, D.; Nikanjam, M.; et al. ATR inhibition facilitates targeting of leukemia dependence on convergent nucleotide biosynthetic pathways. *Nat. Commun.* **2017**, *8*, 241. [CrossRef]
132. Kwon, Y.; Kim, M.; Jung, H.S.; Kim, Y.; Jeoung, D. Targeting Autophagy for Overcoming Resistance to Anti-EGFR Treatments. *Cancers* **2019**, *11*, 1374. [CrossRef]
133. Lamoureux, F.; Zoubeidi, A. Dual inhibition of autophagy and the AKT pathway in prostate cancer. *Autophagy* **2013**, *9*, 1119–1120. [CrossRef]
134. Reyes-Castellanos, G.; Abdel Hadi, N.; Carrier, A. Autophagy Contributes to Metabolic Reprogramming and Therapeutic Resistance in Pancreatic Tumors. *Cells* **2022**, *11*, 1374. [CrossRef]
135. Senthebane, D.A.; Rowe, A.; Thomford, N.E.; Shipanga, H.; Munro, D.; Mazeedi, M.; Almazyadi, H.A.M.; Kallmeyer, K.; Dandara, C.; Pepper, M.S.; et al. The Role of Tumor Microenvironment in Chemoresistance: To Survive, Keep Your Enemies Closer. *Int. J. Mol. Sci.* **2017**, *18*, 1586. [CrossRef]
136. Cerezo, M.; Rocchi, S. Cancer cell metabolic reprogramming: A keystone for the response to immunotherapy. *Cell Death Dis.* **2020**, *11*, 964. [CrossRef]
137. Santos, J.C.; Lima, N.D.S.; Sarian, L.O.; Matheu, A.; Ribeiro, M.L.; Derchain, S.F.M. Exosome-mediated breast cancer chemoresistance via miR-155 transfer. *Sci. Rep.* **2018**, *8*, 829. [CrossRef]
138. Yoshida, G.J.; Saya, H. Therapeutic strategies targeting cancer stem cells. *Cancer Sci.* **2016**, *107*, 5–11. [CrossRef]
139. Tsai, S.T.; Chien, I.H.; Shen, W.H.; Kuo, Y.Z.; Jin, Y.T.; Wong, T.Y.; Hsiao, J.R.; Wang, H.P.; Shih, N.Y.; Wu, L.W. ENO1, a potential prognostic head and neck cancer marker, promotes transformation partly via chemokine CCL20 induction. *Eur. J. Cancer* **2010**, *46*, 1712–1723. [CrossRef]
140. Song, Y.; Luo, Q.; Long, H.; Hu, Z.; Que, T.; Zhang, X.; Li, Z.; Wang, G.; Yi, L.; Liu, Z.; et al. Alpha-enolase as a potential cancer prognostic marker promotes cell growth, migration, and invasion in glioma. *Mol. Cancer* **2014**, *13*, 65. [CrossRef]
141. Mahammedi, H.; Planchat, E.; Pouget, M.; Durando, X.; Cure, H.; Guy, L.; Van-Praagh, I.; Savareux, L.; Atger, M.; Bayet-Robert, M.; et al. The New Combination Docetaxel, Prednisone and Curcumin in Patients with Castration-Resistant Prostate Cancer: A Pilot Phase II Study. *Oncology* **2016**, *90*, 69–78. [CrossRef]
142. Huang, K.; Liang, Q.; Zhou, Y.; Jiang, L.L.; Gu, W.M.; Luo, M.Y.; Tang, Y.B.; Wang, Y.; Lu, W.; Huang, M.; et al. A Novel Allosteric Inhibitor of Phosphoglycerate Mutase 1 Suppresses Growth and Metastasis of Non-Small-Cell Lung Cancer. *Cell Metab.* **2019**, *30*, 1107–1119.e8. [CrossRef] [PubMed]
143. Tu, S.H.; Chang, C.C.; Chen, C.S.; Tam, K.W.; Wang, Y.J.; Lee, C.H.; Lin, H.W.; Cheng, T.C.; Huang, C.S.; Chu, J.S.; et al. Increased expression of enolase alpha in human breast cancer confers tamoxifen resistance in human breast cancer cells. *Breast Cancer Res. Treat.* **2010**, *121*, 539–553. [CrossRef] [PubMed]
144. Principe, M.; Borgoni, S.; Cascione, M.; Chattaragada, M.S.; Ferri-Borgogno, S.; Capello, M.; Bulfamante, S.; Chapelle, J.; Di Modugno, F.; Defilippi, P.; et al. Alpha-enolase (ENO1) controls alpha v/beta 3 integrin expression and regulates pancreatic cancer adhesion, invasion, and metastasis. *J. Hematol. Oncol.* **2017**, *10*, 16. [CrossRef] [PubMed]
145. Wang, L.; Qu, M.; Huang, S.; Fu, Y.; Yang, L.; He, S.; Li, L.; Zhang, Z.; Lin, Q.; Zhang, L. A novel alpha-enolase-targeted drug delivery system for high efficacy prostate cancer therapy. *Nanoscale* **2018**, *10*, 13673–13683. [CrossRef]
146. Santana-Rivera, Y.; Rabelo-Fernandez, R.J.; Quinones-Diaz, B.I.; Grafals-Ruiz, N.; Santiago-Sanchez, G.; Lozada-Delgado, E.L.; Echevarria-Vargas, I.M.; Apiz, J.; Soto, D.; Rosado, A.; et al. Reduced expression of enolase-1 correlates with high intracellular glucose levels and increased senescence in cisplatin-resistant ovarian cancer cells. *Am. J. Transl. Res.* **2020**, *12*, 1275–1292.
147. Jiang, Y.X.; Siu, M.K.Y.; Wang, J.J.; Leung, T.H.Y.; Chan, D.W.; Cheung, A.N.Y.; Ngan, H.Y.S.; Chan, K.K.L. PFKFB3 Regulates Chemoresistance, Metastasis and Stemness via IAP Proteins and the NF-kappaB Signaling Pathway in Ovarian Cancer. *Front. Oncol.* **2022**, *12*, 748403. [CrossRef]
148. Mondal, S.; Roy, D.; Sarkar Bhattacharya, S.; Jin, L.; Jung, D.; Zhang, S.; Kalogera, E.; Staub, J.; Wang, Y.; Xuyang, W.; et al. Therapeutic targeting of PFKFB3 with a novel glycolytic inhibitor PFK158 promotes lipophagy and chemosensitivity in gynecologic cancers. *Int. J. Cancer* **2019**, *144*, 178–189. [CrossRef]
149. Meng, M.B.; Wang, H.H.; Guo, W.H.; Wu, Z.Q.; Zeng, X.L.; Zaorsky, N.G.; Shi, H.S.; Qian, D.; Niu, Z.M.; Jiang, B.; et al. Targeting pyruvate kinase M2 contributes to radiosensitivity of non-small cell lung cancer cells in vitro and in vivo. *Cancer Lett.* **2015**, *356*, 985–993. [CrossRef]
150. Sizemore, S.T.; Zhang, M.; Cho, J.H.; Sizemore, G.M.; Hurwitz, B.; Kaur, B.; Lehman, N.L.; Ostrowski, M.C.; Robe, P.A.; Miao, W.; et al. Pyruvate kinase M2 regulates homologous recombination-mediated DNA double-strand break repair. *Cell Res.* **2018**, *28*, 1090–1102. [CrossRef]
151. Sfakianaki, M.; Papadaki, C.; Tzardi, M.; Trypaki, M.; Manolakou, S.; Messaritakis, I.; Saridaki, Z.; Athanasakis, E.; Mavroudis, D.; Tsiaoussis, J.; et al. PKM2 Expression as Biomarker for Resistance to Oxaliplatin-Based Chemotherapy in Colorectal Cancer. *Cancers* **2020**, *12*, 2058. [CrossRef]
152. Anastasiou, D.; Poulogiannis, G.; Asara, J.M.; Boxer, M.B.; Jiang, J.K.; Shen, M.; Bellinger, G.; Sasaki, A.T.; Locasale, J.W.; Auld, D.S.; et al. Inhibition of pyruvate kinase M2 by reactive oxygen species contributes to cellular antioxidant responses. *Science* **2011**, *334*, 1278–1283. [CrossRef]
153. Zhu, H.; Wu, J.; Zhang, W.; Luo, H.; Shen, Z.; Cheng, H.; Zhu, X. PKM2 enhances chemosensitivity to cisplatin through interaction with the mTOR pathway in cervical cancer. *Sci. Rep.* **2016**, *6*, 30788. [CrossRef]

154. Sun, Q.; Chen, X.; Ma, J.; Peng, H.; Wang, F.; Zha, X.; Wang, Y.; Jing, Y.; Yang, H.; Chen, R.; et al. Mammalian target of rapamycin up-regulation of pyruvate kinase isoenzyme type M2 is critical for aerobic glycolysis and tumor growth. *Proc. Natl. Acad. Sci. USA* **2011**, *108*, 4129–4134. [CrossRef]
155. Lin, Y.; Lv, F.; Liu, F.; Guo, X.; Fan, Y.; Gu, F.; Gu, J.; Fu, L. High Expression of Pyruvate Kinase M2 is Associated with Chemosensitivity to Epirubicin and 5-Fluorouracil in Breast Cancer. *J. Cancer* **2015**, *6*, 1130–1139. [CrossRef]
156. Yoo, B.C.; Ku, J.L.; Hong, S.H.; Shin, Y.K.; Park, S.Y.; Kim, H.K.; Park, J.G. Decreased pyruvate kinase M2 activity linked to cisplatin resistance in human gastric carcinoma cell lines. *Int. J. Cancer* **2004**, *108*, 532–539. [CrossRef]
157. Martinez-Balibrea, E.; Plasencia, C.; Gines, A.; Martinez-Cardus, A.; Musulen, E.; Aguilera, R.; Manzano, J.L.; Neamati, N.; Abad, A. A proteomic approach links decreased pyruvate kinase M2 expression to oxaliplatin resistance in patients with colorectal cancer and in human cell lines. *Mol. Cancer Ther.* **2009**, *8*, 771–778. [CrossRef]
158. Al-Sawaf, O.; Bazeos, A.; Robrecht, S.; Bahlo, J.; Gower, C.; Fink, A.M.; Tresckow, J.; Cramer, P.; Langerbeins, P.; Kutsch, N.; et al. Mode of progression after first line treatment correlates with outcome of chronic lymphocytic leukemia (CLL). *Am. J. Hematol.* **2019**, *94*, 1002–1006. [CrossRef]
159. Cohen, E.E.W.; Bell, R.B.; Bifulco, C.B.; Burtness, B.; Gillison, M.L.; Harrington, K.J.; Le, Q.T.; Lee, N.Y.; Leidner, R.; Lewis, R.L.; et al. The Society for Immunotherapy of Cancer consensus statement on immunotherapy for the treatment of squamous cell carcinoma of the head and neck (HNSCC). *J. Immunother. Cancer* **2019**, *7*, 184. [CrossRef]
160. Reck, M.; Remon, J.; Hellmann, M.D. First-Line Immunotherapy for Non-Small-Cell Lung Cancer. *J. Clin. Oncol.* **2022**, *40*, 586–597. [CrossRef]
161. Bagchi, S.; Yuan, R.; Engleman, E.G. Immune Checkpoint Inhibitors for the Treatment of Cancer: Clinical Impact and Mechanisms of Response and Resistance. *Annu. Rev. Pathol.* **2021**, *16*, 223–249. [CrossRef]
162. Hegde, P.S.; Chen, D.S. Top 10 Challenges in Cancer Immunotherapy. *Immunity* **2020**, *52*, 17–35. [CrossRef] [PubMed]
163. DeBerardinis, R.J.; Chandel, N.S. Fundamentals of cancer metabolism. *Sci. Adv.* **2016**, *2*, e1600200. [CrossRef] [PubMed]
164. DePeaux, K.; Delgoffe, G.M. Metabolic barriers to cancer immunotherapy. *Nat. Rev. Immunol.* **2021**, *21*, 785–797. [CrossRef] [PubMed]
165. Chang, C.H.; Qiu, J.; O'Sullivan, D.; Buck, M.D.; Noguchi, T.; Curtis, J.D.; Chen, Q.; Gindin, M.; Gubin, M.M.; van der Windt, G.J.; et al. Metabolic Competition in the Tumor Microenvironment Is a Driver of Cancer Progression. *Cell* **2015**, *162*, 1229–1241. [CrossRef]
166. Andrejeva, G.; Rathmell, J.C. Similarities and Distinctions of Cancer and Immune Metabolism in Inflammation and Tumors. *Cell Metab.* **2017**, *26*, 49–70. [CrossRef]
167. Watson, M.J.; Vignali, P.D.A.; Mullett, S.J.; Overacre-Delgoffe, A.E.; Peralta, R.M.; Grebinoski, S.; Menk, A.V.; Rittenhouse, N.L.; DePeaux, K.; Whetstone, R.D.; et al. Metabolic support of tumour-infiltrating regulatory T cells by lactic acid. *Nature* **2021**, *591*, 645–651. [CrossRef]
168. Ho, P.C.; Bihuniak, J.D.; Macintyre, A.N.; Staron, M.; Liu, X.; Amezquita, R.; Tsui, Y.C.; Cui, G.; Micevic, G.; Perales, J.C.; et al. Phosphoenolpyruvate Is a Metabolic Checkpoint of Anti-tumor T Cell Responses. *Cell* **2015**, *162*, 1217–1228. [CrossRef]
169. Scharping, N.E.; Rivadeneira, D.B.; Menk, A.V.; Vignali, P.D.A.; Ford, B.R.; Rittenhouse, N.L.; Peralta, R.; Wang, Y.; Wang, Y.; DePeaux, K.; et al. Mitochondrial stress induced by continuous stimulation under hypoxia rapidly drives T cell exhaustion. *Nat. Immunol.* **2021**, *22*, 205–215. [CrossRef]
170. Jancewicz, I.; Szarkowska, J.; Konopinski, R.; Stachowiak, M.; Swiatek, M.; Blachnio, K.; Kubala, S.; Oksinska, P.; Cwiek, P.; Rusetska, N.; et al. PD-L1 Overexpression, SWI/SNF Complex Deregulation, and Profound Transcriptomic Changes Characterize Cancer-Dependent Exhaustion of Persistently Activated CD4(+) T Cells. *Cancers* **2021**, *13*, 4148. [CrossRef]
171. Martins, C.P.; New, L.A.; O'Connor, E.C.; Previte, D.M.; Cargill, K.R.; Tse, I.L.; Sims-Lucas, S.; Piganelli, J.D. Glycolysis Inhibition Induces Functional and Metabolic Exhaustion of CD4(+) T Cells in Type 1 Diabetes. *Front. Immunol.* **2021**, *12*, 669456. [CrossRef]
172. Gerriets, V.A.; Kishton, R.J.; Nichols, A.G.; Macintyre, A.N.; Inoue, M.; Ilkayeva, O.; Winter, P.S.; Liu, X.; Priyadharshini, B.; Slawinska, M.E.; et al. Metabolic programming and PDHK1 control CD4+ T cell subsets and inflammation. *J. Clin. Investig.* **2015**, *125*, 194–207. [CrossRef]
173. Gerriets, V.A.; Kishton, R.J.; Johnson, M.O.; Cohen, S.; Siska, P.J.; Nichols, A.G.; Warmoes, M.O.; de Cubas, A.A.; MacIver, N.J.; Locasale, J.W.; et al. Foxp3 and Toll-like receptor signaling balance Treg cell anabolic metabolism for suppression. *Nat. Immunol.* **2016**, *17*, 1459–1466. [CrossRef]
174. Kelly, B.; O'Neill, L.A. Metabolic reprogramming in macrophages and dendritic cells in innate immunity. *Cell Res.* **2015**, *25*, 771–784. [CrossRef]
175. Netea-Maier, R.T.; Smit, J.W.A.; Netea, M.G. Metabolic changes in tumor cells and tumor-associated macrophages: A mutual relationship. *Cancer Lett.* **2018**, *413*, 102–109. [CrossRef]
176. Bohn, T.; Rapp, S.; Luther, N.; Klein, M.; Bruehl, T.J.; Kojima, N.; Aranda Lopez, P.; Hahlbrock, J.; Muth, S.; Endo, S.; et al. Tumor immunoevasion via acidosis-dependent induction of regulatory tumor-associated macrophages. *Nat. Immunol.* **2018**, *19*, 1319–1329. [CrossRef]
177. Wang, L.; Liu, Y.; Dai, Y.; Tang, X.; Yin, T.; Wang, C.; Wang, T.; Dong, L.; Shi, M.; Qin, J.; et al. Single-cell RNA-seq analysis reveals BHLHE40-driven pro-tumour neutrophils with hyperactivated glycolysis in pancreatic tumour microenvironment. *Gut* **2022**, 1–14. [CrossRef]

178. Li, W.; Tanikawa, T.; Kryczek, I.; Xia, H.; Li, G.; Wu, K.; Wei, S.; Zhao, L.; Vatan, L.; Wen, B.; et al. Aerobic Glycolysis Controls Myeloid-Derived Suppressor Cells and Tumor Immunity via a Specific CEBPB Isoform in Triple-Negative Breast Cancer. *Cell Metab.* **2018**, *28*, 87–103.e6. [CrossRef]
179. Yang, X.; Lu, Y.; Hang, J.; Zhang, J.; Zhang, T.; Huo, Y.; Liu, J.; Lai, S.; Luo, D.; Wang, L.; et al. Lactate-Modulated Immunosuppression of Myeloid-Derived Suppressor Cells Contributes to the Radioresistance of Pancreatic Cancer. *Cancer Immunol. Res.* **2020**, *8*, 1440–1451. [CrossRef]
180. Cascone, T.; McKenzie, J.A.; Mbofung, R.M.; Punt, S.; Wang, Z.; Xu, C.; Williams, L.J.; Wang, Z.; Bristow, C.A.; Carugo, A.; et al. Increased Tumor Glycolysis Characterizes Immune Resistance to Adoptive T Cell Therapy. *Cell Metab.* **2018**, *27*, 977–987.e974. [CrossRef]
181. Song, B.S.; Moon, J.S.; Tian, J.; Lee, H.Y.; Sim, B.C.; Kim, S.H.; Kang, S.G.; Kim, J.T.; Nga, H.T.; Benfeitas, R.; et al. Mitoribosomal defects aggravate liver cancer via aberrant glycolytic flux and T cell exhaustion. *J. Immunother. Cancer* **2022**, *10*, e004337. [CrossRef]
182. Yu, Y.R.; Imrichova, H.; Wang, H.; Chao, T.; Xiao, Z.; Gao, M.; Rincon-Restrepo, M.; Franco, F.; Genolet, R.; Cheng, W.C.; et al. Disturbed mitochondrial dynamics in CD8(+) TILs reinforce T cell exhaustion. *Nat. Immunol.* **2020**, *21*, 1540–1551. [CrossRef] [PubMed]
183. Kooshki, L.; Mahdavi, P.; Fakhri, S.; Akkol, E.K.; Khan, H. Targeting lactate metabolism and glycolytic pathways in the tumor microenvironment by natural products: A promising strategy in combating cancer. *Biofactors* **2022**, *48*, 359–383. [CrossRef] [PubMed]
184. Fischer, K.; Hoffmann, P.; Voelkl, S.; Meidenbauer, N.; Ammer, J.; Edinger, M.; Gottfried, E.; Schwarz, S.; Rothe, G.; Hoves, S.; et al. Inhibitory effect of tumor cell-derived lactic acid on human T cells. *Blood* **2007**, *109*, 3812–3819. [CrossRef] [PubMed]
185. Robert, C.; Lewis, K.D.; Gutzmer, R.; Stroyakovskiy, D.; Gogas, H.; Protsenko, S.; Pereira, R.P.; Eigentler, T.; Rutkowski, P.; Demidov, L.; et al. Biomarkers of treatment benefit with atezolizumab plus vemurafenib plus cobimetinib in BRAF(V600) mutation-positive melanoma. *Ann. Oncol.* **2022**, *33*, 544–555. [CrossRef] [PubMed]
186. Calcinotto, A.; Filipazzi, P.; Grioni, M.; Iero, M.; De Milito, A.; Ricupito, A.; Cova, A.; Canese, R.; Jachetti, E.; Rossetti, M.; et al. Modulation of microenvironment acidity reverses anergy in human and murine tumor-infiltrating T lymphocytes. *Cancer Res.* **2012**, *72*, 2746–2756. [CrossRef]
187. Mendler, A.N.; Hu, B.; Prinz, P.U.; Kreutz, M.; Gottfried, E.; Noessner, E. Tumor lactic acidosis suppresses CTL function by inhibition of p38 and JNK/c-Jun activation. *Int. J. Cancer* **2012**, *131*, 633–640. [CrossRef]
188. Shime, H.; Yabu, M.; Akazawa, T.; Kodama, K.; Matsumoto, M.; Seya, T.; Inoue, N. Tumor-secreted lactic acid promotes IL-23/IL-17 proinflammatory pathway. *J. Immunol.* **2008**, *180*, 7175–7183. [CrossRef]
189. Ma, S.; Zhao, Y.; Lee, W.C.; Ong, L.T.; Lee, P.L.; Jiang, Z.; Oguz, G.; Niu, Z.; Liu, M.; Goh, J.Y.; et al. Hypoxia induces HIF1alpha-dependent epigenetic vulnerability in triple negative breast cancer to confer immune effector dysfunction and resistance to anti-PD-1 immunotherapy. *Nat. Commun.* **2022**, *13*, 4118. [CrossRef]
190. Ford, B.R.; Vignali, P.D.A.; Rittenhouse, N.L.; Scharping, N.E.; Peralta, R.; Lontos, K.; Frisch, A.T.; Delgoffe, G.M.; Poholek, A.C. Tumor microenvironmental signals reshape chromatin landscapes to limit the functional potential of exhausted T cells. *Sci. Immunol.* **2022**, *7*, eabj9123. [CrossRef]
191. Clambey, E.T.; McNamee, E.N.; Westrich, J.A.; Glover, L.E.; Campbell, E.L.; Jedlicka, P.; de Zoeten, E.F.; Cambier, J.C.; Stenmark, K.R.; Colgan, S.P.; et al. Hypoxia-inducible factor-1 alpha-dependent induction of FoxP3 drives regulatory T-cell abundance and function during inflammatory hypoxia of the mucosa. *Proc. Natl. Acad. Sci. USA* **2012**, *109*, E2784–E2793. [CrossRef]
192. Deng, J.; Li, J.; Sarde, A.; Lines, J.L.; Lee, Y.C.; Qian, D.C.; Pechenick, D.A.; Manivanh, R.; Le Mercier, I.; Lowrey, C.H.; et al. Hypoxia-Induced VISTA Promotes the Suppressive Function of Myeloid-Derived Suppressor Cells in the Tumor Microenvironment. *Cancer Immunol. Res.* **2019**, *7*, 1079–1090. [CrossRef]
193. Zheng, J.B.; Wong, C.W.; Liu, J.; Luo, X.J.; Zhou, W.Y.; Chen, Y.X.; Luo, H.Y.; Zeng, Z.L.; Ren, C.; Xie, X.M.; et al. Glucose metabolism inhibitor PFK-015 combined with immune checkpoint inhibitor is an effective treatment regimen in cancer. *Oncoimmunology* **2022**, *11*, 2079182. [CrossRef]
194. Bader, J.E.; Voss, K.; Rathmell, J.C. Targeting Metabolism to Improve the Tumor Microenvironment for Cancer Immunotherapy. *Mol. Cell* **2020**, *78*, 1019–1033. [CrossRef]
195. Chow, E.; Yang, A.; Chung, C.H.L.; Chan, J.C.N. A Clinical Perspective of the Multifaceted Mechanism of Metformin in Diabetes, Infections, Cognitive Dysfunction, and Cancer. *Pharmaceuticals* **2022**, *15*, 442. [CrossRef]
196. Rubiño, M.E.; Carrillo, E.; Alcalá, G.; Domínguez-Martín, A.; Marchal, J.A.; Boulaiz, H. Phenformin as an Anticancer Agent: Challenges and Prospects. *Int. J. Mol. Sci.* **2019**, *20*, 3316. [CrossRef]
197. Shackelford, D.B.; Shaw, R.J. The LKB1-AMPK pathway: Metabolism and growth control in tumour suppression. *Nat. Rev. Cancer* **2009**, *9*, 563–575. [CrossRef]
198. MacIver, N.J.; Blagih, J.; Saucillo, D.C.; Tonelli, L.; Griss, T.; Rathmell, J.C.; Jones, R.G. The liver kinase B1 is a central regulator of T cell development, activation, and metabolism. *J. Immunol.* **2011**, *187*, 4187–4198. [CrossRef]
199. Denko, N.C. Hypoxia, HIF1 and glucose metabolism in the solid tumour. *Nat. Rev. Cancer* **2008**, *8*, 705–713. [CrossRef]
200. Icard, P.; Simula, L.; Fournel, L.; Leroy, K.; Lupo, A.; Damotte, D.; Charpentier, M.C.; Durdux, C.; Loi, M.; Schussler, O.; et al. The strategic roles of four enzymes in the interconnection between metabolism and oncogene activation in non-small cell lung cancer: Therapeutic implications. *Drug Resist. Updat.* **2022**, *63*, 100852. [CrossRef]

201. Xu, K.; Yin, N.; Peng, M.; Stamatiades, E.G.; Shyu, A.; Li, P.; Zhang, X.; Do, M.H.; Wang, Z.; Capistrano, K.J.; et al. Glycolysis fuels phosphoinositide 3-kinase signaling to bolster T cell immunity. *Science* **2021**, *371*, 405–410. [CrossRef]
202. Jacobs, S.R.; Herman, C.E.; Maciver, N.J.; Wofford, J.A.; Wieman, H.L.; Hammen, J.J.; Rathmell, J.C. Glucose uptake is limiting in T cell activation and requires CD28-mediated Akt-dependent and independent pathways. *J. Immunol.* **2008**, *180*, 4476–4486. [CrossRef] [PubMed]
203. Miyamoto, S.; Murphy, A.N.; Brown, J.H. Akt mediates mitochondrial protection in cardiomyocytes through phosphorylation of mitochondrial hexokinase-II. *Cell Death Differ.* **2008**, *15*, 521–529. [CrossRef] [PubMed]
204. Chae, Y.C.; Vaira, V.; Caino, M.C.; Tang, H.Y.; Seo, J.H.; Kossenkov, A.V.; Ottobrini, L.; Martelli, C.; Lucignani, G.; Bertolini, I.; et al. Mitochondrial Akt Regulation of Hypoxic Tumor Reprogramming. *Cancer Cell* **2016**, *30*, 257–272. [CrossRef] [PubMed]
205. Araki, K.; Turner, A.P.; Shaffer, V.O.; Gangappa, S.; Keller, S.A.; Bachmann, M.F.; Larsen, C.P.; Ahmed, R. mTOR regulates memory CD8 T-cell differentiation. *Nature* **2009**, *460*, 108–112. [CrossRef]
206. Rao, R.R.; Li, Q.; Odunsi, K.; Shrikant, P.A. The mTOR kinase determines effector versus memory CD8+ T cell fate by regulating the expression of transcription factors T-bet and Eomesodermin. *Immunity* **2010**, *32*, 67–78. [CrossRef]
207. Madden, M.Z.; Rathmell, J.C. The Complex Integration of T-cell Metabolism and Immunotherapy. *Cancer Discov.* **2021**, *11*, 1636–1643. [CrossRef]
208. Maude, S.L.; Laetsch, T.W.; Buechner, J.; Rives, S.; Boyer, M.; Bittencourt, H.; Bader, P.; Verneris, M.R.; Stefanski, H.E.; Myers, G.D.; et al. Tisagenlecleucel in Children and Young Adults with B-Cell Lymphoblastic Leukemia. *N. Engl. J. Med.* **2018**, *378*, 439–448. [CrossRef]
209. Huang, J.; Huang, X.; Huang, J. CAR-T cell therapy for hematological malignancies: Limitations and optimization strategies. *Front. Immunol.* **2022**, *13*, 1019115. [CrossRef]
210. Qin, X.; Wu, F.; Chen, C.; Li, Q. Recent advances in CAR-T cells therapy for colorectal cancer. *Front. Immunol.* **2022**, *13*, 904137. [CrossRef]
211. Kawalekar, O.U.; O'Connor, R.S.; Fraietta, J.A.; Guo, L.; McGettigan, S.E.; Posey, A.D., Jr.; Patel, P.R.; Guedan, S.; Scholler, J.; Keith, B.; et al. Distinct Signaling of Coreceptors Regulates Specific Metabolism Pathways and Impacts Memory Development in CAR T Cells. *Immunity* **2016**, *44*, 380–390. [CrossRef]
212. Guedan, S.; Madar, A.; Casado-Medrano, V.; Shaw, C.; Wing, A.; Liu, F.; Young, R.M.; June, C.H.; Posey, A.D., Jr. Single residue in CD28-costimulated CAR-T cells limits long-term persistence and antitumor durability. *J. Clin. Investig.* **2020**, *130*, 3087–3097. [CrossRef]
213. Zeng, H.; Cohen, S.; Guy, C.; Shrestha, S.; Neale, G.; Brown, S.A.; Cloer, C.; Kishton, R.J.; Gao, X.; Youngblood, B.; et al. mTORC1 and mTORC2 Kinase Signaling and Glucose Metabolism Drive Follicular Helper T Cell Differentiation. *Immunity* **2016**, *45*, 540–554. [CrossRef]
214. Sabharwal, S.S.; Rosen, D.B.; Grein, J.; Tedesco, D.; Joyce-Shaikh, B.; Ueda, R.; Semana, M.; Bauer, M.; Bang, K.; Stevenson, C.; et al. GITR Agonism Enhances Cellular Metabolism to Support CD8(+) T-cell Proliferation and Effector Cytokine Production in a Mouse Tumor Model. *Cancer Immunol. Res.* **2018**, *6*, 1199–1211. [CrossRef]
215. Pacella, I.; Procaccini, C.; Focaccetti, C.; Miacci, S.; Timperi, E.; Faicchia, D.; Severa, M.; Rizzo, F.; Coccia, E.M.; Bonacina, F.; et al. Fatty acid metabolism complements glycolysis in the selective regulatory T cell expansion during tumor growth. *Proc. Natl. Acad. Sci. USA* **2018**, *115*, E6546–E6555. [CrossRef]
216. Kishton, R.J.; Sukumar, M.; Restifo, N.P. Metabolic Regulation of T Cell Longevity and Function in Tumor Immunotherapy. *Cell Metab.* **2017**, *26*, 94–109. [CrossRef]
217. Leone, R.D.; Zhao, L.; Englert, J.M.; Sun, I.M.; Oh, M.H.; Sun, I.H.; Arwood, M.L.; Bettencourt, I.A.; Patel, C.H.; Wen, J.; et al. Glutamine blockade induces divergent metabolic programs to overcome tumor immune evasion. *Science* **2019**, *366*, 1013–1021. [CrossRef]
218. Shen, L.; Xiao, Y.; Zhang, C.; Li, S.; Teng, X.; Cui, L.; Liu, T.; Wu, N.; Lu, Z. Metabolic reprogramming by ex vivo glutamine inhibition endows CAR-T cells with less-differentiated phenotype and persistent antitumor activity. *Cancer Lett.* **2022**, *538*, 215710. [CrossRef]
219. Sukumar, M.; Liu, J.; Ji, Y.; Subramanian, M.; Crompton, J.G.; Yu, Z.; Roychoudhuri, R.; Palmer, D.C.; Muranski, P.; Karoly, E.D.; et al. Inhibiting glycolytic metabolism enhances CD8+ T cell memory and antitumor function. *J. Clin. Investig.* **2013**, *123*, 4479–4488. [CrossRef]
220. Klebanoff, C.A.; Crompton, J.G.; Leonardi, A.J.; Yamamoto, T.N.; Chandran, S.S.; Eil, R.L.; Sukumar, M.; Vodnala, S.K.; Hu, J.; Ji, Y.; et al. Inhibition of AKT signaling uncouples T cell differentiation from expansion for receptor-engineered adoptive immunotherapy. *JCI Insight* **2017**, *2*, e95103. [CrossRef]
221. Klein Geltink, R.I.; Edwards-Hicks, J.; Apostolova, P.; O'Sullivan, D.; Sanin, D.E.; Patterson, A.E.; Puleston, D.J.; Ligthart, N.A.M.; Buescher, J.M.; Grzes, K.M.; et al. Metabolic conditioning of CD8(+) effector T cells for adoptive cell therapy. *Nat. Metab.* **2020**, *2*, 703–716. [CrossRef]
222. Gemta, L.F.; Siska, P.J.; Nelson, M.E.; Gao, X.; Liu, X.; Locasale, J.W.; Yagita, H.; Slingluff, C.L., Jr.; Hoehn, K.L.; Rathmell, J.C.; et al. Impaired enolase 1 glycolytic activity restrains effector functions of tumor-infiltrating CD8(+) T cells. *Sci. Immunol.* **2019**, *4*, eaap9520. [CrossRef] [PubMed]
223. Philip, M.; Fairchild, L.; Sun, L.; Horste, E.L.; Camara, S.; Shakiba, M.; Scott, A.C.; Viale, A.; Lauer, P.; Merghoub, T.; et al. Chromatin states define tumour-specific T cell dysfunction and reprogramming. *Nature* **2017**, *545*, 452–456. [CrossRef] [PubMed]

224. Scharping, N.E.; Menk, A.V.; Moreci, R.S.; Whetstone, R.D.; Dadey, R.E.; Watkins, S.C.; Ferris, R.L.; Delgoffe, G.M. The Tumor Microenvironment Represses T Cell Mitochondrial Biogenesis to Drive Intratumoral T Cell Metabolic Insufficiency and Dysfunction. *Immunity* **2016**, *45*, 374–388. [CrossRef] [PubMed]
225. Siska, P.J.; Beckermann, K.E.; Mason, F.M.; Andrejeva, G.; Greenplate, A.R.; Sendor, A.B.; Chiang, Y.J.; Corona, A.L.; Gemta, L.F.; Vincent, B.G.; et al. Mitochondrial dysregulation and glycolytic insufficiency functionally impair CD8 T cells infiltrating human renal cell carcinoma. *JCI Insight* **2017**, *2*, e93411. [CrossRef] [PubMed]
226. Bengsch, B.; Johnson, A.L.; Kurachi, M.; Odorizzi, P.M.; Pauken, K.E.; Attanasio, J.; Stelekati, E.; McLane, L.M.; Paley, M.A.; Delgoffe, G.M.; et al. Bioenergetic Insufficiencies Due to Metabolic Alterations Regulated by the Inhibitory Receptor PD-1 Are an Early Driver of CD8(+) T Cell Exhaustion. *Immunity* **2016**, *45*, 358–373. [CrossRef]
227. Patsoukis, N.; Bardhan, K.; Chatterjee, P.; Sari, D.; Liu, B.; Bell, L.N.; Karoly, E.D.; Freeman, G.J.; Petkova, V.; Seth, P.; et al. PD-1 alters T-cell metabolic reprogramming by inhibiting glycolysis and promoting lipolysis and fatty acid oxidation. *Nat. Commun.* **2015**, *6*, 6692. [CrossRef]
228. Lee, M.J.; Yun, S.J.; Lee, B.; Jeong, E.; Yoon, G.; Kim, K.; Park, S. Association of TIM-3 expression with glucose metabolism in Jurkat T cells. *BMC Immunol.* **2020**, *21*, 48. [CrossRef]
229. Tsurutani, N.; Mittal, P.; St Rose, M.C.; Ngoi, S.M.; Svedova, J.; Menoret, A.; Treadway, F.B.; Laubenbacher, R.; Suarez-Ramirez, J.E.; Cauley, L.S.; et al. Costimulation Endows Immunotherapeutic CD8 T Cells with IL-36 Responsiveness during Aerobic Glycolysis. *J. Immunol.* **2016**, *196*, 124–134. [CrossRef]
230. Previte, D.M.; Martins, C.P.; O'Connor, E.C.; Marre, M.L.; Coudriet, G.M.; Beck, N.W.; Menk, A.V.; Wright, R.H.; Tse, H.M.; Delgoffe, G.M.; et al. Lymphocyte Activation Gene-3 Maintains Mitochondrial and Metabolic Quiescence in Naive CD4(+) T Cells. *Cell Rep.* **2019**, *27*, 129–141.e4. [CrossRef]
231. Xie, J.; Wang, X.; Proud, C.G. mTOR inhibitors in cancer therapy. *F1000Res* **2016**, *5*, 2078. [CrossRef]
232. Jia, W.; Luo, S.; Guo, H.; Kong, D. Development of PI3Kalpha inhibitors for tumor therapy. *J. Biomol. Struct. Dyn.* **2022**, 1–18. [CrossRef]
233. Kopf, H.; de la Rosa, G.M.; Howard, O.M.; Chen, X. Rapamycin inhibits differentiation of Th17 cells and promotes generation of FoxP3+ T regulatory cells. *Int. Immunopharmacol.* **2007**, *7*, 1819–1824. [CrossRef]
234. Shi, L.Z.; Wang, R.; Huang, G.; Vogel, P.; Neale, G.; Green, D.R.; Chi, H. HIF1alpha-dependent glycolytic pathway orchestrates a metabolic checkpoint for the differentiation of TH17 and Treg cells. *J. Exp. Med.* **2011**, *208*, 1367–1376. [CrossRef]
235. Siska, P.J.; van der Windt, G.J.; Kishton, R.J.; Cohen, S.; Eisner, W.; MacIver, N.J.; Kater, A.P.; Weinberg, J.B.; Rathmell, J.C. Suppression of Glut1 and Glucose Metabolism by Decreased Akt/mTORC1 Signaling Drives T Cell Impairment in B Cell Leukemia. *J. Immunol.* **2016**, *197*, 2532–2540. [CrossRef]
236. Chapman, N.M.; Zeng, H.; Nguyen, T.M.; Wang, Y.; Vogel, P.; Dhungana, Y.; Liu, X.; Neale, G.; Locasale, J.W.; Chi, H. mTOR coordinates transcriptional programs and mitochondrial metabolism of activated Treg subsets to protect tissue homeostasis. *Nat. Commun.* **2018**, *9*, 2095. [CrossRef]
237. Nozhat, Z.; Mohammadi-Yeganeh, S.; Azizi, F.; Zarkesh, M.; Hedayati, M. Effects of metformin on the PI3K/AKT/FOXO1 pathway in anaplastic thyroid Cancer cell lines. *Daru* **2018**, *26*, 93–103. [CrossRef]
238. Shao, S.; Zhao, L.; An, G.; Zhang, L.; Jing, X.; Luo, M.; Li, W.; Meng, D.; Ning, Q.; Zhao, X.; et al. Metformin suppresses HIF-1alpha expression in cancer-associated fibroblasts to prevent tumor-stromal cross talk in breast cancer. *FASEB J.* **2020**, *34*, 10860–10870. [CrossRef]
239. Pearce, E.L.; Walsh, M.C.; Cejas, P.J.; Harms, G.M.; Shen, H.; Wang, L.S.; Jones, R.G.; Choi, Y. Enhancing CD8 T-cell memory by modulating fatty acid metabolism. *Nature* **2009**, *460*, 103–107. [CrossRef]
240. Chao, R.; Nishida, M.; Yamashita, N.; Tokumasu, M.; Zhao, W.; Kudo, I.; Udono, H. Nutrient Condition in the Microenvironment Determines Essential Metabolisms of CD8(+) T Cells for Enhanced IFNgamma Production by Metformin. *Front. Immunol.* **2022**, *13*, 864225. [CrossRef]
241. Chen, S.; Zhou, X.; Yang, X.; Li, W.; Li, S.; Hu, Z.; Ling, C.; Shi, R.; Liu, J.; Chen, G.; et al. Dual Blockade of Lactate/GPR81 and PD-1/PD-L1 Pathways Enhances the Anti-Tumor Effects of Metformin. *Biomolecules* **2021**, *11*, 1373. [CrossRef]
242. Chung, Y.M.; Khan, P.P.; Wang, H.; Tsai, W.B.; Qiao, Y.; Yu, B.; Larrick, J.W.; Hu, M.C. Sensitizing tumors to anti-PD-1 therapy by promoting NK and CD8+ T cells via pharmacological activation of FOXO3. *J. Immunother. Cancer* **2021**, *9*, e002772. [CrossRef] [PubMed]
243. Song, C.W.; Kim, H.; Cho, H.; Kim, M.S.; Paek, S.H.; Park, H.J.; Griffin, R.J.; Terezakis, S.; Cho, L.C. HIF-1alpha Inhibition Improves Anti-Tumor Immunity and Promotes the Efficacy of Stereotactic Ablative Radiotherapy (SABR). *Cancers* **2022**, *14*, 3273. [CrossRef] [PubMed]
244. Mostafavi, S.; Zalpoor, H.; Hassan, Z.M. The promising therapeutic effects of metformin on metabolic reprogramming of cancer-associated fibroblasts in solid tumors. *Cell. Mol. Biol. Lett.* **2022**, *27*, 58. [CrossRef] [PubMed]
245. Pouyssegur, J.; Marchiq, I.; Parks, S.K.; Durivault, J.; Zdralevic, M.; Vucetic, M. 'Warburg effect' controls tumor growth, bacterial, viral infections and immunity—Genetic deconstruction and therapeutic perspectives. *Semin. Cancer Biol.* **2022**, *86*, 334–346. [CrossRef]
246. Luo, Y.; Li, Y.; Huang, Z.; Li, X.; Wang, Y.; Hou, J.; Zhou, S. A Nanounit Strategy Disrupts Energy Metabolism and Alleviates Immunosuppression for Cancer Therapy. *Nano Lett.* **2022**, *22*, 6418–6427. [CrossRef]

247. Xu, R.; Wu, M.; Liu, S.; Shang, W.; Li, R.; Xu, J.; Huang, L.; Wang, F. Glucose metabolism characteristics and TLR8-mediated metabolic control of CD4(+) Treg cells in ovarian cancer cells microenvironment. *Cell Death Dis.* **2021**, *12*, 22. [CrossRef]
248. Guo, D.; Tong, Y.; Jiang, X.; Meng, Y.; Jiang, H.; Du, L.; Wu, Q.; Li, S.; Luo, S.; Li, M.; et al. Aerobic glycolysis promotes tumor immune evasion by hexokinase2-mediated phosphorylation of IkappaBalpha. *Cell Metab.* **2022**, *34*, 1312–1324.e6. [CrossRef]
249. Wu, L.; Hollinshead, K.E.R.; Hao, Y.; Au, C.; Kroehling, L.; Ng, C.; Lin, W.Y.; Li, D.; Silva, H.M.; Shin, J.; et al. Niche-Selective Inhibition of Pathogenic Th17 Cells by Targeting Metabolic Redundancy. *Cell* **2020**, *182*, 641–654.e20. [CrossRef]
250. Lei, J.; Yang, Y.; Lu, Z.; Pan, H.; Fang, J.; Jing, B.; Chen, Y.; Yin, L. Taming metabolic competition via glycolysis inhibition for safe and potent tumor immunotherapy. *Biochem. Pharmacol.* **2022**, *202*, 115153. [CrossRef]
251. Shan, Y.; Ni, Q.; Zhang, Q.; Zhang, M.; Wei, B.; Cheng, L.; Zhong, C.; Wang, X.; Wang, Q.; Liu, J.; et al. Targeting tumor endothelial hyperglycolysis enhances immunotherapy through remodeling tumor microenvironment. *Acta Pharm. Sin. B* **2022**, *12*, 1825–1839. [CrossRef]
252. Chen, D.P.; Ning, W.R.; Jiang, Z.Z.; Peng, Z.P.; Zhu, L.Y.; Zhuang, S.M.; Kuang, D.M.; Zheng, L.; Wu, Y. Glycolytic activation of peritumoral monocytes fosters immune privilege via the PFKFB3-PD-L1 axis in human hepatocellular carcinoma. *J. Hepatol.* **2019**, *71*, 333–343. [CrossRef] [PubMed]
253. Liu, Y.; Yan, H.; Gu, H.; Zhang, E.; He, J.; Cao, W.; Qu, J.; Xu, R.; Cao, L.; He, D.; et al. Myeloma-derived IL-32gamma induced PD-L1 expression in macrophages facilitates immune escape via the PFKFB3-JAK1 axis. *Oncoimmunology* **2022**, *11*, 2057837. [CrossRef] [PubMed]
254. Yu, Y.; Liang, Y.; Li, D.; Wang, L.; Liang, Z.; Chen, Y.; Ma, G.; Wu, H.; Jiao, W.; Niu, H. Glucose metabolism involved in PD-L1-mediated immune escape in the malignant kidney tumour microenvironment. *Cell Death Discov.* **2021**, *7*, 15. [CrossRef] [PubMed]
255. Palsson-McDermott, E.M.; Dyck, L.; Zaslona, Z.; Menon, D.; McGettrick, A.F.; Mills, K.H.G.; O'Neill, L.A. Pyruvate Kinase M2 Is Required for the Expression of the Immune Checkpoint PD-L1 in Immune Cells and Tumors. *Front. Immunol.* **2017**, *8*, 1300. [CrossRef]
256. Xia, Q.; Jia, J.; Hu, C.; Lu, J.; Li, J.; Xu, H.; Fang, J.; Feng, D.; Wang, L.; Chen, Y. Tumor-associated macrophages promote PD-L1 expression in tumor cells by regulating PKM2 nuclear translocation in pancreatic ductal adenocarcinoma. *Oncogene* **2022**, *41*, 865–877. [CrossRef]
257. Xu, W.; Yang, W.; Wu, C.; Ma, X.; Li, H.; Zheng, J. Enolase 1 Correlated with Cancer Progression and Immune-Infiltrating in Multiple Cancer Types: A Pan-Cancer Analysis. *Front. Oncol.* **2020**, *10*, 593706. [CrossRef]
258. Cappello, P.; Tomaino, B.; Chiarle, R.; Ceruti, P.; Novarino, A.; Castagnoli, C.; Migliorini, P.; Perconti, G.; Giallongo, A.; Milella, M.; et al. An integrated humoral and cellular response is elicited in pancreatic cancer by alpha-enolase, a novel pancreatic ductal adenocarcinoma-associated antigen. *Int. J. Cancer* **2009**, *125*, 639–648. [CrossRef]
259. Cappello, P.; Rolla, S.; Chiarle, R.; Principe, M.; Cavallo, F.; Perconti, G.; Feo, S.; Giovarelli, M.; Novelli, F. Vaccination with ENO1 DNA prolongs survival of genetically engineered mice with pancreatic cancer. *Gastroenterology* **2013**, *144*, 1098–1106. [CrossRef]
260. Cappello, P.; Principe, M.; Bulfamante, S.; Novelli, F. Alpha-Enolase (ENO1), a potential target in novel immunotherapies. *Front. Biosci. (Landmark Ed.)* **2017**, *22*, 944–959. [CrossRef]
261. Mandili, G.; Curcio, C.; Bulfamante, S.; Follia, L.; Ferrero, G.; Mazza, E.; Principe, M.; Cordero, F.; Satolli, M.A.; Spadi, R.; et al. In pancreatic cancer, chemotherapy increases antitumor responses to tumor-associated antigens and potentiates DNA vaccination. *J. Immunother. Cancer* **2020**, *8*, e001071. [CrossRef]
262. Curcio, C.; Brugiapaglia, S.; Bulfamante, S.; Follia, L.; Cappello, P.; Novelli, F. The Glycolytic Pathway as a Target for Novel Onco-Immunology Therapies in Pancreatic Cancer. *Molecules* **2021**, *26*, 1642. [CrossRef]
263. Chen, M.L.; Yuan, T.T.; Chuang, C.F.; Huang, Y.T.; Chung, I.C.; Huang, W.C. A Novel Enolase-1 Antibody Targets Multiple Interacting Players in the Tumor Microenvironment of Advanced Prostate Cancer. *Mol. Cancer Ther.* **2022**, *21*, 1337–1347. [CrossRef]
264. Hanker, A.B.; Sudhan, D.R.; Arteaga, C.L. Overcoming Endocrine Resistance in Breast Cancer. *Cancer Cell* **2020**, *37*, 496–513. [CrossRef]
265. Reeves, G.K.; Beral, V.; Green, J.; Gathani, T.; Bull, D.; Million Women Study, C. Hormonal therapy for menopause and breast-cancer risk by histological type: A cohort study and meta-analysis. *Lancet Oncol.* **2006**, *7*, 910–918. [CrossRef]
266. Rugo, H.S.; Rumble, R.B.; Macrae, E.; Barton, D.L.; Connolly, H.K.; Dickler, M.N.; Fallowfield, L.; Fowble, B.; Ingle, J.N.; Jahanzeb, M.; et al. Endocrine Therapy for Hormone Receptor-Positive Metastatic Breast Cancer: American Society of Clinical Oncology Guideline. *J. Clin. Oncol.* **2016**, *34*, 3069–3103. [CrossRef]
267. Pan, H.; Gray, R.; Braybrooke, J.; Davies, C.; Taylor, C.; McGale, P.; Peto, R.; Pritchard, K.I.; Bergh, J.; Dowsett, M.; et al. 20-Year Risks of Breast-Cancer Recurrence after Stopping Endocrine Therapy at 5 Years. *N. Engl. J. Med.* **2017**, *377*, 1836–1846. [CrossRef]
268. Fiorillo, M.; Sanchez-Alvarez, R.; Sotgia, F.; Lisanti, M.P. The ER-alpha mutation Y537S confers Tamoxifen-resistance via enhanced mitochondrial metabolism, glycolysis and Rho-GDI/PTEN signaling: Implicating TIGAR in somatic resistance to endocrine therapy. *Aging* **2018**, *10*, 4000–4023. [CrossRef]
269. Kulkoyluoglu-Cotul, E.; Arca, A.; Madak-Erdogan, Z. Crosstalk between Estrogen Signaling and Breast Cancer Metabolism. *Trends Endocrinol. Metab.* **2019**, *30*, 25–38. [CrossRef]
270. Toy, W.; Shen, Y.; Won, H.; Green, B.; Sakr, R.A.; Will, M.; Li, Z.; Gala, K.; Fanning, S.; King, T.A.; et al. ESR1 ligand-binding domain mutations in hormone-resistant breast cancer. *Nat. Genet.* **2013**, *45*, 1439–1445. [CrossRef]

271. Li, S.; Shen, D.; Shao, J.; Crowder, R.; Liu, W.; Prat, A.; He, X.; Liu, S.; Hoog, J.; Lu, C.; et al. Endocrine-therapy-resistant ESR1 variants revealed by genomic characterization of breast-cancer-derived xenografts. *Cell Rep.* **2013**, *4*, 1116–1130. [CrossRef]
272. Robinson, D.R.; Wu, Y.M.; Vats, P.; Su, F.; Lonigro, R.J.; Cao, X.; Kalyana-Sundaram, S.; Wang, R.; Ning, Y.; Hodges, L.; et al. Activating ESR1 mutations in hormone-resistant metastatic breast cancer. *Nat. Genet.* **2013**, *45*, 1446–1451. [CrossRef] [PubMed]
273. Ambrosio, M.R.; D'Esposito, V.; Costa, V.; Liguoro, D.; Collina, F.; Cantile, M.; Prevete, N.; Passaro, C.; Mosca, G.; De Laurentiis, M.; et al. Glucose impairs tamoxifen responsiveness modulating connective tissue growth factor in breast cancer cells. *Oncotarget* **2017**, *8*, 109000–109017. [CrossRef] [PubMed]
274. Rivenzon-Segal, D.; Boldin-Adamsky, S.; Seger, D.; Seger, R.; Degani, H. Glycolysis and glucose transporter 1 as markers of response to hormonal therapy in breast cancer. *Int. J. Cancer* **2003**, *107*, 177–182. [CrossRef] [PubMed]
275. Zhang, P.; Yang, Y.; Qian, K.; Li, L.; Zhang, C.; Fu, X.; Zhang, X.; Chen, H.; Liu, Q.; Cao, S.; et al. A novel tumor suppressor ZBTB1 regulates tamoxifen resistance and aerobic glycolysis through suppressing HER2 expression in breast cancer. *J. Biol. Chem.* **2020**, *295*, 14140–14152. [CrossRef] [PubMed]
276. Lv, L.; Yang, S.; Zhu, Y.; Zhai, X.; Li, S.; Tao, X.; Dong, D. Relationship between metabolic reprogramming and drug resistance in breast cancer. *Front. Oncol.* **2022**, *12*, 942064. [CrossRef]
277. Woo, Y.M.; Shin, Y.; Lee, E.J.; Lee, S.; Jeong, S.H.; Kong, H.K.; Park, E.Y.; Kim, H.K.; Han, J.; Chang, M.; et al. Inhibition of Aerobic Glycolysis Represses Akt/mTOR/HIF-1alpha Axis and Restores Tamoxifen Sensitivity in Antiestrogen-Resistant Breast Cancer Cells. *PLoS ONE* **2015**, *10*, e0132285. [CrossRef]
278. Steifensand, F.; Gallwas, J.; Bauerschmitz, G.; Grundker, C. Inhibition of Metabolism as a Therapeutic Option for Tamoxifen-Resistant Breast Cancer Cells. *Cells* **2021**, *10*, 2398. [CrossRef]
279. Chen, Z.; Wang, Y.; Warden, C.; Chen, S. Cross-talk between ER and HER2 regulates c-MYC-mediated glutamine metabolism in aromatase inhibitor resistant breast cancer cells. *J. Steroid Biochem. Mol. Biol.* **2015**, *149*, 118–127. [CrossRef]
280. Sengupta, S.; Sevigny, C.M.; Liu, X.; Jin, L.; Pohlmann, P.R.; Clarke, R. Targeting glycolysis enzyme, PFKFB3, in endocrine therapy resistant breast cancers. *Cancer Res.* **2018**, *78*, 907. [CrossRef]
281. Jones, B.C.; Sengupta, S.; Sevigny, C.M.; Jin, L.; Pohlmann, P.R.; Shajahan-Haq, A.; Clarke, R. Pfkfb3 inhibition significantly decreases endocrine-resistant breast cancer growth and induces necroptotic cell death. In Proceedings of the San Antonio Breast Cancer Symposium, San Antonio, TX, USA, 7–10 December 2021.
282. Bacci, M.; Giannoni, E.; Fearns, A.; Ribas, R.; Gao, Q.; Taddei, M.L.; Pintus, G.; Dowsett, M.; Isacke, C.M.; Martin, L.A.; et al. miR-155 Drives Metabolic Reprogramming of ER+ Breast Cancer Cells Following Long-Term Estrogen Deprivation and Predicts Clinical Response to Aromatase Inhibitors. *Cancer Res.* **2016**, *76*, 1615–1626. [CrossRef]
283. Zuo, Q.; Mogol, A.N.; Liu, Y.J.; Santaliz Casiano, A.; Chien, C.; Drnevich, J.; Imir, O.B.; Kulkoyluoglu-Cotul, E.; Park, N.H.; Shapiro, D.J.; et al. Targeting Metabolic Adaptations in the Breast Cancer-Liver Metastatic Niche Using Dietary Approaches to Improve Endocrine Therapy Efficacy. *Mol. Cancer Res.* **2022**, *20*, 923–937. [CrossRef]
284. Blundon, M.A.; Dasgupta, S. Metabolic Dysregulation Controls Endocrine Therapy-Resistant Cancer Recurrence and Metastasis. *Endocrinology* **2019**, *160*, 1811–1820. [CrossRef]
285. Pertega-Gomes, N.; Felisbino, S.; Massie, C.E.; Vizcaino, J.R.; Coelho, R.; Sandi, C.; Simoes-Sousa, S.; Jurmeister, S.; Ramos-Montoya, A.; Asim, M.; et al. A glycolytic phenotype is associated with prostate cancer progression and aggressiveness: A role for monocarboxylate transporters as metabolic targets for therapy. *J. Pathol.* **2015**, *236*, 517–530. [CrossRef]
286. Wang, J.; Xu, W.; Wang, B.; Lin, G.; Wei, Y.; Abudurexiti, M.; Zhu, W.; Liu, C.; Qin, X.; Dai, B.; et al. GLUT1 is an AR target contributing to tumor growth and glycolysis in castration-resistant and enzalutamide-resistant prostate cancers. *Cancer Lett.* **2020**, *485*, 45–55. [CrossRef]
287. Gonzalez-Menendez, P.; Hevia, D.; Alonso-Arias, R.; Alvarez-Artime, A.; Rodriguez-Garcia, A.; Kinet, S.; Gonzalez-Pola, I.; Taylor, N.; Mayo, J.C.; Sainz, R.M. GLUT1 protects prostate cancer cells from glucose deprivation-induced oxidative stress. *Redox Biol.* **2018**, *17*, 112–127. [CrossRef]
288. Gonzalez-Menendez, P.; Hevia, D.; Mayo, J.C.; Sainz, R.M. The dark side of glucose transporters in prostate cancer: Are they a new feature to characterize carcinomas? *Int. J. Cancer* **2018**, *142*, 2414–2424. [CrossRef]
289. Cui, Y.; Nadiminty, N.; Liu, C.; Lou, W.; Schwartz, C.T.; Gao, A.C. Upregulation of glucose metabolism by NF-kappaB2/p52 mediates enzalutamide resistance in castration-resistant prostate cancer cells. *Endocr. Relat. Cancer* **2014**, *21*, 435–442. [CrossRef]
290. Efstathiou, E.; Titus, M.; Wen, S.; Hoang, A.; Karlou, M.; Ashe, R.; Tu, S.M.; Aparicio, A.; Troncoso, P.; Mohler, J.; et al. Molecular characterization of enzalutamide-treated bone metastatic castration-resistant prostate cancer. *Eur. Urol.* **2015**, *67*, 53–60. [CrossRef]
291. Atif, F.; Yousuf, S.; Stein, D.G. Anti-tumor effects of progesterone in human glioblastoma multiforme: Role of PI3K/Akt/mTOR signaling. *J. Steroid Biochem. Mol. Biol.* **2015**, *146*, 62–73. [CrossRef]
292. Agostinis, P.; Berg, K.; Cengel, K.A.; Foster, T.H.; Girotti, A.W.; Gollnick, S.O.; Hahn, S.M.; Hamblin, M.R.; Juzeniene, A.; Kessel, D.; et al. Photodynamic therapy of cancer: An update. *CA Cancer J. Clin.* **2011**, *61*, 250–281. [CrossRef]
293. Cengel, K.A.; Simone, C.B., 2nd; Glatstein, E. PDT: What's Past Is Prologue. *Cancer Res.* **2016**, *76*, 2497–2499. [CrossRef] [PubMed]
294. Hamblin, M.R. Photodynamic Therapy for Cancer: What's Past is Prologue. *Photochem. Photobiol.* **2020**, *96*, 506–516. [CrossRef] [PubMed]
295. Trachootham, D.; Alexandre, J.; Huang, P. Targeting cancer cells by ROS-mediated mechanisms: A radical therapeutic approach? *Nat. Rev. Drug Discov.* **2009**, *8*, 579–591. [CrossRef] [PubMed]

296. Ogura, S.; Maruyama, K.; Hagiya, Y.; Sugiyama, Y.; Tsuchiya, K.; Takahashi, K.; Abe, F.; Tabata, K.; Okura, I.; Nakajima, M.; et al. The effect of 5-aminolevulinic acid on cytochrome c oxidase activity in mouse liver. *BMC Res. Notes* **2011**, *4*, 66. [CrossRef]
297. Hara, T.; Koda, A.; Nozawa, N.; Ota, U.; Kondo, H.; Nakagawa, H.; Kamiya, A.; Miyashita, K.; Itoh, H.; Nakajima, M.; et al. Combination of 5-aminolevulinic acid and ferrous ion reduces plasma glucose and hemoglobin A1c levels in Zucker diabetic fatty rats. *FEBS Open Bio* **2016**, *6*, 515–528. [CrossRef]
298. Grigalavicius, M.; Ezzatpanah, S.; Papakyriakou, A.; Raabe, T.T.H.; Yannakopoulou, K.; Theodossiou, T.A. 5-ALA Is a Potent Lactate Dehydrogenase Inhibitor but Not a Substrate: Implications for Cell Glycolysis and New Avenues in 5-ALA-Mediated Anticancer Action. *Cancers* **2022**, *14*, 4003. [CrossRef]
299. Kaur, P.; Nagar, S.; Bhagwat, M.; Uddin, M.; Zhu, Y.; Vancurova, I.; Vancura, A. Activated heme synthesis regulates glycolysis and oxidative metabolism in breast and ovarian cancer cells. *PLoS ONE* **2021**, *16*, e0260400. [CrossRef]
300. Simcox, J.A.; Mitchell, T.C.; Gao, Y.; Just, S.F.; Cooksey, R.; Cox, J.; Ajioka, R.; Jones, D.; Lee, S.H.; King, D.; et al. Dietary iron controls circadian hepatic glucose metabolism through heme synthesis. *Diabetes* **2015**, *64*, 1108–1119. [CrossRef]
301. Sato, T.; Yasuzawa, T.; Uesaka, A.; Izumi, Y.; Kamiya, A.; Tsuchiya, K.; Kobayashi, Y.; Kuwahata, M.; Kido, Y. Type 2 diabetic conditions in Otsuka Long-Evans Tokushima Fatty rats are ameliorated by 5-aminolevulinic acid. *Nutr. Res.* **2014**, *34*, 544–551. [CrossRef]
302. Handschin, C.; Lin, J.; Rhee, J.; Peyer, A.K.; Chin, S.; Wu, P.H.; Meyer, U.A.; Spiegelman, B.M. Nutritional regulation of hepatic heme biosynthesis and porphyria through PGC-1alpha. *Cell* **2005**, *122*, 505–515. [CrossRef]
303. McNicholas, K.; MacGregor, M.N.; Gleadle, J.M. In order for the light to shine so brightly, the darkness must be present-why do cancers fluoresce with 5-aminolaevulinic acid? *Br. J. Cancer* **2019**, *121*, 631–639. [CrossRef] [PubMed]
304. Golding, J.P.; Wardhaugh, T.; Patrick, L.; Turner, M.; Phillips, J.B.; Bruce, J.I.; Kimani, S.G. Targeting tumour energy metabolism potentiates the cytotoxicity of 5-aminolevulinic acid photodynamic therapy. *Br. J. Cancer* **2013**, *109*, 976–982. [CrossRef] [PubMed]
305. Nakano, A.; Tsuji, D.; Miki, H.; Cui, Q.; El Sayed, S.M.; Ikegame, A.; Oda, A.; Amou, H.; Nakamura, S.; Harada, T.; et al. Glycolysis inhibition inactivates ABC transporters to restore drug sensitivity in malignant cells. *PLoS ONE* **2011**, *6*, e27222. [CrossRef] [PubMed]
306. Hagiya, Y.; Fukuhara, H.; Matsumoto, K.; Endo, Y.; Nakajima, M.; Tanaka, T.; Okura, I.; Kurabayashi, A.; Furihata, M.; Inoue, K.; et al. Expression levels of PEPT1 and ABCG2 play key roles in 5-aminolevulinic acid (ALA)-induced tumor-specific protoporphyrin IX (PpIX) accumulation in bladder cancer. *Photodiagnosis Photodyn. Ther.* **2013**, *10*, 288–295. [CrossRef]
307. Kobuchi, H.; Moriya, K.; Ogino, T.; Fujita, H.; Inoue, K.; Shuin, T.; Yasuda, T.; Utsumi, K.; Utsumi, T. Mitochondrial localization of ABC transporter ABCG2 and its function in 5-aminolevulinic acid-mediated protoporphyrin IX accumulation. *PLoS ONE* **2012**, *7*, e50082. [CrossRef]
308. Alkarakooly, Z.; Al-Anbaky, Q.A.; Kannan, K.; Ali, N. Metabolic reprogramming by Dichloroacetic acid potentiates photodynamic therapy of human breast adenocarcinoma MCF-7 cells. *PLoS ONE* **2018**, *13*, e0206182. [CrossRef]
309. Kiesslich, T.; Plaetzer, K.; Oberdanner, C.B.; Berlanda, J.; Obermair, F.J.; Krammer, B. Differential effects of glucose deprivation on the cellular sensitivity towards photodynamic treatment-based production of reactive oxygen species and apoptosis-induction. *FEBS Lett.* **2005**, *579*, 185–190. [CrossRef]
310. Jiang, F.; Lilge, L.; Belcuig, M.; Singh, G.; Grenier, J.; Li, Y.; Chopp, M. Photodynamic therapy using Photofrin in combination with buthionine sulfoximine (BSO) to treat 9L gliosarcoma in rat brain. *Lasers Surg. Med.* **1998**, *23*, 161–166. [CrossRef]
311. Perotti, C.; Casas, A.; Del, C.B.A.M. Scavengers protection of cells against ALA-based photodynamic therapy-induced damage. *Lasers Med. Sci.* **2002**, *17*, 222–229. [CrossRef]
312. Mascaraque-Checa, M.; Gallego-Rentero, M.; Nicolas-Morala, J.; Portillo-Esnaola, M.; Cuezva, J.M.; Gonzalez, S.; Gilaberte, Y.; Juarranz, A. Metformin overcomes metabolic reprogramming-induced resistance of skin squamous cell carcinoma to photodynamic therapy. *Mol. Metab.* **2022**, *60*, 101496. [CrossRef]
313. Xie, R.; Xu, T.; Zhu, J.; Wei, X.; Zhu, W.; Li, L.; Wang, Y.; Han, Y.; Zhou, J.; Bai, Y. The Combination of Glycolytic Inhibitor 2-Deoxyglucose and Microbubbles Increases the Effect of 5-Aminolevulinic Acid-Sonodynamic Therapy in Liver Cancer Cells. *Ultrasound Med. Biol.* **2017**, *43*, 2640–2650. [CrossRef]
314. Feng, X.; Zhang, Y.; Wang, P.; Liu, Q.; Wang, X. Energy metabolism targeted drugs synergize with photodynamic therapy to potentiate breast cancer cell death. *Photochem. Photobiol. Sci.* **2014**, *13*, 1793–1803. [CrossRef]
315. Yang, X.; Palasuberniam, P.; Myers, K.A.; Wang, C.; Chen, B. Her2 oncogene transformation enhances 5-aminolevulinic acid-mediated protoporphyrin IX production and photodynamic therapy response. *Oncotarget* **2016**, *7*, 57798–57810. [CrossRef]
316. Zhong, G.; Li, Q.; Luo, Y.; Liu, Y.; Liu, D.; Li, B.; Wang, T. FECH Expression Correlates with the Prognosis and Tumor Immune Microenvironment in Clear Cell Renal Cell Carcinoma. *J. Oncol.* **2022**, *2022*, 8943643. [CrossRef]
317. Chelakkot, V.S.; Liu, K.; Yoshioka, E.; Saha, S.; Xu, D.; Licursi, M.; Dorward, A.; Hirasawa, K. MEK reduces cancer-specific PpIX accumulation through the RSK-ABCB1 and HIF-1alpha-FECH axes. *Sci. Rep.* **2020**, *10*, 22124. [CrossRef]
318. Yoshioka, E.; Chelakkot, V.S.; Licursi, M.; Rutihinda, S.G.; Som, J.; Derwish, L.; King, J.J.; Pongnopparat, T.; Mearow, K.; Larijani, M.; et al. Enhancement of Cancer-Specific Protoporphyrin IX Fluorescence by Targeting Oncogenic Ras/MEK Pathway. *Theranostics* **2018**, *8*, 2134–2146. [CrossRef]
319. Kierans, S.J.; Taylor, C.T. Regulation of glycolysis by the hypoxia-inducible factor (HIF): Implications for cellular physiology. *J. Physiol.* **2021**, *599*, 23–37. [CrossRef]

320. Franklin, D.A.; Sharick, J.T.; Ericsson-Gonzalez, P.I.; Sanchez, V.; Dean, P.T.; Opalenik, S.R.; Cairo, S.; Judde, J.G.; Lewis, M.T.; Chang, J.C.; et al. MEK activation modulates glycolysis and supports suppressive myeloid cells in TNBC. *JCI Insight* **2020**, *5*, e134290. [CrossRef]
321. Warburg, O. On respiratory impairment in cancer cells. *Science* **1956**, *124*, 269–270. [CrossRef]
322. Warburg, O. On the origin of cancer cells. *Science* **1956**, *123*, 309–314. [CrossRef]
323. Yang, M.; Soga, T.; Pollard, P.J. Oncometabolites: Linking altered metabolism with cancer. *J. Clin. Investig.* **2013**, *123*, 3652–3658. [CrossRef] [PubMed]
324. Vasan, K.; Werner, M.; Chandel, N.S. Mitochondrial Metabolism as a Target for Cancer Therapy. *Cell Metab.* **2020**, *32*, 341–352. [CrossRef] [PubMed]
325. Vander Heiden, M.G. Targeting cancer metabolism: A therapeutic window opens. *Nat. Rev. Drug Discov.* **2011**, *10*, 671–684. [CrossRef] [PubMed]
326. Groves, A.M.; Win, T.; Haim, S.B.; Ell, P.J. Non-[18F] FDG PET in clinical oncology. *Lancet Oncol.* **2007**, *8*, 822–830. [CrossRef] [PubMed]
327. Zhang, Y.; Li, Q.; Huang, Z.; Li, B.; Nice, E.C.; Huang, C.; Wei, L.; Zou, B. Targeting Glucose Metabolism Enzymes in Cancer Treatment: Current and Emerging Strategies. *Cancers* **2022**, *14*, 4568. [CrossRef]
328. Zdralevic, M.; Brand, A.; Di Ianni, L.; Dettmer, K.; Reinders, J.; Singer, K.; Peter, K.; Schnell, A.; Bruss, C.; Decking, S.M.; et al. Double genetic disruption of lactate dehydrogenases A and B is required to ablate the "Warburg effect" restricting tumor growth to oxidative metabolism. *J. Biol. Chem.* **2018**, *293*, 15947–15961. [CrossRef]
329. Boudreau, A.; Purkey, H.E.; Hitz, A.; Robarge, K.; Peterson, D.; Labadie, S.; Kwong, M.; Hong, R.; Gao, M.; Del Nagro, C.; et al. Metabolic plasticity underpins innate and acquired resistance to LDHA inhibition. *Nat. Chem. Biol.* **2016**, *12*, 779–786. [CrossRef]
330. Stine, Z.E.; Schug, Z.T.; Salvino, J.M.; Dang, C.V. Targeting cancer metabolism in the era of precision oncology. *Nat. Rev. Drug Discov.* **2022**, *21*, 141–162. [CrossRef]
331. Cassim, S.; Vucetic, M.; Zdralevic, M.; Pouyssegur, J. Warburg and Beyond: The Power of Mitochondrial Metabolism to Collaborate or Replace Fermentative Glycolysis in Cancer. *Cancers* **2020**, *12*, 1119. [CrossRef]
332. Noble, R.A.; Thomas, H.; Zhao, Y.; Herendi, L.; Howarth, R.; Dragoni, I.; Keun, H.C.; Vellano, C.P.; Marszalek, J.R.; Wedge, S.R. Simultaneous targeting of glycolysis and oxidative phosphorylation as a therapeutic strategy to treat diffuse large B-cell lymphoma. *Br. J. Cancer* **2022**, *127*, 937–947. [CrossRef]
333. Hay, N. Reprogramming glucose metabolism in cancer: Can it be exploited for cancer therapy? *Nat. Rev. Cancer* **2016**, *16*, 635–649. [CrossRef]
334. Yang, L.; Yan, X.; Chen, J.; Zhan, Q.; Hua, Y.; Xu, S.; Li, Z.; Wang, Z.; Dong, Y.; Zuo, D.; et al. Hexokinase 2 discerns a novel circulating tumor cell population associated with poor prognosis in lung cancer patients. *Proc. Natl. Acad. Sci. USA* **2021**, *118*, e2012228118. [CrossRef]

International Journal of *Molecular Sciences*

MDPI

*Review*

# Taste and Smell Disorders in Cancer Treatment: Results from an Integrative Rapid Systematic Review

**Tania Buttiron Webber [1], Irene Maria Briata [1], Andrea DeCensi [1,2], Isabella Cevasco [3] and Laura Paleari [4,*]**

1 Division of Medical Oncology, E.O. Galliera Hospital, 16128 Genoa, Italy
2 Wolfson Institute of Preventive Medicine, Queen Mary University of London, London E1 4NS, UK
3 Department of Health Professions, E.O. Galliera Hospital, 16128 Genoa, Italy
4 Research, Innovation and HTA Unit, (A.Li.Sa.) Liguria Health Authority, 16121 Genoa, Italy
* Correspondence: laura.paleari@alisa.liguria.it; Tel.: +39-010-5484-243

**Abstract:** Taste and smell disorders (TSDs) are common side effects in patients undergoing cancer treatments. Knowing which treatments specifically cause them is crucial to improve patients' quality of life. This review looked at the oncological treatments that cause taste and smell alterations and their time of onset. We performed an integrative rapid review. The PubMed, PROSPERO, and Web of Science databases were searched in November 2022. The article screening and study selection were conducted independently by two reviewers. Data were analyzed narratively. Fourteen studies met the inclusion criteria and were included. A high heterogeneity was detected. Taste disorders ranged between 17 and 86%, while dysosmia ranged between 8 and 45%. Docetaxel, paclitaxel, nab-paclitaxel, capecitabine, cyclophosphamide, epirubicin, anthracyclines, and oral 5-FU analogues were found to be the drugs most frequently associated with TSDs. This review identifies the cancer treatments that mainly lead to taste and smell changes and provides evidence for wider studies, including those focusing on prevention. Further studies are warranted to make conclusive indication possible.

**Keywords:** dysgeusia; dysosmia; taste; smell; cancer treatment; rapid review

**Citation:** Buttiron Webber, T.; Briata, I.M.; DeCensi, A.; Cevasco, I.; Paleari, L. Taste and Smell Disorders in Cancer Treatment: Results from an Integrative Rapid Systematic Review. *Int. J. Mol. Sci.* **2023**, *24*, 2538. https://doi.org/10.3390/ijms24032538

Academic Editor: Takeo Nakanishi

Received: 23 December 2022
Revised: 18 January 2023
Accepted: 20 January 2023
Published: 28 January 2023



## 1. Introduction

Global cancer statistics estimate that around 19.3 million new cancer diagnoses occurred in 2020. Lung cancer remained the leading cause of death, with an estimated 1.8 million deaths (18%), followed by colorectal (9.4%), liver (8.3%), stomach (7.7%), and female breast (6.9%) cancers [1].

Taste and smell disorders (TSD) are common side effects in cancer patients undergoing chemotherapy (CT) treatments but often are described as single entities and patients may have difficulty in identifying them [2]. The reported prevalence of taste disturbance ranged from 20% to 86% [3], and its development occurs approximately 2–3 weeks after the start of cancer treatment and persists throughout the duration of the therapy [4]. The prevalence of patients with dysosmia is in the range of 5–60% [5].

Interestingly, in the literature, it is reported that only a few patients report taste and smell alterations spontaneously and these symptoms are often underestimated by oncologists and nurses [6]. A study by Gill et al. reported a discrepancy in the importance given to retaining a normal sense of taste and smell, as reported by patients and by the multidisciplinary team involved in their care ($p < 0.013$) [7]. We previously hypothesized that the hesitancy of physicians in approaching these disorders may be due to a "cultural aspect" where the physician tends to underestimate and leave untreated the adverse events (AEs) related to therapies that do not have a clinical implication [8]. However, it is important to consider these disorders as they can lead to reduced food enjoyment and, most importantly, an inappropriate nutrient intake, with a high impact on the nutritional status, quality of life, and possibly on the efficacy of therapy itself [9].

Literature describes five basic tastes: sweet, sour, bitter, salty, and umami [10,11]. The sense of taste starts with the activation of the taste receptors located on the microvilli or taste receptor cells. These cells are clustered and together they form taste buds. Taste receptor cells are modified epithelial cells that can detect and process gustatory, olfactory, and trigeminal stimulation [12]. Dysgeusia can be classified as follows: (1) ageusia, which is a complete lack of taste; (2) hypogeusia, which is a decreased taste sensitivity; (3) hypergeusia, which is a heightened taste sensitivity; and (4) phantageusia, which is the perception of an unpleasant taste in the absence of a corresponding stimulus in the environment [9]. The sense of smell is even more complex than the sense of taste. Over a trillion different smells can be identified. There are two ways for odors to reach the olfactory epithelium: via the ortho-nasal passage or via the retro-nasal passage [13]. In general, four categories of smell disorders are classified depending on how they impact odor perception. (1) Anosmia is the absence of smell perception; (2) hyposmia is a quantitatively reduced ability to perceive scent; (3) parosmia is a qualitative distortion of an ordinarily detected smell; and (4) phantosmia is the perception of odors when none are present [14].

In addition to cancer treatment, including radiotherapy and surgical treatments, other factors may contribute to taste and smell disorders, such as age, oral infections, smoking, alcohol abuse, chronic diseases such as diabetes, hypertension, and chronic rhinosinusitis, and the type of cancer [15–17]. In fact, the study by Dhuibhir and colleagues showed a high prevalence of taste and smell disorders in newly diagnosed cancer patients before treatment [18].

Currently, TSD can be assessed through clinical methods (objective) or self-reported by patients (subjective). The objective methods assess the oral sensitivity to taste agents through the thresholds of the five taste qualities. The numerical results facilitate the comparison of the taste perception abilities between populations [19] but do not reflect the 'real-world' taste experience [20] as they do not capture dimensions of taste that are important to patients, such as flavor, food enjoyment, or hedonic changes [21]. For this reason, patient-reported questionnaires and qualitative research methods that capture patients' individual experience are recommended [6,22].

A review by Enriquez-Fernandez et al. [23] reports a growing interest in the assessment of taste and smell changes in cancer patients but presents limitations in terms of the heterogeneity in the number of items, assessment range, and in the domains of taste changes. They suggest developing a standardized tool validated by patients to ensure that the terms associated with sensory changes are understood and reliably used by clinicians and researchers.

The aforementioned papers have mainly highlighted the pathophysiology, prevalence, clinical features, and assessment tools of chemosensory alterations. However, there are limited literature reviews highlighting the oncological therapies that lead to these alterations. The aim of this rapid review was to examine the existing and current literature on cancer treatments that can cause TSDs to develop prevention and education strategies in the future.

## 2. Materials and Methods

### 2.1. Study Design

Between October and November 2022, an integrative rapid review was conducted as a knowledge synthesis strategy to provide timely information [24]. Despite being thoroughly studied, the field of TSD is evolving due to novel cancer treatments. As a result, timely reviews can describe current research and report on clinical and organizational levels [25].

### 2.2. Needs Assessment and Topic Selection

The primary need was to map the most recent data on cancer drugs that cause TSD, to summarize the knowledge and enable nurses and oncologists to continuously improve the quality of care and patient management. Thus, the review question was: which oncological drugs cause taste and smell disorders? To this review, we decided to include only studies focused on cancer treatment for solid tumors in adults.

### 2.3. Study Development

According to the methodological process inspired firstly by Tricco and colleagues in 2017 [25], which was then further developed by Langlois et al. in 2019 [26], the following seven-stage process was implemented: (1) a needs assessment and topic selection; (2) study development; (3) literature search; (4) screening and study selection; (5) data extraction; and (6) risk-of-bias assessment. In addition, the Preferred Reporting Items for Systematic reviews and Meta-Analysis (PRISMA) guidelines were applied [27].

### 2.4. Literature Search

The literature search was performed independently by 2 reviewers (TBW and IMB) through the MEDLINE (via PubMed), PROSPERO, and Web of Science databases between October and November 2022. The inclusion criteria were: (a) studies on adult patients with solid tumors undergoing treatment with oncological drugs, (b) studies designed to detect the incidence and prevalence of TSD and/or assess the time of onset; (c) quantitative and qualitative primary studies; and (d) studies published in English within the past 10 years. The exclusion criteria were: (a) studies on patients with hematologic malignancies and (b) studies on patients undergoing radiation therapy. Therefore, the Medical Subject Headings [MeSH] and free-text words used were: "dysgeusia", "taste alteration", "anosmia", "olfaction disorders", "smell alteration", "therapeutics", "therapies", "treatments", "cancer", and "neoplasm". The research was limited to the last 10 years.

### 2.5. Screening and Study Selection

The screening of titles and abstracts was performed by three researchers (TBW, IMB, and LP) to identify the articles' eligibility in relation to the inclusion criteria. Then, an independent full-text evaluation was performed by the same researchers to determine if the studies fully met the inclusion criteria. When disagreements occurred, the final decision to include or exclude an article was made by a consensus. As reported in the PRISMA flow diagram (Figure 1), the flow of a study's inclusion is summarized together with the reasons for exclusion. At the end of the process, 14 papers were retrieved.

### 2.6. Data Extraction

The data were extracted and reported in a Microsoft Excel® spreadsheet. The following data were collected from each selected study and reported in the grid: the (a) author(s), year, country; (b) study design; (c) aims; (d) participants; (e) assessment tool used for TSD detection; (f) cancer treatment; and (g) key findings. The full grid is available as Table 1.

### 2.7. Risk of Bias Assessment

The review team shared in advance the decisions about inclusion and exclusion to prevent information and a selection bias. In addition, to ensure the consistency of the results, the following methodological requirements [28] were respected: (a) the verification of the study selection and data extraction were performed by three reviewers [TBW, IMB, LP]) and (b) an additional independent researcher (=a fourth reviewer [ADC]) contributed and reviewed the narrative synthesis and the summary table.

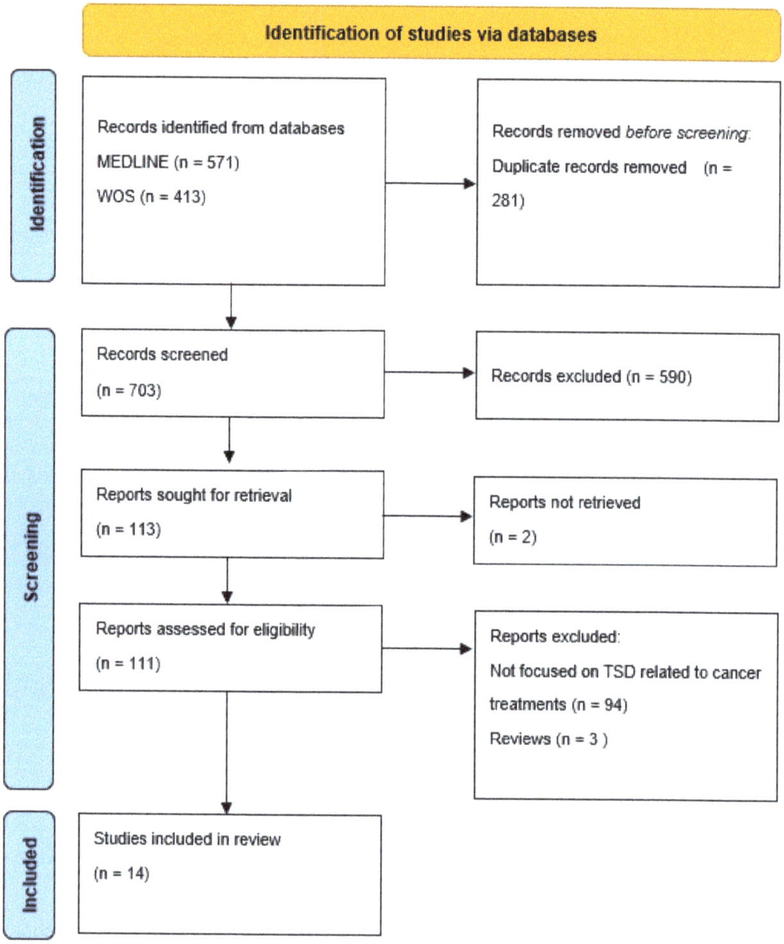


**Figure 1.** Flowchart of the data collection and selection process in accordance with PRISMA-ScR guidelines.

**Table 1.** Characteristics of studies that evaluated cancer therapy-related dysgeusia.

| Ref. n. | Study Design | Aim (s) | Participants (N) | Assessment Tool (s) | Cancer Treatments | Key Findings |
|---|---|---|---|---|---|---|
| [9] | Prospective | Prevalence | 75 Prostate cancer | Survey regarding the taste and smell of food, appetite, and nausea. | CT and/or HT (regimen not specified) | • TAs: 17%; SAs: 8%. • TAs most frequent in patients treated with denosumab (35.0% vs. 10.9%, OR = 4.40, $p = 0.020$) or docetaxel (41.7% vs. 12.7%, OR = 4.91, $p = 0.022$). • Poor taste of food associated with poor appetite and ≥10% weight loss. |

**Table 1.** *Cont.*

| Ref. n. | Study Design | Aim (s) | Participants (N) | Assessment Tool (s) | Cancer Treatments | Key Findings |
|---|---|---|---|---|---|---|
| [29] | Prospective | Self-reported TSDs based on the type of CT treatment. Impact of CT on the severity of the TSDs. | 151 | Questionnaire structured in three sections: eating habits; sensory changes (taste/smell changes and thermal sensitivity); and other clinical disorders (nausea, vomiting, dry mouth, mucositis, and dysphagia). | CT (regimen, Paclitaxel, oxaliplatin, docetaxel, carboplatin, anthracyclines, cisplatin, irinotecan, 5-FU, vinorelbine) | • TAs: 76%.<br>• SAs: 45%.<br>• TAs in patients treated with anthracyclines, paclitaxel, carboplatin, and docet-axel.<br>• Cisplatin and 5-FU are the CT resulting in the lowest complaints.<br>• Xerostomia is strongly associated with bad taste in mouth (OR = 5.96; CI = 2.37–14.94; $p$ value = 0.000) and taste loss (OR = 5.96; CI = 2.37–14.94; $p$ value = 0.000). |
| [30] | Cross-sectional | Prevalence, severity and self-reported characteristics of TAs induced by CT. TAs across CT regimes. | 243 | Validated TA Scale Self-reported TAs duration CiTAS. | CT (regimen: FOLFOX, paclitaxel, docetaxel, cisplatin, pemetrexed, FEC, EC, FOLFIRI, Gemcitabina, TJ, TPF, Gemcarbo, Cisgem, Gemox) | • TAs ranged from 51–86%.<br>• 43% of participants complained of TAs with the start of Ct and 75% reported TAs within the fourth week of treatment.<br>• TAs in patients treated with gemcitabine, cisplatin plus pemetrexed and epirubi-cin plus cyclophospha-mide (EC).<br>• Low levels of TAs were found among participants receiving GEMCARBO and CISGEM.<br>• 55% of participants reported some difficulties in tasting food. Tasting saltiness was the most affected ability. |
| [31] | Prospective | Incidence of TAs | 41 BC | Not validated Questionnaire, filter paper disk method, CTCAE v. 4.0. | CT (regimen: Epirubicin, cyclophos-phamide) | • TA on the 4th day after CT was 53%.<br>• TAs decreased to about 9% immediately before new cycles.<br>• Age and body surface area influenced Tas. |

**Table 1.** *Cont.*

| Ref. n. | Study Design | Aim (s) | Participants (N) | Assessment Tool (s) | Cancer Treatments | Key Findings |
|---|---|---|---|---|---|---|
| [32] | Prospective | Prevalence TAs across CT regimes. | 109 BC, gyneco-logical | Validated TAs scale. | CT (regimen, Gemcitabina, epirubicin, doc-etaxel, capecitabine, epirubicin/docetaxel) | • TAs was 76.1%.<br>• The highest TAs with epiru-bicin, docetaxel, capecita-bine.<br>• Lowest TAs with gem-citabine. |
| [33] | Prospective | To provide new data about TSDs. | 33 head/neck | Sniffin' Sticks test (Determination of threshold, discrimination, and identification, TDI). | CT (regimen: Cisplatin, carboplatin, 5-FU, docetaxel) | • In normosmic or hyposmic, the mean decrease in TDI-score was significant lower during the second cycle.<br>• Age (>55 years) and smokers had a significant (negative) impact. |
| [34] | Prospective | Prevalence of dysgeusia. | 31 males 15 females (9 did not undergo CT) | Salt-impregnated taste strips with 6 concentrations of Sodium chloride. | CT (regimen: 5-FU, platinum, Tx) | • TAs in 38.8%.<br>• 48% in 5-FU or its oral analogues.<br>• 55.6% of patients receiving oral 5-FU analogues.<br>• Patients aged ≥70 years also tended to experience dysgeusia (75%). |
| [35] | Prospective | Effect of cisplatin CT on odor perception. | 15 bronchial cancer patients and 15 control subjects | European Test of Olfactory Capabilities (ETOC). | CT (regimen: cisplatin) | • Odor detection and odor identification abilities were not influenced by the administration of cisplatin, a decrease in pleasantness was observed only for food odors, and not for non-food odors. |
| [36] | Cross-sectional | TAs charac-teristics | 100 | Taste recognition thresholds (TRTs) via a taste disc kit PRO-CTCAE CiTAS. | CT (regimen: Tx based) | • TAs in 59%.<br>• DTX associated with a higher prevalence of more severe and longer TAs than PTX or nab-PTX regimens.<br>• Significantly elevated taste recognition thresholds (hypogeusia) for sweet, sour, and bitter tastes in the taste alteration group receiving nab-paclitaxel ($p$ = 0.022, 0.020, and 0.039, respectively).<br>• Docetaxel, previous CT, dry mouth, and peripheral neuropathy were significantly associated with Tas. |

**Table 1.** *Cont.*

| Ref. n. | Study Design | Aim (s) | Participants (N) | Assessment Tool (s) | Cancer Treatments | Key Findings |
| --- | --- | --- | --- | --- | --- | --- |
| [37] | Observational | Prevalence and clinical therapeutic risk factors. | 7425 | CTCAE v5.0 | CT (not specified) | • TAs in 19.0%; 15.0% grade 1 dysgeusia and 6.0% grade 2.<br>• CT duration ($p < 0.001$), female sex ($p < 0.001$), location of the primary tumor in the uterus ($p = 0.008$), head and neck ($p = 0.012$), and testicles ($p = 0.011$), and use of ifosfamide ($p = 0.009$), docetaxel ($p = 0.001$), paclitaxel ($p < 0.001$), pertuzumab ($p = 0.005$), bevacizumab ($p < 0.001$), and dacarbazine ($p = 0.002$) independently increased the risk of dysgeusia. |
| [38] | Mixed methods | To investigate whether mycotoxic and/or neurotoxic drugs compromise olfactory performance. | 44 | Sniffin' Sticks test (Determination of threshold, discrimination, and identification, TDI). | CT (regimen: Oxaliplatin, 5-FU, capecitabine, gemcitabine, carboplatin, cisplatin, doxorubicin, liposomal doxorubicin, taxanes) | • TDI scores were significantly lower after chemotherapy in all age groups.<br>• Patients older than 50 years were more susceptible to olfactory toxicity. |
| [39] | Case control | Changes in the perception of tastes. | 43 | Taste strips | CT (regimen: platinum based) | • Salty and sour were the most affected tastes in the study group ($p = 0.001$ and 0.05). |
| [40] | Observational | Changes in the detection (DT) and recognition (RT) thresholds of umami, sweet, and bitter tastes. | 40 (NSCLC) | Rinsing technique. Not validated Questionnaire | CT (regimen: Cisplatin, paclitaxel) | • TAs 34% after treatment.<br>• 42% reported a bitter taste in the mouth.<br>• 57% reported dry mouth.<br>• 35% reported food being tasteless, and 12% reported food having an unpleasant taste. |

**Table 1.** *Cont.*

| Ref. n. | Study Design | Aim (s) | Participants (N) | Assessment Tool (s) | Cancer Treatments | Key Findings |
|---|---|---|---|---|---|---|
| [41] | Qualitative | Patient and carer de-scriptions, experiences and conse-quences of taste and flavor changes. | 10 patients 4 carers | Semi-structured interviews | Ct (regimen: Oxaliplatin) | • TAs were apparent by the third CT cycle. • Worse symptoms in the 5–7 days immediately post CT infusion, relief toward the end of a CT cycle. • Full resolution of symp-toms by 6–8 weeks fol-lowing the completion of oxaliplatin treatment. • Most common oral sensations: 'metallic' or a 'slick' or 'coating' in the mouth. |

Abbreviations. CT: chemotherapy; HT: hormone therapy; 5-FU: 5-fluorouracil; TAs: taste alterations; FOLFOX: folinic acid, fluorouracil, oxaliplatin; FEC: epirubicin, fluorouracil, cyclophosphamide; EC: epirubicin, cyclophosphamide; FOLFIRI: folinic acid, fluorouracil, irinotecan; TA: taste alteration; TJ: carboplatin, paclitaxel; TPF: docetaxel, cisplatin, fluorouracil; GEMCARBO: gemcitabine, carboplatin; GEMOX: gemcitabine, oxaliplatin; CISGEM: cisplatin, gemcitabine; BC: breast cancer; CTCAE: Common Terminology Criteria for Adverse Events; Tx: taxane; CiTAS: Chemotherapy induced Taste Alteration Scale; PRO-CTCAE: patient-reported outcome; DTX: docetaxel; PTX: paclitaxel; nab-PTX: nab-paclitaxel; NSCLC: non-small cell lung cancer.

## 3. Results

### *3.1. Study Characteristics*

A preliminary search aimed at exploring knowledge on cancer treatments causing TSDs was conducted by examining the MEDLINE database (via PubMed), Web of Science, and PROSPERO between October and November 2022. There were not recently published or ongoing reviews on this specific topic; meanwhile, a large number of articles on TSDs in cancer patients related to eating habits and quality of life have been published in recent years. The database searches, after the duplicates were removed, returned 703 articles, of which 113 were screened. Out of the 113 articles assessed for their eligibility, 14 studies met the inclusion criteria (Figure 1). The specific characteristics of each selected study that met the inclusion criteria are presented in Table 1. Most of the selected studies used a quantitative method [9,29–40], and only one was a qualitative study [41]. Nine studies focused on changes in the taste [30–32,34,36,37,39–41], 3 focused on smell [33,35,38], and 2 focused on a combination of both [9,29]. The study population among the studies was very different in terms of the cancer diagnosis, stage, treatment, line of therapy, and sample size. Moreover, a high degree of heterogeneity in the tools to assess TSD was observed, even within the study itself. In fact, some studies used a subjective evaluation [9,29,30] while others used validated questionnaires (e.g., CITAS) [30,36] or standardized measurement scales such as CTCAE [31,36,37]. Two studies used the Sniffin' Sticks test [33,38] while one study used taste recognition thresholds (TRTs) [36].

### *3.2. Prevalence, Onset, Resolution of Taste and Smell Disorders*

Taste disorders were found in 17% to 86% of people and were linked to a poor appetite and a 10% weight loss [29,30]. Campagna et al. reported that 43% of participants complained of TSDs at the start of CT, and 75% reported it by the fourth week of treatment [30]. The worsening of symptoms occurs within 5–7 days immediately following the CT infusion and decreases by about 9% immediately before the next cycle [31]. The complete resolution of symptoms (e.g., from oxaliplatin) occurs within 6–8 weeks after the completion of treatment [31,41]. Xerostomia is strongly associated with a bad taste in the mouth (OR = 5.96; CI = 2.37–14.94; $p$-value = 0.000) and a loss of taste (OR = 5.96; CI = 2.37–14.94; $p$-value = 0.000) [29]. Salty and sour were the most affected tastes ($p = 0.001$ and 0.05, respectively) [40]. Body surface, smokers, and people over the age of 70 had a significant negative impact on taste and smell [31,33,34]. The duration of CT ($p = 0.001$), female gender ($p = 0.001$), and location of the primary tumor in the uterus ($p = 0.008$), head and neck ($p = 0.012$), and testicles ($p = 0.011$) independently increased the risk of dysgeusia [37]. The prevalence of olfactory disorders ranged from 8% to 45%; in normosmics or hyposmics, the mean decrease in the threshold determination, discrimination, and identification (TDI score) was significant during the second cycle of cancer treatment and smoking and being over 50 years old were risk factors for smell alterations [9,29,33].

### 3.3. Cancer Treatment

Docetaxel is the main drug related to the occurrence of TSDs and is associated with a higher prevalence of more severe and longer taste alterations than paclitaxel or nab paclitaxel [36]. Anthracyclines, carboplatin, epirubicin, cyclophosphamide, capecitabine, and cisplatin/pemetrexed are also frequently related to TSDs [9,30]. Low levels of taste alterations were found in gemcitabine/carboplatin (GEMCARBO) and cisplatin/gemcitabine (CISGEM) combinations [30], as well as in cisplatin and gemcitabine administered individually [32]. Additionally, 5-fluorouracil (5-FU) or its oral analogues showed a high prevalence of TAs [34]. Docetaxel, previous CT, dry mouth, and peripheral neuropathy were significantly associated with taste alterations [36]. Regarding the effect of cisplatin on odor, detection, and identification abilities were unaffected by an administration of cisplatin; a decrease in pleasantness was observed only for food odors and not for non-food odors [35].

## 4. Discussion

In addition to oncology drugs, the literature reports hundreds of drugs from all major therapeutic classes that have been clinically reported to cause unpleasant and altered taste sensations when administered alone or in combination with other drugs. These unpleasant sensations include metallic and bitter tastes, a partial or complete loss of taste, and distortions and perversions of taste [42].

As suggested in the review by Schiffman et al., there are a number of topics which are useful for understanding the biological basis of drug-induced taste disorders: (1) the interaction of drugs with taste receptors on the apical side of the tongue in the oral cavity; (2) genetic differences among patients that affect the taste perception of drugs; (3) taste sensations caused by injectable drugs; (4) drug interactions caused by the use of multiple drugs; and (5) potential biochemical causes.

It is also important to recognize which groups are most vulnerable to this alteration. These include (1) the elderly, who use a disproportionate number of drugs [43], (2) people with certain genetic polymorphisms related to the perception of a bitter taste [44], (3) people with a reduced drug clearance [45], and (4) people with a drug metabolism [46].

TSDs in cancer patients are an often underestimated and underreported problem that may result from the disease and/or its treatment; this is probably because physicians and nurses do not regularly use standardized taste tests to verify and validate drug-related taste disorders in patients. Additionally, many disorders of taste cannot be categorized according to conventional tastes such as sweet (sucrose), sour (citric acid), salty (NaCl), bitter (quinine), or umami (monosodium glutamate). Dysgeusia and dysosmia alter the pleasure of eating and reduce appetite, which, especially in compromised patients, can lead to malnutrition, increased treatment-related toxicities, and a worsened quality of life. Therefore, the identification of risk factors, such as the use of a specific oncological treatment, that may promote the development of TSDs, is an important aspect to reduce the impact of this condition on these frail patients. Our research identified 14 articles published in the last 10 years that investigated cancer treatments leading to TSDs. In accordance with the existing literature, the range of taste alterations varies between 17% and 86% [29,30] and the severity of the symptoms varies during the cycle [30,31,35]. In fact, the symptoms severity tends to worsen 5–7 days after the CT cycle and then diminish by about 10% before the following cycle [35]. Moreover, the taste alterations tend to persist for a long time [31,41], suggesting that the risk of malnutrition and a worsened quality of life may continue even after the end of the cancer treatment. Salty and sour tastes seem to be the tastes which are most affected by cancer treatments [40], so it might be useful to provide patients with specific nutritional guidance aimed to minimize the alterations. Dysosmia is less investigated but still its prevalence ranges from 8 to 45% of cancer patients [9,29]. According to the results of the current study, dysgeusia and dysosmia were more strongly associated with breast, gynecological, and colorectal cancer [32]. Docetaxel, paclitaxel, nab-paclitaxel, capecitabine, cyclophosphamide, epirubicin, anthracyclines, and oral 5-FU analogues were found to be the drugs most frequently associated with TSDs [9,30,34,36]. Other important risk factors for TSDs included the number of chemotherapy cycles, the female sex, the presence of distant metastases, and the primary tumor's location in the uterus, testicles, or head and neck [37]. An interesting correlation emerged between dysgeusia and peripheral neuropathy; numbness or tingling in the hands or feet (OR, 2.04; 95% CI, 1.25–3.57; $p = 0.004$) were significantly associated with TAs [36]. Knowing the factors most associated with TSDs is crucial for physicians and nurses to carefully monitor their occurrence and severity and to implement adequate prevention strategies. Sevryugin et al., in a recent review [3], summarized a wide range of therapy alternatives, including zinc and polaprezinc, radioprotectors, vitamins and supplements, anti-xerostomia agents, active swallowing exercises, nutritional interventions, delta-9-tetrahydrocannabinol, and photobiomodulation that can be used as a strategy to reduce TSDs.

The high heterogeneity among the selected studies in terms of the diagnosis, stage of disease, treatment, the instruments used to assess the TSDs, and the sample size makes it difficult to make firm conclusions. The limited number of studies exploring specifically which cancer therapies cause alterations in taste and smell leads us to hypothesize that these disorders have not yet been given due attention. In addition, none of the studies included in the present review considered new therapies, such as immunotherapy, suggesting that further studies are needed to investigate the impact of cancer therapies more comprehensively on TSDs. Most published studies relate taste and smell alterations to quality of life, so interventions in a preventive context would be necessary; although, there is no consensus on the prevention strategies to be used in this setting, so an algorithm for selecting the best treatment for TSDs was developed [3]. The algorithm can help the clinician to provide a therapeutic solution for chemosensory disorders or it can help the researcher to design an appropriate clinical trial to increase the knowledge on this underestimated problem.

### *4.1. Study Limitations*

This rapid review aimed to highlight current cancer drugs that can cause changes in taste and smell; however, we know that there are no studies that take into consideration other therapies such as hormone therapy, target therapies, immunotherapies, and monoclonal therapy. Furthermore, the wide heterogeneity of the evaluation tools used, and the different moments of detection do not allow for an accurate generalizability of the results. Our results are not to be considered conclusive, as another limitation is that we explored a limited number of databases.

### *4.2. Implications for Clinical Practice and Research*

Oncologists and nurses should be trained on treatments that induce taste and smell disorders to educate patients about proper nutrition and reduce the impact of these symptoms on their quality of life.

## 5. Conclusions

Taste and smell disorders are not life-threatening events for patients but have a significant impact on their quality of life. Oncologists, nurses, and nutritionists play an important role in the management of these chemotherapy-related symptoms, so further studies are needed to provide specific information to patients on which oncology drugs cause dysgeusia or anosmia, the time of their onset and duration, and to support clinical governance strategies as well.

**Author Contributions:** T.B.W., I.M.B. and L.P. designed the literature search. T.B.W., I.M.B. and L.P. screened and reviewed articles, extracted and analyzed data. T.B.W. and I.M.B. wrote the first draft of the manuscript. L.P., I.C. and A.D. critically reviewed and revised the manuscript and approved the final draft. All authors have read and agreed to the published version of the manuscript.

**Funding:** This research received no external funding.

**Institutional Review Board Statement:** Not applicable.

**Informed Consent Statement:** Not applicable.

**Acknowledgments:** To my beloved son Rami.

**Conflicts of Interest:** The authors declare no conflict of interest.

## References

1. Sung, H.; Ferlay, J.; Siegel, R.L.; Laversanne, M.; Soerjomataram, I.; Jemal, A.; Bray, F. Global cancer statistics 2020: GLOBOCAN estimates of incidence and mortality worldwide for 36 cancers in 185 countries. *CA Cancer J. Clin.* **2021**, *71*, 209–249. [CrossRef] [PubMed]
2. Wrobel, B.B.; Leopold, D.A. Clinical assessment of patients with smell and taste disorders. *Otolaryngol. Clin. N. Am.* **2004**, *37*, 1127–1142. [CrossRef] [PubMed]
3. Sevryugin, O.; Kasvis, P.; Vigano, M.; Vigano, A. Taste and smell disturbances in cancer patients: A scoping review of available treatments. *Support Care Cancer.* **2021**, *29*, 49–66. [CrossRef] [PubMed]
4. Takei, M.; Okada, N.; Nakamura, S.; Kagawa, K.; Fujii, S.; Miki, H.; Ishizawa, K.; Abe, M.; Sato, Y. A Genome-Wide Association Study Predicts the Onset of Dysgeusia Due to Anti-cancer Drug Treatment. *Biol. Pharm. Bull.* **2022**, *45*, 114–117. [CrossRef] [PubMed]
5. Gamper 0745, E.M.; Zabernigg, A.; Wintner, L.M.; Giesinger, J.M.; Oberguggenberger, A.; Kemmler, G.; Sperner-Unterweger, B.; Holzner, B. Coming to your senses: Detecting taste and smell alterations in chemotherapy patients. A systematic review. *J. Pain Symptom. Manage.* **2012**, *44*, 880–895. [CrossRef] [PubMed]
6. Bernhardson, B.M.; Tishelman, C.; Rutqvist, L.E. Taste and smell changes in patients receiving cancer chemotherapy: Distress, impact on daily life, and self-care strategies. *Cancer Nurs.* **2009**, *32*, 45–54. [CrossRef]
7. Gill, S.S.; Frew, J.; Fry, A.; Adam, J.; Paleri, V.; Dobrowsky, W.; Chatterjee, S.; Kelly, C.G. Priorities for the head and neck cancer patient, their companion and members of the multidisciplinary team and decision regret. *Clin. Oncol. (R Coll. Radiol).* **2011**, *23*, 518–524. [CrossRef]
8. Buttiron Webber, T.; Marra, D.; Puntoni, M.; Giuliano, S.; Briata, I.M.; Cevasco, I.; Clavarezza, M.; D'Amico, M.; Defferrari, C.; Gozza, A.; et al. Patient-versus physician-reported outcomes in a low-dose tamoxifen trial in noninvasive breast cancer. *Breast. J.* **2021**, *27*, 817–823. [CrossRef] [PubMed]
9. Amézaga 4296, J.; Alfaro, B.; Ríos, Y.; Larraioz, A.; Ugartemendia, G.; Urruticoechea, A.; Tueros, I. Assessing taste and smell alterations in cancer patients undergoing chemotherapy according to treatment. *Support Care Cancer* **2018**, *26*, 4077–4086. [CrossRef]
10. Hartley, I.E.; Liem, D.G.; Keast, R. Umami as an 'alimentary' taste. A new perspective on taste classification. *Nutrients* **2019**, *11*, 182. [CrossRef]
11. Witt, M. Anatomy and development of the human taste system. *Handb. Clin. Neurol.* **2019**, *164*, 147–171.
12. Corremans, M.; Mortelmans, D.; Geurden, B.; Luyten, S.; Bekkering, G. Prevalence and incidence of chemotherapy-induced taste alterations in adult cancer patients: A systematic review protocol. *JBI Evid. Synth.* **2022**, *20*, 1338. [CrossRef] [PubMed]
13. van Oort, S.; Kramer, E.; de Groot, J.-W.; Visser, O. Taste alterations and cancer treatment. *Curr. Opin. Support Palliat* **2018**, *12*, 162–167. [CrossRef] [PubMed]
14. Mainland, J.D.; Barlow, L.A.; Munger, S.D.; Millar, S.E.; Vergara, M.N.; Jiang, P.; Schwob, J.E.; Goldstein, B.J.; Boye, S.E.; Martens, J.R.; et al. Identifying Treatments for Taste and Smell Disorders: Gaps and Opportunities. *Chem. Senses.* **2020**, *45*, 493–502. [CrossRef] [PubMed]
15. Leopold, D.; Holbrook, E.H.; Noell, C.A.; Mabry, R. Disorders of Taste and Smell Medscape; 2021 [Updated Aug, 2022]. Available online: https://emedicine.medscape.com/article/861242-overview (accessed on 22 November 2022).
16. Liu, G.; Zong, G.; Doty, R.L.; Sun, Q. Prevalence and risk factors of taste and smell impairment in a nationwide representative sample of the US population: A cross-sectional study. *BMJ Open* **2016**, *6*, 1–10. [CrossRef] [PubMed]
17. Fokkens, W.J.; Lund, V.J.; Hopkins, C.; Hellings, P.W.; Kern, R.; Reitsma, S.; Toppila-Salmi, S.; Bernal-Sprekelsen, M.; Mullol, J. Executive summary of EPOS 2020 including integrated care pathways. *Rhinology* **2020**, *58*, 82–111. [CrossRef]

18. Uí Dhuibhir, P.; Barrett, M.; O'Donoghue, N.; Gillham, C.; El Beltagi, N.; Walsh, D. Self-reported and objective taste and smell evaluation in treatment-naive solid tumour patients. *Support Care Cancer* **2020**, *28*, 2389. [CrossRef]
19. Snyder 2396, D.J.; Prescott, J.; Bartoshuk, L.M. Modern psychophysics and the assessment of human oral sensation. *Adv. Oto-Rhino-Laryngol.* **2006**, *63*, 221–241. [CrossRef]
20. Boltong, A.; Keast, R. The influence of chemotherapy on taste perception and food hedonics: A systematic review. *Canc. Treat. Rev.* **2012**, *38*, 152–163. [CrossRef]
21. McGowan, D. Chemotherapy-induced oral dysfunction: A literature review. *Br. J. Nurs.* **2008**, *17*, 1422–1426. [CrossRef]
22. Wismer, W.V. Assessing alterations in taste and their impact on cancer care. *Curr. Opin. Support. Palliat. Care* **2008**, *2*, 282–287. [CrossRef] [PubMed]
23. Enriquez-Fernandez, B.E.; Martinez-Michel, L.; Thorlakson, J.; Wismer, W.V. Patient-reported taste change assessment questionnaires used in the oncology setting: A narrative review. *Eur. J. Oncol. Nurs.* **2020**, *47*, 10. [CrossRef] [PubMed]
24. O'Leary 1775, D.F.; Casey, M.; O'Connor, L.; Stokes, D.; Fealy, G.M.; O'Brien, D.; Smith, R.; McNamara, M.S.; Egan, C. Using rapid reviews: An example from a study conducted to inform policy-making. *J. Adv. Nurs.* **2017**, *73*, 742–752. [CrossRef] [PubMed]
25. Tricco, A.C.; Langlois, E.V.; Straus, S.E. *Rapid Reviews to Strengthen Health Policy and Systems: A Practical Guide*; World Health Organization, Alliance for Health Policy and Systems Research: Geneva, Switzerland, 2017; ISBN 978 92 4 151276 3.
26. Langlois, E.V.; Straus, S.E.; Antony, J.; King, V.J.; Tricco, A.C. Using rapid reviewto strengthen health policy and systems and progress towards universal health coverage. *BMJ Glob Health* **2019**, *4*, e001178. [CrossRef] [PubMed]
27. Page, M.J.; McKenzie, J.E.; Bossuyt, P.M.; Boutron, I.; Hoffmann, T.C.; Mulrow, C.D.; Shamseer, L.; Tetzlaff, J.M.; Akl, E.A.; Brennan, S.E.; et al. The PRISMA 2020 statement: An updated guideline for reporting systematic reviews. *BMJ* **2021**, *372*, 1.
28. Pluddemann, A.; Aronson, J.K.; Onakpoya, I.; Heneghan, C.; Mahtani, K.R. Redefining rapid reviews: A flexible framework for restricted systematic reviews. *BMJ Evid. Based Med.* **2018**, *23*, 201–203. [CrossRef]
29. Alonzi, S.; Hoerger, M.; Perry, L.M.; Chow, L.D.; Manogue, C.; Cotogno, P.; Ernst, E.M.; Ledet, E.M.; Sartor, O. Changes in taste and smell of food during prostate cancer treatment. *Support Care Cancer* **2021**, *29*, 2807–2809. [CrossRef]
30. Campagna, S.; Gonella, S.; Sperlinga, R.; Giuliano, P.L.; Marchese, R.; Pedersini, R.; Berchialla, P.; Dimonte, V. Prevalence, Severity, and Self-Reported Characteristics of Taste Alterations in Patients Receiving Chemotherapy. *Oncol. Nurs. Forum.* **2018**, *45*, 342–353. [CrossRef]
31. Denda, Y.; Niikura, N.; Satoh-Kuriwada, S.; Yokoyama, K.; Terao, M.; Morioka, T.; Tsuda, B.; Okamura, T.; Ota, Y.; Tokuda, Y.; et al. Taste alterations in patients with breast cancer following chemotherapy: A cohort study. *Breast Cancer* **2020**, *27*, 954–962. [CrossRef]
32. Gamper, E.M.; Giesinger, J.M.; Oberguggenberger, A.; Kemmler, G.; Wintner, L.M.; Gattringer, K.; Sperner-Unterweger, B.; Holzner, B.; Zabernigg, A. Taste alterations in breast and gynaecological cancer patients receiving chemotherapy: Prevalence, course of severity, and quality of life correlates. *Acta. Oncol.* **2012**, *51*, 490–496. [CrossRef]
33. Haxel, B.R.; Berg, S.; Boessert, P.; Mann, W.J.; Fruth, K. Olfaction in chemotherapy for head and neck malignancies. *Auris. Nasus. Larynx.* **2016**, *43*, 74–78. [CrossRef]
34. Imai, H.; Soeda, H.; Komine, K.; Otsuka, K.; Shibata, H. Preliminary estimation of the prevalence of chemotherapy-induced dysgeusia in Japanese patients with cancer. *BMC Palliat. Care* **2013**, *12*, 38. [CrossRef]
35. Joussain, P.; Giboreau, A.; Fontas, M.; Laville, M.; Hummel, T.; Souquet, P.J.; Bensafi, M. Cisplatin chemotherapy induces odor perception changes in bronchial cancer patients. *Lung. Cancer* **2013**, *82*, 168–170. [CrossRef]
36. Kaizu, M.; Komatsu, H.; Yamauchi, H.; Yamauchi, T.; Sumitani, M.; Doorenbos, A.Z. Characteristics of taste alterations in people receiving taxane-based chemotherapy and their association with appetite, weight, and quality of life. *Support Care Cancer* **2021**, *29*, 5103–5114. [CrossRef]
37. Malta, C.E.N.; de Lima Martins, J.O.; Carlos, A.C.A.M.; Freitas, M.O.; Magalhães, I.A.; de Vasconcelos, H.C.A.; de Lima Silva-Fernandes, I.J.; de Barros Silva, P.G. Risk factors for dysgeusia during chemotherapy for solid tumors: A retrospective cross-sectional study. *Support Care Cancer* **2022**, *5114*, 313–325. [CrossRef]
38. Riga, M.; Chelis, L.; Papazi, T.; Danielides, V.; Katotomichelakis, M.; Kakolyris, S. Hyposmia: An underestimated and frequent adverse effect of chemotherapy. *Support Care Cancer* **2015**, *23*, 3053–3058. [CrossRef]
39. Sicchieri, J.M.F.; Peria, F.M.; Sartorelli, D.S.; Diez-Garcia, R.W. Recognition of taste in patients during antineoplastic therapy with platinum drugs. *Nutrition* **2019**, *67–68*, 11. [CrossRef]
40. Turcott, J.G.; Juárez-Hernández, E.; De la Torre-Vallejo, M.; Sánchez-Lara, K.; Luvian-Morales, J.; Arrieta, O. Value: Changes in the Detection and Recognition Thresholds of Three Basic Tastes in Lung Cancer Patients Receiving Cisplatin and Paclitaxel and Its Association with Nutritional and Quality of Life Parameters. *Nutr. Cancer* **2016**, *68*, 241–249. [CrossRef]
41. Boltong, A.; Keast, R.; Aranda, S. Experiences and consequences of altered taste, flavour and food hedonics during chemotherapy treatment. *Support Care Cancer* **2012**, *20*, 2765–2774. [CrossRef]
42. Shiffman, S.S. Influence of Drugs on Taste Function. In *Handbook of Olfaction and Gustation*, 3rd ed.; Doty, R.L., Ed.; John Wiley & Sons: Hoboken, NJ, USA, 2015; pp. 911–926.
43. Qato, D.M.; Alexander, G.C.; Conti, R.M.; Johnson, M.; Schumm, P.; Lindau, S.T. Use of prescription and over-the-counter medications and dietary supplements among older adults in the United States. *JAMA* **2008**, *300*, 2867–2878. [CrossRef]
44. Hamor, G.H.; Lafdjian, A. Dualistic thiourea moiety taste response of methimazole. *J. Pharm. Sci.* **1967**, *56*, 777–778. [CrossRef]

45. Shi, S.; Klotz, U. Age-related changes in pharmacokinetics. *Curr. Drug Metab.* **2011**, *12*, 601–610. [CrossRef] [PubMed]
46. Lee, L.S.; Nafziger, A.N.; Bertino, J.S., Jr. Evaluation of inhibitory drug interactions during drug development: Genetic polymorphisms must be considered. *Clin. Pharmacol. Ther.* **2005**, *78*, 1–6. [CrossRef] [PubMed]

International Journal of
*Molecular Sciences*

MDPI

*Review*

# Liquid Biopsy for Lung Cancer: Up-to-Date and Perspectives for Screening Programs

**Giovanna Maria Stanfoca Casagrande [1,†], Marcela de Oliveira Silva [1,†], Rui Manuel Reis [1,2,3] and Letícia Ferro Leal [1,4,*]**

1 Molecular Oncology Research Center, Barretos Cancer Hospital, 1331 Rua Antenor Duarte Vilela, Barretos 14784-400, Brazil
2 Life and Health Sciences Research Institute (ICVS), School of Medicine, Campus de Gualtar, University of Minho, 4710-057 Braga, Portugal
3 ICVS/3B's—PT Government Associate Laboratory, 4710-057 Braga, Portugal
4 Barretos School of Medicine Dr. Paulo Prata—FACISB, Barretos 14785-002, Brazil
* Correspondence: leticiaferro@hotmail.com
† These authors contributed equally to this work.

**Abstract:** Lung cancer is the deadliest cancer worldwide. Tissue biopsy is currently employed for the diagnosis and molecular stratification of lung cancer. Liquid biopsy is a minimally invasive approach to determine biomarkers from body fluids, such as blood, urine, sputum, and saliva. Tumor cells release cfDNA, ctDNA, exosomes, miRNAs, circRNAs, CTCs, and DNA methylated fragments, among others, which can be successfully used as biomarkers for diagnosis, prognosis, and prediction of treatment response. Predictive biomarkers are well-established for managing lung cancer, and liquid biopsy options have emerged in the last few years. Currently, detecting EGFR p.(Tyr790Met) mutation in plasma samples from lung cancer patients has been used for predicting response and monitoring tyrosine kinase inhibitors (TKi)-treated patients with lung cancer. In addition, many efforts continue to bring more sensitive technologies to improve the detection of clinically relevant biomarkers for lung cancer. Moreover, liquid biopsy can dramatically decrease the turnaround time for laboratory reports, accelerating the beginning of treatment and improving the overall survival of lung cancer patients. Herein, we summarized all available and emerging approaches of liquid biopsy—techniques, molecules, and sample type—for lung cancer.

**Keywords:** liquid biopsy; lung cancer; biomarkers; precision medicine

**Citation:** Casagrande, G.M.S.; Silva, M.d.O.; Reis, R.M.; Leal, L.F. Liquid Biopsy for Lung Cancer: Up-to-Date and Perspectives for Screening Programs. *Int. J. Mol. Sci.* **2023**, *24*, 2505. https://doi.org/10.3390/ijms24032505

Academic Editor: Laura Paleari

Received: 17 November 2022
Revised: 9 December 2022
Accepted: 19 December 2022
Published: 28 January 2023



## 1. Introduction

Lung cancer has the highest incidence rate for cancer type and the second-highest mortality rate in the world [1]. The high mortality rates associated with lung cancer are mostly due to its late detection and diagnosis, leading to a decrease in these overall survival rates [2,3]. However, even patients diagnosed in the early stages of the disease and still during cancer's contained stage still show a poor five-years overall survival rate, which is lower than 60% compared to other cancer types. Despite the early detection, the poor prognosis is attributed to the disease progression [4].

Cigarette smoking and other tobacco exposure habits are the major risk factor for lung cancer, and it is associated with approximately 80% of all lung cancer cases [5,6]. However, additional risk factors associated with environmental and occupational exposures have also been linked to lung cancer development [7]. In addition to environmental risk factors, single nucleotide polymorphisms (SNPs) have also been associated with an increased risk for lung cancer development [8–10].

The biological processes behind lung cancer are complex, and the tumors are highly heterogeneous [11]. Histologically, lung cancer is divided into two major subtypes: small-cell lung cancer (SCLC) and non-small-cell lung cancer (NSCLC) [12]. The SCLC accounts

for approximately 15–20% of all lung cancer cases, and SCLC patients show extremely low survival rates [13]. The NSCLC accounts for approximately 80–85% of all lung cancer patients, and it is further classified into three main histological subtypes: adenocarcinoma, squamous cell carcinoma (SCC), and large cell carcinoma [14]. NSCLC patients present heterogeneous survival rates depending on several features, including the stage of the disease at the time of diagnosis, the patient's performance status, smoking habits, histological subtypes, and molecular characteristics [15,16].

In addition to the histological categorization, molecular characteristics are a critical additional component of the biological study of lung cancer due to the advances in precision medicine for both methodological procedural aspects and molecular biomarkers [17]. Currently, the number of FDA-approved drugs targeting molecular biomarkers has dramatically increased in the market, creating a more efficient route for treating NSCLC patients [18]. On the other hand, tumor tissue biopsy has continued to be mandatory for histological evaluation and diagnosis, and in the last decades, it has also been employed in the detection of molecular biomarkers [18,19]. However, the methods commonly used for tissue procurement are highly invasive and may have many side effects, which depending on the performance status of a patient, may not even be eligible for such an invasive procedure. Moreover, tumor tissue biopsy represents a small fragment of the tumor as a whole, which may result in unviable tissue biopsies for assessing tumor heterogeneity [20]. Thus, the development of new approaches is pivotal, given that the current approaches have some disadvantages, especially when it comes to offering a viable option for those who are not eligible for the currently existent methods of sample procurement [20–23].

In the last few years, a new approach has emerged, attracting various efforts for its implementation in the diagnostic routine of lung cancer [24,25]. Liquid biopsy is a minimally invasive approach for sample procurement, mostly body fluids, and it is used to detect molecular alterations, tumor cells, and metabolites. Furthermore, liquid biopsy, the emerging approach attracting substantial attention, has currently been employed for clinical management, specifically for guiding treatment and monitoring disease. In addition, to being a minimally invasive approach, the liquid biopsy also enables the serial collection of samples allowing early detection of residual disease and relapses and resistance to treatment [20,21,23,26,27]. For NSCLC, this approach is particularly valuable for patients who are not eligible for the conventional tissue biopsy, mainly due to patients' individual conditions and the tumor location, as previously discussed. As a result, liquid biopsy has been employed for disease and treatment monitoring and for creating a targeted treatment for NSCLC patients within the field of precision medicine [21,28].

This review aims to gather the most up-to-date information for the procedure of liquid biopsy, given the context of NSCLC patients. This review will focus on information such as the various types of body fluids used for liquid biopsy, the currently available and promising biomarkers in the field, and a discussion of the major challenges involved in analyzing and securing samples for liquid biopsy.

## 2. Sample Types and Analytes for Liquid Biopsy from NSCLC

Body fluids such as plasma, sputum, saliva, urine, stool, cerebrospinal fluid, and pleural effusions, among others, are suitable sources for detecting diagnostic, prognostic, and predictive biomarkers for NSCLC [21,29–33]. As a result, liquid biopsy has been employed due to its minimally invasive sample procurement. However, it is important to point out that not all body fluids are collected by a minimally invasive sample procurement (e.g., cerebrospinal fluid). Thus, the type of body fluid selected for liquid biopsy must be carefully chosen based on the type of cancer, especially because some body fluids do not properly represent tumor origin (Table 1 and Figure 1). Moreover, the most important benefit of liquid biopsy is the minimally invasive sample procurement, and we should avail this advantage.

**Table 1.** Liquid biopsy analytes for NSCLC: major clinical applications, biofluids, and methods.

| Clinical Application | | Biofluids | Methodologies | Reference |
|---|---|---|---|---|
| ctDNA/cfDNA | Diagnosis, Tumor burden, Treatment response, Prognosis. | Peripheral blood, Sputum. | qPCR, dPCR, ddPCR, ARMS, BEAMing | [32–35] |
| CTCs | Diagnosis, Tumor burden, Prognosis. | Peripheral blood | RT-qPCR, Ep-CAM, NGS | [36–38] |
| Extracellular vesicles (exosomes) | Diagnosis, Prognosis. | Peripheral blood | ultracentrifugation, exosomes immunoprecipitation, immune beads precipitation | [39] |
| miRNAs | Diagnosis, Disease, Progression. | Plasma, Serum, Sputum. | RT-qPCR | [40] |
| DNA Methylation biomarkers | Diagnosis, Disease progression. | Plasma | Immunoprecipitation, methyl-sensitive restriction enzymes, sodium bisulfite conversion, q-PCR, and Next-Generation Techniques | [41,42] |
| Metabolites/Proteins | Progression, Predictive, Diagnosis, Prognosis. | Serum | HRMAS MRS UPLC–MS and immunoradiometric assay | [43–46] |
| Autoantibodies tumor-associated antigens | Diagnosis, Predictive. | Serum | ELISA | [47,48] |

ARMS: Scorpion amplification-refractory mutation system; qPCR: quantitative Polymerase Chain Reaction; dPCR: digital Polymerase Chain Reaction; ddPCR: droplet digital Polymerase Chain Reaction; BEAMing: beads, emulsion, amplification, and magnetics; Ep-CAM: Epithelial cell adhesion molecule, NGS: Next-Generation Sequencing; HRMAS: High-Resolution Magic Angle Spinning; MRS: magnetic resonance spectroscopy; UPLC–MS: ultra-performance liquid chromatography-tandem mass spectrometry.

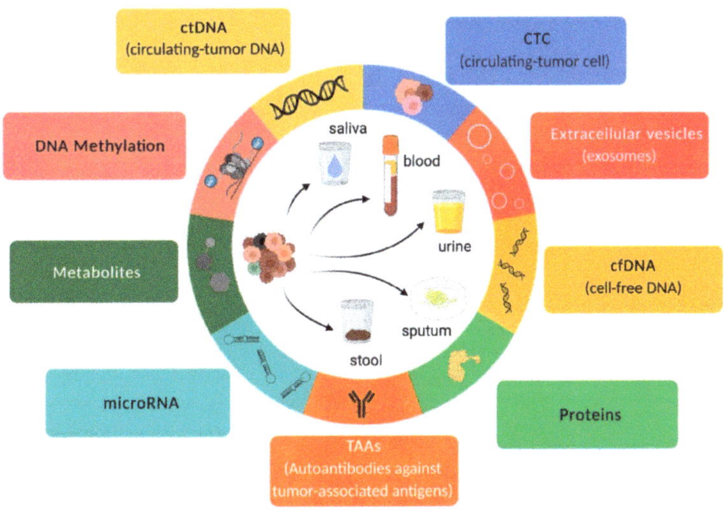


**Figure 1.** Overview of liquid biopsy. Inspired by Hanahan & Weinberg (192).

Liquid biopsy samples are mostly composed of cell-free DNA (cfDNA), cell tumor DNA (ctDNA), circulating cell-free microRNAs (ccfmiRNAs), circulating tumor cells (CTCs), metabolites, and proteins as well as extracellular vesicles such as exosomes, which contain proteins and cell-free nucleic acids (cfNA) such as miRNAs (Figure 1). These components are released into body fluids through processes such as apoptosis, necrosis, and secretion [26]. Therefore, liquid biopsy has been an important "supporting actor" for guiding therapeutic strategies for NSCLC patients and has emerged as a "leading title-role actor" in the routine setting of precision medicine.

## 3. NSCLC Biomarkers for Detection in Liquid Biopsy Samples

Clinical biomarkers can be found in different biofluids; currently, the most frequently used in precision medicine are plasma and serum, both of which originate from the peripheral blood [49]. Both blood-derived biofluids can be used to analyze different biomarkers from distinctive sources, such as CTCs, ctDNA, cfDNA, ccfmiRNAs, and metabolites, among others [50]. The major applications of each type of molecule in the field of lung cancer are summarized hereafter.

### *3.1. Circulating-Tumor DNA (ctDNA) and Cell-Free DNA (cfDNA)*

Fragments of cell-free DNA (cfDNA) are freely available throughout the blood. CfDNA is released into the bloodstream due to natural body mechanisms, such as apoptosis, necrosis, and active secretion [26,31]. CfDNA can be found in both healthy subjects and cancer patients, although cfDNA levels tend to be higher in cancer patients (Figure 2).

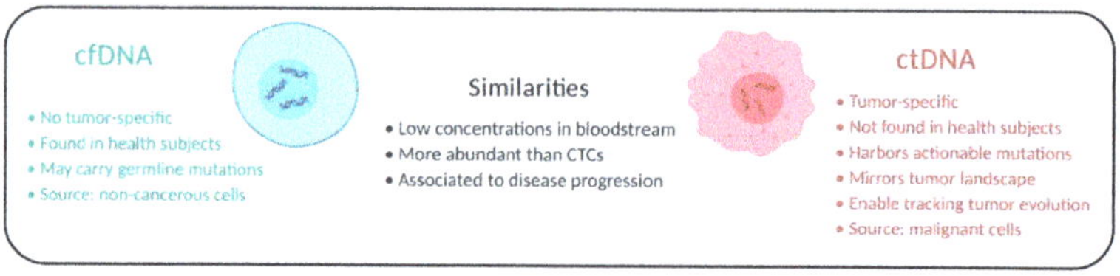


**Figure 2.** Comparison of cfDNA and ctDNA. Individual aspects and similarities of both molecules are described.

Circulating tumor DNA (CtDNA), different from cfDNA, can harbor somatic mutations reflecting the tumor dynamics [51,52]. Tissue biopsy is a single "snapshot" of the tumor and, therefore, can mirror a unique subclone or a few subclones, while ctDNA can be more representative of the tumor's entire tissue composition. Moreover, it is currently possible to trace the subclone's origin of relapses and metastases through the phylogenetic profile of ctDNA. Tracing the clonal origin of the tumor can provide a collage of a tumor's evolution, opening possibilities for the translation of such information into clinical practice [34].

Using liquid biopsy to detect ctDNA is extremely effective, as it is possible to collect a sample at any time during the disease and therapy course, assessing the progression of the disease in real-time [51]. However, liquid biopsy can become challenging when the concentration of ctDNA is measured to be extremely low (<1%) compared to the concentration of cfDNA in a patient's bloodstream. Thence, highly sensitive and specific techniques are required for detecting mutations to track, detect, and monitor genomic alterations as cancer progresses [52–54]. The current standard methods for detecting somatic mutations, such as RT-PCR, Sanger sequencing, and next-generation sequencing (NGS), may not be sensitive enough for the detection of mutational ctDNA, specifically mutations presented with a low variant allele frequency (VAF) [51].

The main advantage of analyzing ctDNAs is the high specificity of the molecule, as it is proven that any molecular alterations present in these molecules are identical to those in the tumor tissue. In addition, advanced-stage cancer patients usually present with higher levels of ctDNA, which allows for easy monitoring of the disease's course, tumor heterogeneity and dynamics, tumor evolution, and any tumor-acquired resistance to targeted treatments.

Recently, ctDNA was associated with shorter survival, and actionable alterations in ctDNA were not detectable in time-matched tissue [55]. In addition, the turnaround time was decreased for the release of a biopsy report to guide therapeutic decisions that can be made earlier and safer [55].

Currently, highly sensitive PCR-based techniques, such as droplet digital PCR (ddPCR) and BEAMing, are considered sensitive enough for detecting mutations at low frequencies and have emerged as potential tools not only for advanced cancers but also for early detection [53]. NGS is also an extremely sensitive technique for detecting somatic mutations and identifying mutational frequencies as low as 0.02% VAF; however, the NGS technique produces an error rate of about 0.5–2.0% [56].

Therefore, the detection of rare variants by NGS remains challenging due to its limit of detection (LoD) and errors incorporated during sequencing. Furthermore, random errors can be incorporated into the DNA molecules during the library preparation or the sequencing processes, which could be mistaken for true variants. Strategies using molecular barcoding principles, such as hybrid capture and amplicon-based NGS, have also been used to detect low-frequency mutations in both liquid and conventional biopsy samples to decrease false negative results [51,53]. Each molecule in the sequencing library is marked with a small sequence of random nucleotides (8–16 N), which can be called a Unique Molecular Identifier (UMI), Tag Sequencing, or molecular barcodes and to decrease the chances of detecting false variants, the sequenced reads are grouped according to these UMI. As a result, sequencing artifacts can then be detected as they are not present in all reads with the same UMI, which increases the reliability of naming the true variants. Therefore, molecular barcoding technology allows for the correction of sequencing errors [57,58].

Due to the advances in sensitive technologies, ctDNA and mutational analysis are now possible for NSCLC patients. In addition, the detection rate of ctDNA can be higher than 80% in the plasma from NSCLC patients, suggesting that ctDNA analysis is an adequate alternative when sampling tissue biopsy is not an option [59]. *EGFR, KRAS, ERBB2*, and *BRAF* mutations, gene rearrangements (*EML4—ALK, ROS1, NTRK1/2*, and *RET*), exon skipping alterations, and gene amplifications (MET) are routinely evaluated in the management of care for NSCLC patients [51]. All these molecular alterations have been adapted into the clinical practice for guiding and monitoring patients' treatment and disease state [53]. Currently, broader NGS panels have been employed in clinical practices, such as MSK-IMPACT (tissue) and MSK-ACCESS (plasma) [55]. A quarter of the patients presented alterations in ctDNA not detected in tissues [55], which suggests that plasma samples may present a higher specificity then previously reported, supporting the use of NGS as a sensitive, specific tool for plasma samples analysis [60].

Unfortunately, not all of the actionable alterations mentioned above have been translated for the clinical management of NSCLC patients through liquid biopsy sampling.

The FDA recently approved two IVD (in vitro diagnosis) tests (Cobas *EGFR* Mutation Test v2, Roche, and Idylla TM ctEGFR Mutation Assay) for NSCLC patients, which employs plasma samples for the detection of *EGFR* resistance mutation p.(Tyr790Met), exon 19 deletions and p/L858R mutations [61]. Although the IVD test shows a lower sensitivity when compared with other methods of variant detection (e.g., ddPCR and NGS), its approval was a great addition to the present tools used in guiding therapeutic decisions and monitoring treatment results for NSCLC patients [62].

### 3.2. *cfDNA Methylation Biomarkers*

Genetic mutations and epigenetic modifications, such as DNA methylation, have been detected in cfDNA and ctDNA. It has been considered a promising approach for translation into clinical applications to be used for diagnosis, prognosis, and predictive purposes.

Methylation is an incorporation of a methyl group (CH3) into a Cytosine in regions enriched with CG bases, also known as CpG islands [63]. Methylation occurs in CpG islands when found at the promoter regions of several genes, and it often results in gene silencing commonly found in tumor suppressor genes. On the other hand, the transcriptional activation of genes with the methylation present in the gene body is associated with various cancer types, including lung cancer [41,64]. Methylome analysis has yielded highly successful results on tumor tissues, especially when the analysis is focused on molecular subtyping and biomarkers discovery of several tumor types, including lung cancer [65,66]. In addition to plasma and serum, other body fluids deserve attention, such as sputum and bronchoalveolar lavage fluids, specifically due to their proximity to the tumor's location [41].

Analysis of methylation-based biomarkers in cfDNA can also be used for diagnostic purposes for the management of lung cancer patients. Plasma cfDNA showed significant differences in DNA methylation level for early-stage NSCLC patients, including stage IA patients, indicating this type of biomarker could be a valuable tool for screening and early detection of NSCLC combined with imaging tests to improve the detection of early-stage pulmonary nodules [63,67–70]. Multi-cancer early detection (MCED) test may be a promising approach for cancer screening and early detection. MCED is a targeted methylation-based assay for complementary use in screening programs [71–73]. However, results for stage I lung cancer have not shown to be sensitive enough for being employed in a screening setting [72]. A recent study screened asymptomatic subjects from the NHS-Galleri trial (ISRCTN91431511) but results about the clinical utility of the MCED test for lung cancer still need to be addressed [73].

The combined methylation analysis of the CDO1 and HOXA9 was associated with unfavorable outcomes, while the combination of PTGDR and AJAP1 methylation was associated with favorable outcomes. A prognostic risk based on these methylation-based biomarkers may be useful to refine risk stratification [65,67]. In addition to prognostication, methylation-based biomarkers can also be useful as predictive biomarkers. Deregulation of methylation in cfDNA was also associated with EGFR-TKI resistance in early-stage NSCLC patients [74,75].

### 3.3. Circulating Tumor Cells (CTCs)

Circulating tumor cells (CTCs) are cells derived from primary tumors that were dissociated from tumor mass by either mechanical motion, the loss of adhesion molecules on the surface of cells entering the circulatory system, or a combination of both. CTCs are commonly detected in lower concentrations in the peripheral blood compared to other types of analytes (e.g., ctDNA), and it may be a challenge requiring the sampling of CTC-rich peripheral blood [38,52,54]. Although studies have shown that aggressive tumors can release thousands of these cells into the bloodstream every day, and it may be associated with the mechanism of distant metastasis, such a result would require completing a complex process [38,54]. Once these cells enter the bloodstream, they are capable of planting metastatic sites through active trans-endothelial migration while remaining inactive; however, no studies have yet been able to show how the biological process of transition from dormancy to active growth happens [52]. As a result, the mechanisms through which cancer spreads from one organ to another using CTCs are of great interest to the scientific community [54].

Furthermore, the methods of isolating CTCs include identifying these cells based on the presence of specific markers, such as the epithelial cell adhesion molecule (EpCAM), the cytokeratin of epithelial CTCs, and the N-Cadherin or vimentin of mesenchymal CTCs [76,77]. The FDA has confirmed an IVD method, based on anti-EpCAM ferromagnetic microbeads, called CellSearch CTC kit® (Veridex LLC, Raritan, NJ, USA) for prognostic assessment. This IVD method detects anti-EpCAM ferromagnetic microbeads on circulating tumor cells (CTC) in peripheral blood samples [78]. As previously mentioned, the characterization of CTCs is difficult since the detection method requires high sensitivity to detect these molecules' low concentrations in extracorporeal fluids.

The presence of CTCs is considered a prognostic biomarker since it can help predict disease progression in cancer patients, including the progression of NSCLC [54]. One study showed that patients with metastatic lung cancer and a state of progressive disease presented a higher expression of the *PIK3CA, AKT2, TWIST*, and *ALDH1* genes in CTCs compared to patients with non-metastatic disease. Thus, it is possible to correlate the presence of CTCs with the diagnostic scope of disease burden, including possible metastasis and disease progression [37]. The presence of CTCs was found to be independent of the tumor stage at diagnosis (49% stage I, 48% stage II, 48% stage III, and 52% stage IV patients) and histology (47% adenocarcinoma and 40% squamous cell carcinoma) when using a size-based filtration method for CTC detection [79]. Moreover, of the cells with the highest glucose uptake, hypermetabolic CTCs were isolated to analyze *EGFR* and *KRAS* mutations using ddPCR. The comparison between the primary tumor's *EGFR* and *KRAS* mutations and that of CTCs showed a match in 70% of the cases [80].

### 3.4. Extracellular Vesicles

Extracellular vesicles (EVs) can be divided into microvesicles, vesicles, and exosomes. Exosomes are released by various methods and are detectable in several cell types and bodily fluids, including cancer cells. Moreover, exosomes are released during the process of exocytosis following the fusion of multivesicular bodies (MVBs) and cell membranes, and they can also be detected in body fluids such as blood (plasma and serum), urine, pleural effusions, saliva, cerebrospinal fluid, and semen [27,54]. In addition, exosomes have been shown to mediate intercellular communication between the tumor and the stroma, resulting in a rich source of molecular information that enables the location of origin cells for these exosomes [50,81].

Furthermore, ultracentrifugation or differential centrifugation (DC) and size-exclusion chromatography, based on size selection and chemical isolation during polymeric-based precipitation (PBP), are all commonly used for EV isolation [82–84].

Tumor cells can release more exosomes than non-tumor cells, making exosomes a potential biomarker in liquid biopsies for various types of tumors [27]. In lung cancer, exosomal RNA, DNA, and proteins can be used to detect molecular alterations, including actionable mutations [27,85]. Qu et al. (2019) identified exosomes harboring EGFR mutation, showcasing exosomes as a valuable component in the analysis's tumor progression, pre-metastatic niche formation, and resistance to treatment [86]. The comparison between exosome-derived and tumor-derived mutations showed a sensitivity of 100% in detecting EGFR mutations and a specificity greater than 96% [86].

Furthermore, exosomal miRNAs (exo-miRNA) have also been generating interest from the scientific community as a potential biomarker in several types of cancer, including lung cancer. Exo-miRNA is protected from RNAse degradation in the bloodstream due to the lipid bilayer membrane, which results in their stability in body fluids and easily detectable form in body fluids [27]. As a result, exo-miRNA load was reported as a prognostic biomarker [87]. Additionally, exo-miRNAs miR-564 and miR-659 switched sensitive to resistant cells to gefitinib [88], which proved to be a predictive phenotype for exo-miRNA. On the other hand, the exo-miRNA miR-302-b is associated with the suppression of lung cancer cell proliferation and migration via the TGFβRII/ERK pathways, which has established the miR-302-b as a potential therapeutic target for lung cancer patients [89]. Thus, exosomal miRNAs may be employed in the process and measurements of diagnosis, prognosis, and monitoring of lung cancer patients, which reveals the need for further studies of exosomes and their clinical application in the context of precision medicine.

### 3.5. MicroRNA/CircRNA

MicroRNAs (miRNAs) are small RNAs (19–24 nucleotides) responsible for regulating gene expression. However, most miRNAs have unknown biological functions [90,91]. MiRNAs differ from mRNAs due to their stability, small size, and regulatory control in gene expression, rendering them a new generation of biomarkers [40]. Circulating-free miRNAs (cfmiRNAs) have already been reported in body fluids from cancer patients, including plasma, serum, urine, and saliva [92–94].

The detection of miRNAs in liquid biopsy samples was reported to distinguish NSCLC patients from healthy subjects paving the way for cfmiRNAs as promising early detection biomarkers [95]. Reis and collaborators (2020) recently reported differential expression of a set of miRNAs in plasma samples from early-stage lung cancer patients [96]. Similarly, plasma samples from early-stage NSCLC patients paired with non-cancer patients were screened for 754 circulating miRNAs, and a highly accurate 24-miRNAs panel was proposed for early detection of lung cancer in addition to the known risk factors [95]. In addition, dysregulation of miRNA expression can also be associated with exposure factors.

For example, the downregulation of let-7i-3p and miR-154-5p was found in serum from smokers, both non-cancer subjects and lung cancer patients. These miRNAs are involved in lung cancer development and progression, rendering these circulating miRNAs a potential role as a diagnostic and prognostic biomarker in lung cancer for tobacco-exposed patients [97,98]. Moreover, high expression of miR-34 and miR34-c in plasma samples from surgically resected NSCLC patients was associated with increased survival [98].

Recently, some publications have explored the potential of circular RNA (circRNA) as a biomarker in liquid biopsy [99]. This is because circRNA are resistant to RNases and other exonucleases due to the lack of final 5′ or 3′ regions [100,101], having a longer half-life than linear RNA [102]. Moreover, its specific tissue expression and stage of development increase this interest as a potential biomarker [98,103–105]. Studies show that circRNA expression varies in different tumor types, including lung cancer [104,106–108]. In adenocarcinoma patients, circRNA can be over-expressed, such as circ_0013958, which was upregulated [107], and circFARSA, a circRNA derived from exon 5–7 of the FARSA gene, in the plasma of patients with NSCLC compared to controls without cancer [109,110].

In addition, studies with circRNAs reported that they could detect specific signature profiles in tissue samples, capable of distinguishing histological groups in addition to NCLSC patients from healthy ones [105]. However, studies are being carried out to analyze the profile of circRNAs signatures in liquid biopsy samples, such as plasma, exosomes, and lymphocytes, to implement these molecules as diagnostic and predictive biomarkers.

Altogether, the detection of miRNAs/circRNAs in minimally invasive samples has emerged as a promising approach for the early detection of NSCLC to be incorporated in lung cancer screening programs and for prognostic purposes to be incorporated in the clinical management of these patients.

### 3.6. Metabolites and Proteins

In pathological conditions, including cancer, circulating metabolites can be detected in body fluids [45]. For detecting metabolites in biological fluids, Gas chromatography-mass spectrometry (GC-MS), nuclear magnetic resonance (NMR), and high-performance liquid chromatography UV detector (LC-UV) along with software (Metaboanalyst, Metabolite Set Enrichment Analysis [MSEA], Metlin, BioStatFlow and Human metabolome database [HMDB]) have been employed for detecting, processing, and analyzing metabolic data [111].

Lung cancer patients show changes in metabolic pathways compared to non-cancer patients; starch and sucrose metabolism, galactose metabolism, fructose, mannose degradation, purine metabolism, and tryptophan metabolism are among the unbalanced metabolic pathways [43]. In addition, some amino acids such as valine, leucine, and isoleucine are related to stress and energy production, regulating many signaling pathways, such as protein synthesis, lipid synthesis, cell growth, and autophagy, and can be found at higher levels in lung cancer patients compared with non-cancer patients [44]. However, these previously reported data must be clinically tested and reproduced in other series.

Although metabolomics has been a promising approach for managing and monitoring cancer patients, some limitations have still been experienced, such as identifying unknown compounds, detecting cancer-specific metabolites, and standardization of cutoffs [111].

In addition to metabolites, serum proteins have currently been used as cancer biomarkers, such as carcinoembryonic antigen (CEA), cytokeratin 19fragment (CYFRA 21-1), cancer antigen 125 (CA 125), neuron-specific enolase (NSE) and squamous cell carcinoma antigen (SCCA) [112,113]. Immunoassay (e.g., ELISA) and mass spectrometry can detect these serum proteins. In the last few years, considerable efforts have been made to find serum protein biomarkers for the early detection of lung cancer that will potentially be employed in screening programs soon.

Acute-phase reactant proteins (APRPs) are produced in response to inflammation caused by cancer and can also be used as potential biomarkers for the diagnosis of different types of cancer [114]. Lung cancer patients presented higher serum haptoglobin β (HP- β) chain levels than healthy controls. However, patients with other respiratory diseases also presented increased levels of Hp-β chain; thus, medical history, and radiological images should also be considered [115]. Lung cancer patients also presented higher serum amyloid A (SAA), another APRP, compared with healthy controls [116]. With higher levels of SAA1 and SAA2, lung adenocarcinoma patients also presented decreased serum levels of Apo A-1—a protein responsible for removing endogenous cholesterol from inflammatory sites. Thus, SAA1, SAA2, and Apo A-1 could also be considered potential biomarkers for the early detection of lung cancer [117].

In addition, to early detection applications, some proteins involved with metastasis and tumor progression are promising prognostic biomarkers. For example, plasma and pleural effusions from NSCLC patients with high levels of the S100A6 protein—a member of the S100 family with pro-apoptotic function—presented longer survival time compared with S100A6-negative cases [118,119]. On the other hand, NSCLC patients show high serum Cytokeratins (CKs) levels, such as CK 8, 18, and 19, which were associated with unfavorable prognoses [119,120]. In addition, the combined analysis of regulators of actin—calmodulin, thymosin β4, cofilin-1, and thymosin β10—was suitable for predicting patients' outcomes [121].

The CancerSEEK reported forty-one potential protein biomarkers detectable at least in one out of the eight tumor types. The authors selected the best eight biomarkers for composing the final bead-based immunoassay test (CA-125, CA19-9, CEA, HGF, Myeloperoxidase, OPN, Prolactin, TIMP-1), which were highly effective in distinguishing cancer patients from healthy controls [122]. The positivity for the CancerSEEK test was about 70% for all cancer types. According to this study, combining protein biomarkers with cfDNA mutations increased sensitivity without significantly decreasing specificity [122].

Although proteomic approaches have been promising tools for precision medicine in the lung cancer field, there are some limitations regarding using proteins as biomarkers. Many of these proteins are associated with different tumor types. They present poor sensitivity, are not organ-specific, and can also be detected in non-malignant diseases [112,113,123]. Thus, clinical history and radiological exams should always be considered. Moreover, some proteins are presented with low serum levels, precluding their employment for early lung cancer diagnosis [123,124]. Finally, the analysis of these biomarkers can be influenced by the contamination of intracellular proteins caused by cell lysis during sample procurement and processing. Thus, pre-analytical steps are pivotal [118].

Scalable proteomics has also emerged as a tool for the identification of high-risk subjects for lung cancer. The most recent example is Olink Proteomics, which was created to detect circulating proteins (https://olink.com/, accessed on 16 November 2022). A 36-protein multiplexed assay was developed for risk assessment of lung cancer development, and this set of proteins includes growth factors, tumor necrosis factor receptors, and chemokines and cytokines [125]. Scalable proteomic approaches may not be affordable for low-middle-income countries, but when feasible, they should still be considered as a complementary tool along with LDCT in lung cancer screening programs. However, more studies should be conducted to prove these proteomic approaches are cost-effective for lung cancer screening, especially considering limited resources and minorities.

### 3.7. Autoantibodies against Tumor-Associated Antigens

Autoantibodies against tumor-associated antigens (TAAs) are circulating antibodies to autologous cellular antigens [126–128]. Tumor tissues can release cellular proteins, leading to the activation of the immune system and the production of autoantibodies [129,130]. Therefore, cancer patients produce autoantibodies against aberrant or overexpressed proteins produced by these cancer cells [131]. TAAs are stable in the serum and have been studied in several types of tumors, including lung cancer. TAAs can be detected through the immunoenzymatic assay (ELISA) [48,129,132]. Moreover, they can also provide information for early detection and disease monitoring [47,48,132–135].

In the 90s, Lubin et al. (1995) analyzed the presence of p53 antibodies in the serum of patients with lung cancer [136]. These antibodies were 30% higher in cancer patients than in non-cancer patients, associated with TP53 mutations [136]. TAAs panels have also been developed, combining many different TAAs to improve test sensitivity [48,126,132,137,138]. The panels developed include TAAs for p53, NY-ESO-1, CAGE, GBU4–5, Annexin 1, SOX2, c-Myc, MDM2, NPM1, p16, cyclin B1, among others [48,132,139]. Although the levels of most biomarkers increase proportionally to tumor burden, TAAs levels do not differ among different stages of lung cancer, possibly due to the humoral immune response—present from the beginning of tumorigenesis [48,132,139]. This feature renders TAAs a potential tool for early detection [48,132,140,141]. However, studies including proper sample size and validation set are required for greater reliability of TAAs in the routine lung cancer setting.

## 4. Clinical Application of Liquid Biopsy for NSCLC

The absence of pathognomonic symptoms in lung cancer leads to late diagnosis—many patients may mistakenly receive another diagnosis, such as pneumonia, Chronic obstructive pulmonary disease (COPD), among others, especially from low-middle income countries, where resources are limited, and health systems are recurrently offscourings. The late diagnosis in limited therapeutic options and curative intent is no longer available, culminating in the highest lethality rate among all cancer types [3].

### 4.1. Prognosis

Lung cancer in the early stage has an excellent prognostic after surgery due to early screening but needs more information about the disease is needed to understand metastatic lung cancer [142]. After the liquid biopsy advent, the prediction of outcomes for NSCLC patients employing molecular biomarkers became more feasible for those who may not be eligible for conventional biopsy [21,27].

In NSCLC and SCLC, more CTCs were associated with adverse prognostic factors for survival [54,143,144]. A study compared the number of CTCs present in blood in patients with small-cell lung cancer. The CTC count was made at the beginning, before, and after treatment. The authors show that patients with eight or more CTCs, per 7.5 mL of blood had a worse survival than those with less than 8 CTCs at the pretreatment, and patients whose baseline CTC levels remained lower than 8 CTC at the posttreatment showed better survival [143].

Undetectable ctDNA blood levels may serve as a prognostic marker for targeted therapy and chemotherapy in patients with NSCLC [142,145]. CtDNA has been proposed as a non-invasive, real-time biomarker to provide prognostic information to monitor treatment since patients with the same mutation may differ in response to treatment [146]. Based on this, techniques can detect recurrent point mutations in controller genes. When using ddPCR, plasma G12/G13 status was associated with an unfavorable prognostic in progression-free survival and overall survival in NSCLC patients [147]. The prognostic value of TP53 in lung cancer is being debated. Several changes in this gene are found in patients with advanced NSCLC. Patients carrying pathogenic plasma mutations in TP53 have lower overall survival when compared to wild TP53. In addition, the risk of extrathoracic metastases in patients with TP53 ctDNA alteration is greater [148].

In addition to point mutations, the methylation status of critical genes can also be analyzed in minimally invasive samples, and the methylation status can serve as a prognostic biomarker [135,142]. For example, Wen et al. (2021) showed that circulating metHOXA9 is an adverse prognostic factor in patients with advanced NSCLC. In this study, the authors analyzed the levels of metHOXA9 in patients' blood before starting treatment and before each cycle of chemotherapy. As a result, it was observed that the levels of metHOXA9 increased after the first treatment cycle; that is, the treatment changed the status of the biomarker, decreasing the overall survival of these patients [149].

The hypermethylation of some genes as p16, CDH1, FHIT, and APC, can be considered a prognostic factor. In addition to being classified by stages and histological type, genetic alterations among each stratified group may reveal additional and more accurate prognostic factors [150]. Such as, the methylation of genes p16, CDH13, RASSF1A, and APC in patients with early-stage NSCLC-treated surgery was associated with an early recurrence [151,152]. More studies are needed to analyze the best genes to use as prognostic biomarkers in lung cancer. By considering post-transcriptional mechanisms, miRNAs can also serve as prognostic biomarkers. For example, NSCLC patients with downregulated miR-590-5p had significantly lower median survival rates when compared to patients expressing high miR-590-5p, and it was associated as a potential prognostic marker for the progression of NSCLC [152].

High expression levels of miR-18a, miR-20a, miR- 92a, miR-126, miR-210, and miR-19a correlated with worse disease-free survival (DFS), in addition to a shorter overall survival (OS) than patients with low expression levels of these miRNAs. So, these results suggested that these miRNAs can be potential biomarkers for the prognosis of NSCLC patients [153]. In addition, elevated levels of miR-34a correlated with a prolonged DFS and OS compared to low levels of expression in 196 NSCLC patients, i.e., miR-34a has potential prognostic value for lung cancer patients [98].

Boeri et al. (2011) and Tian et al. (2016), with a short cohort, reported that miR-486-5p was downregulated in adenocarcinoma patients' plasma, and miR-181b-5p was upregulated in squamous cell carcinoma patients' plasma. In addition, the two miRNAs negatively regulate their targets—RASSF1 and PIK3R1 [148,149], while that elevated levels of miR-21 in plasma samples predicted poorer overall survival of NSCLC patients [154,155]. However, more studies with increased sample sizes must incorporate microRNAs into clinical practice as prognostic biomarkers [154].

### *4.2. Precision Medicine and Disease Monitoring*

Positron Emission Tomography/Computed Tomography (PET/CT) and tissue biopsy are currently employed for NSCLC disease staging and monitoring and for guiding therapeutic strategies either based on disease staging or molecular alterations [43,156]. However, not all patients are eligible for tissue biopsy, depending on tumor location and the patient's clinical performance [157]. Using liquid biopsy for clinical management, including treatment and disease monitoring, can better represent tumor heterogeneity, and predictive biomarkers can be successfully detected to guide therapeutic options for NSCLC. Unfortunately, the successful detection of predictive biomarkers in this sample depends on assay sensitivity and the variant allele frequency (VAF). Fortunately, susceptible technologies have emerged as suitable approaches for cfDNA analysis, making rare detecting alleles feasible [126,158,159].

There are two approved tests for the search for actionable mutations in liquid biopsy, the Idylla TM ctEGFR Mutation Assay and the Cobas®EGFR Mutation Test v2, equally real-time PCR-based [160,161]. Identifying EGFR mutations has become essential since EGFR-TKI therapy has become the standard treatment choice for EGFR mutant patients [162]. The acquisition of resistance to treatment is recurrent in up to 60% of patients, and disease recurrence undergoing EGFR-TKI therapy is generally mediated by the p.(Tyr790Met) mutation [163]. As Cabanero and Tsao (2018) suggest, in patients diagnosed with NSCLC, ctDNA analysis can provide tumor resistance responses acquired in real-time to TKIs for EGFR [164]. For them, the clinical application of these tests is about to reach reality because the current digital platforms approach greater sensitivity and precision to ctDNA [165]. Some patients presented with the EGFR p.(Tyr790Met) mutation early as 344 days before disease recurrence [160]. EGFR-mutated patients presented undetectable ctDNA after four weeks of TKi therapy, associated with a 12-week radiological response and progression-free and overall survival [66].

A prospective study was carried out in 2015 using available serial samples of blood—collected and follow-up for ten months—of 41 patients with lung cancer. All samples of blood were analyzed cfDNA observing EGFR mutation and p.(Tyr790Met) mutation, and the authors observed that the appearance or increase in a unit of the p.(Tyr790Met) allele frequency almost triples the risk of death and progression, and this information can be used for to estimate whether p.(Tyr790Met) positive patients should start second-line treatment based on molecular data rather than imaging data [166].

In addition to mutations, the decreased concentration of CTCs has been associated with the radiographic response of the tumor during the different treatments in patients with NSCLC (surgery, chemotherapy, radiotherapy, target therapy, immunotherapy), while the increased number of CTCs has been correlated with the disease progression [167].

Studies have also investigated the expression of PD-L1 in CTC and white blood cells (WBC) in NSCLC [68,168–170]. Expression of PD-L1 in CTC and WBC was highly correlated with the tumor tissue expression, pointing out the importance of this evaluation of the "liquid microenvironment" to assist in the immunotherapy stratification and the monitoring of disease [170]. Studies with tumor tissue from patients with NSCLC have shown a relationship between high tumor mutational burden (TMB) and a longer duration of response and survival in patients treated with anti-PD-1 or anti-PD-L1 therapy [171–173]. For example, Gandara et al. (2018) analyzed 797 plasma samples from NSCLC patients, and the authors observed that TMB in plasma (bTMB)$\geq$ 16 had a higher benefit for progression-free survival with atezolizumab therapy [174]. However, few studies have been conducted with a blood-based assay to measure bTMB in lung cancer, and the effectiveness of bTMB remains unclear [175].

Gene fusions can also be detected in blood samples of NSCLC patients, such as anaplastic lymphoma kinase (ALK, ROS1, RET, and NTRK) [176]. The detection and permanence of this fusion were associated in this study with shorter progression-free survival to crizotinib [177].

### 4.3. Early Detection as an Emerging Application of Liquid Biopsy for NSCLC

The high mortality rate of lung cancer patients is related chiefly to late diagnosis when curative treatments are ineffective. In order to decrease lung cancer mortality, screening programs have been implemented worldwide [178,179].

There are currently cancer screening programs, such as breast cancer and lung cancer, which use X-ray and low-dose computed tomography (LDCT) emission, respectively, to increase the chances of early diagnosis and increase patient survival. However, these tests are performed only in the percentage of the population considered at high risk; that is, the young do not have the opportunity to undergo the screening procedure because they also have a chance of developing cancer. In addition, developing countries, associated with low income and the lack of easy access to public health programs, make screening programs difficult to occur due to the cost of equipment and the need for mobile units, which lead to late diagnosis of cancer [180]. For this reason, implementing public policies to aid research and development in the context of liquid biopsy is significant.

However, many challenges have been experienced in lung cancer screening programs, and additional strategies are required to increase the early detection of lung cancer [181]. Plasma levels of cfDNA from NSCLC patients are higher than controls with no cancer, rendering cfDNA plasma levels an exciting approach as a diagnostic biomarker [107,182]. CTCs can be eliminated by the primary tumor, even in the early stages of tumor development. However, they are available at low abundance, requiring precision and empathetic methodologies to detect them [183]. Improvements in the CTC isolation methods have emerged, making CTCs in clinical practice feasible to increase the patient's possibility of care and quality of life [76]. Moreover, the use of biomarkers to distinguish benign from malignant lesions, and to identify molecules that can complement imaging tests during lung cancer screening, can reduce the number of false positives and false negatives [179,180].

MiRNAs released in plasma and serum may also be used for lung cancer screening [184]. The use of microRNAs for lung cancer screening and early detection should be considered to be used along with CT images in order to improve the accuracy of screening programs when CT images do not show clinically detectable disease. The Multicenter Italian Lung Detection (BioMILD) trial showed a miRNA signature classifier (MSC) combined with CT was more effective for risk stratification than only CT or only MSC [185]. Subjects with CT positive and MSC negative (CT+/MSC-) showed an HR of 13.73, and subjects with CT positive and MSC positive (CT+/MSC+) showed an HR of 30.71. However, validation and reproducibility studies are necessary for the implementation of this strategy proposed by the BioMILD trial. In addition, the employment of miRNAs as a minimally invasive tool together with CT in lung cancer screening programs still needs to be better addressed, especially due to conflict data about these biomarkers in blood samples [186].

Technologies are under development to improve the detection of rare and/or small molecules present in the human body. circRNAs-based signatures have emerged as a promising tool for the early detection of lung cancer [187,188].

The CancerSEEK improved the early detection of several types of tumors, including lung cancer, by combining several biomarkers in plasma [123]. The authors analyzed plasma samples from 1005 patients with eight different tumor types, and none of the patients received neoadjuvant chemotherapy nor harbored evident metastases before blood collection. CtDNA isolated from plasma for mutation analysis was initially submitted to PCR amplification using primers designed to amplify regions of interest in the 16 genes. Simultaneously, eight proteins were analyzed in the same samples by the Luminex bead-based immunoassays technique (Millipore, Bilerica, NY). The accuracy of the prediction of this approach was 39% for lung cancer. The sensitivity of CancerSEEK for lung cancer was approximately 60% [123].

The INTEGRAL Risk Biomarker and Nodule Malignancy project analyzes a proteomic panel, based on Olink assay, with 21 protein relevant for lung cancer using a minimum quantity of blood (<50 uL) to optimize LDCT screening [189]. This project was initiated with case-control cohorts—training and validation sets, totalizing more than four thousand people—of several screening programs, main in EUA, Australia, Singapura, and Canada. However, we expect results from this trial to prove the effectiveness of using proteomics panels in lung cancer screening scenarios.

On the other hand, a multiple cancer early detection test (MCED) developed a biomarker based on cfDNA methylation in the plasma of 4077 subjects. This biomarker showed an overall sensitivity and specificity of 51.5% and 99.5%, respectively [72]. For lung cancer, this same methylation panel has shown a sensitivity of 79.5% in cases with stage II (95% CI).

The DETECT-A study (Detecting cancers Earlier Through Elective mutation-based blood Collection and Testing) was performed with 10,000 women, 65 to 75 years old, with no personal history of cancer, and was analyze mutations of 16 genes in cfDNA and nine protein highly validated biomarkers in blood samples [190]. Twenty-six cancer patients were detected by blood testing, and nine were lung cancer. In addition, a PET-CT scan was performed for 15 patients to exclude distant metastasis [190]. Only 1.2% of the individuals tested in the blood test were submitted to PET-CT, decreasing the costs of imaging scans [190]. These data demonstrate the possible relevance of combining multi-cancer blood testing with PET-CT in the clinical routine, improving early detection of lung cancer in a non-invasive way [191]. In addition, lung cancer screening programs can shift the diagnosis scenario from metastatic to early-stage disease. Although overdiagnosis remains a concern, a recent study reported no increase in the overall incidence rate of lung cancer but a decrease in metastatic cases and an increased number of stage I cases compared with an unscreened population [192]. Furthermore, the development of alternative techniques for screening and early detection, such as liquid biopsy, can help the conventional methods already used, reducing the costs of public coffers and reducing exposure to radiation and patient discomfort.

## 5. Challenges and Limitations

Liquid biopsy has been a promising approach for detecting biomarkers in NSCLC patients. This minimally invasive approach better represents tumor heterogeneity and can also be effective for lung cancer screening.

The lack of standardization prevents liquid biopsy implementation in the clinical routine. For this reason, more studies involving protocol standardization and a more significant number of cases become necessary to obtain a more excellent population representation, generating accurate and applicable results. Another limitation is that some biomarkers are fragile, requiring great pre-analytical care. The interaction between genetics and environment is challenging to control. Additionally, specific and sensitive methodologies are necessary to isolate and analyze these biomarkers, mainly due to the low concentration of some molecules in the body fluid.

## 6. Open Issues e Future Perspectives

Although many studies and advances have been made over the last few years concerning liquid biopsy, little has been translated into clinical practice, probably due to the several challenges already mentioned in this review. In addition, we must consider that the vast majority of studies analyze populations of European and North American, opening a gap for the genomic analyses referring to more heterogeneous populations, such as the population of South America and Africa [193,194].

The development and improvement of specific techniques, both to isolate analytes from liquid biopsy and to analyze them, are necessary to increase the sensitivity and specificities of the tests so that they are safe to be used for both early detection, prognosis, and monitoring. Methodologies that increase confidence to detect rare variants or allow a small initial sample input is welcome in this area. Training professionals in this area is also critical for reliable results. In addition, improving new mathematical and computational methods based on machine learning can also improve liquid biopsy methods making this approach even closer to the routine setting.

In this review, we summarized promising biomarkers that can be used for lung cancer screening associated with gold standard methods—such as LDCT—improving detection rates of screening programs (Figure 3). In the near future, a liquid biopsy will hopefully increase the early detection of lung cancer.

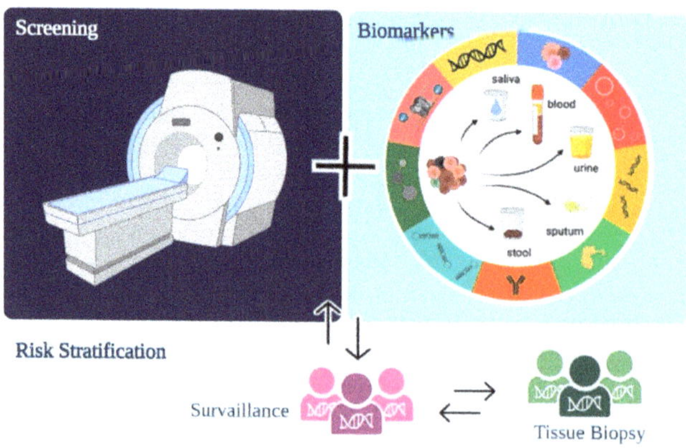


**Figure 3.** Purposed workflow for lung cancer screening programs associating low-dose computed tomography (LDCT) and minimally invasive biomarkers.

**Author Contributions:** Wrote the manuscript: G.M.S.C. and M.d.O.S. (both equally contribute to this review). Review, Editing, and Supervising: L.F.L. and R.M.R. All authors have read and agreed to the published version of the manuscript.

**Funding:** GMSC is supported by Sao Paulo Research Foundation (FAPESP), and MOS is supported by the National Ministry of Health (Brazil). LFL was supported by the Public Ministry of Labor Campinas (Research, Prevention, and Education of Occupational Cancer). RMR is a recipient of a CNPq Productivity (Brazil) fellowship. This research was funded by PRONON—PRONON/MS (Abordagens móveis e de tecnologia para prevenção primária e secundária de câncer—NUP: 25000.015000/2019-53).

**Institutional Review Board Statement:** Review manuscripts do not require IRB approval.

**Informed Consent Statement:** Not applicable due to the nature of the article (Review article). We confirmed no patients data were collected for the current article.

**Data Availability Statement:** No original data were generated by the current manuscript. All information were gathered from open access and/or publicly available manuscripts.

**Acknowledgments:** Funding sources have no contribution to filling out authorship for the present study. We thank all members of the GTOP group (Translational Group of Pulmonary Oncology—Barretos Cancer Hospital, Brazil) for scientific discussion and suggestions and Nicole Mayara Dias de Souza for English editing.

**Conflicts of Interest:** The authors declare no conflict of interest.

## References

1. Siegel, R.L.; Miller, K.D.; Jemal, A. Cancer statistics, 2019. *CA Cancer J. Clin.* **2019**, *69*, 7–34. [CrossRef]
2. Henschke, C.I.; International Early Lung Cancer Action Program Investigators. Survival of patients with clinical stage I lung cancer diagnosed by computed tomography screening for lung cancer. *Clin. Cancer Res. AACR* **2007**, *13*, 4949–4950. [CrossRef] [PubMed]
3. Warren, G.W.; Ostroff, J.S.; Goffin, J.R. Lung cancer screening, cancer treatment, and addressing the continuum of health risks caused by tobacco. *Am. Soc. Clin. Oncol. Educ. Book* **2016**, *36*, 223–229. [CrossRef] [PubMed]
4. Howlader, N.N.; Noone, A.M.; Krapcho, M.; Miller, D.; Brest, A.; Yu, M.; Ruhl, J.; Tatalovich, Z.; Mariotto, A.; Lewis, D.R.; et al. *SEER Cancer Statistics Review, 1975–2017*; National Cancer Institute: Bethesda, MD, USA, 2020; Volume 4.
5. Thun, M.J.; Henley, S.J.; Calle, E.E. Tobacco use and cancer: An epidemiologic perspective for geneticists. *Oncogene* **2002**, *21*, 7307–7325. [CrossRef]
6. Ridge, C.A.; McErlean, A.M.; Ginsberg, M.S. (Eds.) Epidemiology of lung cancer. In *Seminars in Interventional Radiology*; Thieme Medical Publishers: New York, NY, USA, 2013.
7. Markowitz, S.B.; Levin, S.M.; Miller, A.; Morabia, A. Asbestos, Asbestosis, Smoking, and Lung Cancer. New Findings from the North American Insulator Cohort. *Am. J. Respir. Crit. Care Med.* **2013**, *188*, 90–96. [CrossRef]
8. Shankar, A.; Dubey, A.; Saini, D.; Singh, M.; Prasad, C.P.; Roy, S.; Bharati, S.J.; Rinki, M.; Singh, N.; Seth, T.; et al. Environmental and occupational determinants of lung cancer. *Transl. Lung Cancer Res.* **2019**, *8*, S31–S49. [CrossRef] [PubMed]
9. McKay, J.D.; EPIC Study; Hung, R.J.; Gaborieau, V.; Boffetta, P.; Chabrier, A.; Byrnes, G.; Zaridze, D.; Mukeria, A.; Szeszenia-Dabrowska, N.; et al. Lung cancer susceptibility locus at 5p15.33. *Nat. Genet.* **2008**, *40*, 1404–1406. [CrossRef] [PubMed]
10. McKay, J.D.; SpiroMeta Consortium; Hung, R.J.; Han, Y.; Zong, X.; Carreras-Torres, R.; Christiani, D.C.; Caporaso, N.E.; Johansson, M.; Xiao, X.; et al. Large-scale association analysis identifies new lung cancer susceptibility loci and heterogeneity in genetic susceptibility across histological subtypes. *Nat. Genet.* **2017**, *49*, 1126–1132. [CrossRef] [PubMed]
11. Hanahan, D. Hallmarks of Cancer: New Dimensions. *Cancer Discov.* **2022**, *12*, 31–46. [CrossRef]
12. Cruz, C.S.D.; Tanoue, L.T.; Matthay, R.A. Lung cancer: Epidemiology, etiology, and prevention. *Clin. Chest Med.* **2011**, *32*, 605–644. [CrossRef]
13. Howlader, N.N.A.M.; Noone, A.M.; Krapcho, M.E.; Miller, D.; Brest, A.; Yu, M.E.; Ruhl, J.; Tatalovich, Z.; Mariotto, A.; Lewis, D.R.; et al. *SEER Cancer Statistics Review, 1975–2016*; National Cancer Institute: Bethesda, MD, USA, 2019; Volume 1.
14. Planchard, D.; Smit, E.F.; Groen, H.J.M.; Mazieres, J.; Besse, B.; Helland, Å.; Giannone, V.; D'Amelio, A.M., Jr.; Zhang, P.; Mookerjee, B.; et al. Dabrafenib plus trametinib in patients with previously untreated BRAFV600E-mutant metastatic non-small-cell lung cancer: An open-label, phase 2 trial. *Lancet Oncol.* **2017**, *18*, 1307–1316. [CrossRef] [PubMed]
15. Duma, N.; Santana-Davila, R.; Molina, J.R. (Eds.) *Non–Small Cell Lung Cancer: Epidemiology, Screening, Diagnosis, and Treatment*; Mayo Clinic Proceedings; Elsevier: Amsterdam, The Netherlands, 2019.
16. Leal, L.F.; de Paula, F.E.; De Marchi, P.; Viana, L.D.S.; Pinto, G.D.J.; Carlos, C.D.; Berardinelli, G.N.; Miziara, J.E.; da Silva, C.M.; Silva, E.C.A.; et al. Mutational profile of Brazilian lung adenocarcinoma unveils association of EGFR mutations with high Asian ancestry and independent prognostic role of KRAS mutations. *Sci. Rep.* **2019**, *9*, 3209. [CrossRef] [PubMed]
17. Kanwal, M.; Ding, X.-J.; Cao, Y. Familial risk for lung cancer. *Oncol. Lett.* **2017**, *13*, 535–542. [CrossRef]
18. Tan, A.C.; Tan, D.S.W. Targeted Therapies for Lung Cancer Patients With Oncogenic Driver Molecular Alterations. *J. Clin. Oncol.* **2022**, *40*, 611–625. [CrossRef]
19. Harrison, S.; Judd, J.; Chin, S.; Ragin, C. Disparities in Lung Cancer Treatment. *Curr. Oncol. Rep.* **2022**, *24*, 241–248. [CrossRef] [PubMed]

20. Lim, M.; Kim, C.-J.; Sunkara, V.; Kim, M.-H.; Cho, Y.-K. Liquid Biopsy in Lung Cancer: Clinical Applications of Circulating Biomarkers (CTCs and ctDNA). *Micromachines* **2018**, *9*, 100. [CrossRef] [PubMed]
21. Lu, J.; Han, B. Liquid Biopsy Promotes Non-Small Cell Lung Cancer Precision Therapy. *Technol. Cancer Res. Treat.* **2018**, *17*, 1533033818801809. [CrossRef] [PubMed]
22. Pisapia, P.; Malapelle, U.; Troncone, G. Liquid Biopsy and Lung Cancer. *Acta Cytol.* **2019**, *63*, 489–496. [CrossRef]
23. Hoseok, I.; Cho, J.-Y. Lung cancer biomarkers. *Adv. Clin. Chem.* **2015**, *72*, 107–170.
24. Buder, A.; Tomuta, C.; Filipits, M. The potential of liquid biopsies. *Curr. Opin. Oncol.* **2016**, *28*, 130–134. [CrossRef]
25. Castro-Giner, F.; Gkountela, S.; Donato, C.; Alborelli, I.; Quagliata, L.; Ng, C.K.Y.; Piscuoglio, S.; Aceto, N. Cancer Diagnosis Using a Liquid Biopsy: Challenges and Expectations. *Diagnostics* **2018**, *8*, 31. [CrossRef] [PubMed]
26. Wan, J.C.M.; Massie, C.; Garcia-Corbacho, J.; Mouliere, F.; Brenton, J.D.; Caldas, C.; Pacey, S.; Baird, R.; Rosenfeld, N. Liquid biopsies come of age: Towards implementation of circulating tumour DNA. *Nat. Rev. Cancer* **2017**, *17*, 223–238. [CrossRef] [PubMed]
27. Bracht, J.W.P.; Mayo-De-Las-Casas, C.; Berenguer, J.; Karachaliou, N.; Rosell, R. The Present and Future of Liquid Biopsies in Non-Small Cell Lung Cancer: Combining Four Biosources for Diagnosis, Prognosis, Prediction, and Disease Monitoring. *Curr. Oncol. Rep.* **2018**, *20*, 70. [CrossRef] [PubMed]
28. Trombetta, D.; Sparaneo, A.; Fabrizio, F.P.; Muscarella, L.A. Liquid biopsy and NSCLC. *Lung Cancer Manag.* **2016**, *5*, 91–104. [CrossRef]
29. Chan, M.H.M.; Chow, K.M.; Chan, A.T.C.; Leung, C.B.; Chan, L.Y.S.; Chow, K.C.K.; Lam, C.W.; Lo, Y.M.D. Quantitative Analysis of Pleural Fluid Cell-free DNA as a Tool for the Classification of Pleural Effusions. *Clin. Chem.* **2003**, *49*, 740–745. [CrossRef]
30. Pu, D.; Liang, H.; Wei, F.; Akin, D.; Feng, Z.; Yan, Q.; Li, Y.; Zhen, Y.; Xu, L.; Dong, G.; et al. Evaluation of a novel saliva-based epidermal growth factor receptor mutation detection for lung cancer: A pilot study. *Thorac. Cancer* **2016**, *7*, 428–436. [CrossRef]
31. Wang, Y.; Springer, S.; Zhang, M.; McMahon, K.W.; Kinde, I.; Dobbyn, L.; Ptak, J.; Brem, H.; Chaichana, K.; Gallia, G.L.; et al. Detection of tumor-derived DNA in cerebrospinal fluid of patients with primary tumors of the brain and spinal cord. *Proc. Natl. Acad. Sci. USA* **2015**, *112*, 9704–9709. [CrossRef]
32. Fujii, T.; Barzi, A.; Sartore-Bianchi, A.; Cassingena, A.; Siravegna, G.; Karp, D.D.; Piha-Paul, S.A.; Subbiah, V.; Tsimberidou, A.M.; Huang, H.J.; et al. Mutation-Enrichment Next-Generation Sequencing for Quantitative Detection of KRAS Mutations in Urine Cell-Free DNA from Patients with Advanced CancersKRAS Mutations in Urine cfDNA. *Clin. Cancer Res.* **2017**, *23*, 3657–3666. [CrossRef]
33. Benlloch, S.; Martí-Ciriquián, J.L.; Galbis-Caravajal, J.M.; Martín, C.; Sánchez-Payá, J.; Rodríguez-Paniagua, J.M.; Romero, S.; Massutí, B. Cell-Free DNA Concentration in Pleural Fluid and Serum: Quantitative Approach and Potential Prognostic Factor in Patients with Cancer and Pleural Effusions. *Clin. Lung Cancer* **2006**, *8*, 140–145. [CrossRef]
34. Abbosh, C.; Birkbak, N.J.; Wilson, G.A.; Jamal-Hanjani, M.; Constantin, T.; Salari, R.; Le Quesne, J.; Moore, D.A.; Veeriah, S.; Rosenthal, R.; et al. Phylogenetic ctDNA analysis depicts early-stage lung cancer evolution. *Nature* **2017**, *545*, 446–451. [CrossRef]
35. Yamamoto, H.; Watanabe, Y.; Itoh, F. Cell-Free DNA. In *Biomarkers in Cancer Therapy*; Springer: Berlin/Heidelberg, Germany, 2019; pp. 11–24.
36. Cescon, D.W.; Bratman, S.V.; Chan, S.M.; Siu, L.L. Circulating tumor DNA and liquid biopsy in oncology. *Nat. Cancer* **2020**, *1*, 276–290. [CrossRef] [PubMed]
37. Hanssen, A.; Wagner, J.; Gorges, T.M.; Taenzer, A.; Uzunoglu, F.G.; Driemel, C.; Stoecklein, N.H.; Knoefel, W.T.; Angenendt, S.; Hauch, S.; et al. Characterization of different CTC subpopulations in non-small cell lung cancer. *Sci. Rep.* **2016**, *6*, 28010. [CrossRef] [PubMed]
38. Shaffer, C. Circulating Tumor Cells and the Liquid Biopsy: Processed appropriately, liquid biopsy samples and circulating biomarkers may aid cancer diagnosis and predict treatment outcomes. *Genet. Eng. Biotechnol. News* **2019**, *39*, 42–44. [CrossRef]
39. Jia, S.; Zhang, R.; Li, Z.; Li, J. Clinical and biological significance of circulating tumor cells, circulating tumor DNA, and exosomes as biomarkers in colorectal cancer. *Oncotarget* **2017**, *8*, 55632. [CrossRef]
40. Iqbal, M.A.; Arora, S.; Prakasam, G.; Calin, G.A.; Syed, M.A. MicroRNA in lung cancer: Role, mechanisms, pathways and therapeutic relevance. *Mol. Asp. Med.* **2019**, *70*, 3–20. [CrossRef]
41. Li, L.; Fu, K.; Zhou, W.; Snyder, M. Applying circulating tumor DNA methylation in the diagnosis of lung cancer. *Precis. Clin. Med.* **2019**, *2*, 45–56. [CrossRef]
42. Moss, J.; Magenheim, J.; Neiman, D.; Zemmour, H.; Loyfer, N.; Korach, A.; Samet, Y.; Maoz, M.; Druid, H.; Arner, P.; et al. Comprehensive human cell-type methylation atlas reveals origins of circulating cell-free DNA in health and disease. *Nat. Commun.* **2018**, *9*, 5068. [CrossRef] [PubMed]
43. Sun, Q.; Zhao, W.; Wang, L.; Guo, F.; Song, D.; Zhang, Q.; Zhang, D.; Fan, Y.; Wang, J. Integration of metabolomic and transcriptomic profiles to identify biomarkers in serum of lung cancer. *J. Cell. Biochem.* **2019**, *120*, 11981–11989. [CrossRef] [PubMed]
44. Tang, Y.; Li, Z.; Lazar, L.; Fang, Z.; Tang, C.; Zhao, J. Metabolomics workflow for lung cancer: Discovery of biomarkers. *Clin. Chim. Acta* **2019**, *495*, 436–445. [CrossRef] [PubMed]
45. Seow, W.J.; Shu, X.-O.; Nicholson, J.; Holmes, E.; Walker, D.I.; Hu, W.; Cai, Q.; Gao, Y.-T.; Xiang, Y.-B.; Moore, S.C.; et al. Association of Untargeted Urinary Metabolomics and Lung Cancer Risk Among Never-Smoking Women in China. *JAMA Netw. Open* **2019**, *2*, e1911970. [CrossRef]
46. Yokota, H.; Guo, J.; Matoba, M.; Higashi, K.; Tonami, H.; Nagao, Y. Lactate, choline, and creatine levels measured by vitro 1H-MRS as prognostic parameters in patients with non-small-cell lung cancer. *J. Magn. Reson. Imaging Off. J. Int. Soc. Magn. Reson. Med.* **2007**, *25*, 992–999.

47. Chapman, C.; Murray, A.; Chakrabarti, J.; Thorpe, A.; Woolston, C.; Sahin, U.; Barnes, A.; Robertson, J. Autoantibodies in breast cancer: Their use as an aid to early diagnosis. *Ann. Oncol.* **2007**, *18*, 868–873. [CrossRef] [PubMed]
48. Du, Q.; Yu, R.; Wang, H.; Yan, N.; Yuan, Q.; Ma, Y.; Slamon, D.; Hou, D.; Wang, H.; Wang, Q. Significance of tumor-associated autoantibodies in the early diagnosis of lung cancer. *Clin. Respir. J.* **2018**, *12*, 2020–2028. [CrossRef] [PubMed]
49. Luo, W.; Rao, M.; Qu, J.; Luo, D. Applications of liquid biopsy in lung cancer-diagnosis, prognosis prediction, and disease monitoring. *Am. J. Transl. Res.* **2018**, *10*, 3911. [PubMed]
50. Miyanaga, A.; Masuda, M.; Yamada, T. Biomarkers of Lung Cancer: Liquid Biopsy Comes of Age. In *Biomarkers in Cancer Therapy*; Springer: Berlin/Heidelberg, Germany, 2019; pp. 105–113.
51. Diaz, L.A., Jr.; Bardelli, A. Liquid biopsies: Genotyping circulating tumor DNA. *J. Clin. Oncol. Off. J. Am. Soc. Clin. Oncol.* **2014**, *32*, 579. [CrossRef]
52. Davis, A.A.; Cristofanilli, M. Detection of Predictive Biomarkers Using Liquid Biopsies. In *Predictive Biomarkers in Oncology*; Springer: Berlin/Heidelberg, Germany, 2019; pp. 107–117.
53. Gale, D.; Lawson, A.R.; Howarth, K.; Madi, M.; Durham, B.; Smalley, S.; Calaway, J.; Blais, S.; Jones, G.; Clark, J.; et al. Development of a highly sensitive liquid biopsy platform to detect clinically-relevant cancer mutations at low allele fractions in cell-free DNA. *PLoS ONE* **2018**, *13*, e0194630. [CrossRef]
54. Revelo, A.E.; Martin, A.; Velasquez, R.; Kulandaisamy, P.C.; Bustamante, J.; Keshishyan, S.; Otterson, G. Liquid biopsy for lung cancers: An update on recent developments. *Ann. Transl. Med.* **2019**, *7*, 349. [CrossRef]
55. Jee, J.; Lebow, E.S.; Murciano-Goroff, Y.R.; Jayakumaran, G.; Shen, R.; Brannon, A.R.; Benayed, R.; Namakydoust, A.; Offin, M.; Paik, P.K.; et al. *Overall Survival with Circulating Tumor DNA-Guided Therapy in Advanced Non-Small Cell Lung Cancer*; Wolters Kluwer Health: Philadelphia, PA, USA, 2021.
56. Shendure, J.; Ji, H. Next-generation DNA sequencing. *Nat. Biotechnol.* **2008**, *26*, 1135–1145. [CrossRef]
57. Wong, W.H.; Tong, R.S.; Young, A.L.; Druley, T.E. Rare Event Detection Using Error-corrected DNA and RNA Sequencing. *J. Vis. Exp.* **2018**, e57509. [CrossRef]
58. Teder, H.; Koel, M.; Paluoja, P.; Jatsenko, T.; Rekker, K.; Laisk-Podar, T.; Kukuškina, V.; Velthut-Meikas, A.; Fjodorova, O.; Peters, M.; et al. TAC-seq: Targeted DNA and RNA sequencing for precise biomarker molecule counting. *NPJ Genom. Med.* **2018**, *3*, 34. [CrossRef]
59. Villaflor, V.; Won, B.; Nagy, R.; Banks, K.; Lanman, R.B.; Talasaz, A.; Salgia, R. Biopsy-free circulating tumor DNA assay identifies actionable mutations in lung cancer. *Oncotarget* **2016**, *7*, 66880. [PubMed]
60. Sabari, J.K.; Offin, M.; Stephens, D.; Ni, A.; Lee, A.; Pavlakis, N.; Clarke, S.; Diakos, C.I.; Datta, S.; Tandon, N.; et al. A Prospective Study of Circulating Tumor DNA to Guide Matched Targeted Therapy in Lung Cancers. *Gynecol. Oncol.* **2019**, *111*, 575–583. [CrossRef] [PubMed]
61. FuFaD, A. Aprovação Pré-Comercialização P150044, Teste de Mutação Cobas EGFR V2. 2017 FD. 2017. Available online: https://www.accessdata.fda.gov/cdrh_docs/pdf12/p120019s007c.pdf (accessed on 1 September 2022).
62. Torres, S.; González, Á.; Tomás, A.J.C.; Fariñas, S.C.; Ferrero, M.; Mirda, D.; Sirera, R.; Jantus-Lewintre, E.; Camps, C. A profile on cobas®EGFR Mutation Test v2 as companion diagnostic for first-line treatment of patients with non-small cell lung cancer. *Expert Rev. Mol. Diagn.* **2020**, *20*, 575–582.
63. Vrba, L.; Oshiro, M.M.; Kim, S.S.; Garland, L.L.; Placencia, C.; Mahadevan, D.; Nelson, M.A.; Futscher, B.W. DNA methylation biomarkers discovered in silico detect cancer in liquid biopsies from non-small cell lung cancer patients. *Epigenetics* **2020**, *15*, 419–430. [PubMed]
64. Duruisseaux, M.; Esteller, M. (Eds.) Lung cancer epigenetics: From knowledge to applications. In *Seminars in Cancer Biology*; Elsevier: Amsterdam, The Netherlands, 2018.
65. Bjaanaes, M.M.; Fleischer, T.; Halvorsen, A.R.; Daunay, A.; Busato, F.; Solberg, S.; Jørgensen, L.; Kure, E.; Edvardsen, H.; Børresen-Dale, A.-L.; et al. Genome-wide DNA methylation analyses in lung adenocarcinomas: Association with EGFR, KRAS and TP53 mutation status, gene expression and prognosis. *Mol. Oncol.* **2016**, *10*, 330–343. [PubMed]
66. Wang, J.; Duan, Y.; Meng, Q.-H.; Gong, R.; Guo, C.; Zhao, Y.; Zhang, Y. Integrated analysis of DNA methylation profiling and gene expression profiling identifies novel markers in lung cancer in Xuanwei, China. *PLoS ONE* **2018**, *13*, e0203155.
67. Ooki, A.; Maleki, Z.; Tsay, J.-C.J.; Goparaju, C.; Brait, M.; Turaga, N.; Nam, H.-S.; Rom, W.N.; Pass, H.I.; Sidransky, D.; et al. A Panel of Novel Detection and Prognostic Methylated DNA Markers in Primary Non–Small Cell Lung Cancer and Serum DNA. *Clin. Cancer Res.* **2017**, *23*, 7141–7152.
68. Yang, Z.; Qi, W.; Sun, L.; Zhou, H.; Zhou, B.; Hu, Y. DNA methylation analysis of selected genes for the detection of early-stage lung cancer using circulating cell-free DNA. *Adv. Clin. Exp. Med. Off. Organ Wroc. Med. Univ.* **2019**, *28*, 355–360. [CrossRef]
69. Pérez-Ramírez, C.; Cañadas-Garre, M.; Robles, A.I.; Molina, M.; Faus-Dáder, M.J.; Calleja-Hernández, M. Liquid biopsy in early stage lung cancer. *Transl. Lung Cancer Res.* **2016**, *5*, 517. [CrossRef]
70. Hulbert, A.; Jusue-Torres, I.; Stark, A.; Chen, C.; Rodgers, K.; Lee, B.; Griffin, C.; Yang, A.; Huang, P.; Wrangle, J.; et al. Early Detection of Lung Cancer Using DNA Promoter Hypermethylation in Plasma and SputumEpigenetic Lung Cancer Screening. *Clin. Cancer Res.* **2017**, *23*, 1998–2005.
71. Liu, M.C.; Oxnard, G.R.; Klein, E.A.; Swanton, C.; Seiden, M.V.; CCGA Consortium. Sensitive and specific multi-cancer detection and localization using methylation signatures in cell-free DNA. *Ann. Oncol.* **2020**, *31*, 745–759. [PubMed]

72. Klein, E.; Richards, D.; Cohn, A.; Tummala, M.; Lapham, R.; Cosgrove, D.; Chung, G.; Clement, J.; Gao, J.; Hunkapiller, N.; et al. Clinical validation of a targeted methylation-based multi-cancer early detection test using an independent validation set. *Ann. Oncol.* **2021**, *32*, 1167–1177. [PubMed]
73. Neal, R.D.; Johnson, P.; Clarke, C.A.; Hamilton, S.A.; Zhang, N.; Kumar, H.; Swanton, C.; Sasieni, P. Cell-Free DNA–Based Multi-Cancer Early Detection Test in an Asymptomatic Screening Population (NHS-Galleri): Design of a Pragmatic, Prospective Randomised Controlled Trial. *Cancers* **2022**, *14*, 4818. [PubMed]
74. Poggiana, C.; Rossi, E.; Zamarchi, R. Possible role of circulating tumor cells in early detection of lung cancer. *J. Thorac. Dis.* **2020**, *12*, 3821. [CrossRef]
75. Santarpia, M.; Liguori, A.; Karachaliou, N.; Gonzalez-Cao, M.; Daffinà, M.G.; D'Aveni, A.; Marabello, G.; Altavilla, G.; Rosell, R. Osimertinib in the treatment of non-small-cell lung cancer: Design, development and place in therapy. *Lung Cancer Targets Ther.* **2017**, *8*, 109. [CrossRef]
76. He, Y.; Shi, J.; Schmidt, B.; Liu, Q.; Shi, G.; Xu, X.; Liu, C.; Gao, Z.; Guo, T.; Shan, B. Circulating Tumor Cells as a Biomarker to Assist Molecular Diagnosis for Early Stage Non-Small Cell Lung Cancer. *Cancer Manag. Res.* **2020**, *12*, 841. [CrossRef]
77. Correnti, M.; Raggi, C. Stem-like plasticity and heterogeneity of circulating tumor cells: Current status and prospect challenges in liver cancer. *Oncotarget* **2017**, *8*, 7094.
78. Zhang, Y.; Wang, Y.; Du, Z.; Wu, M.; Zhang, G. Detection of micrometastases in lung cancer with magnetic nanoparticles and quantum dots. *Int. J. Nanomed.* **2012**, *7*, 2315. [CrossRef]
79. Tu, Q.; Wu, X.; Le Rhun, E.; Blonski, M.; Wittwer, B.; Taillandier, L.; Bittencourt, M.D.C.; Faure, G.C. CellSearch® technology applied to the detection and quantification of tumor cells in CSF of patients with lung cancer leptomeningeal metastasis. *Lung Cancer* **2015**, *90*, 352–357.
80. Hofman, V.; Long, E.; Ilie, M.; Bonnetaud, C.; Vignaud, J.M.; Fléjou, J.F.; Lantuejoul, S.; Piaton, E.; Mourad, N.; Butori, C.; et al. Morphological analysis of circulating tumour cells in patients undergoing surgery for non-small cell lung carcinoma using the isolation by size of epithelial tumour cell (ISET) method. *Cytopathology* **2012**, *23*, 30–38. [CrossRef]
81. Turetta, M.; Bulfoni, M.; Brisotto, G.; Fasola, G.; Zanello, A.; Biscontin, E.; Mariuzzi, L.; Steffan, A.; Di Loreto, C.; Cesselli, D.; et al. Assessment of the Mutational Status of NSCLC Using Hypermetabolic Circulating Tumor Cells. *Cancers* **2018**, *10*, 270. [PubMed]
82. Reclusa, P.; Taverna, S.; Pucci, M.; Durendez, E.; Calabuig, S.; Manca, P.; Serrano, M.J.; Sober, L.; Pauwels, P.; Russo, A.; et al. Exosomes as diagnostic and predictive biomarkers in lung cancer. *J. Thorac. Dis.* **2017**, *9* (Suppl. S13), S1373–S1382. [CrossRef] [PubMed]
83. Vanni, I.; Alama, A.; Grossi, F.; Dal Bello, M.G.; Coco, S. Exosomes: A new horizon in lung cancer. *Drug Discov. Today* **2017**, *22*, 927–936. [PubMed]
84. Momen-Heravi, F.; Balaj, L.; Alian, S.; Mantel, P.-Y.; Halleck, A.E.; Trachtenberg, A.J.; Soria, C.E.; Oquin, S.; Bonebreak, C.M.; Saracoglu, E.; et al. Current methods for the isolation of extracellular vesicles. *Biol. Chem.* **2013**, *394*, 1253–1262.
85. Böing, A.N.; van der Pol, E.; Grootemaat, A.E.; Coumans, F.A.W.; Sturk, A.; Nieuwland, R. Single-step isolation of extracellular vesicles by size-exclusion chromatography. *J. Extracell. Vesicles* **2014**, *3*, 23430.
86. Qu, X.; Li, Q.; Yang, J.; Zhao, H.; Wang, F.; Zhang, F.; Zhang, S.; Zhang, H.; Wang, R.; Wang, Q.; et al. Double-Stranded DNA in Exosomes of Malignant Pleural Effusions as a Novel DNA Source for EGFR Mutation Detection in Lung Adenocarcinoma. *Front. Oncol.* **2019**, *9*, 931.
87. Siravegna, G.; Marsoni, S.; Siena, S.; Bardelli, A. Integrating liquid biopsies into the management of cancer. *J. Clin. Oncol. Off. J. Am. Soc. Clin. Oncol.* **2017**, *14*, 531–548.
88. Rodríguez, M.; Silva, J.; López Alfonso, A.; López-Muñiz, M.B.; Peña, C.; Domínguez, G.; García, J.M.; López-González, A.; Méndez, M.; Provencio, M.; et al. Different exosome cargo from plasma/bronchoalveolar lavage in non-small-cell lung cancer. *Genes Chromosom. Cancer* **2014**, *53*, 713–724.
89. Rabinowits, G.; Gerçel-Taylor, C.; Day, J.M.; Taylor, D.D.; Kloecker, G.H. Exosomal MicroRNA: A Diagnostic Marker for Lung Cancer. *Clin. Lung Cancer* **2009**, *10*, 42–46. [CrossRef]
90. Tomasetti, M.; Lee, W.; Santarelli, L.; Neuzil, J. Exosome-derived microRNAs in cancer metabolism: Possible implications in cancer diagnostics and therapy. *Exp. Mol. Med.* **2017**, *49*, e285.
91. Galka-Marciniak, P.; Urbanek-Trzeciak, M.O.; Nawrocka, P.M.; Dutkiewicz, A.; Giefing, M.; Lewandowska, M.A.; Kozlowski, P. Somatic Mutations in miRNA Genes in Lung Cancer—Potential Functional Consequences of Non-Coding Sequence Variants. *Cancers* **2019**, *11*, 793. [PubMed]
92. Almeida, M.I.; Reis, R.M.; Calin, G.A. MicroRNA history: Discovery, recent applications, and next frontiers. *Mutat. Res./Fundam. Mol. Mech. Mutagen.* **2011**, *717*, 1–8.
93. Mitchell, P.S.; Parkin, R.K.; Kroh, E.M.; Fritz, B.R.; Wyman, S.K.; Pogosova-Agadjanyan, E.L.; Peterson, A.; Noteboom, J.; O'Briant, K.C.; Allen, A.; et al. Circulating microRNAs as stable blood-based markers for cancer detection. *Proc. Natl. Acad. Sci. USA* **2008**, *105*, 10513–10518. [CrossRef] [PubMed]
94. Puerta-Gil, P.; García-Baquero, R.; Jia, A.Y.; Ocaña, S.; Alvarez-Múgica, M.; Alvarez-Ossorio, J.L.; Cordon-Cardo, C.; Cava, F.; Sánchez-Carbayo, M. miR-143, miR-222, and miR-452 Are Useful as Tumor Stratification and Noninvasive Diagnostic Biomarkers for Bladder Cancer. *Am. J. Pathol.* **2012**, *180*, 1808–1815. [CrossRef]
95. Zhi, F.; Cao, X.; Xie, X.; Wang, B.; Dong, W.; Gu, W.; Ling, Y.; Wang, R.; Yang, Y.; Liu, Y. Identification of Circulating MicroRNAs as Potential Biomarkers for Detecting Acute Myeloid Leukemia. *PLoS ONE* **2013**, *8*, e56718.

96. Reis, P.P.; Drigo, S.A.; Carvalho, R.F.; Lapa, R.M.L.; Felix, T.F.; Patel, D.; Cheng, D.; Pintilie, M.; Liu, G.; Tsao, M.-S. Circulating miR-16-5p, miR-92a-3p, and miR-451a in Plasma from Lung Cancer Patients: Potential Application in Early Detection and a Regulatory Role in Tumorigenesis Pathways. *Cancers* **2020**, *12*, 2071. [CrossRef]
97. Wozniak, M.B.; Scelo, G.; Muller, D.C.; Mukeria, A.; Zaridze, D.; Brennan, P. Circulating MicroRNAs as Non-Invasive Biomarkers for Early Detection of Non-Small-Cell Lung Cancer. *PLoS ONE* **2015**, *10*, e0125026.
98. Huang, J.; Wu, J.; Li, Y.; Li, X.; Yang, T.; Yang, Q.; Jiang, Y. Deregulation of Serum MicroRNA Expression Is Associated with Cigarette Smoking and Lung Cancer. *BioMed Res. Int.* **2014**, *2014*, 364316. [CrossRef]
99. Zhao, K.; Cheng, J.; Chen, B.; Liu, Q.; Xu, D.; Zhang, Y. Circulating microRNA-34 family low expression correlates with poor prognosis in patients with non-small cell lung cancer. *J. Thorac. Dis.* **2017**, *9*, 3735. [CrossRef]
100. Arnaiz, E.; Sole, C.; Manterola, L.; Iparraguirre, L.; Otaegui, D.; Lawrie, C.H. (Eds.) CircRNAs and cancer: Biomarkers and master regulators. In *Seminars in Cancer Biology*; Elsevier: Amsterdam, The Netherlands, 2019.
101. Suzuki, H.; Zuo, Y.; Wang, J.; Zhang, M.Q.; Malhotra, A.; Mayeda, A. Characterization of RNase R-digested cellular RNA source that consists of lariat and circular RNAs from pre-mRNA splicing. *Nucleic Acids Res.* **2006**, *34*, e63. [CrossRef]
102. Vincent, H.A.; Deutscher, M.P. Substrate Recognition and Catalysis by the Exoribonuclease RNase R. *J. Biol. Chem.* **2006**, *281*, 29769–29775. [CrossRef] [PubMed]
103. Salzman, J.; Chen, R.E.; Olsen, M.N.; Wang, P.L.; Brown, P.O. Cell-type specific features of circular RNA expression. *PLoS Genet.* **2013**, *9*, e1003777. [CrossRef]
104. Jeck, W.R.; Sharpless, N.E. Detecting and characterizing circular RNAs. *Nat. Biotechnol.* **2014**, *32*, 453–461. [CrossRef]
105. De Fraipont, F.; Gazzeri, S.; Cho, W.C.; Eymin, B. Circular RNAs and RNA splice variants as biomarkers for prognosis and therapeutic response in the liquid biopsies of lung cancer patients. *Front. Genet.* **2019**, *10*, 390. [PubMed]
106. Pedraz-Valdunciel, C.; Giannoukakos, S.; Potie, N.; Giménez-Capitán, A.; Huang, C.; Hackenberg, M.; Fernandez-Hilario, A.; Bracht, J.; Filipska, M.; Aldeguer, E.; et al. Digital multiplexed analysis of circular RNAs in FFPE and fresh non-small cell lung cancer specimens. *Mol. Oncol.* **2022**, *16*, 2367–2383. [CrossRef]
107. Chen, K.-Z.; Lou, F.; Yang, F.; Zhang, J.-B.; Ye, H.; Chen, W.; Guan, T.; Zhao, M.-Y.; Su, X.-X.; Shi, R.; et al. Circulating Tumor DNA Detection in Early-Stage Non-Small Cell Lung Cancer Patients by Targeted Sequencing. *Sci. Rep.* **2016**, *6*, 31985. [CrossRef]
108. Zhu, X.; Wang, X.; Wei, S.; Chen, Y.; Chen, Y.; Fan, X.; Han, S.; Wu, G. hsa_circ_0013958: A circular RNA and potential novel biomarker for lung adenocarcinoma. *FEBS J.* **2017**, *284*, 2170–2182. [CrossRef]
109. Zhang, X.; Zhou, H.; Jing, W.; Luo, P.; Qiu, S.; Liu, X.; Zhu, M.; Liang, C.; Yu, M.; Tu, J. The Circular RNA hsa_circ_0001445 Regulates the Proliferation and Migration of Hepatocellular Carcinoma and May Serve as a Diagnostic Biomarker. *Dis. Markers* **2018**, *2018*, 3073467. [CrossRef]
110. Hang, D.; Zhou, J.; Qin, N.; Zhou, W.; Ma, H.; Jin, G.; Hu, Z.; Dai, J.; Shen, H. A novel plasma circular RNA circFARSA is a potential biomarker for non-small cell lung cancer. *Cancer Med.* **2018**, *7*, 2783–2791. [CrossRef]
111. Yu, L.; Li, K.; Zhang, X. Next-generation metabolomics in lung cancer diagnosis, treatment and precision medicine: Mini review. *Oncotarget* **2017**, *8*, 115774. [CrossRef]
112. Kumar, A.; Misra, B.B. Challenges and Opportunities in Cancer Metabolomics. *Proteomics* **2019**, *19*, e1900042. [CrossRef] [PubMed]
113. Molina, R.; Filella, X.; Augé, J.; Fuentes, R.; Bover, I.; Rifa, J.; Moreno, V.; Canals, E.; Viñolas, N.; Marquez, A.; et al. Tumor Markers (CEA, CA 125, CYFRA 21-1, SCC and NSE) in Patients with Non-Small Cell Lung Cancer as an Aid in Histological Diagnosis and Prognosis. *Tumor Biol.* **2003**, *24*, 209–218. [CrossRef] [PubMed]
114. Barlési, F.; Gimenez, C.; Torre, J.-P.; Doddoli, C.; Mancini, J.; Greillier, L.; Roux, F.; Kleisbauer, J.-P. Prognostic value of combination of Cyfra 21-1, CEA and NSE in patients with advanced non-small cell lung cancer. *Respir. Med.* **2004**, *98*, 357–362. [CrossRef]
115. Pang, W.W.; Abdul-Rahman, P.S.; Wan-Ibrahim, W.I.; Hashim, O.H. Can the acute-phase reactant proteins be used as cancer biomarkers? *Int. J. Biol. Markers* **2010**, *25*, 1–11. [CrossRef]
116. Kang, S.-M.; Sung, H.-J.; Ahn, J.-M.; Park, J.-Y.; Lee, S.-Y.; Park, C.-S.; Cho, J.-Y. The Haptoglobin β chain as a supportive biomarker for human lung cancers. *Mol. Biosyst.* **2011**, *7*, 1167–1175. [CrossRef] [PubMed]
117. Sung, H.-J.; Ahn, J.-M.; Yoon, Y.-H.; Rhim, T.-Y.; Park, C.-S.; Park, J.-Y.; Lee, S.-Y.; Kim, J.-W.; Cho, J.-Y. Identification and Validation of SAA as a Potential Lung Cancer Biomarker and its Involvement in Metastatic Pathogenesis of Lung Cancer. *J. Proteome Res.* **2011**, *10*, 1383–1395. [CrossRef] [PubMed]
118. Jensen, L.; Whitehead, A.S. Regulation of serum amyloid A protein expression during the acute-phase response. *Biochem. J.* **1998**, *334*, 489–503. [CrossRef]
119. De Petris, L.; Orre, L.M.; Kanter, L.; Pernemalm, M.; Koyi, H.; Lewensohn, R.; Lehtiö, J. Tumor expression of S100A6 correlates with survival of patients with stage I non-small-cell lung cancer. *Lung Cancer* **2009**, *63*, 410–417. [CrossRef] [PubMed]
120. Pujol, J.L.; Grenier, J.; Daurès, J.P.; Daver, A.; Pujol, H.; Michel, F.B. Serum fragment of cytokeratin subunit 19 measured by CYFRA 21-1 immunoradiometric assay as a marker of lung cancer. *Cancer Res.* **1993**, *53*, 61–66.
121. Bergqvist, M.; Brattström, D.; Hesselius, P.; Wiklund, B.; Silen, A.; Wagenius, G.; Brodin, O. Cytokeratin 8 and 18 fragments measured in serum and their relation to survival in patients with non-small cell lung cancer. *Anticancer Res.* **1999**, *19*, 1833–1836.
122. Xu, B.J.; Gonzalez, A.L.; Kikuchi, T.; Yanagisawa, K.; Massion, P.P.; Wu, H.; Mason, S.E.; Olson, S.J.; Shyr, Y.; Carbone, D.P.; et al. MALDI-MS derived prognostic protein markers for resected non-small cell lung cancer. *Proteom. Clin. Appl.* **2008**, *2*, 1508–1517. [CrossRef] [PubMed]

123. Cohen, J.D.; Li, L.; Wang, Y.; Thoburn, C.; Afsari, B.; Danilova, L.; Douville, C.; Javed, A.A.; Wong, F.; Mattox, A.; et al. Detection and localization of surgically resectable cancers with a multi-analyte blood test. *Science* **2018**, *359*, 926–930. [CrossRef] [PubMed]
124. Zamay, T.N.; Zamay, G.S.; Kolovskaya, O.S.; Zukov, R.A.; Petrova, M.M.; Gargaun, A.; Berezovski, M.V.; Kichkailo, A.S. Current and Prospective Protein Biomarkers of Lung Cancer. *Cancers* **2017**, *9*, 155. [CrossRef]
125. Alcala, K.; Guida, F.; Johansson, M.; Johansson, M.; Robbins, H.A.; Smith-Byrne, K.; Stevens, V.; Zahed, H. Lung Cancer Cohort Consortium The Blood Proteome of Imminent Lung Cancer Diagnosis. *medRxiv* **2022**. [CrossRef]
126. Mehan, M.R.; Williams, S.A.; Siegfried, J.M.; Bigbee, W.L.; Weissfeld, J.L.; Wilson, D.O.; Pass, H.I.; Rom, W.N.; Muley, T.; Meister, M.; et al. Validation of a blood protein signature for non-small cell lung cancer. *Clin. Proteom.* **2014**, *11*, 32. [CrossRef] [PubMed]
127. Boyle, P.; Chapman, C.J.; Holdenrieder, S.; Murray, A.; Robertson, C.; Wood, W.C.; Maddison, P.; Healey, G.; Fairley, G.H.; Barnes, A.C.; et al. Clinical validation of an autoantibody test for lung cancer. *Ann. Oncol.* **2011**, *22*, 383–389. [CrossRef]
128. Tan, E.M. Autoantibodies as reporters identifying aberrant cellular mechanisms in tumorigenesis. *J. Clin. Investig.* **2001**, *108*, 1411–1415. [CrossRef]
129. Houghton, A.N. Cancer antigens: Immune recognition of self and altered self. *J. Exp. Med.* **1994**, *180*, 1–4. [CrossRef]
130. Tan, E.M.; Zhang, J. Autoantibodies to tumor-associated antigens: Reporters from the immune system. *Immunol. Rev.* **2008**, *222*, 328–340. [CrossRef]
131. Zhang, J.-Y.; Tan, E.M. Autoantibodies to tumor-associated antigens as diagnostic biomarkers in hepatocellular carcinoma and other solid tumors. *Expert Rev. Mol. Diagn.* **2010**, *10*, 321–328. [CrossRef]
132. Veronesi, G.; Bianchi, F.; Infante, M.; Alloisio, M. The challenge of small lung nodules identified in CT screening: Can biomarkers assist diagnosis? *Biomark. Med.* **2016**, *10*, 137–143. [CrossRef] [PubMed]
133. Dai, L.; Tsay, J.-C.J.; Li, J.; Yie, T.-A.; Munger, J.S.; Pass, H.; Rom, W.N.; Zhang, Y.; Tan, E.M.; Zhang, J.-Y. Autoantibodies against tumor-associated antigens in the early detection of lung cancer. *Lung Cancer* **2016**, *99*, 172–179. [CrossRef] [PubMed]
134. Burotto, M.; Thomas, A.; Subramaniam, D.; Giaccone, G.; Rajan, A. Biomarkers in Early-Stage Non–Small-Cell Lung Cancer: Current Concepts and Future Directions. *J. Thorac. Oncol.* **2014**, *9*, 1609–1617. [CrossRef] [PubMed]
135. Chen, H.; Werner, S.; Tao, S.; Zörnig, I.; Brenner, H. Blood autoantibodies against tumor-associated antigens as biomarkers in early detection of colorectal cancer. *Cancer Lett.* **2014**, *346*, 178–187. [CrossRef]
136. Lubin, R.; Zalcman, G.; Bouchet, L.; Trédaniel, J.; Legros, Y.; Cazals, D.; Hirsch, A.; Soussi, T. Serum p53 antibodies as early markers of lung cancer. *Nat. Med.* **1995**, *1*, 701–702. [CrossRef] [PubMed]
137. Murray, A.; Chapman, C.; Healey, G.; Peek, L.; Parsons, G.; Baldwin, D.; Barnes, A.; Sewell, H.; Fritsche, H.; Robertson, J. Technical validation of an autoantibody test for lung cancer. *Ann. Oncol.* **2010**, *21*, 1687–1693. [CrossRef]
138. Farlow, E.C.; Patel, K.; Basu, S.; Lee, B.S.; Kim, A.W.; Coon, J.S.; Faber, L.P.; Bonomi, P.; Liptay, M.J.; Borgia, J.A. Development of a Multiplexed Tumor-Associated Autoantibody-Based Blood Test for the Detection of Non–Small Cell Lung CancerEarly Detection Serum Test for NSCLC. *Clin. Cancer Res.* **2010**, *16*, 3452–3462. [CrossRef]
139. Wu, L.; Chang, W.; Zhao, J.; Yu, Y.; Tan, X.; Su, T.; Zhao, L.; Huang, S.; Liu, S.; Cao, G. Development of Autoantibody Signatures as Novel Diagnostic Biomarkers of Non–Small Cell Lung CancerNew Serum Diagnostic Biomarkers for NSCLC. *Clin. Cancer Res.* **2010**, *16*, 3760–3768. [CrossRef]
140. Lam, S.; Boyle, P.; Healey, G.F.; Maddison, P.; Peek, L.; Murray, A.; Chapman, C.J.; Allen, J.; Wood, W.C.; Sewell, H.F.; et al. *Early*CDT-Lung: An Immunobiomarker Test as an Aid to Early Detection of Lung Cancer. *Cancer Prev. Res.* **2011**, *4*, 1126–1134. [CrossRef]
141. Wang, T.; Liu, H.; Pei, L.; Wang, K.; Song, C.; Wang, P.; Ye, H.; Zhang, J.; Ji, Z.; Ouyang, S.; et al. Screening of tumor-associated antigens based on Oncomine database and evaluation of diagnostic value of autoantibodies in lung cancer. *Clin. Immunol.* **2020**, *210*, 108262. [CrossRef]
142. Jiang, D.; Wang, Y.; Liu, M.; Si, Q.; Wang, T.; Pei, L.; Wang, P.; Ye, H.; Shi, J.; Wang, X.; et al. A panel of autoantibodies against tumor-associated antigens in the early immunodiagnosis of lung cancer. *Immunobiology* **2020**, *225*, 151848. [CrossRef]
143. Li, R.-Y.; Liang, Z.-Y. Circulating tumor DNA in lung cancer: Real-time monitoring of disease evolution and treatment response. *Chin. Med. J.* **2020**, *133*, 2476–2485. [CrossRef] [PubMed]
144. Zhang, Z.; Xiao, Y.; Zhao, J.; Chen, M.; Xu, Y.; Zhong, W.; Xing, J.; Wang, M. Relationship between circulating tumour cell count and prognosis following chemotherapy in patients with advanced non-small-cell lung cancer. *Respirology* **2016**, *21*, 519–525. [CrossRef] [PubMed]
145. Hou, J.-M.; Krebs, M.G.; Lancashire, L.; Sloane, R.; Backen, A.; Swain, R.K.; Priest, L.J.C.; Greystoke, A.; Zhou, C.; Morris, K.; et al. Clinical Significance and Molecular Characteristics of Circulating Tumor Cells and Circulating Tumor Microemboli in Patients With Small-Cell Lung Cancer. *J. Clin. Oncol.* **2012**, *30*, 525–532. [CrossRef] [PubMed]
146. Kim, T.; Kim, E.Y.; Lee, S.H.; Kwon, D.S.; Kim, A.; Chang, Y.S. Presence of mEGFR ctDNA predicts a poor clinical outcome in lung adenocarcinoma. *Thorac. Cancer* **2019**, *10*, 2267–2273. [CrossRef] [PubMed]
147. Hur, J.Y.; Lee, J.S.; Kim, I.A.; Kim, H.J.; Kim, W.S.; Lee, K.Y. Extracellular vesicle-based EGFR genotyping in bronchoalveolar lavage fluid from treatment-naive non-small cell lung cancer patients. *Transl. Lung Cancer Res.* **2019**, *8*, 1051. [CrossRef] [PubMed]
148. Michaelidou, K.; Koutoulaki, C.; Mavridis, K.; Vorrias, E.; Papadaki, M.A.; Koutsopoulos, A.V.; Mavroudis, D.; Agelaki, S. Detection of *KRAS* G12/G13 Mutations in Cell Free-DNA by Droplet Digital PCR, Offers Prognostic Information for Patients with Advanced Non-Small Cell Lung Cancer. *Cells* **2020**, *9*, 2514. [CrossRef]
149. Wen, S.W.C.; Andersen, R.F.; Hansen, T.F.; Nyhus, C.H.; Hager, H.; Hilberg, O.; Jakobsen, A. The prognostic impact of circulating homeobox A9 methylated DNA in advanced non-small cell lung cancer. *Transl. Lung Cancer Res.* **2021**, *10*, 855. [CrossRef]

150. Pavan, A.; Bragadin, A.B.; Calvetti, L.; Ferro, A.; Zulato, E.; Attili, I.; Nardo, G.; Maso, A.D.; Frega, S.; Menin, A.G.; et al. Role of next generation sequencing-based liquid biopsy in advanced non-small cell lung cancer patients treated with immune checkpoint inhibitors: Impact of STK11, KRAS and TP53 mutations and co-mutations on outcome. *Transl. Lung Cancer Res.* **2021**, *10*, 202. [CrossRef]
151. Suzuki, M.; Yoshino, I. Aberrant methylation in non-small cell lung cancer. *Surg. Today* **2010**, *40*, 602–607. [CrossRef]
152. Brock, M.V.; Hooker, C.M.; Ota-Machida, E.; Han, Y.; Guo, M.; Ames, S.; Glöckner, S.; Piantadosi, S.; Gabrielson, E.; Pridham, G.; et al. DNA Methylation Markers and Early Recurrence in Stage I Lung Cancer. *N. Engl. J. Med.* **2008**, *358*, 1118–1128. [CrossRef] [PubMed]
153. Khandelwal, A.; Seam, R.K.; Gupta, M.; Rana, M.K.; Prakash, H.; Vasquez, K.M.; Jain, A. Circulating micro RNA-590-5p functions as a liquid biopsy marker in non-small cell lung cancer. *Cancer Sci.* **2020**, *111*, 826–839. [CrossRef] [PubMed]
154. Boeri, M.; Verri, C.; Conte, D.; Roz, L.; Modena, P.; Facchinetti, F.; Calabrò, E.; Croce, C.M.; Pastorino, U.; Sozzi, G. MicroRNA signatures in tissues and plasma predict development and prognosis of computed tomography detected lung cancer. *Proc. Natl. Acad. Sci. USA* **2011**, *108*, 3713–3718. [CrossRef]
155. Tian, F.; Shen, Y.; Chen, Z.; Li, R.; Lu, J.; Ge, Q. Aberrant miR-181b-5p and miR-486-5p expression in serum and tissue of non-small cell lung cancer. *Gene* **2016**, *591*, 338–343. [CrossRef]
156. Wang, Z.X.; Bian, H.B.; Wang, J.R.; Cheng, Z.X.; Wang, K.M.; De, W. Prognostic significance of serum miRNA-21 expression in human non-small cell lung cancer. *J. Surg. Oncol.* **2011**, *104*, 847–851. [CrossRef]
157. Bianconi, F.; Palumbo, I.; Spanu, A.; Nuvoli, S.; Fravolini, M.L.; Palumbo, B. PET/CT Radiomics in Lung Cancer: An Overview. *Appl. Sci.* **2020**, *10*, 1718. [CrossRef]
158. Yoneda, K.; Imanishi, N.; Ichiki, Y.; Tanaka, F. A liquid biopsy in primary lung cancer. *Surg. Today* **2019**, *49*, 1–14. [CrossRef]
159. Kessler, M.D.; Pawar, N.R.; Martin, S.S.; Antalis, T.M.; O'Connor, T.D. Improving Cancer Detection and Treatment with Liquid Biopsies and ptDNA. *Trends Cancer* **2018**, *4*, 643–654. [CrossRef]
160. Sorensen, B.S.; Wu, L.; Wei, W.; Tsai, J.; Weber, B.; Nexo, E.; Meldgaard, P. Monitoring of epidermal growth factor receptor tyrosine kinase inhibitor-sensitizing and resistance mutations in the plasma DNA of patients with advanced non–small cell lung cancer during treatment with erlotinib. *Cancer* **2014**, *120*, 3896–3901. [CrossRef]
161. Reijans, M.; Gestel, S.; Haes, E.; Vandesteene, C.; Gomez, J.C.; Gouedard, C.; Patera, S.; Murray, S.; Maertens, G. Feasibility study of a ctEGFR prototype assay on the fully automated Idylla™ platform. *Ann. Oncol.* **2019**, *30*, v577. [CrossRef]
162. Heeke, S.; Benzaquen, J.; Hofman, V.; Ilié, M.; Allegra, M.; Long-Mira, E.; Lassalle, S.; Tanga, V.; Salacroup, C.; Bonnetaud, C.; et al. Critical Assessment in Routine Clinical Practice of Liquid Biopsy for EGFR Status Testing in Non–Small-Cell Lung Cancer: A Single-Laboratory Experience (LPCE, Nice, France). *Clin. Lung Cancer* **2020**, *21*, 56–65.e8. [CrossRef] [PubMed]
163. Lee, D.H. Treatments for EGFR-mutant non-small cell lung cancer (NSCLC): The road to a success, paved with failures. *Pharmacol. Ther.* **2017**, *174*, 1–21. [CrossRef] [PubMed]
164. Cabanero, M.; Tsao, M.S. Circulating Tumour DNA in EGFR-Mutant Non-Small-Cell Lung Cancer. *Curr. Oncol.* **2018**, *25*, 38–44. [CrossRef]
165. Ding, P.; Becker, T.; Bray, V.; Chua, W.; Ma, Y.; Lynch, D.; Po, J.; Luk, A.; Caixeiro, N.; de Souza, P.; et al. The predictive and prognostic significance of liquid biopsy in advanced epidermal growth factor receptor-mutated non-small cell lung cancer: A prospective study. *Lung Cancer* **2019**, *134*, 187–193. [CrossRef] [PubMed]
166. Provencio, M.; Torrente, M.; Calvo, V.; Pérez-Callejo, D.; Gutiérrez, L.; Franco, F.; Pérez-Barrios, C.; Barquín, M.; Royuela, A.; García-García, F.; et al. Prognostic value of quantitative ctDNA levels in non small cell lung cancer patients. *Oncotarget* **2018**, *9*, 488. [CrossRef]
167. Pérez-Callejo, D.; Torrente, M.; Romero, A.; Provencio, M. Clinical and Investigational Applications of Liquid Biopsy in Non-Small Cell Lung Cancer. *J. Tumor* **2016**, *4*, 461–468. [CrossRef]
168. Nicolazzo, C.; Raimondi, C.; Mancini, M.; Caponnetto, S.; Gradilone, A.; Gandini, O.; Mastromartino, M.; Del Bene, G.; Prete, A.A.; Longo, F.; et al. Monitoring PD-L1 positive circulating tumor cells in non-small cell lung cancer patients treated with the PD-1 inhibitor Nivolumab. *Sci. Rep.* **2016**, *6*, 31726. [CrossRef]
169. Boffa, D.J.; Graf, R.P.; Salazar, M.C.; Hoag, J.; Lu, D.; Krupa, R.; Louw, J.; Dugan, L.; Wang, Y.; Landers, M.; et al. Cellular Expression of PD-L1 in the Peripheral Blood of Lung Cancer Patients is Associated with Worse Survival. *Cancer Epidemiol. Biomark. Prev.* **2017**, *26*, 1139–1145. [CrossRef]
170. Ilié, M.; Szafer-Glusman, E.; Hofman, V.; Chamorey, E.; Lalvée, S.; Selva, E.; Leroy, S.; Marquette, C.-H.; Kowanetz, M.; Hedge, P.; et al. Detection of PD-L1 in circulating tumor cells and white blood cells from patients with advanced non-small-cell lung cancer. *Ann. Oncol.* **2018**, *29*, 193–199. [CrossRef]
171. Rizvi, N.A.; Hellmann, M.D.; Snyder, A.; Kvistborg, P.; Makarov, V.; Havel, J.J.; Lee, W.; Yuan, J.; Wong, P.; Ho, T.S.; et al. Cancer immunology. Mutational landscape determines sensitivity to PD-1 blockade in non–small cell lung cancer. *Science* **2015**, *348*, 124–128. [CrossRef]
172. Goodman, A.M.; Kato, S.; Bazhenova, L.; Patel, S.P.; Frampton, G.M.; Miller, V.; Stephens, P.J.; Daniels, G.A.; Kurzrock, R. Tumor Mutational Burden as an Independent Predictor of Response to Immunotherapy in Diverse CancersTMB Predicts Response to Immunotherapy in Diverse Cancers. *Mol. Cancer Ther.* **2017**, *16*, 2598–2608. [CrossRef] [PubMed]
173. Singal, G.; Miller, P.; Agarwala, V.; Li, G.; Kaushik, G.; Backenroth, D.; Gossai, A.; Frampton, G.M.; Torres, A.Z.; Lehnert, E.M.; et al. Association of Patient Characteristics and Tumor Genomics With Clinical Outcomes Among Patients With Non–Small Cell Lung Cancer Using a Clinicogenomic Database. *JAMA* **2019**, *321*, 1391–1399. [CrossRef] [PubMed]
174. Gandara, D.R.; Paul, S.M.; Kowanetz, M.; Schleifman, E.; Zou, W.; Li, Y.; Rittmeyer, A.; Fehrenbacher, L.; Otto, G.; Malboeuf, C.; et al. Blood-based tumor mutational burden as a predictor of clinical benefit in non-small-cell lung cancer patients treated with atezolizumab. *Nat. Med.* **2018**, *24*, 1441–1448. [CrossRef]

175. Zill, O.A.; Banks, K.C.; Fairclough, S.R.; Mortimer, S.A.; Vowles, J.V.; Mokhtari, R.; Gandara, D.R.; Mack, P.C.; Odegaard, J.I.; Nagy, R.J.; et al. The Landscape of Actionable Genomic Alterations in Cell-Free Circulating Tumor DNA from 21,807 Advanced Cancer PatientsSomatic Genomic Landscape of Circulating Tumor DNA. *Clin. Cancer Res.* **2018**, *24*, 3528–3538. [CrossRef]
176. Shaw, A.T.; Yeap, B.Y.; Mino-Kenudson, M.; Digumarthy, S.R.; Costa, D.B.; Heist, R.S.; Solomon, B.; Stubbs, H.; Admane, S.; McDermott, U.; et al. Clinical Features and Outcome of Patients With Non–Small-Cell Lung Cancer Who Harbor EML4-ALK. *J. Clin. Oncol.* **2009**, *27*, 4247. [CrossRef] [PubMed]
177. Nilsson, R.J.A.; Karachaliou, N.; Berenguer, J.; Gimenez-Capitan, A.; Schellen, P.; Teixido, C.; Tannous, J.; Kuiper, J.L.; Drees, E.; Grabowska, M.; et al. Rearranged EML4-ALK fusion transcripts sequester in circulating blood platelets and enable blood-based crizotinib response monitoring in non-small-cell lung cancer. *Oncotarget* **2016**, *7*, 1066. [CrossRef]
178. de Koning, H.J.; van der Aalst, C.M.; de Jong, P.A.; Scholten, E.T.; Nackaerts, K.; Heuvelmans, M.A.; Lammers, J.W.J.; Weenink, C.; Yousaf-Khan, U.; Horeweg, N.; et al. Reduced lung-cancer mortality with volume CT screening in a randomized trial. *N. Engl. J. Med.* **2020**, *382*, 503–513. [CrossRef]
179. Hochhegger, B.; Camargo, S.; Teles, G.B.D.S.; Chate, R.C.; Szarf, G.; Guimarães, M.D.; Gross, J.L.; Barbosa, P.N.V.P.; Chiarantano, R.S.; Reis, R.M.; et al. Challenges of Implementing Lung Cancer Screening in a Developing Country: Results of the Second Brazilian Early Lung Cancer Screening Trial (BRELT2). *JCO Glob. Oncol.* **2022**, *8*, e2100257. [CrossRef]
180. Chiarantano, R.S.; Vazquez, F.L.; Franco, A.; Ferreira, L.C.; da Costa, M.C.; Talarico, T.; Oliveira, N.; Miziara, J.E.; Mauad, E.C.; da Silva, E.C.; et al. Implementation of an Integrated Lung Cancer Prevention and Screening Program Using a Mobile Computed Tomography (CT) Unit in Brazil. *Cancer Control* **2022**, *29*, 10732748221121385. [CrossRef]
181. Blandin Knight, S.; Crosbie, P.A.; Balata, H.; Chudziak, J.; Hussell, T.; Dive, C. Progress and prospects of early detection in lung cancer. *Open Biol.* **2017**, *7*, 170070. [CrossRef] [PubMed]
182. Phallen, J.; Sausen, M.; Adleff, V.; Leal, A.; Hruban, C.; White, J.; Anagnostou, V.; Fiksel, J.; Cristiano, S.; Papp, E.; et al. Direct detection of early-stage cancers using circulating tumor DNA. *Sci. Transl. Med.* **2017**, *9*, eaan2415. [CrossRef] [PubMed]
183. Zhang, Z.; Ramnath, N.; Nagrath, S. Current status of CTCs as liquid biopsy in lung cancer and future directions. *Front. Oncol.* **2015**, *5*, 209. [CrossRef] [PubMed]
184. Foss, K.M.; Sima, C.; Ugolini, D.; Neri, M.; Allen, K.E.; Weiss, G.J. miR-1254 and miR-574-5p: Serum-Based microRNA Biomarkers for Early-Stage Non-small Cell Lung Cancer. *J. Thorac. Oncol.* **2011**, *6*, 482–488. [CrossRef]
185. Pastorino, U.; Boeri, M.; Sestini, S.; Sabia, F.; Milanese, G.; Silva, M.; Suatoni, P.; Verri, C.; Cantarutti, A.; Sverzellati, N.; et al. Baseline computed tomography screening and blood microRNA predict lung cancer risk and define adequate intervals in the BioMILD trial. *Ann. Oncol.* **2022**, *33*, 395–405. [CrossRef] [PubMed]
186. Kumar, S.; Gaur, V.; Mir, I.A.; Saikia, J.; Kumar, S. MicroRNA-3692-3p is overexpressed in lung tumors but may not serve as a prognostic biomarker in lung cancer patients. *Mol. Biol. Rep.* **2022**, 1–10. [CrossRef] [PubMed]
187. Lian, X.; Cao, D.; Hu, X.; Mo, W.; Yao, X.; Mo, J.; Wang, H. Circular RNAs Hsa_circ_101555 and Hsa_circ_008068 as Diagnostic Biomarkers for Early-Stage Lung Adenocarcinoma. *Int. J. Gen. Med.* **2022**, *15*, 5579. [CrossRef]
188. Zhang, Q.; Qin, S.; Peng, C.; Liu, Y.; Huang, Y.; Ju, S. Circulating circular RNA hsa_circ_0023179 acts as a diagnostic biomarker for non-small-cell lung cancer detection. *J. Cancer Res. Clin. Oncol.* **2022**, 1–12. [CrossRef]
189. Robbins, H.A.; Alcala, K.; Moez, E.K.; Guida, F.; Thomas, S.; Zahed, H.; Warkentin, M.T.; Smith-Byrne, K.; Brhane, Y.; Muller, D.; et al. Design and methodological considerations for biomarker discovery and validation in the Integrative Analysis of Lung Cancer Etiology and Risk (INTEGRAL) Program. *Ann. Epidemiol.* **2023**, *77*, 1–12. [CrossRef]
190. Lennon, A.M.; Buchanan, A.H.; Kinde, I.; Warren, A.; Honushefsky, A.; Cohain, A.T.; Ledbetter, D.H.; Sanfilippo, F.; Sheridan, K.; Rosica, D.; et al. Feasibility of blood testing combined with PET CT to screen for cancer and guide intervention. *Science* **2020**, *369*, eabb9601. [CrossRef]
191. LaRose, T.L.; Meheus, F.; Brennan, P.; Johansson, M.; Robbins, H.A. Assessment of Biomarker Testing for Lung Cancer Screening Eligibility. *JAMA Netw. Open* **2020**, *3*, e200409. [CrossRef]
192. Vachani, A.; Carroll, N.M.; Simoff, M.J.; Neslund-Dudas, C.; Honda, S.; Greenlee, R.T.; Rendle, K.A.; Burnett-Hartman, A.; Ritzwoller, D.P. Stage Migration and Lung Cancer Incidence After Initiation of Low-Dose Computed Tomography Screening. *J. Thorac. Oncol.* **2022**, *17*, 1355–1364. [CrossRef] [PubMed]
193. Fatumo, S.; Chikowore, T.; Choudhury, A.; Ayub, M.; Martin, A.R.; Kuchenbaecker, K. A roadmap to increase diversity in genomic studies. *Nat. Med.* **2022**, *28*, 243–250. [CrossRef] [PubMed]
194. Saulsberry, L.; Olopade, O.I. Precision oncology: Directing genomics and pharmacogenomics toward reducing cancer inequities. *Cancer Cell* **2021**, *39*, 730–733. [CrossRef] [PubMed]

International Journal of *Molecular Sciences*

MDPI

*Article*

# ONC201 Suppresses Neuroblastoma Growth by Interrupting Mitochondrial Function and Reactivating Nuclear ATRX Expression While Decreasing MYCN

**Jian-Ching Wu [1], Chao-Cheng Huang [2,3], Pei-Wen Wang [1,4], Ting-Ya Chen [1], Wen-Ming Hsu [5], Jiin-Haur Chuang [1,6,†] and Hui-Ching Chuang [1,7,*,†]**

1 Center for Mitochondrial Research and Medicine, Kaohsiung Chang Gung Memorial Hospital, Kaohsiung 83301, Taiwan
2 Department of Pathology, Kaohsiung Chang Gung Memorial Hospital and Chang Gung University College of Medicine, Kaohsiung 83301, Taiwan
3 Biobank and Tissue Bank, Kaohsiung Chang Gung Memorial Hospital, Kaohsiung 83301, Taiwan
4 Department of Internal Medicine, Kaohsiung Chang Gung Memorial Hospital and Chang Gung University College of Medicine, Kaohsiung 83301, Taiwan
5 Department of Surgery, National Taiwan University Hospital, National Taiwan University College of Medicine, Taipei 10617, Taiwan
6 Department of Pediatric Surgery, Kaohsiung Chang Gung Memorial Hospital and Chang Gung University College of Medicine, Kaohsiung 83301, Taiwan
7 Department of Otolaryngology, Kaohsiung Chang Gung Memorial Hospital and Chang Gung University College of Medicine, Kaohsiung 83301, Taiwan
* Correspondence: entjulia@cgmh.org.tw; Tel.: +886-7-7317123 (ext. 8896); Fax: +886-7-7311696
† These authors contributed equally to this work.

**Abstract:** Neuroblastoma (NB) is characterized by several malignant phenotypes that are difficult to treat effectively without combination therapy. The therapeutic implication of mitochondrial ClpXP protease ClpP and ClpX has been verified in several malignancies, but is unknown in NB. Firstly, we observed a significant increase in ClpP and ClpX expression in immature and mature ganglion cells as compared to more malignant neuroblasts and less malignant Schwannian-stroma-dominant cell types in human neuroblastoma tissues. We used ONC201 targeting ClpXP to treat NB cells, and found a significant suppression of mitochondrial protease, i.e., ClpP and ClpX, expression and downregulation of mitochondrial respiratory chain subunits SDHB and NDUFS1. The latter was associated with a state of energy depletion, increased reactive oxygen species, and decreased mitochondrial membrane potential, consequently promoting apoptosis and suppressing cell growth of NB. Treatment of NB cells with ONC201 as well as the genetic attenuation of ClpP and ClpX through specific short interfering RNA (siRNA) resulted in the significant upregulation of the tumor suppressor alpha thalassemia/mental retardation X-linked (ATRX) and promotion of neurite outgrowth, implicating mitochondrial ClpXP proteases in *MYCN*-amplified NB cell differentiation. Furthermore, ONC201 treatment significantly decreased MYCN protein expression and suppressed tumor formation with the reactivation of ATRX expression in *MYCN*-amplified NB-cell-derived xenograft tumors. Taken together, ONC201 could be the potential agent to provide diversified therapeutic application in NB, particularly in NB with *MYCN* amplification.

**Keywords:** neuroblastoma; ATRX; MYCN; ClpPX protease; ONC201; apoptosis; mitochondria

**Citation:** Wu, J.-C.; Huang, C.-C.; Wang, P.-W.; Chen, T.-Y.; Hsu, W.-M.; Chuang, J.-H.; Chuang, H.-C. ONC201 Suppresses Neuroblastoma Growth by Interrupting Mitochondrial Function and Reactivating Nuclear ATRX Expression While Decreasing MYCN. *Int. J. Mol. Sci.* **2023**, *24*, 1649. https://doi.org/10.3390/ijms24021649

Academic Editor: Laura Paleari

Received: 22 August 2022
Revised: 3 January 2023
Accepted: 7 January 2023
Published: 13 January 2023



## 1. Introduction

Neuroblastoma (NB) is a genetically heterogeneous tumor characterized by pleomorphic cells, including those with *MYCN* amplification that are highly glycolytic, presenting the opportunity to use glycolytic inhibitor 2-deoxyglucose (2DG) to suppress NB cell growth [1,2]. Significant changes in mitochondrial membrane potential, impaired mitochondrial function, increased oxidative stress, and subsequent cell death are involved in the process [3,4], indicating the importance of studying mitochondrial function.

The proper execution of mitochondrial functions depends on the maintenance of the integrated mitochondrial proteome. Mitochondrial unfolded protein response ($UPR^{mt}$) is a feedback loop, wherein the dysfunction of the mitochondrial proteome is sensed and communicated to the nucleus, which subsequently launches an extensive transcriptional program to repair the damage and to rescue mitochondrial function [5]. $UPR^{mt}$ is maintained primarily by the chaperones HSP10, HSP60, HSP70, and HSP90 and by proteases of the AAA+ superfamily, specifically the Lon, ClpXP, and m-AAA proteases. Dysregulation of the proteases in cancer may serve as a novel therapeutic target in malignancies which are highly dependent on mitochondrial functions for survival, including acute myeloid leukemias and a subset of melanomas [6]. ClpXP protease is composed of ClpP, which forms a multimeric complex with ClpX, leading to proteolytic activity. ClpXP is overexpressed in hematologic malignancies and solid tumors. Notably, both the inhibition and hyperactivation of ClpXP was reported to lead to impaired respiratory chain activity and cause cancer cell death [7].

ONC201 was discovered initially as a first-in-class TNF-related apoptosis-inducing ligand (TRAIL) compound (known as TIC10). Based on the central core structure of ONC201, several compounds of the imipridone family have been developed, including ONC201, ONC206, and ONC212. There are more than 15 ongoing Phase I and Phase II clinical trials of ONC201, either as a single agent or in combination, to treat cancers including glioblastoma, acute leukemia, breast, colorectal, endometrial high-grade gliomas, multiple myeloma, neuroendocrine, and other solid cancers [8,9].

Many studies emphasize the importance of ClpP in mediating cancer cell death, including in breast cancer tissue, SK-N-SH neuroblastoma cells, and acute myeloid leukemia [10–13]. Moreover, the coordination of ClpP and ClpX was important in a study that showed boron-based peptidomimetics as potent inhibitors of human ClpP in the presence of human ClpX [10]. ClpX is suppressed upon ONC212 treatment in pancreatic cancer cells and in diffuse intrinsic pontine glioma (DIPG) cells treated with ONC206 [11,12]. However, both ClpP and ClpX are suppressed when DIPG cells are treated with ONC212 [12]. Results of these studies clearly indicate the importance of studying both ClpP and ClpX when treating cancer cells with imipridones because their diversified response is obviously dependent on the cell types as well as on different members of the imipridone family.

In addition to cell proliferation and death, ClpXP protease is also involved in cell differentiation, including the morphological differentiation of Streptomyces lividans [13,14]. Preclinical testing of a combined treatment with the ClpP activator ONC201/TIC10 and 2DG resulted in a dual metabolic reprogramming and synergistic anti-proliferative and anti-migratory effects on glioblastoma cells [15]. In that respect, ClpP-mediated degradation of mitochondrial energy-related components is additive to the effects of 2DG as a glycolytic inhibitor [16]. However, our previous study has revealed that 2DG alone can simultaneously target cancer and endothelial cells to significantly suppress NB growth in mice regardless of the status of *MYCN* amplification [2]. The latter means that if we use 2DG in combination with ONC201 to treat NB cells, it might mask the efficacy of ONC201 for the treatment of NB. In this study, we decided to test the effect of ONC201 alone on NB cell differentiation and tumor growth. The results offer promise for the development of future therapy for NB.

## 2. Results

### 2.1. The Expressions of ClpP and ClpX Are Associated with NB Differentiation

Elevated ClpP expression is associated with poor survival in patients with other malignancies, but remains unclear with regard to NB tumors. Firstly, we evaluated the expression of ClpP and ClpX from the resected tumors of patients with NB. Immunostaining of ClpP and its multimeric partner ClpX was herein performed in NB tumors. A total of 23 patients with 5 different International Neuroblastoma Pathology Classification (INPC) statuses were included. The staining intensity of ClpP and ClpX varied significantly according to the cell type, even in different tissue blocks of the same patient. Accordingly, the expressions of ClpP and ClpX in the NB tumors were classified according to the predominant cell type, categorized as primitive neuroblast, differentiated neuroblast, immature ganglion cell, mature ganglion cell, and Schwannian stroma. The expressions of ClpP and ClpX were clearly different among the various cell types (Figure 1A). The primitive neuroblast and Schwannian stroma cell types were either devoid of ClpP expression or were very weak; on the contrary, the expression of ClpP in the immature ganglion cells was significantly higher than in the former two cell types, followed by that of mature ganglion cells and differentiated neuroblasts (Figure 1B). The expression of ClpX in the NB cells essentially followed the same trend as that of ClpP, except that the highest expression was identified in mature ganglion cells, followed by immature ganglion cells, and differentiated neuroblasts. The primitive neuroblast and Schwannian stroma were essentially devoid of ClpX expression (Figure 1C). These results suggest that both ClpP and ClpX may involve the differentiation process in NB.

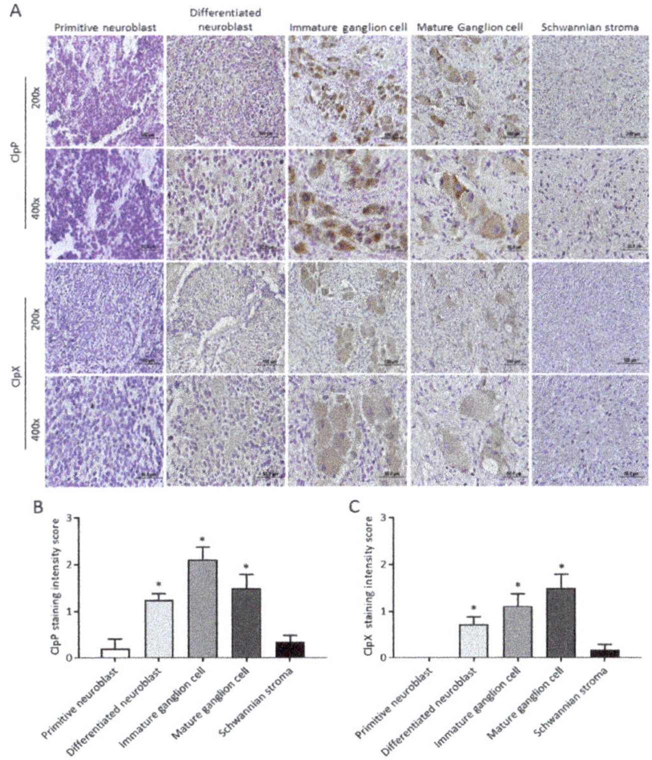


**Figure 1.** Comparison of ClpP and ClpX expressions in subtypes of human NBs: (**A**) Representative profiles of ClpP and ClpX immunostaining in differentiated NBs. (**B**,**C**) Histological scoring for ClpP and ClpX intensity in NBs tissues. Data in each bar chart are represented as mean ± SD. Primitive neuroblast ($n$ = 4), differentiated neuroblast ($n$ = 12), immature ganglion cell ($n$ = 10), mature ganglion cell ($n$ = 4), and Schwannian stroma ($n$ = 12). Upper panel, ×200 magnification, scale bar 100 μm; lower panel, ×400 magnification, scale bar 50 μm. * $p < 0.05$.

*2.2. Administration of ONC201 Inhibits NB Cell Growth and Promotes NB Cell Death*

To investigate the efficacy of ONC201 alone in the treatment of NB cells, *MYCN*-nonamplified SK-N-AS and *MYCN*-amplified BE(2)M17 cell lines were treated with various concentrations of ONC201 (from 0 to 40 μM) for 48 h and 96 h. As shown in Figure 2A, treatment of ONC201 alone displayed about 20% of growth suppression in two NB cell lines at 48 h; however, more than 60% was suppressed in both cell lines at 96 h, suggesting that the low concentration of ONC201 is sustainable for the treatment of NB. The effect of ONC201 on the ClpXP complex in the four NB cell lines was measured by Western blot. Interestingly, the results of the short treatment period demonstrate that ONC201 (5 μM) did not affect the protein levels of ClpP, but was associated with the significant downregulation of ClpX expression in the four NB cell lines (Supplementary Figure S1A–D). For the longer treatment period, ONC201 almost eliminated the protein expression of ClpX at both 24 h and 48 h, whereas ClpP was notably decreased by ONC201 at 48 h in the four NB cell lines (Supplementary Figure S1A–D). The phase contrast images show that the cell density was markedly decreased with ONC201 (5 μM) treatment of SK-N-AS cells and BE(2)M17 cells (Figure 2B).

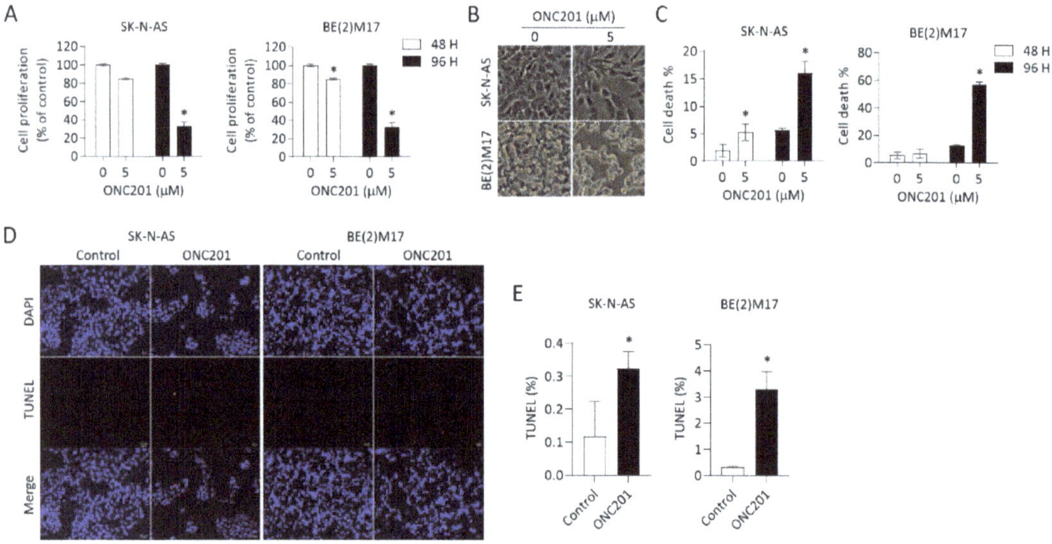


**Figure 2.** Effects of ONC201 on cell survival in NB cells: (**A**) SK-N-AS and BE(2)M17 cells were treated with ONC201 (5 μM) for 48 h and 96 h. The cell survival was detected by Resazurin assay. (**B**) Representative phase contrast images demonstrating cell density and morphological changes in NBs exposed to ONC201 for 48 h. Scale bar 50 μm. (**C**) After incubation with ONC201 (5 μM) for 48 h, cells were collected and stained by trypan blue solution. The percentage of trypan-blue-positive cells was quantified using flow cytometry. (**D**,**E**) Representative profiles of TUNEL staining (red) in control and ONC201-treated NB cells. DAPI (blue) was used as a nuclear counterstain. Scale bar 20 μm. (**D**) The apoptotic cells are expressed as the percentage of the number of TUNEL-positive cells over the number of DAPI-positive cells. All data are expressed as mean ± SD from triplicate experiments. * $p < 0.05$.

We further evaluated cell viability by trypan blue staining using flow cytometry analysis as well as TUNEL assay. Consistent with the previous results of cell growth inhibition, the percentage of trypan-blue-positive cells was less significant at 48 h of treatment with ONC201, but more significant at 96 h in both SK-N-AS and BE(2)M17 NB cells (Figure 2C).

An additional assay using TUNEL staining to detect DNA breaks also revealed significantly increased cell death in both NB cell lines treated with ONC201 for 48 h (Figure 2D,E). These results suggest that monotherapy with ONC201 alone is effective to suppress NB cell growth and promote cell death within the time frame of the study, regardless the status of *MYCN* amplification.

We also studied the effect of siRNA targeting ClpP and ClpX on the proliferation and death of SK-N-AS and BE(2)M17 cells. The data show that siClpP and sClpX could significantly decrease cell proliferation while increase cell death in both NB cells (Supplementary Figure S2A,B).

### 2.3. ONC201 Alone Is Capable of Disturbing Respiratory Chain Proteins and Mitochondrial Functions in NB Cells

Respiratory chain proteins have previously been identified as putative ClpP substrates, and the inhibition of ClpP is known to accumulate degraded or misfolded subunits. Thus, we herein investigated two representative respiratory chain subunits, SDHB and NDUFS1. The results reveal that treatment with ONC201 alone remarkably reduced ClpP and ClpX expressions in representative SK-N-AS cells (Figure 3A). Moreover, the protein levels of SDHB and NDUFS1 were significantly decreased by ONC201 in SK-N-AS cells (Figure 3A). We next investigated whether mitochondrial functions were changed in NB cells. To achieve this, the Seahorse XF analyzer was applied to evaluate OCR (quantity of mitochondrial respiration) and ECAR (indicator of glycolysis) in ONC201-treated NB cells. The OCR of SK-N-AS cells treated with ONC201 (5 μM) significantly diminished to less than 30 pMoles/min/2 × $10^4$ cells of basal respiration (Figure 3B,C). However, ONC201 treatment did not boost the inhibitory effect on ECAR (Figure 3D,E). These results reveal that ONC201 inhibits the oxidative phosphorylation chain but not glycolysis in the NB cells.

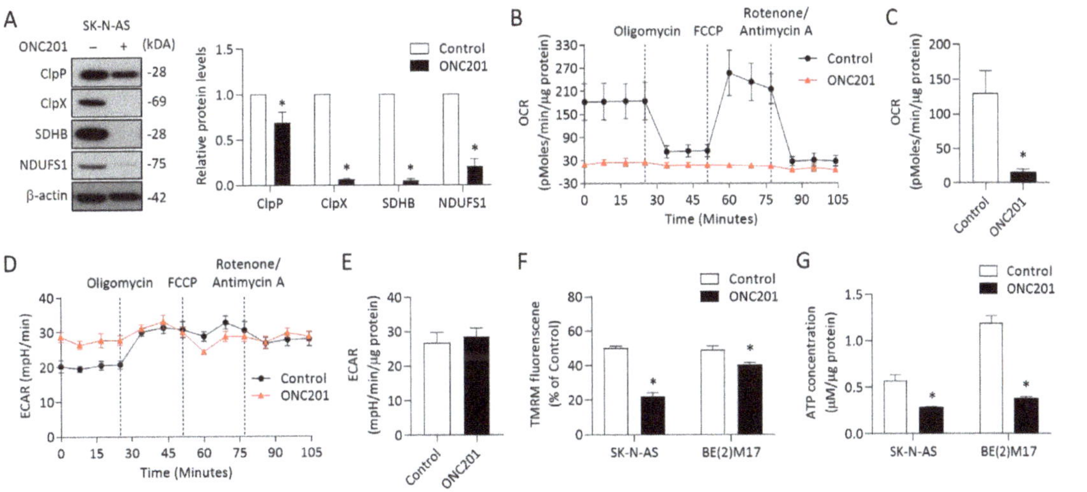


**Figure 3.** Effects of ONC201 on mitochondrial functions in NB cells: (**A**) The relative protein levels of ClpP, ClpX, SDHB, and NDUFS1 were determined by Western blot and quantified by Image Pro-plus analysis software. (**B**,**C**) OCR in ONC201-treated cells were detected by the Seahorse XF24 analyzer in the absence or presence of oligomycin (1 μM), FCCP (0.5 μM), and rotenone (1 μM) by counting 1 × $10^5$ cells. (**D**,**E**) Profiling of ECAR in control and ONC201-treated cells was measured by the Seahorse XF24 analyzer. (**F**) Cells were labeled with TMRM (100 nM) to examine mitochondrial membrane potential and analyzed by flow cytometry. (**G**) Total cellular ATP levels were measured using the luciferase-based luminescence assay kit. All data were obtained from three independent experiments, and are expressed as mean ± SD. * $p < 0.05$.

We next examined other mitochondrial functions, including MMP and ATP production, in NB cells treated with ONC201. The MMP detected by TMRM revealed that ONC201 treatment significantly decreased the MMP in both SK-N-AS and BE(2)17 cells (Figure 3F).

The baseline ATP contents were higher in the *MYCN*-amplified BE(2)17 cells than in the *MYCN*-nonamplified SK-N-AS cells. However, single treatment with ONC201 significantly decreased of the ATP production in both NB cells (Figure 3G). Collectively, these results indicate that ONC201 treatment alone could disturb mitochondrial functions regardless of the status of *MYCN* amplification.

### 2.4. *ONC201 Treatment Induces ROS Generation and Apoptosis in NB Cells*

Excess ROS has been implicated in the regulation of apoptosis in several tumor types. We thus measured the ROS content by DHE and MitoSox Red assays in ONC201-treated NB cells. A significant increase in DHE fluorescence was detected in the ONC201-treated SK-N-AS and BE(2)17 cells (Figure 4A). Subsequently, MitoSox Red staining revealed an elevated mitochondrial ROS accumulation in ONC201-treated SK-N-AS and BE(2)17 cells (Figure 4B).

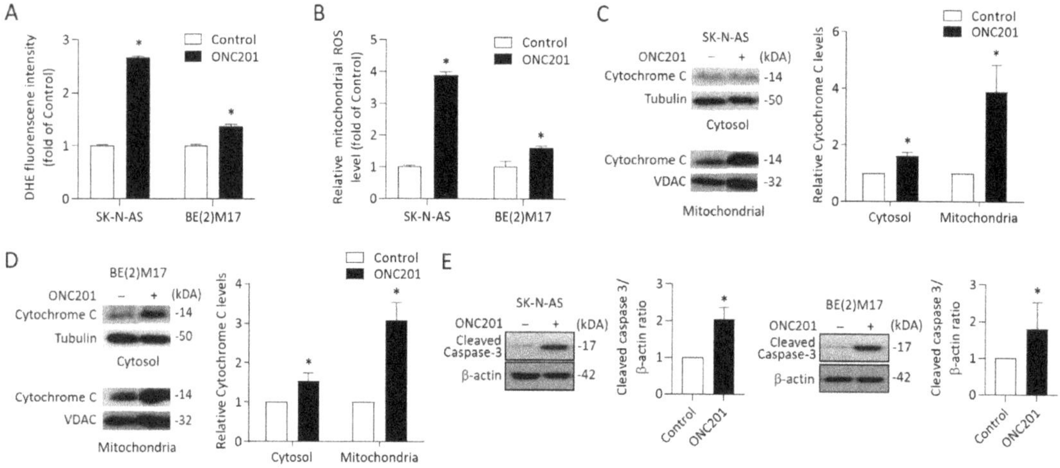


**Figure 4.** Effects of ONC201 on mitochondria-mediated apoptosis signaling in NB cells: (**A**) The intracellular ROS production of ONC201-treated cells was detected by DHE staining. The relative intensity of DHE fluorescence was analyzed by flow cytometry. (**B**) Mitochondrial ROS was detected by MitoSOX Red and immediately analyzed by flow cytometry. Data are expressed as fold change compared with control group. (**C**,**D**) To detect cytochrome C release, cytosolic and mitochondrial fractions of NB cells were subjected into 15% SDS-PAGE and probed with anti-cytochrome c antibody by Western blot analysis. Alpha-tubulin and VDAC1 were used as the cytosolic and mitochondrial internal control, respectively. (**E**) The protein levels of cleaved caspase 3 were detected by Western blot. The relative protein ratio was normalized with β-actin using Image Pro-plus analysis software. All data were obtained from three independent experiments, and are expressed as mean ± SD. * $p < 0.05$.

To investigate whether the treatment triggered mitochondria-mediated intrinsic apoptosis in NB cells, NB cells were treated with ONC201 for 48 h before being harvested for cytosol and mitochondria isolation. Western blot analysis indicated that ONC201 significantly increased the expression of cytochrome C in the cytosol and mitochondria fraction in both SK-N-AS and BE(2)17 cells (Figure 4C,D).

To further verify the apoptotic signaling activation, we evaluated the expression of cleaved caspase 3 by Western blot. The results reveal that ONC201 alone significantly enhanced the protein levels of cleaved caspase 3 in both SK-N-AS and BE(2)17 cells (Figure 4E). These results suggest that single treatment with ONC201 could provoke ROS generation and mitochondria-mediated apoptosis in NB cells, regardless of the status of *MYCN* amplification.

### 2.5. Administration of ONC201 Induces Neurite Outgrowth and Promotes ATRX Expression and Nuclear Translocation in MYCN-Amplified NB Cells

Via microscopy observations, we found that neurite outgrowth was observed when NB cells were treated with ONC201, which was obvious in BE(2)M17 cells, but not in SK-N-AS cells (Figure 5A). Given that ATRX has been reported to exhibit involvement in chromatin remodeling and neuronal differentiation [17], we assessed the levels of ATRX in NB cells when exposed to ONC201. Western blot analysis showed that ONC201 significantly increased the protein levels of ATRX in BE(2)M17 cells, which was associated with a significant decrease in MYCN (Figure 5B). To further investigate the distribution of increased ATRX in NB cells, immunofluorescence staining was performed to label ATRX (red) and mitochondria using an anti-TOM20 antibody (green) under confocal microscopy observation. Of note, stronger intensity of red fluorescence was found in the nuclei of ONC201-treated SK-N-AS and BE(2)M17 cells compared with the control group (Figure 5C). Consistent with the above findings, there was an increase in ATRX in the nucleus, but a decrease in ATRX in the cytosol (Figure 5D).

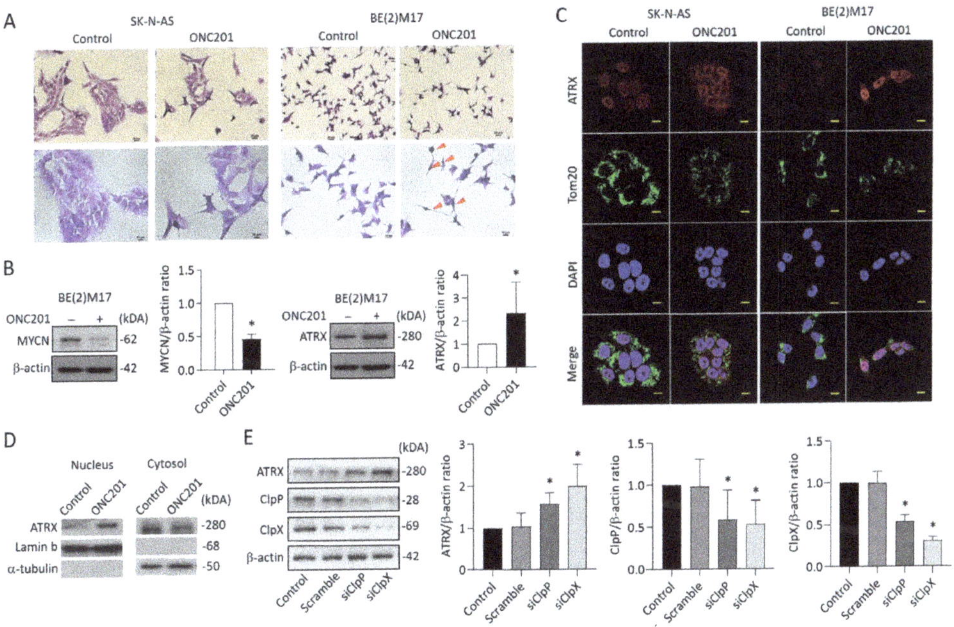


**Figure 5.** Effects of ONC201 on neurite outgrowth and the express profiles of ATRX in NB cells: (**A**) Representative of Giemsa staining in ONC201-treated SK-N-AS and BE(2)M17 cells. Scale bar 20 μm (**upper panel**) and 10 μm (**lower panel**), respectively. Red arrow indicates the neurite extension. (**B**) Western blot analysis of MYCN and ATRX protein in ONC201-treated BE(2)M17 cells. The relative protein ratio was normalized with β-actin using Image Pro-plus analysis software. (**C**) Representative images demonstrating ATRX enhancement (red) in the nucleus of SK-N-AS and BE(2)M17 cells when exposed to ONC201 (5 μM) for 48 h. The mitochondria were labeled with anti-TOM20 antibody (green). The nuclei were labeled with DAPI (blue). Scale bar 10 μm. (**D**) Western blot analysis of ATRX protein in nuclear and cytosolic fractions in BE(2)M17 cells. The relative ATRX ratio was normalized with lamin b and alpha-tubulin, respectively, and expressed as mean ± SD. (**E**) Analysis of ATRX expression in siClpP- and siClpX-treated SK-N-AS cells by Western blot. The relative protein ratio was normalized with β-actin, and is expressed as mean ± SD. All data were obtained from four to six independent experiments. * $p < 0.05$.

To investigate whether ClpXP downregulation can modulate ATRX expression, BE(2)M17 cells were transfected with siClpP or siClpX for 48 h and then harvested for Western blot analysis. The results indicate that the protein levels of ClpP and ClpX were reduced up to 50% in both siClpP- and siClpX-treated cells compared with the control group. As expected, the ATRX expression was significantly increased in BE(2)M17 cells receiving siClpP or siClpX, implying that ATRX upregulation might be dependent on the suppression of ClpXP protease (Figure 5E).

### 2.6. *ONC201 Significantly Suppresses the Development of MYCN-Amplified NB Xenograft Tumors*

Next, we evaluated the therapeutic efficacy of ONC201 in a xenograft NB animal model. Administration of ONC201 (50 μg/g) significantly inhibited the tumor growth. Compared to the control group, the tumor weight of ONC201-treated group (0.28 ± 0.27 g) decreased down to only around 20% of the control group (1.44 ± 0.45 g) (Figure 6A). Further immunohistochemistry analysis demonstrated a significantly higher percentage of nuclear ATRX-positive cells in ONC201-treated NB tissues than the control group (Figure 6B). Interestingly, the immunoreactive ClpP and ClpX protein staining was also significantly increased in the high-dose group, indicating that the administration of ONC201 (50 μg/g) for a longer period may reverse the decreased protein expression of both ClpP and ClpX in the in vitro studies (Figure 6C,D). We also verified the effect of ONC201 on MYCN expression in *MYCN*-amplified NB cells. Surprisingly, the ONC201 treatment suppressed the expression of MYCN in BE(2)M17 cells in a dose-dependent manner (Figure 6E).

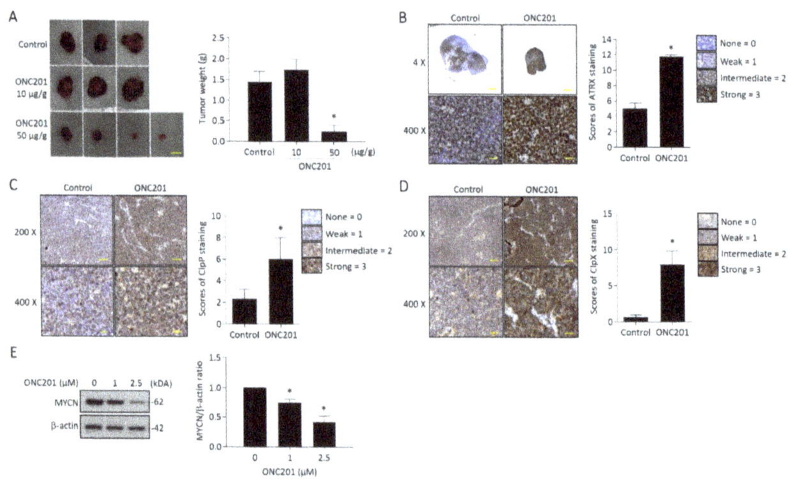

**Figure 6.** Effect of ONC201 administration on tumor growth in NB xenograft animal model and ClpXP and ATRX expression in tumor tissues: (**A**) Images of the resected tumors in the three groups. Tumor weight (g) was measured at sacrifice by microbalance and expressed as mean ± SD. Control = 3, treatment = 5. Scale bar 1 cm. (**B**) Representative pictures of ATRX expression in control and ONC201-treated neuroblastoma tissues. Scale bar 100 μm (**upper panel**) and 30 μm (**lower panel**), respectively. The percentage of nuclear ATRX-positive cells was calculated from five random images at 400x magnification per tissue, and are expressed as mean ± SD ($n$ = 3). (**C**,**D**) Immunohistochemistry analysis of ClpP and ClpX expression in controls and ONC201-treated neuroblastoma tissues. Scale bar 30 μm. The intensities of ClpP and ClpX staining were calculated from five random images at 400× magnification per tissue, and are expressed as mean ± SD ($n$ = 3). (**E**) Analysis of MYCN expression in ONC201-treated BE(2)M17 cells by Western blot. The relative protein ratio was normalized with β-actin using Image Pro-plus analysis software. * $p < 0.05$.

Since the molecular mechanism of the upregulation of ATRX by ONC201 is not clear at present, we proceeded to study EZH2 expression. The rationale is that MYCN upregulates EZH2, leading to the inactivation of a tumor suppressor program in neuroblastoma [18], while ATRX in-frame fusion neuroblastoma is sensitive to EZH2 inhibition [19]. The results of our study reveal that ONC201 did not change the expression of EZH2 protein. Knock-down of EZH2 also failed to change the ATRX protein expression (Supplementary Figure S3).

Finally, we also evaluated the potentially toxic effects of ONC201 on the mice by measuring the body weight, organ weight, and histology of two major organs, the liver and kidney, which showed no significant difference between groups (Supplementary Figure S4).

Taken together, ONC201 achieves the significant tumor inhibitory effect through interrupting the mitochondria function and modulating the expression of ATRX and MYCN, the two critical clinical risk factors for patients with NB.

## 3. Discussion

The present study first confirms a role for ClpXP protease in the differentiation of human NB tissues. Using ONC201 to suppress ClpXP protease, we observed significant decreases in mitochondrial proteases ClpP and ClpX in NB cells, regardless of the *MYCN* amplification status, which is associated with impaired mitochondrial respiratory chain function. ONC201 alone can induce a state of energy depletion, increased ROS, and decreased mitochondrial membrane potential in NB cells. The overall cascade effect is significantly enhanced cytochrome c accumulation in the mitochondria, release into the cytosol, and consequent suppression of NB cell growth through mitochondria-mediated apoptosis, which is particularly significant in *MYCN*- amplified NB cells. The latter is also vulnerable to glycolytic inhibition by 2DG than *MYCN*-nonamplified NB cells [1,2]. We also found that MYCN is suppressed in *MYCN*-amplified NB cells, while ATRX is activated and accumulates in the nucleus after suppression of ClpP and ClpX with ONC201 or by siRNA targeting the genes in both NB cell types. The findings are critical since inactivation of ATRX has been correlated with a high risk of the disease and poor prognosis [20,21]. Our findings indicate that ATRX could be reactivated by ONC201 treatment or by suppressing of ClpP/ClpX expression to improve the survival of the patients with high-risk *MYCN*-amplified NB.

We found that the highest immunoreactive staining of both ClpP and ClpX proteins occurred in immature or mature ganglion cells rather than more malignant neuroblasts in human NB tumors. The expression is also low in the Schwannian-stroma-dominant cell type, as they are deficient in ganglion cells and also a favorable histological finding, according to the Shimada classification [22]. The implication of ClpP and ClpX in neuronal differentiation and maturation in NB seems contrary to previous studies reporting that ClpXP is overexpressed in hematologic malignancies and solid tumors, and is necessary for the viability of a subset of tumors [7]. Both inhibition and hyperactivation of ClpXP lead to impaired respiratory chain activity and causes cell death in cancer cells [7]. The data provided in our study clearly show that ONC201 suppresses ClpXP and disrupts mitochondrial function in vitro, but repeated treatment with a high dose of ONC201 in *MYCN*-amplified NB xenograft promote the expression of immunoreactive ClpP and ClpX proteins. The latter is consistent with several other cancers, such as acute myeloid leukemia, in which ONC201 functions as an activator of ClpXP [9,23].

Respiratory chain subunits have been identified as putative ClpP substrates, while the inhibition of ClpP leads to the accumulation of degraded or misfolded subunits. Our findings of notable downregulations of SDHB and NDUFS1 are consistent with a study reporting that ClpP hyperactivation induces lethality in leukemias and lymphomas, while selective proteolysis of subsets of the mitochondrial proteome involving mitochondrial respiration and oxidative phosphorylation are responsible [23]. As one of the subunits of mitochondrial respiratory chain complex II, SDHB has been shown to significantly increase the carcinogenesis of oral squamous cell carcinoma [24]. By contrast, SDHB has been reported to suppress tumorigenesis in clear cell renal cell carcinoma, while decreased

SDHB accelerates growth in hepatocellular carcinoma (HCC) [25,26], which is ascribed to a switch from aerobic respiration to glycolysis in HCC. In this study, ONC201 decreased the expression of SDHB with the interruption of aerobic respiration in NB cells, eventually leading to NB cell death.

Repression of genes, including NDUFS1, by a sulfonamide anticancer agent, indisulam, has been shown to improve the survival of patients with metastatic melanoma [27]. Granzyme B, a caspase and cytotoxic lymphocyte protease, can increase ROS in target cells by directly cleaving complex I subunits, including NDUFV1, NDUFS1, and NDUFS2, to promote apoptosis [28]. Interestingly, oncostatin M, a pleiotropic cytokine belonging to the IL-6 family, may suppress NDUFS1/2 and improve the glioblastoma response to ionizing radiation, thereby prolonging lifespan [29]. The above findings are consistent with our results of a profound decrease in NDUFS1 using ONC201, leading to increased ROS and promotion of apoptosis in susceptible NB cells.

ROS increases were detected in the NB cells treated with ONC201. ONC201-induced reduction in ClpP has been shown to blunt $UPR^{mt}$ induction, leading to increased generation of ROS, and decreased membrane potential [14]. Interestingly, both superoxide and hydrogen peroxide production, and mitochondrial ROS generation are highest in SK-N-AS, which is also known to have a high expression of c-Myc. The latter is known to regulate ROS generation along with HIF-1 alpha [30], which may be ascribed to the c-Myc induction of nuclear-encoded mitochondrial gene expression and mitochondrial biogenesis [31]. Gene expression analysis has revealed that MYCN is associated with increased ROS, downregulated mitophagy, and poor prognosis [32]. Forced overexpression of MYCN in neural crest progenitor cells enhances glutaminolysis, leading to ROS production and rendering NB cells sensitive to ROS augmentation [33].

Downregulation of ClpXP by ONC201 may be responsible for a decrease in ClpP in myoblasts, the mouse islet β-cell line (Min6), and human trophoblast cells, which are all associated with decreased MMP [14,34,35]. Contrary to a previous study reporting that the downregulation of ClpP in muscle cells impairs myoblast differentiation [14], ONC201 treatment or siRNA targeting ClpP and ClpX in our study significantly increased ATRX expression in the nucleus of BE(2)M17. ATRX has been implicated in neuronal differentiation, which may be impaired by the silencing of ATRX [36]. ATRX as a tumor suppressor has been associated with protection from DNA replication stress through a resolution of difficult-to-replicate G-quadruplex (G4) DNA structures [37]. Meanwhile, ATRX mutation has been regarded as an unfavorable histology [38]. Our finding of an increased expression of ATRX in the nucleus is of interest and significant. A recent report of a case with ATRX mutation associated with complex I deficiency suggests that target genes of the ATRX protein include those responsible for mitochondrial function [39]. ONC201 and ClpXP protease target mitochondria, which concurrently upregulate ATRX expression in the nucleus, indicating that some form of feedback loop exists in this particular scenario. A previous study revealed that MYCN upregulates EZH2, leading to the inactivation of a tumor suppressor program in neuroblastoma [18]; meanwhile ATRX in-frame fusion neuroblastoma is sensitive to EZH2 inhibition [19]. Our study, using ONC201 or siRNA targeting EZH2, reveals that even using higher doses that could significantly decrease EZH2 protein expression, there is little change in ATRX expression. Our results fail to duplicate that found in glioma or other cancers [40,41]. Obviously, the genetic background in our NB cell models could be responsible for their difference from diffuse midline glioma. Special manipulation, such as ATRX in-frame fusion [19], may be required for this purpose. We will continue to solve this problem in our future study.

*MYCN* gene amplification is one of the most important prognostic markers for high-risk NB patients. The reported prevalence of NB patients with *MYCN* gene amplification is around 20–30%, with poor overall survival rate of less than 50% [42]. A recent report on targeting of the *MYCN* gene by genetic deletion using DNA alkylating agent, MYCN-A3, induced apoptosis in *MYCN*-amplified cells, but not in nonamplified cells [43]. In our present study, we found that ONC201 treatment could significantly inhibit the protein expression of MYCN in *MYCN*-amplified NB cells with a promising tumor suppressor effect in xenograft tumors. Although the exact mechanism is needed to be elucidated, the tumor inhibitory effect is encouraging for future application to treat high-risk NB patients.

One limitation of the current study is that we did not investigate how ONC201 decreases the expression of MYCN in MYCN-amplified NB cells. Similar findings are shown in a study using MYCN-amplified IMR-32 and MYCN-nonamplified SK-N-SH NB cells to address the issues of differential tumorigenic protein expression [44]. They also failed to verify the mechanism explaining how ONC201 and ONC206 accomplish the specific aim of decreasing the expression of MYCN and several other tumorigenic proteins. Future studies are necessary to clarify the issue.

## 4. Materials and Methods

### 4.1. Cell Cultures and Reagents

Human NB cell lines SK-N-AS and BE (2)-M17 were purchased from the American Type Culture Collection (Manassas, VA, USA). All cell lines were maintained in Dulbecco's modified Eagle's medium (DMEM) (Thermo Fisher Scientific, Waltham, MA, USA) and supplemented with 10% heat-inactivated fetal bovine serum (FBS; Thermo Fisher Scientific, Waltham, MA, USA), GlutaMAX (Thermo Fisher Scientific, Waltham, MA, USA), nonessential amino acids (Thermo Fisher Scientific, Waltham, MA, USA), and an antibiotic–antimycotic (Thermo Fisher Scientific, Waltham, MA, USA) in a 5% $CO_2$ humidified incubator at 37 °C. ONC201 (SML1068), DHE (309800), and Resazurin sodium salt (199303) were purchased from Sigma-Aldrich (St. Louis, MO, USA). Cleaved caspase 3 (Asp175) and N-Myc (9405S) antibodies were purchased from Cell Signaling Technology (Danvers, MA, USA). ClpX (ab168338), ATRX (ab97508), and EZH2 (ab186006) antibodies were purchased from Abcam (Cambridge, MA, USA). β-actin (sc-8432), ClpP (sc-271284), SDHB (sc-271548), VDAC1 (sc-390996), Lamin b (sc-6216), and NDUFS1 (sc-271510) were purchased from Santa Cruz Biotechnology (Santa Cruz, CA, USA). A-tubulin (GTX112141) was purchased from GeneTex (Hsinchu, Taiwan).

### 4.2. Neuroblastoma Xenograft Animal Model

Four-week-old male nonobese diabetic/SCID (NOD/SCID, NOD.CB17-Prkdcscid/NcrCrl) mice were purchased from the National Laboratory Animal Center (Taipei, Taiwan). All animal procedures in this study are approved by the Institute of Animal Care and Use Committee of Kaohsiung Chang Gung Memorial Hospital, Taiwan (IACUC No. 2019091805). To induce human neuroblastoma, SK-N-DZ cells were subcutaneously inoculated into the right flank of mice ($1 \times 10^7$ cells in 0.1 mL of medium). After implantation for 8 days, tumor-bearing mice were randomly divided into control ($n$ = 3), ONC201 (10 μg/g), and ONC201 (50 μg/g) groups ($n$ = 5), respectively. The mice then received intraperitoneal injections of the indicated treatment once weekly for 4 weeks. At the end of the experiment, mice were sacrificed, and the tumors were harvested and weighed using an electronic microbalance.

### 4.3. Cell Survival Assessment

The Resazurin reduction test was utilized to detect the cell survival. Briefly, NB cells were seeded in a 96-well culture plate ($1.5 \times 10^4$ cells/well) overnight. Then, cells were treated with indicated concentrations of ONC201 for 48 h and 96 h. At the end of the experiment, cells were supplemented with 20 μL per well of Resazurin dye (Sigma-Aldrich, St. Louis, MO, USA) and incubated for another 3 h at 37 °C. The optical density of resorufin was measured at 570 nm using a 96-well spectrophotometric plate reader (Hidex Sense, Turku, Finland).

### 4.4. Flow Cytometry to Detect Cell Death

Cell death was detected by trypan blue solution (Thermo Fisher Scientific, Waltham, MA, USA). After ONC201 (5 μM) treatment, cells were dissociated from the plate with trypsin-EDTA and then stained with trypan blue solution for 5 min. The percentage of cell death was detected by FACS caliber 101 flow cytometer (BD Biosciences, San Jose, CA, USA) and analyzed using winMDI software.

### 4.5. Terminal Deoxynucleotidyl Transferase dUTP Nick End Labeling (TUNEL) Assay

ONC201-induced NB cell death was detected by enzyme labeling of DNA strand breaks using a TUNEL assay (In Situ Cell Death Detection Kit; Roche, Germany) according to manufacturer's instructions. Briefly, cells were fixed with 4% paraformaldehyde for 10 min and then incubated with a mixture of terminal deoxynucleotidyl transferase and fluorescein-dUTP at 37 °C to label free 3′OH ends of DNA. After TUNEL staining, cells were washed with PBS and counterstained with DAPI Fluoromount-G (SouthernBiotech, Birmingham, AL, USA) for 10 min and viewed under a fluorescent microscope. For quantifying the immunostaining of TUNEL, each group was randomly captured by microscopy for three independent fields. The apoptotic percentage was estimated by normalizing the number of TUNEL-positive cells to the total number of DAPI-positive cells and calculated from three independent fields.

### 4.6. Isolation of Mitochondrial and Cytosolic Fractions

After ONC201 treatment for 24 h, cells were harvested and resuspended in 1.2 mL RSB Hypo Buffer (10 mM Tris-HCl, pH 7.5, containing 10 mM NaCl and 1.5 mM MgCl2) for 10 min on ice. Subsequently, cells were homogenized using a 1 mL syringe to pass the cell suspension through a 27-gauge needle 10 times, and then added to 0.8 mL 2.5X MS homogenization buffer (12.5 mM Tris-HCl, pH 7.5, containing 2.5 mM EDTA, 525 mM mannitol, and 175 mM sucrose). Nuclei and intact cells were removed by centrifugation at 3000× *g* rpm for 5 min at 4 °C. The supernatants (cytoplasmic fraction) were recentrifuged at 13,000 rpm for 20 min at 4 °C to move pellets, while the pellets containing mitochondria were washed with 1X MS homogenization buffer (5 mM Tris-HCl, pH 7.5, containing 1 mM EDTA, 210 mM mannitol and 70 mM sucrose) twice at 13,000 rpm for 10 min at 4 °C, and then resuspended in PRO-PREP™ Protein Extraction Solution (Intron Biotechnology Inc., Seongnam, Republic of Korea) for 20 min on ice. The separation efficiency was determined by Western blot.

### 4.7. Isolation of Nuclear and Cytosolic Fractions

Nuclear and cytosolic fractions were extracted according to L&W nucleus/cytoplasm fractionation protocol [45]. Cells were harvested and resuspended in 0.5 mL hypotonic solution (20 mM Tris-HCl pH 7.4, 10 mM KCl, 2 mM $MgCl_2$, 1 mM EGTA, 0.5 mM DTT, 0.5 mM PMSF) containing 0.1% NP-40 and incubated for 3 min on ice. Subsequently, cells were homogenized using a 1 mL syringe to pass the cell suspension through a 27-gauge needle 10 times, and then centrifugation at 3000× *g* rpm for 5 min at 4 °C. The supernatants (cytoplasmic fraction) were recentrifuged at 13,000 rpm for 10 min to move pellet debris, while the pellets containing nuclei were washed with isotonic buffer (20 mM Tris-HCl pH 7.4, 150 mM KCl, 2 mM $MgCl_2$, 1 mM EGTA, 0.5 mM DTT, 0.5 mM PMSF) three times, and then resuspended in PRO-PREP™ Protein Extraction Solution (Intron Biotechnology Inc., Seongnam, Republic of Korea) for 20 min on ice. The supernatants (nuclear fractions) were obtained by centrifugation at 13,000× *g* rpm for 10 min at 4 °C. The separation efficiency was determined by Western blot.

### 4.8. Mitochondrial Bioenergetics Analysis

The rate of oxygen consumption (OCR) and extracellular acidification rate (ECAR) in NB cells were analyzed using a Seahorse XF24 extracellular flux analyzer (Seahorse Bioscience Inc.; Chicopee, MA, USA) [46]. Initially, cells were seeded in Seahorse cell culture 24-well plates (2 × $10^4$ cells/well) overnight, and then treated with ONC201 (5 μM) for 48 h. After washing with sodium-bicarbonate-free DMEM medium, cells were refreshed with 500 μL of medium for further examination. The basic OCR was measured four times and plotted as a function of the cells under basal conditions; the inhibitors, including oligomycin (1 μM), FCCP (0.5 μM), and rotenone (1 μM), were added sequentially before experimental analysis. The basal ECAR was measured after injections of glucose (10 mM) and oligomycin (1 μM). At the end of recording, cells were collected and counted using a trypan blue exclusion assay. The OCR and ECAR values were normalized to total protein levels in individual wells using the BCA protein assay (Thermo Fisher Scientific, Waltham, MA, USA).

### 4.9. Measurement of ROS Generation

The intracellular ROS generation was measured by flow cytometry using DHE staining. Cells were treated with ONC201 (5 μM) for 48 h, and these cells were incubated with DHE (10 μM) for 30 min at 37 °C. The ROS production was further measured using a FACS caliber 101 flow cytometer (BD Biosciences, San Jose, CA, USA) and analyzed using winMDI software. For the detection of mitochondrial superoxide anions, the MitoSox Red reagent (Invitrogen; Carlsbad, CA, USA) was applied according to the manufacturer's protocol. Briefly, at the indicated time points after treatment, cells were incubated with MitoSox Red reagent (5 μM) for 30 min at 37 °C. The cells were then collected, washed twice with PBS, and finally resuspended in a flow tube with 1 mL PBS. The fluorescent signal of cell suspension was then measured using a FACS caliber 101 flow cytometer (BD Biosciences, San Jose, CA, USA) and analyzed using winMDI software.

### 4.10. Measurement of Mitochondrial Membrane Potential

Mitochondrial membrane potential (MMP) was determined by Tetramethyl rhodamine methyl ester (TMRM; Sigma-Aldrich, USA) staining. Briefly, at the indicated time points after treatment, cells were incubated with TMRM (100 nM) for 20 min at 37 °C in the dark, washed twice with ice-cold PBS, and finally resuspended in 1 ml PBS. Mitochondrial permeability transition was measured immediately by FACS caliber 101 flow cytometer (BD Biosciences, San Jose, CA, USA) and analyzed using winMDI software.

### 4.11. Measurement of ATP Content

The cellular ATP concentrations were measured using the ATP Colorimetric/Fluorometric Assay Kit (BioVision, Inc., Milpitas, CA, USA) according to the manufacturer's instructions. Briefly, after treatment with ONC201 (5 μM) for 48 h, cells (4 × $10^5$) were collected and resuspended in a reaction buffer containing 20 mM glycine, 50 mM $MgSO_4$, and 4 mM EDTA. Samples were boiled for 2 min at 100 °C and centrifuged for 5 min at 3000× *g* rpm. Supernatants were collected, and 50 μL of each sample was mixed with 50 μL of an ATP solution for fluorescence readings at Ex/Em535/587 nm using a fluorescence meter (FLUOstar OPTIMA; BMG Labtech, Ortenberg, Germany). The level of ATP production was determined from a standard curve constructed with 10–100 pmol ATP.

### 4.12. Immunofluorescence Analysis

NB cells were plated overnight on 6-well culture dishes containing sterilized coverslips. After treatment for 48 h, cells were fixed with 4% (*w*/*v*) paraformaldehyde and permeabilized with PBS containing 0.1% (*w*/*v*) Triton X-100 and 2% (*w*/*v*) BSA at room temperature for 10 min. Cells were labeled with indicated primary antibodies, followed by Alexa Fluor-conjugated secondary antibody (Invitrogen; Carlsbad, CA, USA). Cells were then washed with PBS and mounted in DAPI Fluoromount-G (SouthernBiotech, Birmingham, AL, USA). Labeled cells were visualized with LSM510 (Carl Zeiss, Thornwood, NY, USA).

### 4.13. Western Blot Analysis

NB cells were treated with ONC201 (5 μM) for 48 h before protein analysis. Cell lysates were separated by SDS-PAGE and transferred to a polyvinylidene difluoride (PVDF) membrane. The PVD membrane was blocked with 5% milk in TBS-T for 1 h, then incubated with specific primary antibodies and secondary antibodies conjugated with HRP (1:10,000 dilutions in 5% milk) for 1 h, respectively. The membrane signals were analyzed using an AutoChemi image system (UVP), or exposed to Fuji medical X-ray film, followed by quantification with Alpha View SA 3.4.0 (ProteinSimple, San Jose, CA, USA).

### 4.14. Immunohistochemistry Analysis and Scoring

For the analysis of ClpXP expression in human NB tissues and ATRX profiles in neuroblastoma tissues, the paraffin sections were deparaffinized, blocked with 3% hydrogen peroxide for 10 min, and subjected to antigen retrieval by microwave heating in 0.01 M citrate buffer for 15 min. The slides were then washed twice with PBS, incubated with primary antibodies, followed by incubation with the polymer conjugated with peroxidase for 30 min using a polymer detection system (Zymed Laboratories, San Francisco, CA). Finally, the color was developed with 3, 3-diaminobenzidine (DAB; Sigma, St. Louis, MO). The slides were counterstained with Gill's hematoxylin, dehydrated, and mounted before microscopic reading. The intensity of positively stained tumor cells was scored as 0 = none; 1 = weak; 2 = intermediate; and 3 = strong. For quantification of ClpXP and ATRX staining intensity, we randomly selected five high-power (×400) fields for each section to evaluate each sample. The intensity of positively stained tumor cells was scored as 0 = none; 1 = weak; 2 = intermediate; and 3 = strong. The proportion of each intensity score was further scored as 0 = no positive cells; 1 = 0–20%; 2 = 21–50%; 3 = 51–80%; 4 = 81–100%. Both scores were multiplied and summed to produce a final immunoreactivity score, ranging from 0 to 12.

### 4.15. RNA Interference

For knocking down the gene expression of ClpXP protease, cells were transfected with control siRNA (D-001810-10-50; Dharmacon, Lafayette, Co, USA), siClpP (L-005811-00-0020; Dharmacon, Lafayette, Co, USA) and siClpX (L-008763-00-0020; Dharmacon, Lafayette, Co, USA) for 4 h, respectively, using Lipofectamine 3000 Reagent (Thermo Fisher Scientific, Waltham, MA, USA) according to the manufacturer's instructions, and then incubated with complete medium for 48 h before the ensuing experimental analyses.

### 4.16. Statistical Analysis

Data are presented as the mean ± standard deviation of three independent experiments, unless otherwise indicated. The statistical analysis was performed using GraphPad Prism 8.0 software (GraphPad Software, San Diego, CA, USA). One-way ANOVA and post hoc multiple comparison via the Tukey test were used to compare multiple groups, and $t$-tests were used for two-group comparisons. A $p$-value less than 0.05 was considered statistically significant.

## 5. Conclusions

The dysregulation of mitochondria, suppression of MYCN, reactivation of ATRX, and the implication of ClpP and ClpX in differentiation are key processes involved in the particular process of ONC201 treatment in NB cells. ONC201, as a single-agent therapy, can suppress *MYCN*-amplified NB xenograft growth. Future therapeutic application in NB, in combination with other agents, is promising. The major finding of our study is summarized and illustrated in Figure 7.

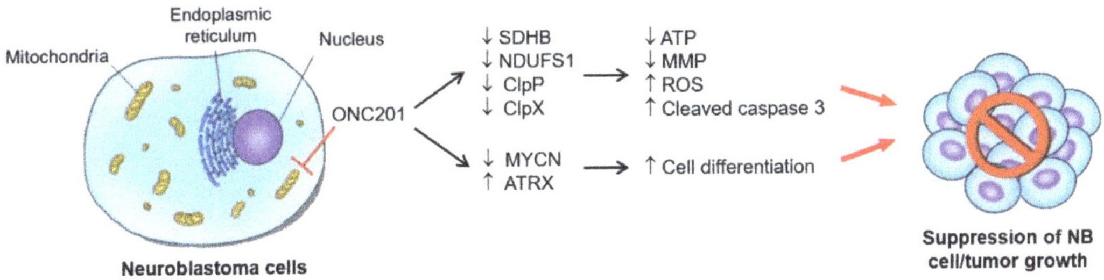


**Figure 7.** The proposed working model for interruption of mitochondrial functions, induced apoptosis, and differentiation in NB cells: ONC201 treatment targets mitochondria to perturb mitochondrial functions, e.g., the treatment disrupts the mitochondrial ClpXP complex, impairs respiratory chain activity, and leads to mitochondrial function loss and ROS accumulation. The enhancement of ROS production finally induces apoptosis in NB cells. In addition to cell death, ONC201 treatment also promotes the nuclear expression of ATRX, which supports neuronal differentiation and the suppression of NB growth.

**Supplementary Materials:** The following supporting information can be downloaded at: https://www.mdpi.com/article/10.3390/ijms24021649/s1.

**Author Contributions:** Conceptualization: J.-H.C., H.-C.C.; Methodology: J.-C.W., C.-C.H., P.-W.W., T.-Y.C.; validation: J.-C.W., H.-C.C.; formal analysis: J.-C.W., C.-C.H., T.-Y.C.; resources: C.-C.H., J.-H.C.; data curation: J.-C.W., T.-Y.C., W.-M.H., J.-H.C., H.-C.C.; writing original draft: J.-C.W.; writing—review and editing: J.-H.C., H.-C.C.; supervision: J.-H.C.; project administration: J.-H.C.; funding acquisition: J.-H.C. All authors have read and agreed to the published version of the manuscript.

**Funding:** This research was supported by (Chang Gung Memorial Hospital, Taiwan) grant numbers of CMRPG8H0181-3 and CMRPG8K0071-3.

**Institutional Review Board Statement:** The study was conducted in accordance with the Declaration of Helsinki, and approved by the Institutional Review Board (IRB) of Chang Gung Medical Foundation (protocol code 201701882B0C503 on 4 October 2021). The animal study protocol was approved by Institutional Animal Care and Use Committee of Chang Gung Memorial Hospital (protocol code 20190918056 on 30 January 2020).

**Informed Consent Statement:** Patient consent was waived under the approval of IRB of Chang Gung Medical Foundation (protocol code 201701882B0C503 on 4 October 2021) using de-identification data.

**Data Availability Statement:** The data presented in this study are available on request from the corresponding author. The data are not publicly available due to privacy.

**Acknowledgments:** The authors thank Chang Gung Medical Foundation Kaohsiung Chang Gung Memorial Hospital Tissue Bank Core Lab and Biobank (CLRPG8L0081) for technical support, and Po-Chin Tsai and Pei-Lin Liao for their invaluable technical assistance.

**Conflicts of Interest:** The authors declare no conflict of interest.

## References

1. Chuang, J.H.; Chou, M.H.; Tai, M.H.; Lin, T.K.; Liou, C.W.; Chen, T.; Hsu, W.M.; Wang, P.W. 2-Deoxyglucose treatment complements the cisplatin- or BH3-only mimetic-induced suppression of neuroblastoma cell growth. *Int. J. Biochem. Cell Biol.* **2013**, *45*, 944–951. [CrossRef] [PubMed]
2. Huang, C.C.; Wang, S.Y.; Lin, L.L.; Wang, P.W.; Chen, T.Y.; Hsu, W.M.; Lin, T.K.; Liou, C.W.; Chuang, J.H. Glycolytic inhibitor 2-deoxyglucose simultaneously targets cancer and endothelial cells to suppress neuroblastoma growth in mice. *Dis. Model Mech.* **2015**, *8*, 1247–1254. [CrossRef] [PubMed]
3. Schibler, J.; Tomanek-Chalkley, A.M.; Reedy, J.L.; Zhan, F.; Spitz, D.R.; Schultz, M.K.; Goel, A. Mitochondrial-Targeted Decyl-Triphenylphosphonium Enhances 2-Deoxy-D-Glucose Mediated Oxidative Stress and Clonogenic Killing of Multiple Myeloma Cells. *PLoS ONE* **2016**, *11*, e0167323. [CrossRef] [PubMed]

4. O'Neill, S.; Porter, R.K.; McNamee, N.; Martinez, V.G.; O'Driscoll, L. 2-Deoxy-D-Glucose inhibits aggressive triple-negative breast cancer cells by targeting glycolysis and the cancer stem cell phenotype. *Sci. Rep.* **2019**, *9*, 3788. [CrossRef] [PubMed]
5. Jovaisaite, V.; Auwerx, J. The mitochondrial unfolded protein response-synchronizing genomes. *Curr. Opin. Cell Biol.* **2015**, *33*, 74–81. [CrossRef]
6. Goard, C.A.; Schimmer, A.D. Mitochondrial matrix proteases as novel therapeutic targets in malignancy. *Oncogene* **2014**, *33*, 2690–2699. [CrossRef]
7. Nouri, K.; Feng, Y.; Schimmer, A.D. Mitochondrial ClpP serine protease-biological function and emerging target for cancer therapy. *Cell Death Dis.* **2020**, *11*, 841. [CrossRef]
8. Allen, J.E.; Kline, C.L.; Prabhu, V.V.; Wagner, J.; Ishizawa, J.; Madhukar, N.; Lev, A.; Baumeister, M.; Zhou, L.; Lulla, A.; et al. Discovery and clinical introduction of first-in-class imipridone ONC201. *Oncotarget* **2016**, *7*, 74380–74392. [CrossRef]
9. Prabhu, V.V.; Morrow, S.; Rahman Kawakibi, A.; Zhou, L.; Ralff, M.; Ray, J.; Jhaveri, A.; Ferrarini, I.; Lee, Y.; Parker, C.; et al. ONC201 and imipridones: Anti-cancer compounds with clinical efficacy. *Neoplasia* **2020**, *22*, 725–744. [CrossRef]
10. Tan, J.; Grouleff, J.J.; Jitkova, Y.; Diaz, D.B.; Griffith, E.C.; Shao, W.; Bogdanchikova, A.F.; Poda, G.; Schimmer, A.D.; Lee, R.E.; et al. De Novo Design of Boron-Based Peptidomimetics as Potent Inhibitors of Human ClpP in the Presence of Human ClpX. *J. Med. Chem.* **2019**, *62*, 6377–6390. [CrossRef]
11. Ferrarini, I.; Louie, A.; Zhou, L.; El-Deiry, W.S. ONC212 is a Novel Mitocan Acting Synergistically with Glycolysis Inhibition in Pancreatic Cancer. *Mol. Cancer Ther.* **2021**, *20*, 1572–1583. [CrossRef]
12. Borsuk, R.; Zhou, L.; Chang, W.I.; Zhang, Y.; Sharma, A.; Prabhu, V.V.; Tapinos, N.; Lulla, R.R.; El-Deiry, W.S. Potent preclinical sensitivity to imipridone-based combination therapies in oncohistone H3K27M-mutant diffuse intrinsic pontine glioma is associated with induction of the integrated stress response, TRAIL death receptor DR5, reduced ClpX and apoptosis. *Am. J. Cancer Res.* **2021**, *11*, 4607–4623.
13. Viala, J.; Mazodier, P. The ATPase ClpX is conditionally involved in the morphological differentiation of Streptomyces lividans. *Mol. Genet. Genomics* **2003**, *268*, 563–569. [CrossRef]
14. Deepa, S.S.; Bhaskaran, S.; Ranjit, R.; Qaisar, R.; Nair, B.C.; Liu, Y.; Walsh, M.E.; Fok, W.C.; Van Remmen, H. Down-regulation of the mitochondrial matrix peptidase ClpP in muscle cells causes mitochondrial dysfunction and decreases cell proliferation. *Free Radic Biol. Med.* **2016**, *91*, 281–292. [CrossRef]
15. Pruss, M.; Dwucet, A.; Tanriover, M.; Hlavac, M.; Kast, R.E.; Debatin, K.M.; Wirtz, C.R.; Halatsch, M.E.; Siegelin, M.D.; Westhoff, M.A.; et al. Dual metabolic reprogramming by ONC201/TIC10 and 2-Deoxyglucose induces energy depletion and synergistic anti-cancer activity in glioblastoma. *Br. J. Cancer* **2020**, *122*, 1146–1157. [CrossRef]
16. Furukawa, T.; Katayama, H.; Oikawa, A.; Negishi, L.; Ichikawa, T.; Suzuki, M.; Murase, K.; Takayama, S.; Sakuda, S. Dioctatin Activates ClpP to Degrade Mitochondrial Components and Inhibits Aflatoxin Production. *Cell Chem. Biol.* **2020**, *27*, 1396–1409.e10. [CrossRef]
17. Marano, D.; Fioriniello, S.; Fiorillo, F.; Gibbons, R.J.; D'Esposito, M.; Della Ragione, F. ATRX Contributes to MeCP2-Mediated Pericentric Heterochromatin Organization during Neural Differentiation. *Int. J. Mol. Sci.* **2019**, *20*, 5371. [CrossRef]
18. Chen, L.; Alexe, G.; Dharia, N.V.; Ross, L.; Iniguez, A.B.; Conway, A.S.; Wang, E.J.; Veschi, V.; Lam, N.; Qi, J.; et al. CRISPR-Cas9 screen reveals a MYCN-amplified neuroblastoma dependency on EZH2. *J. Clin. Invest.* **2018**, *128*, 446–462. [CrossRef]
19. Qadeer, Z.A.; Valle-Garcia, D.; Hasson, D.; Sun, Z.; Cook, A.; Nguyen, C.; Soriano, A.; Ma, A.; Griffiths, L.M., Zeineldin, M.; et al. ATRX In-Frame Fusion Neuroblastoma Is Sensitive to EZH2 Inhibition via Modulation of Neuronal Gene Signatures. *Cancer Cell* **2019**, *36*, 512–527.e9. [CrossRef]
20. Irwin, M.S.; Naranjo, A.; Zhang, F.F.; Cohn, S.L.; London, W.B.; Gastier-Foster, J.M.; Ramirez, N.C.; Pfau, R.; Reshmi, S.; Wagner, E.; et al. Revised Neuroblastoma Risk Classification System: A Report From the Children's Oncology Group. *J. Clin. Oncol.* **2021**, *39*, 3229–3241. [CrossRef]
21. Zeineldin, M.; Federico, S.; Chen, X.; Fan, Y.; Xu, B.; Stewart, E.; Zhou, X.; Jeon, J.; Griffiths, L.; Nguyen, R.; et al. MYCN amplification and ATRX mutations are incompatible in neuroblastoma. *Nat. Commun.* **2020**, *11*, 913. [CrossRef]
22. Shimada, H.; Ambros, I.M.; Dehner, L.P.; Hata, J.; Joshi, V.V.; Roald, B.; Stram, D.O.; Gerbing, R.B.; Lukens, J.N.; Matthay, K.K.; et al. The International Neuroblastoma Pathology Classification (the Shimada system). *Cancer* **1999**, *86*, 364–372. [CrossRef]
23. Ishizawa, J.; Zarabi, S.F.; Davis, R.E.; Halgas, O.; Nii, T.; Jitkova, Y.; Zhao, R.; St-Germain, J.; Heese, L.E.; Egan, G.; et al. Mitochondrial ClpP-Mediated Proteolysis Induces Selective Cancer Cell Lethality. *Cancer Cell* **2019**, *35*, 721–737.e9. [CrossRef] [PubMed]
24. Grimm, M.; Cetindis, M.; Lehmann, M.; Biegner, T.; Munz, A.; Teriete, P.; Kraut, W.; Reinert, S. Association of cancer metabolism-related proteins with oral carcinogenesis—Indications for chemoprevention and metabolic sensitizing of oral squamous cell carcinoma? *J. Transl. Med.* **2014**, *12*, 208. [CrossRef] [PubMed]
25. Fang, Z.; Sun, Q.; Yang, H.; Zheng, J. SDHB Suppresses the Tumorigenesis and Development of ccRCC by Inhibiting Glycolysis. *Front. Oncol.* **2021**, *11*, 639408. [CrossRef] [PubMed]
26. Tseng, P.L.; Wu, W.H.; Hu, T.H.; Chen, C.W.; Cheng, H.C.; Li, C.F.; Tsai, W.H.; Tsai, H.J.; Hsieh, M.C.; Chuang, J.H.; et al. Decreased succinate dehydrogenase B in human hepatocellular carcinoma accelerates tumor malignancy by inducing the Warburg effect. *Sci. Rep.* **2018**, *8*, 3081. [CrossRef] [PubMed]
27. Baur, M.; Gneist, M.; Owa, T.; Dittrich, C. Clinical complete long-term remission of a patient with metastatic malignant melanoma under therapy with indisulam (E7070). *Melanoma Res.* **2007**, *17*, 329–331. [CrossRef]

28. Jacquemin, G.; Margiotta, D.; Kasahara, A.; Bassoy, E.Y.; Walch, M.; Thiery, J.; Lieberman, J.; Martinvalet, D. Granzyme B-induced mitochondrial ROS are required for apoptosis. *Cell Death Differ.* **2015**, *22*, 862–874. [CrossRef]
29. Sharanek, A.; Burban, A.; Laaper, M.; Heckel, E.; Joyal, J.S.; Soleimani, V.D.; Jahani-Asl, A. OSMR controls glioma stem cell respiration and confers resistance of glioblastoma to ionizing radiation. *Nat. Commun.* **2020**, *11*, 4116. [CrossRef]
30. Yu, Y.; Niapour, M.; Zhang, Y.; Berger, S.A. Mitochondrial regulation by c-Myc and hypoxia-inducible factor-1 alpha controls sensitivity to econazole. *Mol. Cancer Ther.* **2008**, *7*, 483–491. [CrossRef]
31. Dang, C.V.; Li, F.; Lee, L.A. Could MYC induction of mitochondrial biogenesis be linked to ROS production and genomic instability? *Cell Cycle* **2005**, *4*, 1465–1466. [CrossRef]
32. Montemurro, L.; Raieli, S.; Angelucci, S.; Bartolucci, D.; Amadesi, C.; Lampis, S.; Scardovi, A.L.; Venturelli, L.; Nieddu, G.; Cerisoli, L.; et al. A Novel MYCN-Specific Antigene Oligonucleotide Deregulates Mitochondria and Inhibits Tumor Growth in MYCN-Amplified Neuroblastoma. *Cancer Res.* **2019**, *79*, 6166–6177. [CrossRef]
33. Wang, T.; Liu, L.; Chen, X.; Shen, Y.; Lian, G.; Shah, N.; Davidoff, A.M.; Yang, J.; Wang, R. MYCN drives glutaminolysis in neuroblastoma and confers sensitivity to an ROS augmenting agent. *Cell Death Dis.* **2018**, *9*, 220. [CrossRef]
34. Yung, H.W.; Colleoni, F.; Dommett, E.; Cindrova-Davies, T.; Kingdom, J.; Murray, A.J.; Burton, G.J. Noncanonical mitochondrial unfolded protein response impairs placental oxidative phosphorylation in early-onset preeclampsia. *Proc. Natl. Acad. Sci. USA* **2019**, *116*, 18109–18118. [CrossRef]
35. Luo, J.; Zeng, B.; Tao, C.; Lu, M.; Ren, G. ClpP regulates breast cancer cell proliferation, invasion and apoptosis by modulating the Src/PI3K/Akt signaling pathway. *PeerJ* **2020**, *8*, e8754. [CrossRef]
36. Ly, J.D.; Grubb, D.R.; Lawen, A. The mitochondrial membrane potential (deltapsi(m)) in apoptosis; an update. *Apoptosis* **2003**, *8*, 115–128. [CrossRef]
37. Teng, Y.C.; Sundaresan, A.; O'Hara, R.; Gant, V.U.; Li, M.; Martire, S.; Warshaw, J.N.; Basu, A.; Banaszynski, L.A. ATRX promotes heterochromatin formation to protect cells from G-quadruplex DNA-mediated stress. *Nat. Commun.* **2021**, *12*, 3887. [CrossRef]
38. Nakazawa, A. Biological categories of neuroblastoma based on the international neuroblastoma pathology classification for treatment stratification. *Pathol. Int.* **2021**, *71*, 232–244. [CrossRef]
39. Aiba, K.; Nakamura, Y.; Sugimoto, M.; Yatsuka, Y.; Okazaki, Y.; Murayama, K.; Ohtake, A.; Yokochi, K.; Saitoh, S. A case of ATR-X syndrome with mitochondrial respiratory chain dysfunction. *Eur. J. Med. Genet.* **2021**, *64*, 104251. [CrossRef]
40. Chi, A.S.; Tarapore, R.S.; Hall, M.D.; Shonka, N.; Gardner, S.; Umemura, Y.; Sumrall, A.; Khatib, Z.; Mueller, S.; Kline, C.; et al. Pediatric and adult H3 K27M-mutant diffuse midline glioma treated with the selective DRD2 antagonist ONC201. *J. Neurooncol.* **2019**, *145*, 97–105. [CrossRef]
41. Zhang, Y.; Zhou, L.; Safran, H.; Borsuk, R.; Lulla, R.; Tapinos, N.; Seyhan, A.A.; El-Deiry, W.S. EZH2i EPZ-6438 and HDACi vorinostat synergize with ONC201/TIC10 to activate integrated stress response, DR5, reduce H3K27 methylation, ClpX and promote apoptosis of multiple tumor types including DIPG. *Neoplasia* **2021**, *23*, 792–810. [CrossRef] [PubMed]
42. Otte, J.; Dyberg, C.; Pepich, A.; Johnsen, J.I. MYCN Function in Neuroblastoma Development. *Front. Oncol.* **2020**, *10*, 624079. [CrossRef] [PubMed]
43. Yoda, H.; Inoue, T.; Shinozaki, Y.; Lin, J.; Watanabe, T.; Koshikawa, N.; Takatori, A.; Nagase, H. Direct Targeting of MYCN Gene Amplification by Site-Specific DNA Alkylation in Neuroblastoma. *Cancer Res.* **2019**, *79*, 830–840. [CrossRef] [PubMed]
44. El-Soussi, S.; Hanna, R.; Semaan, H.; Khater, A.R.; Abdallah, J.; Abou-Kheir, W.; Abou-Antoun, T. A Novel Therapeutic Mechanism of Imipridones ONC201/ONC206 in MYCN-Amplified Neuroblastoma Cells via Differential Expression of Tumorigenic Proteins. *Front Pediatr* **2021**, *9*, 693145. [CrossRef]
45. Senichkin, V.V.; Prokhorova, E.A.; Zhivotovsky, B.; Kopeina, G.S. Simple and Efficient Protocol for Subcellular Fractionation of Normal and Apoptotic Cells. *Cells* **2021**, *10*, 852. [CrossRef]
46. Kuo, H.M.; Weng, S.W.; Chang, A.Y.; Huang, H.T.; Lin, H.Y.; Chuang, J.H.; Lin, T.K.; Liou, C.W.; Tai, M.H.; Lin, C.Y.; et al. Altered mitochondrial dynamics and response to insulin in cybrid cells harboring a diabetes-susceptible mitochondrial DNA haplogroup. *Free Radic Biol. Med.* **2016**, *96*, 116–129. [CrossRef]

International Journal of *Molecular Sciences*

*Article*

# Design and Synthesis of Acridine-Triazole and Acridine-Thiadiazole Derivatives and Their Inhibitory Effect against Cancer Cells

**Lini Huo [1,†], Xiaochen Liu [1,†], Yogini Jaiswal [2], Hao Xu [1], Rui Chen [1,*], Rumei Lu [1,*], Yaqin Nong [1], Leonard Williams [2], Yan Liang [3] and Zhiruo Jia [1]**

1 College of Pharmacy, Guangxi University of Chinese Medicine, Nanning 530222, China
2 Center for Excellence in Post-Harvest Technologies, North Carolina Agricultural and Technical State University, The North Carolina Research Campus, 500 Laureate Way, Kannapolis, NC 28081, USA
3 College of Pharmacy, Guangxi Medical University, Nanning 530021, China
* Correspondence: lurm@gxtcmu.edu.cn (R.C.); chenr@gxtcmu.edu.cn (R.L.)
† These authors contributed equally to this work.

**Abstract:** We report herein the design and synthesis of a series of novel acridine-triazole and acridine-thiadiazole derivatives. The newly synthesized compounds and the key intermediates were all evaluated for their antitumor activities against human foreskin fibroblasts (HFF), human gastric cancer cells-803 (MGC-803), hepatocellular carcinoma bel-7404 (BEL-7404), large cell lung cancer cells (NCI-H460), and bladder cancer cells (T24). Most of the compounds exhibited high levels of antitumor activity against MGC-803 and T24 but low toxicity against human normal liver cells (LO2), and their effect was even better than the commercial anticancer drugs, 5-fluorouracil (5-FU) and cis-platinum. Further, pharmacological mechanisms such as topo I, cell cycle, cell apoptosis, and neovascularization were all evaluated. Only a few compounds exhibited potent topo I inhibitory activity at 100 μM. In addition, the most active compounds with an $IC_{50}$ value of 5.52–8.93 μM were chosen, and they could induce cell apoptosis in the G2 stage of MGC-803 or mainly arrest T24 cells in the S stage. To our delight, most of the compounds exhibited lower zebrafish cytotoxicity but could strongly inhibit the formation of zebrafish sub-intestinal veins, indicating a potential for clinical application.

**Keywords:** acridine-triazole; acridine-thiadiazole; topoisomerase I; anti-angiogenesis; zebrafish

**Citation:** Huo, L.; Liu, X.; Jaiswal, Y.; Xu, H.; Chen, R.; Lu, R.; Nong, Y.; Williams, L.; Liang, Y.; Jia, Z. Design and Synthesis of Acridine-Triazole and Acridine-Thiadiazole Derivatives and Their Inhibitory Effect against Cancer Cells. *Int. J. Mol. Sci.* **2023**, *24*, 64. https://doi.org/10.3390/ijms24010064

Academic Editor: Laura Paleari

Received: 31 October 2022
Revised: 5 December 2022
Accepted: 6 December 2022
Published: 21 December 2022



## 1. Introduction

Today, cancer is one of the major health problems in the world. With the development of molecular biology and molecular pharmacology, the pathogenesis of cancer is being explored at the gene level. Pharmacological mechanisms such as signal transduction, neovascularization, telomerase, topoisomerase, cell cycle and cell apoptosis have major impacts on cancerous cells and can be used as targets in cancer therapy [1].

Acridines are an important classe of nitrogen-containing heterocyclic compounds. Due to their structural characteristics as planar tricyclic aromatic molecules, acridines intercalate tightly but reversibly to the DNA helix [2,3]. These compounds reveal a wide variety of biological activities, including anticancer [4], antimicrobial [5,6], anti-acetylcholinesterase [7], etc. A number of acridine derivatives serve as chemotherapeutic agents, especially in the field of antitumor DNA-binding agents [8]. An example of one such compound is 9-amsacrine, which has been clinically used for the treatment of leukemia [9].

Due to their beneficial characteristics, triazole and thiadiazole derivatives can serve as potential antitumor agents and thus are of pharmaceutical interest. In drug development, the triazole ring is often used to replace the amino group to reduce the resistance of some anticancer drugs and enhance their anticancer activity [10]. Thiadiazole groups are commonly introduced in the design of anticancer drugs because of their high anticancer activity. Kumar et al. recently reported the synthesis and anticancer activity of a series of benzpyrole-thiadiazole derivatives and revealed the important role of the thiadiazole ring in cytotoxicity [11].

Designing hybrid drugs with multiple effects is a common strategy in the recent search for new anticancer drugs [12]. In recent years, many structurally diverse hybrid molecules at the 9-position of the acridine skeleton have been reported for the enhancement of anticancer activity. Examples of such compounds include acridine-mycophenolic acid hybrid (a) [13], acridine-thiazolidinedione hybrid [14] (b), and acridine-chlormethine hybrid (c) [15] (Figure 1).

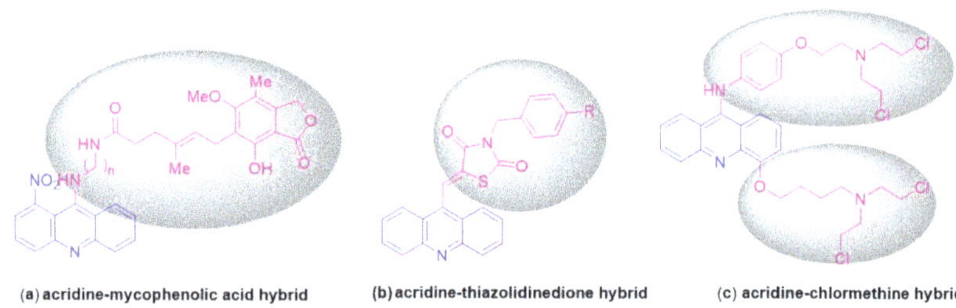


**Figure 1.** Structures of some hybrid molecules (**a**–**c**) at the 9-position of the acridine skeleton.

Considering these facts, our strategy was to couple an acridine and a triazole or thiadiazole nucleus to obtain a new class of compounds such as the acridine-triazole hybrid or acridine-thiadiazole hybrid (Figure 2). The anticancer activities of the synthesized compounds were assessed based on various mechanisms of action and molecular docking.

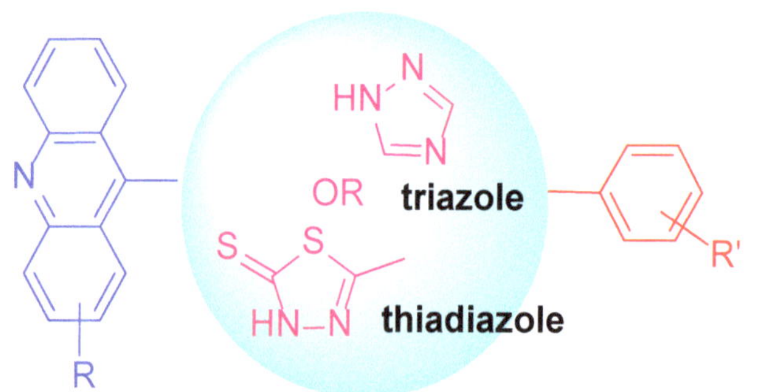


**Figure 2.** Strategy for the design of acridine-triazole or acridine-thiadiazole hybrids.

## 2. Results and Discussion

### 2.1. Chemistry

The general synthetic approach for aroyl thiourea derivatives (**4**), acridinyl 1,2,4-triazole derivatives (**5**) and acridinyl 1,2,4-thiadiazole derivatives (**6**) is illustrated in Scheme 1.

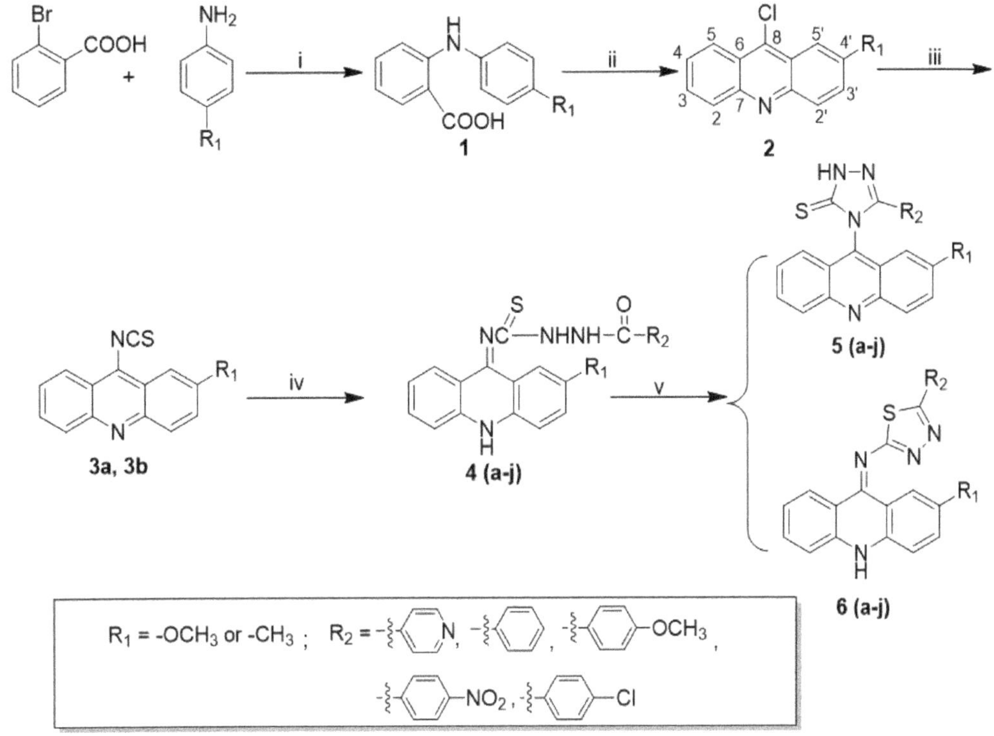


**Scheme 1.** Syntheses of acridinyl derivatives. Reagents and conditions: (i) Cu, K2CO3, 140 °C; (ii) POCl3, 140 °C; (iii) NaSCN/ tetrabutylammonium bromide; (iv) $R_2-\overset{O}{\overset{\|}{C}}-NHNH_2$; (v) Na2CO3, reflux, or 98% H2SO4, 0 °C.

The target compounds of 1,2,4-triazolethiones (**5**) and 1,2,4-thiadiazoles (**6**) were synthesized by means of a ring closure reaction using aroyl thiourea derivatives (**4**) in sodium carbonate or concentrated sulfuric acid conditions, respectively. The synthesis of aroyl thiourea derivatives (**4**) was carried out according to the known procedure of the addition of substituted hydrazides to acridin-9-yl isothiocyanate (**3**). It is important to note that the precipitate **3a** is formed at room temperature, while **3b** needs to be cooled in an ice bath. The key intermediates (**4**) were obtained in 95% EtOH without purification with a yield of 73–92% $^{w}/_{w}$.

As expected, auto-condensation cyclization proceeded effectively in the refluxing condition of 5% $Na_2CO_3$ or 98% concentrated sulfuric acid in an ice bath. It is reported that acridinyl 1,2,4-triazole derivatives (**5**) possibly exist in one of two tautomeric forms (Figure 3), thione (**a**) or thiol (**b**) [16]. And the thione form (**a**) was established by comparison of the HSQC and HMBC spectra and DFT calculations. To further confirm the structure of our synthesized products, a single crystal of compound **5b** was cultivated in absolute ethyl alcohol, and the molecular structure was confirmed as indicated in Figure 3c. The corresponding single crystal structural data for compound **5b** is provided in the supporting information (CCDC 2214949).

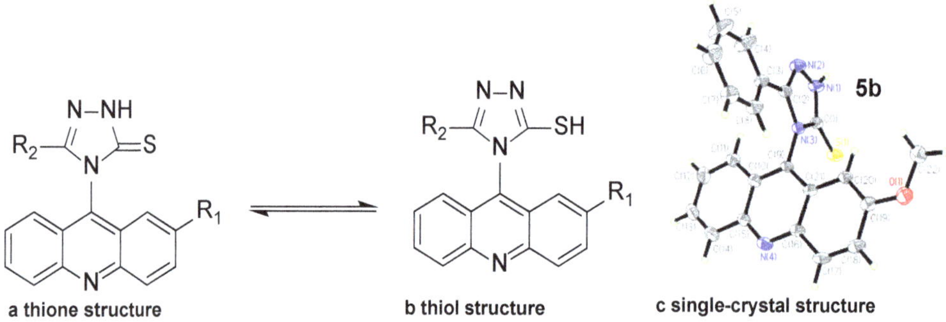


**Figure 3.** The molecular structure of compound **5b**.

The success of the cyclization of compound **6** mainly depended on reaction temperature and reaction time. The reaction temperature had to be maintained below 0 °C. When $R_2$ was an electron-withdrawing group such as pyridyl and nitrophenyl, the reaction time had to be extended almost to 48 h. Interestingly, the final structure of compound **6** was not the desired acridine skeleton (a, Figure 4) for the compound. The N-10 atom of the acridinyl moiety captured a proton and thus resulted in the formation of a 9′,10′-dihydroacridine structure (b, Figure 4), which was verified through X-ray crystallographic analysis (c). The corresponding single crystal structural data of compound **6d** is provided in the supporting information (CCDC 2214923). The exchangeable NH protons of acridine thiosemicarbazides are reported in the literature (Figure 5) [16].

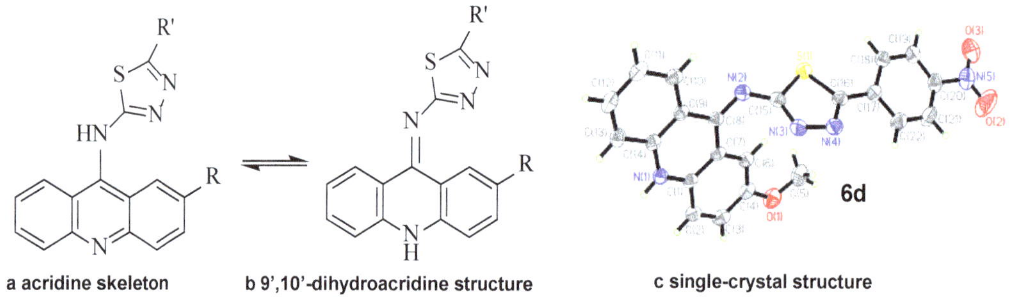


**Figure 4.** The molecular structure of compound **6d**.

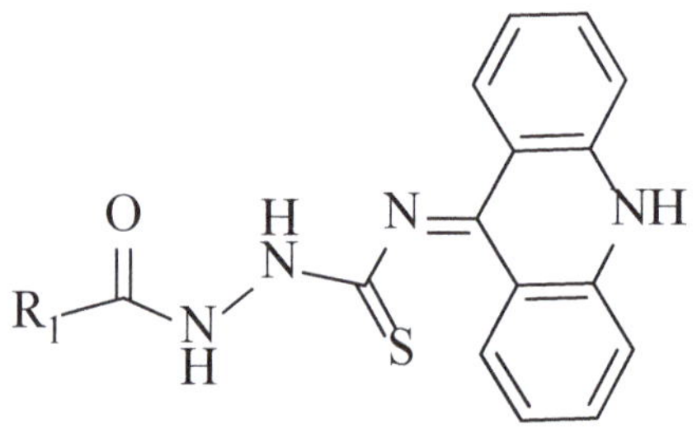

**Figure 5.** Reported structure of acridine thiosemicarbazides [13].

## 2.2. *In-Vitro Anticancer Activity Assay and Structure-Activity Analysis*

All newly synthesized acridinyl derivatives (**4–6**) were screened for their anticancer activities in comparison to the reference compounds, 5-FU and cis-platinum. Compounds

**4–6** were tested for their in vitro antitumor activities against HFF, MGC-803, BEL-7404, NCI-H460, and T24 tumor cell lines, and human normal liver cells (LO2), and the results are shown in Table 1. Most of the compounds had strong selective potency against MGC-803 and T24 cancer cells. In the MGC-803 cell line assay, almost all of the compounds displayed better cytotoxicity than the positive control 5-FU ($IC_{50}$ = 30.45 ± 2.87 μM), with an $IC_{50}$ of 5.52–34.99 μM. This indicates that the introduction of the triazole and thiadiazole groups on the acridine skeleton could improve the antitumor activity against MGC-803. In addition, except for compounds **4c**, **4d**, **4e**, **4i**, **4j**, **5a**, **5g**, **6b**, **6d**, and **6i**, almost all of the compounds demonstrated better cytotoxicity inhibition than cis-platinum ($IC_{50}$ = 15.97 ± 1.53 μM). Particularly, the $IC_{50}$ values of compounds **5d**, **5g**, **5i**, **6g**, **6e**, and **6h** were all below 10 μM, and the $IC_{50}$ of them were 5.52 ± 1.04 μM, 8.5 ± 1.85 μM, 8.92 ± 0.99 μM, 9.01 ± 1.32 μM, 9.95 ± 1.03 μM, and 6.85 ± 0.84 μM, respectively. In the T24 cell line assay, many compounds, especially the series of compound **4**, had significant activity against T24. This implies that there is a significant increase in potency after the introduction of the aroyl thiourea group. Among these compounds, $R_1$ = -$CH_3$ and $R_2$ = -$OCH_3$ might help to improve the antitumor activity of acridine nuclear, such as compounds **4h**, **5h** and **6h**, all of which exhibited the best inhibition compared with other analogues, with $IC_{50}$ values of 8.05 ± 1.06, 11.25 ± 1.16, and 8.93 ± 1.25 μM, respectively. In particular, compounds **4h** and **6h** had better antitumor activities than the two commercial anticancer drugs 5-FU ($IC_{50}$ = 32.04 ± 1.23) and cis-platinum ($IC_{50}$ = 9.13 ± 1.54 μM). To our delight, most 1,2,4-triazolethiones (**5**) and 1,2,4-thiadiazoles (**6**) have low toxicity to LO2 compared with the positive control. Compounds **5d** and **6h** were the most active but had lower toxicities than 5-FU and *cis*-platinum. Therefore, compounds **5d** and **6h** or **4h** and **6h** exhibited good cytotoxicity inhibition against MGC-803 or T24 cancer cells and were selected for further exploration to identify their mechanisms of cancer cell growth inhibition.

**Table 1.** Effect of compounds **4**, **5** and **6** against cell viability of different cell lines # (μM).

| No. | HFF | MGC-803 | BEL-7404 | NCI-H460 | T24 | LO2 |
|---|---|---|---|---|---|---|
| **4a** | 75.79 ± 3.52 | 10.89 ± 1.82 | 26.93 ± 2.58 | 36.41 ± 3.12 | >100 | 21.96 ± 1.71 |
| **4b** | 72.56 ± 3.57 | 14.47 ± 2.06 | 25.78 ± 2.83 | 42.27 ± 2.15 | 29.82 ± 2.85 | 34.37 ± 2.05 |
| **4c** | 63.74 ± 2.59 | 21.04 ± 1.55 | 13.33 ± 1.37 | 52.37 ± 3.67 | 11.23 ± 2.51 | 19.45 ± 1.69 |
| **4d** | 65.28 ± 3.82 | 34.99 ± 3.57 | 43.91 ± 2.59 | 40.54 ± 4.52 | 13.01 ± 1.64 | 25.12 ± 1.33 |
| **4e** | 66.84 ± 4.52 | 25.55 ± 1.97 | 25.95 ± 2.17 | 73.25 ± 3.67 | 10.32 ± 1.07 | 10.23 ± 1.12 |
| **4f** | 55.21 ± 1.36 | 11.24 ± 0.96 | 34.37 ± 2.24 | 53.66 ± 3.04 | 9.66 ± 1.54 | 10.08 ± 0.96 |
| **4g** | 70.11 ± 3.97 | 13.54 ± 1.59 | 20.17 ± 4.13 | 39.01 ± 2.05 | 25.84 ± 1.51 | 14.34 ± 1.52 |
| **4h** | >100 | 11.25 ± 1.46 | 27.10 ± 2.91 | 25.36 ± 3.16 | 8.05 ± 1.06 | 9.01 ± 0.93 |
| **4i** | 60.73 ± 2.31 | 22.34 ± 1.35 | 23.32 ± 1.33 | 60.40 ± 2.95 | 9.89 ± 1.45 | 11.76 ± 1.38 |
| **4j** | 61.53 ± 1.85 | 16.37 ± 1.56 | 24.45 ± 3.53 | 32.22 ± 2.36 | 19.95 ± 1.32 | 20.53 ± 1.39 |
| **5a** | 74.50 ± 4.03 | 22.41 ± 1.32 | 22.06 ± 2.72 | 36.45 ± 2.96 | 21.17 ± 2.72 | 41.99 ± 2.31 |
| **5b** | 53.58 ± 2.78 | 15.81 ± 1.94 | 27.65 ± 2.97 | 42.08 ± 3.74 | 22.05 ± 1.85 | 58.28 ± 3.25 |
| **5c** | 75.51 ± 2.92 | 15.13 ± 0.98 | 28.71 ± 2.24 | 36.45 ± 3.92 | 29.29 ± 1.91 | >100 |
| **5d** | 62.93 ± 1.90 | 5.52 ± 1.04 | 25.07 ± 2.89 | 19.44 ± 1.58 | 15.92 ± 1.38 | 51.79 ± 3.46 |
| **5e** | 74.93 ± 3.35 | 8.50 ± 1.85 | 34.66 ± 2.64 | 35.13 ± 1.94 | 18.45 ± 1.64 | >100 |
| **5f** | 69.22 ± 2.16 | 15.24 ± 1.08 | 44.21 ± 2.68 | 58.79 ± 3.22 | 15.72 ± 1.58 | 46.78 ± 2.93 |
| **5g** | 68.31 ± 2.74 | 19.35 ± 1.38 | 20.54 ± 1.13 | 30.64 ± 2.21 | 19.36 ± 2.17 | 44.55 ± 2.35 |
| **5h** | 68.07 ± 2.64 | 10.88 ± 0.97 | 40.33 ± 2.06 | 26.32 ± 2.51 | 11.25 ± 1.16 | 37.67 ± 2.47 |
| **5i** | 70.85 ± 2.99 | 8.92 ± 0.99 | 31.66 ± 2.36 | 28.31 ± 1.32 | 14.26 ± 1.27 | >100 |
| **5j** | 64.69 ± 3.36 | 13.51 ± 1.91 | 45.87 ± 2.48 | 21.78 ± 2.46 | 13.06 ± 1.70 | 36.44 ± 2.65 |
| **6a** | 68.82 ± 1.87 | 14.31 ± 1.29 | 19.21 ± 1.30 | 25.34 ± 3.57 | 10.18 ± 0.96 | 40.24 ± 2.74 |
| **6b** | 79.32 ± 2.48 | 23.27 ± 1.97 | 32.29 ± 2.82 | 47.51 ± 3.51 | 49.36 ± 4.59 | >100 |
| **6c** | 65.21 ± 3.92 | 12.13 ± 1.22 | 25.11 ± 2.15 | 30.23 ± 2.45 | 24.27 ± 2.34 | >100 |
| **6d** | >100 | 26.66 ± 3.35 | >100 | >100 | >100 | >100 |
| **6e** | 44.71 ± 1.44 | 9.01 ± 1.32 | 21.33 ± 2.81 | 27.88 ± 3.97 | 14.88 ± 1.30 | 33.64 ± 2.01 |

**Table 1.** *Cont.*

| No. | HFF | MGC-803 | BEL-7404 | NCI-H460 | T24 | LO2 |
|---|---|---|---|---|---|---|
| **6f** | 45.38 ± 2.18 | 12.35 ± 1.96 | 40.26 ± 2.19 | 55.72 ± 3.28 | 13.86 ± 1.37 | 37.22 ± 2.12 |
| **6g** | 76.45 ± 2.79 | 9.95 ± 1.03 | 31.25 ± 3.27 | 25.87 ± 1.83 | 19.33 ± 1.05 | >100 |
| **6h** | 33.90 ± 1.28 | 6.85 ± 0.84 | 20.25 ± 1.59 | 13.33 ± 1.39 | 8.93 ± 1.25 | 43.77 ± 2.63 |
| **6i** | 91.95 ± 2.99 | 22.92 ± 1.85 | 43.66 ± 2.36 | 36.41 ± 3.15 | 29.89 ± 2.45 | >100 |
| **6j** | 56.23 ± 3.16 | 12.99 ± 1.89 | 48.47 ± 3.06 | 13.88 ± 1.83 | 15.47 ± 1.98 | 51.17 ± 3.09 |
| **5-FU** | 25.45 ± 1.27 | 30.45 ± 2.87 | 34.52 ± 1.18 | 44.04 ± 0.54 | 32.04 ± 1.23 | 40.15 ± 1.65 |
| **cis-platinum** | 10.85 ± 0.34 | 15.97 ± 1.53 | 10.01 ± 0.52 | 7.126 ± 1.24 | 9.13 ± 1.54 | 21.38 ± 1.25 |

[#] human foreskin fibroblasts (HFF); human gastric cancer cells-803 (MGC-803); hepatocellular carcinoma bel-7404 (BEL-7404); large cell lung cancer cells (NCI-H460); and bladder cancer cells (T24); LO2 human normal liver cells (LO2).

### 2.3. *Antitumor Mechanism Studies*

#### 2.3.1. Apoptosis and Cell-Cycle Analysis

Apoptosis and the cell-cycle play a central role in cancer, since their induction in cancer cells is critical to a successful therapy [17,18]. Therefore, the most active compounds, including **5d** and **6h** or **4h** and **6h** were selected to study their effect on apoptosis and cell cycle profiles in the MGC80-3 or T24 cell lines, respectively.

The apoptosis ratios of MGC80-3 or T24 cell lines induced by the selected compounds at the concentration of $IC_{50}$ and 0.5 $IC_{50}$ were quantitatively determined by flow cytometry. Four quadrant images (Q1, Q2, Q3 and Q4) were observed by flow cytometric analysis. The results of apoptosis ratios (including the early and late apoptosis ratios) after 12 h are presented in Figure 6 (MGC80-3) and Figure 7 (T24). Figure 6 revealed that compounds **5d** and **6h** could induce apoptosis in MGC80-3 cells in a concentration dependent manner. The apoptosis percentage of compound **5d** measured at different concentrations were found to be 6.616% (2.76 μM) and 17.51% (5.52 μM), while the value for control was 0.586%. Treatment was also accompanied by a decrease in the percentage of live cells, with values of 93.0% in control and 81.2% in treated cells. After treatment with compound **6h**, 5.62% (3.43 μM) and 14.25% (6.85 μM) of the cells were apoptotic. These were higher percentages than the one observed in the control (0.586%). These results further demonstrate that apoptosis was induced by compounds **5d** and **6h** in addition to cell proliferation inhibition. From the results of Figure 7, compounds **4h** and **6h** led to an increase in the number of apoptotic cells in T24 with the increase of the concentration (from 0.5 $IC_{50}$ to $IC_{50}$), and their apoptosis ratios at their $IC_{50}$ concentrations were increased to 12.377% and 10.749%, respectively, when compared with the control (1.18%). All compounds had little effect on late apoptosis of MGC80-3 or T24, and some normal cells were found to be necrotic in Q1 region. The results evidently illustrate that representative compounds **5d** and **6h** or **4h** and **6h** could suppress cell proliferation by inducing apoptosis in the early apoptotic period.

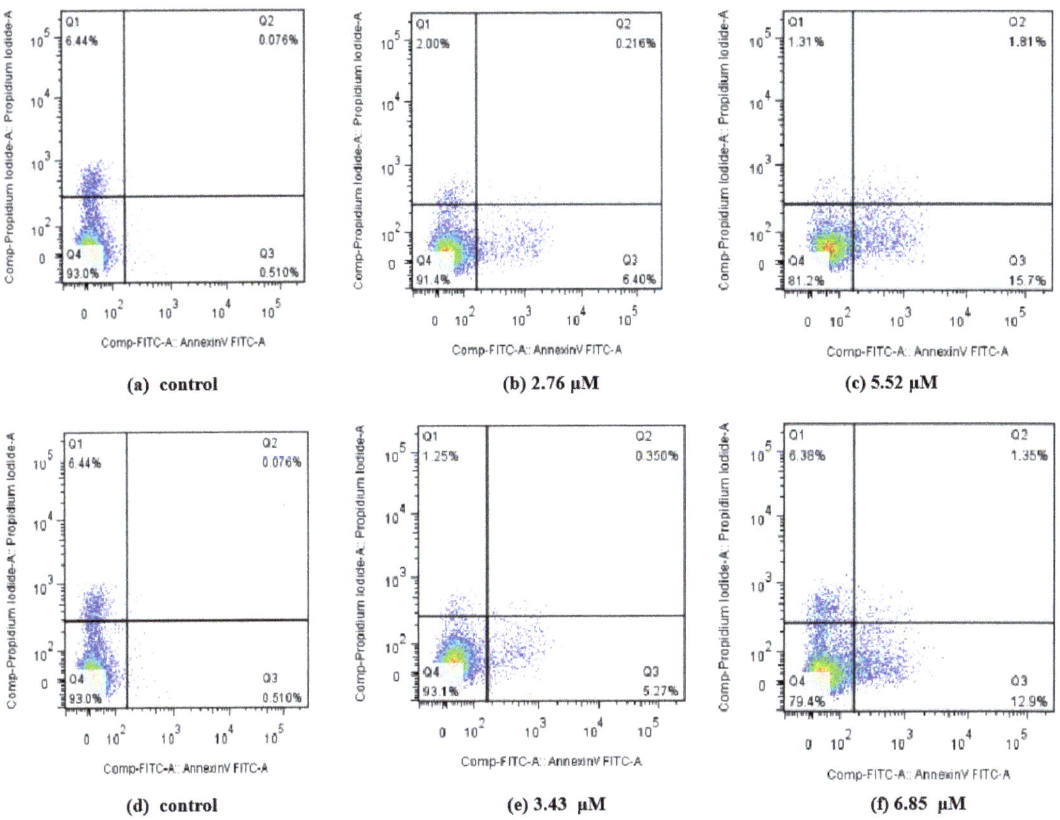


**Figure 6.** Apoptosis ratio detection of compounds **5d** and **6h** by Annexin V-FITC and PI. (**a**,**d**) The MGC80-3 cells not treated with compounds **5d** or **6h** were used as controls; (**b**,**c**) compound **5d** treated MGC80-3 cells for 24 h at concentrations of 2.76 and 5.52 μM, respectively; (**e**,**f**) compound **6h** treated MGC80-3 cells for 24 h at concentrations of 3.34 and 6.85 μM, respectively.

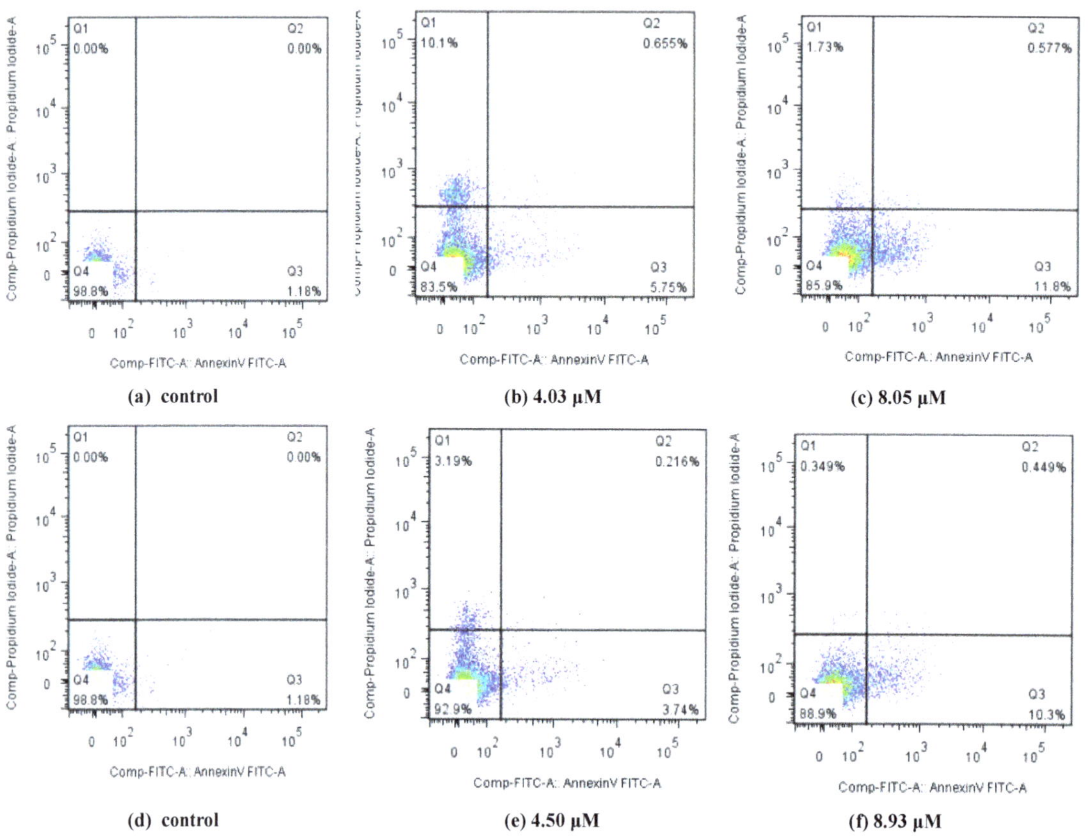


**Figure 7.** Apoptosis ratio detection of compounds **4h** and **6h** by Annexin V-FITC and PI. (**a**,**d**) The T24 cells not treated with compounds **4h** and **6h** were used as controls; (**b**,**c**) compound **4h** treated T24 cells for 24 h at concentrations of 4.03 and 8.05 μM, respectively; (**e**,**f**) compound **6h** treated T24 cells for 24 h at concentrations of 4.50 and 8.93 μM, respectively.

The cell cycle distributions of T24 and MGC80-3cells after 48 h of treatment with the most active compounds, **4h** and **5d**, at their $IC_{50}$ concentrations are shown in Figure 8. Compared to control, both compounds **4h** and **5d** interfered with the cell cycles of T24 and MGC80-3 cells, respectively. As shown in Figure 8a,b, the S-phase population of T24 cells increased by 30.04% compared to the control cells (22.89%), indicating that compound **4h** might inhibit the growth of tumor cells by arresting the cells in S phase during the DNA synthesis period. However, compound **5d** could induce a significant cell cycle arrest in the G2 phase, resulting in a concomitant population increase (13.32%) compared with the control cells (8.91%) at a concentration of 5.52 μM (Figure 8c,d). These results suggest that compound **5d** may inhibit the growth of tumor cells by arresting cells in the G2 phase in the late stage of DNA synthesis.

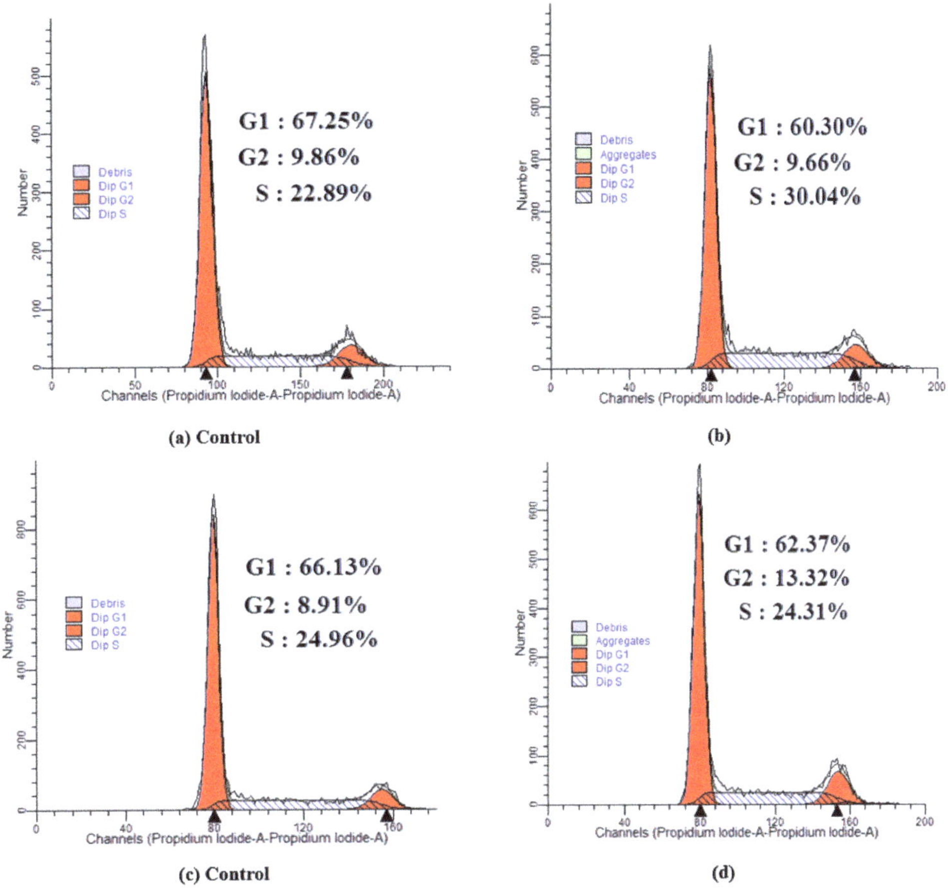


**Figure 8.** Cell cycle analysis of compound **4h** treated T24 cells (**b**) and compound **5d** treated MGC80-3 cells (**d**) at their $IC_{50}$ concentrations (8.05 μM and 5.52 μM) for 48 h. And the T24 and MGC80-3 cells not treated with compounds **4h** and **5d** were used as control, (**a**,**c**). (G1: Prophase of DNA synthesis; S: stage of dna synthesis; G2: Late stages of DNA synthesis).

2.3.2. Evaluation of Topo I Inhibitory Activity

DNA topoisomerase I (topo I) has become the main molecular target in anticancer drugs on account of its significance in all living organisms, participating in replication, transcription, recombination, and repair in many cellular metabolic processes. The topo I inhibitory activity of the compounds with the known topo I inhibitor camptothecin (CPT) is depicted in Figure 9. Only compounds **4e**, **5c**, and **6h** exhibit potent topo I inhibitory activity at 100 μM. Compounds that have little to no inhibitory activity may have other mechanisms for their anticancer effects. Molecular docking studies of the selected compounds were carried out by the Surflex-Dock algorithm of Sybyl-X 2.0 (Tripos Inc., St. Louis, MI, USA). The molecular docking approach was verified by our previously published methods (RMSD (root-mean-square deviation) value was 0.4438 Å) [19]. The binding affinities of protein-ligand complexes were expressed as a total score and shown in Figure 10. compounds **4e**, **5c** and **6h** exhibited good binding affinities, with total scores of 9.79, 7.81 and 9.66, respectively. Potent Topo I inhibitory activity of these compounds may be attributed to the formation of hydrophobic residue, hydrogen bond, and π–π stacking with the same amino acid residue DA113, DC112, TGP11 as CPT.

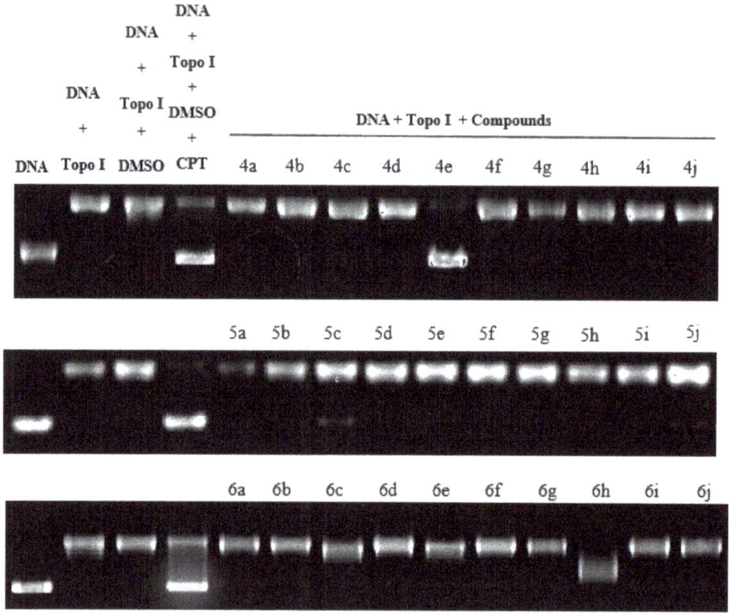


**Figure 9.** DNA topoisomerase I (Topo I) inhibitory activity of CPT and all target compounds (**4a–4j, 5a–5j, 6a–6j**) at 1 mM.

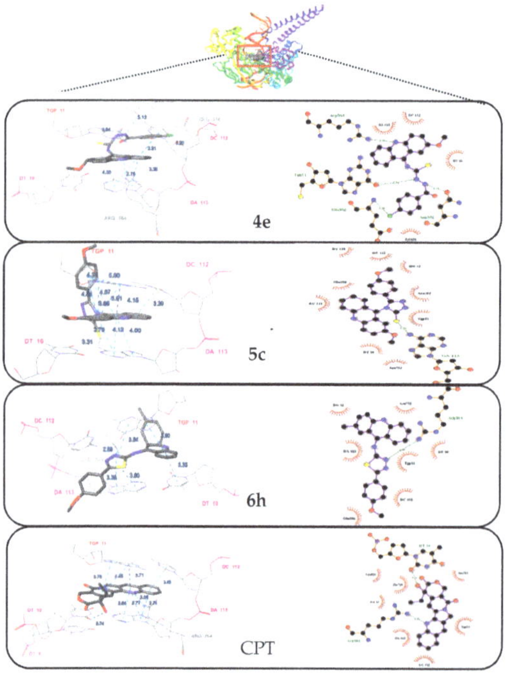


| Compound | Amino Acid | Type | Hydrophobic Residues | Total Score |
|---|---|---|---|---|
| 4e | DA113 | π–π | DA113, DC112, DT10, TYR426 | 9.7926 |
| | DC112 | π–π | | |
| | ASN352 | Hydrogen Bond | | |
| | TGP11 | Hydrogen Bond | | |
| | ARG364 | Hydrogen Bond | | |
| | GLU356 | Hydrogen Bond | | |
| 5c | DA113 | π–π | GLU356, DC112, DT10, ASN722, TGP11, ASN352, DG12, DC111 | 7.8136 |
| | DC112 | π–π | | |
| | TGP11 | π–π | | |
| | DA113 | Hydrogen Bond | | |
| 6h | DA113 | π–π | GLU356, DG12, DA113, DC112, DT10, TGP11, ASN722 | 9.6633 |
| | DT10 | π–π | | |
| | TGP11 | π–π | | |
| | ARG364 | Hydrogen Bond | | |
| CPT | DA113 | π–π | LYS532, THR718, DG12, DA113, DC112, TGP11, ASN722 | 10.2196 |
| | DC112 | π–π | | |
| | TGP11 | π–π | | |
| | ARG364 | Hydrogen Bond | | |
| | DT10 | Hydrogen Bond | | |

**Figure 10.** The best pose of the binding mode of compounds (**4e, 5c, 6h** and **CPT**) with DNA Topo I complex (PDB:1T8I).

2.3.3. Toxicity and Anti-angiogenesis in the Zebrafish Model

Many antitumor drugs inevitably have side effects on normal cells, such as bone marrow suppression, liver and kidney injury, and abnormal blood cells. Therefore, in order to improve the possibility of clinical application of acridine-heterocyclic derivatives, the effective and low-toxicity antitumor drugs were screened using a zebrafish model. In this experiment, 2% DMSO was used to dissolve the target products, and the abnormal rate (MAR) and mortality rate (MOR) of zebrafish embryos (72 hpf) were used for statistics. At different concentrations (1–2 mg/mL) of the selected compounds, various deformities were observed, such as failure to hatch, embryo necrosis, severe angulation of the spine and severe pericardial edema (Figure 11). The mortality and malformation rates of embryos increased with a dose-effect relationship (Table 2). Almost all compounds in the compound **4** series were toxic. Particularly, compounds **4a**, **4b**, **4f** and **4h** had a total mortality and malformation rate of 100% at 2.0 mg.L$^{-1}$, exhibiting the strongest embryonic toxicity. It is worth mentioning that compounds **5d** and **5h** displayed high levels of antitumor activities but were less toxic to zebrafish embryos. At the highest concentration of 2.0 mg.L$^{-1}$, the mortality rate of zebrafish embryos was close to 0% and the malformation rate was less than 15%. Moreover, there was hardly any toxicity observed in compound **6** at lower concentrations (1.0 mg.L$^{-1}$). Compound **6i** exhibited very low toxicity at a high concentration of 2.0 mg.L$^{-1}$ with 0% mortality rate and 25% malformation rate.

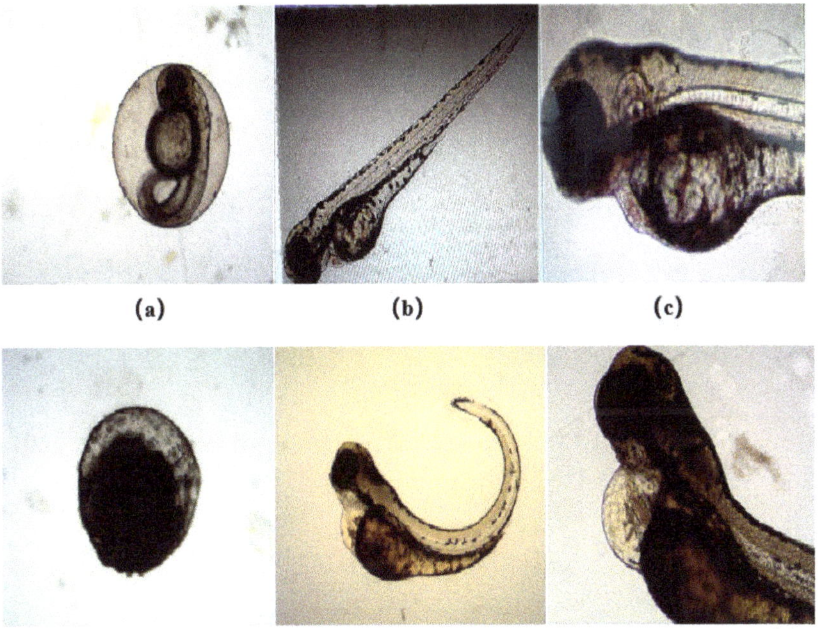

(a) (b) (c)

(d) (e) (f)

**Figure 11.** Normal and abnormal zebrafish embryos (**a**) Normal zebrafish embryos (**b**) Normal zebrafish (**c**) Normal pericardium of zebrafish (**d**) Embryo necrosis (**e**) Severe angulation of the spine (**f**) Severe pericardial edema.

Currently, the zebrafish has emerged as a valuable model organism to substitute traditional models for studying angiogenesis inhibitors [20]. The genes of zebrafish show 70–80% similarity to humans, and the vascular structure of zebrafish has high similarity to that of other vertebrates [21–23]. Therefore, the subintestinal veins (SIVs) in the zebrafish embryos are used as evaluation indicators for anti-angiogenesis inhibitors. In this study,

NBT/BCIP vascular staining was used to observe the angiogenesis effect of representative drugs (**4h**, **4f**, **5d**, **5h**, **6g** and **6h**) in a zebrafish model. As shown in Figure 12, SIVs grew well in the blank group, naturally extending into a network in the abdomen with many branches. The length of SIVs of zebrafish was measured by Image J software and is shown in Figure 10b. Compared with the blank control group, the area of the meshed pattern vessel and the number of vascular branches in the network decreased after the administration of compounds **4h**, **5d**, **5h** and **6h.** Among these compounds, compounds **5d** and **5h** exhibited the strongest antiangiogenic effects that led to a nearly 50% reduction in the vessel length compared to the mean vessel length for the controls. Compounds **4a** and **6g** could reduce the area of blood vessels, but at the same time, additional blood vessels were formed on the blood vessel edge.

**Table 2.** The mortality rate (MOR) and malformation rate (MAR) of some of the selected compounds.

| NO. | Control | | 2% DMSO | | Sample Concentration 2 mg/mL | | 1.5 mg/mL | | 1 mg/mL | |
|---|---|---|---|---|---|---|---|---|---|---|
| | MOR | MAR | MOR | MAR | MOR | MAR | MOR | MAR | MOR | MAR |
| **4a** | 0 | 0 | 0 | 0 | 45% [c] | 55% [c] | 40% [c] | 60% [c] | 15% [c] | 85% [c] |
| **4b** | 0 | 0 | 0 | 0 | 40% [a] | 55% [a] | 15% [a] | 70% [a] | 15% [a] | 70% [a] |
| **4c** | 0 | 0 | 0 | 0 | 35% [a] | 50% [a] | 25% [a] | 60% [a] | 45% [a] | 50% [a] |
| **4f** | 0 | 0 | 0 | 0 | 85% [b] | 15% [b] | 20% [b] | 80% [b] | 10% [b] | 60% [b] |
| **4h** | 0 | 0 | 0 | 0 | 65% [a] | 35% [a] | 15% [a] | 85% [a] | 15% [a] | 35% [a] |
| **5d** | 0 | 0 | 0 | 0 | 0 | 10% [c] | 0 | 10% [a] | 0 | 5% [b] |
| **5h** | 0 | 0 | 0 | 0 | 0 | 15% [a] | 0 | 10% [a] | 0 | 5% [a] |
| **5j** | 0 | 0 | 0 | 0 | 0 | 15% [a] | 0 | 5% [a] | 0 | 0 |
| **6a** | 0 | 0 | 0 | 0 | 5% [b] | 90% [b] | 0 | 95% [b] | 0 | 15% [b] |
| **6e** | 0 | 0 | 0 | 0 | 65% [c] | 35% [c] | 30% [c] | 70% [c] | 10% [c] | 10% [c] |
| **6h** | 0 | 0 | 0 | 0 | 15% [b] | 65% [b] | 0 | 80% [b] | 0 | 35% [b] |
| **6i** | 0 | 0 | 0 | 0 | 0 | 25% [c] | 0 | 10% [b] | 0 | 0 |

Note: [a] = $p < 0.01$; [b] = $p < 0.001$; [c] = $p < 0.002$, compared with the control group.

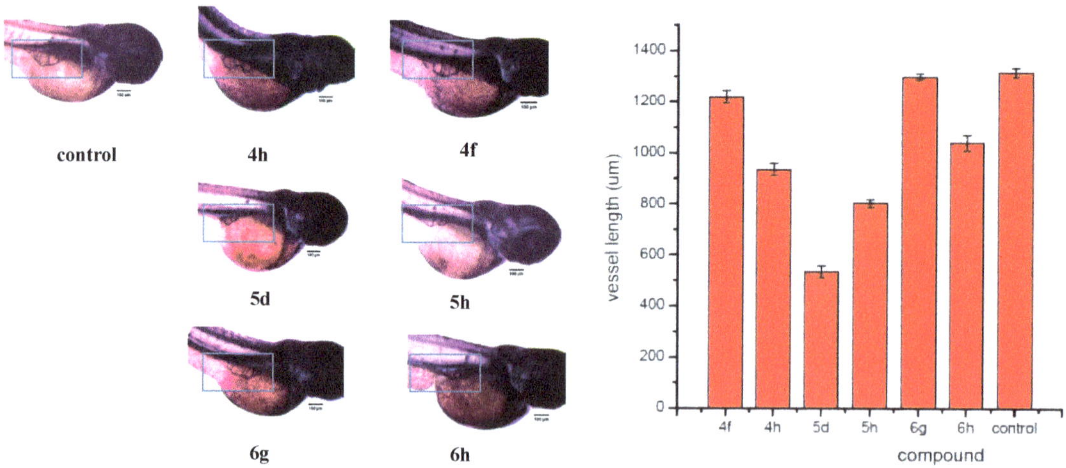


**Figure 12.** Effects of representative compounds (**4f, 4h, 5d, 5h, 6g** and **6h**) and control on the subintestinal veins (SIVs) length of 72 hpf zebrafish embryos ($x \pm s$, $n = 12$), $p < 0.05$.

## 3. Materials and Methods

All commercially available chemicals were reagent grade and bought from Aladdin Reagent Co., LTD (Shanghai, China); NBT/BCIP kit was bought from Tiangen Biochemical Technology Co., LTD (Shanghai, China); AnnexinV-FITC apoptosis detection kit was bought from Nanjing KGI Biotechnology Development Co., Ltd. (Nanjing, China); The spectra such as NMR, MS, and IR were all evaluated and recorded on a Bruker DRX-400 ($^{1}$H: 400 MHz, $^{13}$C: 100 MHz) (Rheinstetten, Germany), a Thermo Fisher LCQ Fleet (ESI) instrument (Waltham, MA, USA), and FT-IR Thermo Nicolet Avatar 360 using a KBr pellet (Waltham, MA, USA). And the melting points were measured by the XT-4 A melting point apparatus (Shanghai, China) without correction. Other instruments include BD FACSAria II Flow cytometer (Franklin Lakes, NJ, USA), MCO96 carbon dioxide incubator (Osaka, Japan) and Bio Tek EL × 800 microplate reader (Winooski, VT, USA), etc.

### 3.1. *Synthesis Methods*

#### 3.1.1. Synthesis of N-phenyl-o-aminobenzoic acid (**1**) and 9-chlorine acridine (**2**)

The synthesis of N-phenyl-o-aminobenzoic acid (**1**) and 9-chlorine acridine (**2**) was carried out according to our previously published procedure, with slight modifications [24]. Compound **1** could proceed to the next step without further purification.

2-methoxyl-9-chlorine acridine (**2a**): Yellow-green needle crystal, yield 85.2%, m.p. 158–159 °C. ESI-MS m/z: 244 ([M + H]$^+$); $^{1}$H NMR ($CDCl_3$, 400 MHz) 8.00 (dd, 2H, *J* = 8.00, ArH), 7.93 (d, 2H, *J* = 8.20, ArH), 7.55–7.60 (m, 2H, ArH), 7.50 (d, 1H, *J* = 8.40, ArH), 7.30 (s, 1H, ArH), and 3.73 (s, 3H, -$OCH_3$).

2-methyl-9-chlorine acridine (**2b**): Pale green needle crystal, yield 78.5%, m.p. 122–123 °C. ESI-MS m/z: 228 ([M + H])$^+$ $^{1}$H NMR ($CDCl_3$, 400 MHz), δ: 8.05 (dd, 2H, *J* = 8.00, ArH), 8.00 (d, 2H, *J* = 9.20, ArH), 7.61–7.68 (m, 2H, ArH), 7.50 (d, 1H, *J* = 5.40, ArH), 7.43 (s, 1H, ArH), 2.35 (s, 3H, -$CH_3$).

#### 3.1.2. Synthesis of 9-acridinyl Isothiocyanate (**3**)

To a solution of chlorine acridine **2** (5 mmol) in acetone (50 mL), NaSCN (0.81 g, 10 mmol) and tetrabutylammonium bromide (0.32 g, 1 mmol) were added, and the mixture was then refluxed at 60 °C for 1 h. After cooling to room temperature, crystals of **3a** were immediately precipitated in the reaction mixture, and crystals of **3b** were precipitated in an ice bath. At the end of the procedure, the crystals were filtered, washed with water, and dried under vacuum, and no further purification was carried out.

2-methoxyl-9-acridinyl isothiocyanate (**3a**): bright yellow crystal, yield 88.0%, m.p. 149–150 °C; ESI-MS m/z: 267 ([M + H]$^+$); $^{1}$H NMR ($CDCl_3$, 400 MHz), δ: 8.25 (d, 2H, *J* = 8.50, ArH), 8.15 (d, 1H, *J* = 9.20, ArH), 7.77–7.81 (q, 1H, ArH), 7.66–7.68 (t, 1H, ArH), 7.51 (d, 1H, *J* = 8.00, ArH), 7.40 (s, 1H, ArH), 4.08(s, 3H, -$OCH_3$); $^{13}$C NMR ($CDCl_3$, 100 MHz) δ: 158.66, 130.48, 127.54, 127.04, 123.48, 122.62, 122.28, 98.50, 55.90; IR (KBr) ν: 2967, 2098 (-N=C=S), 1356–1557 $cm^{-1}$.

2-methyl-9-acridinyl isothiocyanate (**3b**): faint yellow needle crystal, yield 94%, m.p. 128–129 °C; ESI-MS m/z: 351 ([M + H]$^+$); $^{1}$H NMR ($CDCl_3$, 400 MHz), δ: 8.26–8.28 (m, 2H, ArH), 8.15 (d, 1H, *J* = 8.40 Hz, ArH), 8.04 (s, 1H, ArH), 7.83 (t, 1H, ArH), 7.64–7.70 (m, 2H, ArH), 7.40 (s, 1H, ArH), 2.67 (s, 3H, -$CH_3$); $^{13}$C NMR ($CDCl_3$, 100 M Hz) δ: 137.59, 130.46, 127.08, 125.21, 122.92, 122.21, 121.17, 22.12; IR (KBr) ν: 2903, 2143 (-N=C=S), 1411–1630 $cm^{-1}$.

#### 3.1.3. General Procedure for the Synthesis of Acridinyl Aroyl Thiourea Derivatives **4a**–**4f**

To a solution of 9-isothiocyanatoacridine **3** (2mmol) in absolute ethyl alcohol (60 mL), the appropriate substituted hydrazides (2 mmol) were added, and the reaction mixture was refluxed until the reactants had been consumed (monitored by TLC). The precipitate of **4a**–**4f** was prepared, filtered off, washed with 95% ethyl alcohol, and dried at room temperature.

1-2′-methoxyl acridinyl-3-4′-pyridinamide thiourea (**4a**): Yellow powder, Yield 91%, m.p. 200–206 °C; ESI-MS m/z: 404 ([M + H]$^+$); $^{1}$H NMR (400 MHz, DMSO-$d^6$), δ: 11.26 (br, s, 1H, -NH), 10.41 (br, s, 1H, -NH), 10.22 (br, s, 1H, -NH), 8.80 (s, 1H, ArH), 8.05–8.15 (m, 2H, ArH), 7.93 (s, 1H, ArH), 7.77 (t, 1H, ArH), 7.63 (t, 1H, ArH), 8.15 (d, 1H, *J* = 8.40 Hz, ArH), 8.04 (s, 1H, ArH), 7.83 (t, 1H, ArH), 7.53 (d, 1H, *J* = 9.20 Hz, ArH), 7.45 (s, 1H, ArH),

4.02 (s, 3H, -$OCH_3$); $^{13}$C NMR (DMSO-$d^6$, 100 MHz) δ: 183.19, 167.30, 165.41, 158.40, 150.74, 148.72, 141.00, 139.61, 131.03, 129.59, 125.82, 122.40, 113.46, 109.76, 56.09; IR (KBr) ν: 3108, 2948 (N–H), 1695 (-C=O), 1291 (-C=S) $cm^{-1}$.

1-2′-methoxyl acridinyl-3-benzoyl thiosemicarbazides (**4b**): Yellow powder, Yield 76%, m.p. 190–192 °C; ESI-MS m/z: 403 ([M + H]$^+$); $^1$H NMR (400 MHz, DMSO-$d^6$), δ: 10.96 (br, s, 1H, -NH), 10.40 (br, s, 1H, -NH), 10.10 (br, s, 1H, -NH), 8.01–8.14 (m, 4H, ArH), 7.76 (s, 2H, ArH), 7.52–7.58 (m, 6H, ArH), 4.02 (s, 3H, -$OCH_3$); $^{13}$C NMR (101 MHz, DMSO-$d^6$) δ: 183.14, 166.84, 164.80, 157.06, 156.21, 144.80, 141.24, 139.31, 132.72, 129.80, 128.80, 127.28, 125.95, 122.46, 122.11, 112.10, 111.51, 56.22; IR (KBr) ν: 3102, 2941 (-N—H), 1686 (-C=O), 1289 (-C=S) $cm^{-1}$.

1-2′-methoxyl acridinyl-3-4′-methoxy benzoyl thiosemicarbazides (**4c**): Yellow powder, Yield 87%, m.p. 202–203 °C; ESI-MS m/z: 455 ([M + Na]$^+$); $^1$H NMR (400 MHz, DMSO-$d^6$), δ: 10.81 (br, s, 1H, -NH), 10.37 (br, s, 1H, -NH), 10.04 (br, s, 1H, -NH), 8.02–8.09 (m, 4H, ArH), 7.32–7.87 (m, 6H, ArH), 7.06 (s, 1H, ArH), 3.82 (s, 3H, -$OCH_3$), 3.32 (s, 3H, -$OCH_3$); $^{13}$C NMR (101 MHz, DMSO-$d^6$) δ: 182.97, 166.60, 162.03, 156.30, 153.82, 140.77, 136.15, 132.16, 129.58, 128.77, 127.25, 125.82, 121.52, 111.00, 108.32, 55.90; IR (KBr) ν: 3107, 2945 (-N—H), 1677 (-C=O), 1256 (-C=S) $cm^{-1}$.

1-2′-methoxyl acridinyl-3-4′-nitro benzoyl thiosemicarbazides (**4d**): Yellow powder, Yield 94%, m.p. 223–227 °C; ESI-MS m/z: 448 ([M + H]$^+$); $^1$H NMR (400 MHz, DMSO-$d^6$), δ: 11.30 (br, s, 1H, -NH), 10.45 (br, s, 1H, -NH), 10.22 (br, s, 1H, -NH), 8.38 (s, 2H, ArH), 8.26 (s, 2H, ArH), 8.02–8.16 (m, 2H, ArH), 7.53-7.63 (m, 4H, ArH), 7.46 (s, 1H, ArH), 4.02 (s, 3H, -$OCH_3$); $^{13}$C NMR (101 MHz, DMSO-$d^6$) δ: 183.22, 167.59, 162.87, 157.42, 149.87, 147.71, 146.73, 140.26, 138.66, 131.49, 131.11, 129.64, 125.67, 124.67, 124.15, 123.98, 117.74, 110.08, 100.87, 56.20; IR (KBr) ν: 3105, 2947 (-N—H), 1697 (-C=O), 1527 (-C=S) $cm^{-1}$.

1-2′-methoxyl acridinyl-3-4′-nitro benzoyl thiosemicarbazides (**4d**): Yellow powder, Yield 94%, m.p. 223–227 °C; ESI-MS m/z: 448 ([M+H]$^+$); $^1$H NMR (400 MHz, DMSO-$d^6$), δ: 11.30 (br, s, 1H, -NH), 10.45 (br, s, 1H, -NH), 10.22 (br, s, 1H, -NH), 8.38 (s, 2H, ArH), 8.26 (s, 2H, ArH), 8.02-8.16 (m, 2H, ArH), 7.53–7.63 (m, 4H, ArH), 7.46 (s, 1H, ArH), 4.02 (s, 3H, -$OCH_3$); $^{13}$C NMR (101 MHz, DMSO-$d^6$) δ: 183.22, 167.59, 162.87, 157.42, 149.87, 147.71, 146.73, 140.26, 138.66, 131.49, 131.11, 129.64, 125.67, 124.67, 124.15, 123.98, 117.74, 110.08, 100.87, 56.20; IR (KBr) ν: 3105, 2947 (-N—H), 1697 (-C=O), 1527 (-C=S) $cm^{-1}$.

1-2′-methyl acridinyl-3-4′-pyridinamide thiourea (**4f**): Orange powder, Yield 82%, m.p. 176–180 °C; ESI-MS m/z: 410 ([M + Na]$^+$); $^1$H NMR (400 MHz, DMSO-$d^6$), δ: 11.59 (br, s, 1H, -NH), 10.99 (br, s, 1H, -NH), 10.21 (br, s, 1H, -NH), 8.81 (s, 2H, ArH), 7.89 (s, 3H, ArH), 7.42–7.89 (m, 6H, ArH), 2.33 (s, 3H, -$CH_3$); $^{13}$C NMR (101 MHz, DMSO-$d^6$) δ: 183.47, 167.14, 155.61, 150.76, 140.45, 140.16, 138.18, 133.65, 131.24, 126.31, 124.88, 117.52, 111.80, 21.93; IR (KBr) ν: 3104, 2918 (-N—H), 1556 (-C=O), 1471 (-C=S) $cm^{-1}$.

1-2′-methyl acridinyl-3-benzoyl thiosemicarbazides (**4g**): Orange powder, Yield 82%, m.p. 171–173 °C; ESI-MS m/z: 409 [M + Na]$^+$; $^1$H NMR (400 MHz, DMSO-$d^6$), δ: 10.68 (br, s, 1H, -NH), 10.41 (br, s, 1H, -NH), 10.13 (br, s, 1H, -NH), 8.58–8.75 (m, 4H, ArH), 7.42–7.57 (m, 5H, ArH), 7.39 (d, $J$ = 8.5 Hz, 2H, ArH), 7.09 (s, 1H, ArH), 2.42 (s, 3H, -$CH_3$); $^{13}$C NMR (101 MHz, DMSO-$d^6$,) δ: 181.29, 166.81, 153.65, 150.24, 148.96, 140.25, 138.24, 133.58, 130.84, 130.35, 126.22, 125.07, 122.13, 121.38, 117.44, 116.40, 111.44, 21.25; IR (KBr) ν: 3102, 2917 (-N—H), 1569 (-C=O), 1471 (-C=S) $cm^{-1}$.

1-2′-methyl acridinyl-3-4′-methoxy benzoyl thiosemicarbazides (**4h**): Orange-yellow powder, Yield 93%, m.p. 210–212 °C; ESI-MS m/z: 439 ([M + Na]$^+$); $^1$H NMR (400 MHz, DMSO-$d^6$), δ: 10.53 (br, s, 1H, -NH), 10.38 (br, s, 1H, -NH), 10.08 (br, s, 1H, -NH), 7.97–8.18 (m, 4H, ArH), 7.34–7.55 (m, 3H, ArH), 7.34–7.55 (m, 3H, ArH), 7.07 (s, 1H, ArH), 3.44 (s, 3H, -$OCH_3$), 2.42 (s, 3H, -$CH_3$); $^{13}$C NMR (101 MHz, DMSO-$d^6$) δ: 181.09, 166.30, 153.39, 148.81, 140.15, 138.18, 130.79, 130.04, 126.17, 125.32, 121.21, 117.39, 114.12, 111.37, 55.89, 22.24; IR (KBr) ν: 3094, 2914 (-N—H), 1556 (-C=O), 1471 (-C=S) $cm^{-1}$.

1-2′-methyl acridinyl-3-4′-nitro benzoyl thiosemicarbazides (**4i**): Orange-yellow powder, Yield 73%, m.p. 187–188 °C; ESI-MS m/z: 457 ([M + Na]$^+$); $^1$H NMR (400 MHz, DMSO-$d^6$), δ: 11.56 (br, s, 1H, -NH), 10.76 (br, s, 1H, -NH), 10.15 (br, s, 1H, -NH), 8.41 (d, $J$

= 8.5 Hz, 2H), 7.96–8.26 (m, 4H, ArH), 7.26–7.57 (m, 4H, ArH), 7.11 (s, 1H, ArH), 2.44 (s, 3H, $-CH_3$); $^{13}$C NMR (101 MHz, DMSO-$d^6$) δ: 183.42, 166.78, 155.67, 149.20, 140.24, 140.14, 131.87, 130.89, 130.41, 124.68, 117.93, 111.56, 22.50; IR (KBr) ν: 3095, 2918 (-N—H), 1598 (-C=O), 1483 (-C=S) cm$^{-1}$.

1-2′-methyl acridinyl-3-4′-chloro benzoyl thiosemicarbazides (**4j**): Orange-yellow powder, Yield 86%, m.p. 178–179 °C; ESI-MS m/z: 443 ([M + Na]$^+$); $^1$H NMR (400 MHz, DMSO-$d^6$), δ: 10.77 (br, s, 1H, -NH), 10.48 (br, s, 1H, -NH), 10.08 (br, s, 1H, -NH), 8.18–8.34 (m, 4H, ArH), 7.57–7.83 (m, 3H, ArH), 7.33–7.57 (m, 3H, ArH), 7.07 (s, 1H, ArH), 2.42 (s, 3H, $-CH_3$); $^{13}$C NMR (101 MHz, DMSO-$d^6$) δ: 182.61, 166.39, 155.38, 149.09, 140.26, 139.92, 131.03, 130.08, 129.29, 124.20, 117.53, 116.31, 111.58, 104.34, 22.83; IR (KBr) ν: 3094, 2915 (-N—H), 1567 (-C=O), 1480 (-C=S) cm$^{-1}$.

### 3.1.4. General Procedure for the Synthesis of Acridinyl 1,2,4-triazole Derivatives **5a–5f**

The appropriate acyl thiosemicarbazides (**4a–4i**, 1 mmol) and 5% aqueous sodium carbonate (40 mL) were refluxed for 5 h. After cooling, the precipitate was filtered off and the filtrate was acidified by hydrochloric acid to a pH of 2. The precipitates were formed, filtered off and then crystallized from ethyl alcohol.

4-(2-methoxyacridin-9-yl)-5-(pyridin-4-yl)-2,4-dihydro-3H-1,2,4-triazole-3-thione (**5a**): Light yellow powder, Yield 75%, m.p. 272–273 °C; ESI-MS m/z: 384 ([M + H]$^+$); $^1$H NMR (400 MHz, DMSO-$d^6$), δ: 14.79 (br, s, 1H, -NH), 8.38 (dd, $J$ = 4.6, 1.5 Hz, 2H, ArH), 8.33–8.11 (m, 2H, ArH), 7.94–7.76 (m, 1H, ArH), 7.08 (dd, $J$ = 4.6, 1.6 Hz, 2H, ArH), 6.87 (d, $J$ = 2.6 Hz, 1H), 3.86 (s, 3H, $-OCH_3$); $^{13}$C NMR (100 MHz, DMSO-$d^6$) δ: 169.99, 159.21, 157.94, 150.89, 149.46, 134.90, 130.31, 129.15, 126.69, 125.10, 124.89, 120.97, 56.52; IR (KBr) ν: 3069, 2906, 2745 (-N—H), 1505–1633 (-C=N),1480 (-C=S) cm$^{-1}$.

4-(2-methoxyacridin-9-yl)-5-phenyl-2,4-dihydro-3H-1,2,4-triazole-3-thione (**5b**): Light yellow powder, 60%, m.p. 259–261 °C; ESI-MS m/z: 385 ([M + H]$^+$); $^1$H NMR (400 MHz, DMSO-$d^6$), δ: 14.56 (s, 1H, -NH), 8.22 (dd, $J$ = 12.2, 9.1 Hz, 2H, ArH), 7.89–7.75 (m, 1H, ArH), 7.70–7.58 (m, 2H, ArH), 7.56 (d, $J$ = 8.6 Hz, 1H, ArH), 7.26 (t, $J$ = 6.6 Hz, 1H, ArH), 7.21–7.07 (m, 4H, ArH), 6.82 (d, $J$ = 2.6 Hz, 1H, ArH), 3.85 (s, 3H, $-OCH_3$); $^{13}$C NMR (100 MHz, DMSO-$d^6$) δ: 169.50, 159.01, 151.69, 147.45, 146.68, 132.13, 131.27, 130.24, 130.13, 129.26, 128.94, 127.42, 126.58, 125.76, 124.95, 123.68, 122.58, 98.59, 56.44; IR (KBr) ν: 3056, 2912, 2749 (-N—H), 1500–1632 (-C=N), 1476 (-C=S) cm$^{-1}$.

4-(2-methoxyacridin-9-yl)-5-(4-methoxyphenyl)-2,4-dihydro-3H-1,2,4-triazole-3-thione (**5c**): Light yellow powder, Yield 58%, m.p. 262–267 °C; ESI-MS m/z: 415 ([M + H]$^+$); $^1$H NMR (400 MHz, DMSO-$d^6$), δ: 14.45 (s, 1H, -NH), 8.23 (dd, $J$ = 11.8, 9.1 Hz, 2H, ArH), 7.93–7.73 (m, 1H, ArH), 7.64 (s, 2H, ArH), 7.54 (s, 1H, ArH), 7.08 (d, $J$ = 8.9 Hz, 2H, ArH), 6.80 (s, 1H, ArH), 6.70 (d, $J$ = 8.9 Hz, 2H, ArH), 3.85 (s, 3H, $-OCH_3$), 3.58 (s, 3H, $-OCH_3$); $^{13}$C NMR (100 MHz, DMSO-$d^6$) δ: 169.30, 161.33, 159.01, 151.58, 147.51, 146.73, 134.08, 132.17, 130.20, 128.95, 126.56, 125.02, 123.76, 122.59, 117.89, 114.76, 98.57, 56.41, 55.60; IR (KBr) ν: 3066, 2883, 2726 (N—H), 1500–1632 (C=N), 1476 (C=S) cm$^{-1}$.

4-(2-methoxyacridin-9-yl)-5-(4-nitrophenyl)-2,4-dihydro-3H-1,2,4-triazole-3-thione (**5d**): Light yellow powder, Yield 40%, m.p. 253–255 °C; ESI-MS m/z: 430 ([M + H]$^+$); $^1$H NMR (400 MHz, DMSO-$d^6$), δ: 14.79 (s, 1H, -NH), 8.24 (t, $J$ = 9.3 Hz, 2H, ArH), 8.02 (d, $J$ = 8.9 Hz, 2H, ArH), 7.93–7.75 (m, 1H, ArH), 7.63 (dd, $J$ = 9.5, 2.7 Hz, 2H, ArH), 7.58 (s, 1H, ArH), 7.44 (d, $J$ = 8.9 Hz, 2H, ArH), 6.91 (s, 1H, ArH), 3.87 (s, 3H, $-OCH_3$); $^{13}$C NMR (100 MHz, DMSO-$d^6$) δ: 170.00, 159.22, 149.93, 148.90, 147.47, 146.78, 133.20, 132.22, 131.44, 130.25, 129.15, 128.69, 126.70, 124.93, 124.61, 123.51, 122.38, 98.64, 56.54; IR (KBr) ν: 3066, 2883, 2726 (-N—H), 1421–1633 (-C=N), 1345 (-C=S) cm$^{-1}$.

4-(4-chlorophenyl)-4-(2-methoxyacridin-9-yl)-2,4-dihydro-3H-1,2,4-triazole-3-thione (**5e**): Light yellow powder, Yield 78%, m.p. 286–287 °C; ESI-MS m/z: 419 ([M + H]$^+$); $^1$H NMR (400 MHz, DMSO-$d^6$), δ: 14.61 (s, 1H, -NH), 8.23 (dd, $J$ = 11.4, 9.2 Hz, 2H, ArH), 7.93–7.76 (m, 1H, ArH), 7.73–7.60 (m, 2H, ArH), 7.55 (d, $J$ = 8.6 Hz, 1H, ArH), 7.25 (d, $J$ = 8.6 Hz, 2H, ArH), 7.17 (d, $J$ = 8.6 Hz, 2H, ArH), 6.85 (s, 1H, ArH), 3.86 (s, 3H, $-OCH_3$); $^{13}$C NMR (100 MHz, DMSO-$d^6$) δ: 169.59, 159.10, 150.72, 147.44, 146.71, 136.12, 133.51, 132.17, 130.22,

129.54, 129.11, 126.64, 124.93, 124.62, 123.57, 122.49, 98.61, 56.49; IR (KBr) ν: 3056, 2909, 2748 (-N—H), 1344–1632 (-C=N), 1503 (-C=S) $cm^{-1}$.

4-(2-methylacridin-9-yl)-5-(pyridin-4-yl)-2,4-dihydro-3H-1,2,4-triazole-3-thione (**5f**): Light yellow powder, Yield 48%, m.p. 264–268 °C; ESI-MS m/z: 370 ($[M + H]^+$); $^1H$ NMR (400 MHz, DMSO-$d^6$) δ: 14.84 (s, 1H, -NH), 8.37 (d, *J* = 6.0 Hz, 2H, ArH), 8.29 (d, *J* = 8.7 Hz, 1H, ArH), 8.23 (d, *J* = 8.9 Hz, 1H, ArH), 7.96–7.84 (m, 1H, ArH), 7.84–7.74 (m, 1H, ArH), 7.72–7.59 (m, 2H, ArH), 7.50 (s, 1H, ArH), 7.04 (d, *J* = 6.1 Hz, 2H, ArH), 2.52 (s, 3H, -$CH_3$); $^{13}C$ NMR (100 MHz, DMSO-$d^6$) δ: 170.32, 150.92, 149.15, 148.75, 148.53, 139.37, 134.60, 134.56, 132.85, 131.15 130.21, 130.00, 129.04, 123.54, 123.40, 122.87, 120.84, 120.70, 56.49; IR (KBr) ν: 3066, 2917, 2757 (-N—H), 1279–1600 (-C=N), 1426 (-C=S) $cm^{-1}$.

4-(2-methylacridin-9-yl)-5-phenyl-2,4-dihydro-3H-1,2,4-triazole-3-thione (**5g**): White powder, Yield 78%, m.p. 292–293 °C; ESI-MS m/z: 369 ($[M + H]^+$); $^1H$ NMR (400 MHz, DMSO-$d^6$) δ: 14.60 (s, 1H, -NH), 8.26 (d, *J* = 8.8 Hz, 1H, ArH), 8.20 (d, *J* = 8.9 Hz, 1H, ArH), 7.90–7.82 (m, 1H, ArH), 7.77 (d, *J* = 9.0 Hz, 1H, ArH), 7.72–7.56 (m, 2H, ArH), 7.48 (s, 1H, ArH), 7.30–7.17 (m, 1H, ArH), 7.13 (d, *J* = 4.4 Hz, 4H, ArH), 2.51 (s, 3H, -$CH_3$); $^{13}C$ NMR (100 MHz, DMSO-$d^6$) δ: 169.73, 151.47, 148.71, 148.47, 139.08, 135.20, 134.51, 131.29, 131.07, 130.11, 129.91, 129.28, 128.83, 127.33, 125.66, 123.71, 123.58, 123.02, 120.86, 22.15; IR (KBr) ν: 3066, 2917, 2757 (-N—H), 1279–1600 (-C=N), 1426 (-C=S) $cm^{-1}$.

4-(4-methoxyphenyl)-4-(2-methylacridin-9-yl)-2,4-dihydro-3H-1,2,4-triazole-3-thione (**5h**): White powder, Yield 83%, m.p. 268–269 °C; ESI-MS m/z: 399 ($[M + H]^+$); $^1H$ NMR (400 MHz, DMSO-$d^6$) δ: 14.48 (s, 1H, -NH), 8.27 (d, *J* = 8.8 Hz, 1H), 8.21 (d, *J* = 8.9 Hz, 1H, ArH), 7.86 (t, 1H, ArH), 7.78 (dd, *J* = 9.0, 1.5 Hz, 1H, ArH), 7.65 (d, *J* = 6.5 Hz, 1H, ArH), 7.60 (d, *J* = 8.5 Hz, 1H, ArH), 7.46 (s, 1H, ArH), 7.05 (d, *J* = 8.9 Hz, 2H, ArH), 6.68 (d, *J* = 8.9 Hz, 2H, ArH), 3.57 (s, 3H, -$OCH_3$), 2.52 (s, 3H, -$CH_3$); $^{13}C$ NMR (100 MHz, DMSO-$d^6$) δ: 169.51, 161.33, 151.35, 148.77, 148.53, 139.05, 135.42, 134.49, 131.06, 130.06, 128.85, 123.71 123.02, 120.86, 117.78, 114.78, 55.61, 22.17; IR (KBr) ν: 3095, 2925, 2750 (-N—H), 1360–1613 (-C=N), 1514 (-C=S) $cm^{-1}$.

4-(2-methylacridin-9-yl)-5-(4-nitrophenyl)-2,4-dihydro-3H-1,2,4-triazole-3-thione (**5i**): Light yellow powder, Yield 53%, m.p. 245–247 °C; ESI-MS m/z: 414 ($[M + H]^+$); $^1H$ NMR (400 MHz, DMSO-$d^6$) δ: 14.79 (s, 1H, -NH), 8.24 (t, *J* = 9.3 Hz, 2H, ArH), 8.02 (d, *J* = 8.9 Hz, 2H, ArH), 7.83 (s, 1H, ArH), 7.71–7.59 (m, 2H, ArH), 7.57 (d, *J* = 8.7 Hz, 1H, ArH), 7.44 (d, *J* = 8.9 Hz, 2H, ArH), 6.91 (s, H, ArH), 2.51 (s, 3H, -$CH_3$), $^{13}C$ NMR (100 MHz, DMSO-$d^6$) δ: 169.99, 150.53, 148.55, 139.72, 135.26, 131.39, 130.41, 129.19, 128.69, 126.71, 124.62, 123.78, 123.49, 120.98, 22.19; IR (KBr) ν: 3054, 2918, 2755 (-N—H), 1432-1633 (-C=N), 1376 (-C=S) $cm^{-1}$.

4-(4-chlorophenyl)-4-(2-methylacridin-9-yl)-2,4-dihydro-3H-1,2,4-triazole-3-thione (**5j**): Yellow powder, Yield 80%, m.p. 259–263 °C; ESI-MS m/z: 403 ($[M + H]^+$); $^1H$ NMR (400 MHz, DMSO-$d^6$) δ: 14.65 (s, 1H, -NH), 8.27 (d, *J* = 8.8 Hz, 1H, ArH), 8.20 (d, *J* = 8.9 Hz, 1H, ArH), 7.88 (dd, *J* = 10.5, 4.0 Hz, 1H, ArH), 7.77 (d, *J* = 8.8 Hz, 1H, ArH), 7.73–7.57 (m, 2H, ArH), 7.49 (s, 1H, ArH), 7.24 (d, *J* = 8.5 Hz, 2H, ArH), 7.14 (d, *J* = 8.5 Hz, 2H, ArH); $^{13}C$ NMR (100 MHz, DMSO-$d^6$) δ: 169.85, 150.49, 148.74, 148.52, 139.19, 136.15, 134.86, 134.52, 131.08, 130.17, 129.97, 129.55, 129.00, 124.51, 123.56, 122.98, 120.81, 22.15; IR (KBr) ν: 3097, 2933 (-N—H), 1258-1600 (-C=N), 1497 (-C=S) $cm^{-1}$.

### 3.1.5. General Procedure for the Synthesis of Acridinyl 1,3,4-thiadiazol Derivatives (**6**)

About 3 mL of 98% concentrated sulfuric acid was added to a 50 mL round-bottom flask and stirred in an ice bath for 10 min at 0 °C. Then, intermediate **4** (0.5 mmol) was added into the solution in small portions over the course of 1 h. The reaction was continued at room temperature for 24–48 h, and 10 mL pure water was slowly added to reaction mixture in an ice bath. The final product **6** was precipitated, filtered off, washed with water, dried, and crystallized from ethyl alcohol.

7-methoxy-N-(5-(pyridin-4-yl)-1,3,4-thiadiazol-2-yl)-10,10a-dihydroacridin-9(8aH)-imine (**6a**): Orange solids, Yield 80%, m.p. 272–275 °C; ESI-MS m/z: 386 ($[M + H]^+$); $^1H$ NMR (400 MHz, DMSO-$d^6$), δ: 8.78 (d, *J* = 5.7 Hz, 2H, ArH), 8.22 (d, *J* = 8.6 Hz, 1H, ArH), 8.00–7.94

(m, 3H, ArH), 7.90 (d, $J$ = 6.2 Hz, 2H, ArH), 7.75 (dd, $J$ = 9.3, 2.7 Hz, 1H, ArH), 7.58 (s, 1H, ArH), 7.53 (dt, $J$ = 8.3, 4.0 Hz, 1H, ArH), 3.86 (s, 3H, -$OCH_3$); $^{13}$C NMR (DMSO-$d^6$, 100 MHz) δ: 156.40, 149.19, 140.03, 135.03, 130.65, 128.43, 127.65, 127.49, 127.29, 126.86, 125.02, 121.72, 119.38, 103.95, 56.20; IR (KBr) ν: 3055, 3011, 2837 (-C—H, -N—H), 1379-1631 (-C=N) $cm^{-1}$.

7-methoxy-N-(5-phenyl-1,3,4-thiadiazol-2-yl)-10,10a-dihydroacridin-9(8aH)-imine (**6b**): Orange- yellow solid, Yield 81%, m.p. 263–265 °C; ESI-MS m/z: 385 ($[M + H]^+$); $^1$H NMR (400 MHz, DMSO-$d^6$), δ: 8.19 (d, $J$ = 8.7 Hz, 1H, ArH), 7.92 (q, $J$ = 9.3 Hz, 3H, ArH), 7.80 (dd, $J$ = 6.6, 2.9 Hz, 2H, ArH), 7.68 (dd, $J$ = 9.2, 2.7 Hz, 1H, ArH), 7.58 (s, 1H, ArH), 7.54–7.44 (m, 4H, ArH), 3.83 (s, 3H, -$OCH_3$); $^{13}$C NMR (101 MHz, DMSO-$d^6$) δ: 156.06, 140.22, 136.58, 134.49, 131.24, 130.46, 129.83, 127.72, 126.95, 126.79, 124.37, 121.59, 120.02, 120.33, 119.26, 116.76, 104.39, 56.03; IR (KBr) ν: 2771 (-C—H, N—H), 1329-1632 (-C=N) $cm^{-1}$.

7-methoxy-N-(5-(4-methoxyphenyl)-1,3,4-thiadiazol-2-yl)-10,10a-dihydroacridin-9(8aH)-imine (**6c**): Orange-red solid, Yield 54%; m.p. 230–232 °C; ESI-MS m/z: 415 ($[M + H]^+$); $^1$HNMR (400 MHz, DMSO-$d^6$) δ: 8.35 (d, $J$ = 8.5 Hz, 1H, ArH), 8.10 (d, $J$ = 8.1 Hz, 3H, ArH), 8.01 (d, $J$ = 9.8 Hz, 1H, ArH), 7.96 (s, 1H, ArH), 7.91–7.79 (m, 1H, ArH), 7.76–7.55 (m, 3H, ArH), 7.06 (t, $J$ = 9.1 Hz, 1H, ArH), 3.93 (s, 3H, -$OCH_3$), 3.81 (s, 3H, -$OCH_3$); $^{13}$C NMR (101 MHz, DMSO-$d^6$) δ: 156.91, 149.55, 144.54, 140.78, 136.88, 135.34, 129.01, 128.50, 126.77, 124.39, 121.00, 116.85, 112.99, 103.39, 103.21, 56.65, 56.17; IR (KBr) ν: 2771 (-C—H, -N—H), 1567 (-C=N) $cm^{-1}$.

7-methoxy-N-(5-(4-nitrophenyl)-1,3,4-thiadiazol-2-yl)-10,10a-dihydroacridin-9(8aH)-imine (**6d**): Orange-red solid, Yield 33%; m.p. 258–260 °C; ESI-MS m/z: 430 ($[M + H]^+$); $^1$H NMR (DMSO-$d^6$, 400 MHz) δ: 8.32 (d, $J$ = 8.9 Hz, 2H, ArH), 8.29 (d, $J$ = 8.6 Hz, 1H, ArH), 8.04 (dd, $J$ = 6.9, 4.8 Hz, 5H, ArH), 7.81 (dd, $J$ = 9.3, 2.7 Hz, 1H, ArH), 7.61 (dd, $J$ = 8.5, 2.5 Hz, 2H, ArH), 3.89 (s, 3H, -$OCH_3$); $^{13}$C NMR (101 MHz, DMSO-$d^6$) δ: 156.85, 148.76, 145.74, 142.43, 139.80, 136.85, 135.92, 135.35, 129.32, 127.93, 126.55, 125.67, 125.01, 121.90, 120.38, 119.60, 117.03, 103.59, 56.32; IR (KBr) ν: 2829 (-C—H, -N—H), 1346-1633 (-C=N) $cm^{-1}$.

7-methoxy-N-(5-(4-chlorophenyl)-1,3,4-thiadiazol-2-yl)-10,10a-dihydroacridin-9(8aH)-imine (**6e**): Orange-yellow solid, Yield 51%; m.p. 203–205 °C; ESI-MS m/z: 419 ($[M + H]^+$); $^1$H NMR (400 MHz, DMSO-$d^6$) δ: 8.21 (d, $J$ = 8.6 Hz, 1H, ArH), 7.94 (d, $J$ = 8.1 Hz, 3H, ArH), 7.82 (s, 2H, ArH), 7.72 (dd, $J$ = 9.2, 2.6 Hz, 1H, ArH), 7.58 (dd, $J$ = 5.9, 2.6 Hz, 3H, ArH), 7.55–7.41 (m, 1H, ArH), 3.85 (s, 3H, -$OCH_3$); $^{13}$C NMR (101 MHz, DMSO-$d^6$) δ: 156.36, 140.09, 135.90, 134.8, 129.89, 129.19, 128.58, 128.25, 126.65, 124.87, 121.70, 120.01, 119.45, 116.90, 104.12, 102.60, 56.21; IR (KBr) ν: 2781 (-C—H, -N—H), 1467 (-C=N) $cm^{-1}$.

7-methyl-N-(5-(pyridin-4-yl)-1,3,4-thiadiazol-2-yl)-10,10a-dihydroacridin-9(8aH)-imine (**6f**): Orange-yellow solid, Yield 43%; m.p. 276–278 °C; ESI-MS m/z: 370 ($[M + H]^+$); $^1$H NMR (400 MHz, DMSO-$d^6$) δ: 8.34 (d, $J$ = 8.7 Hz, 2H, ArH), 8.23 (d, $J$ = 8.9 Hz, 1H, ArH), 8.14 (d, $J$ = 8.3 Hz, 1H, ArH), 8.05 (t, $J$ = 9.3 Hz, 3H, ArH), 8.01–7.91 (m, 2H, ArH), 7.88 (s, 1H, ArH), 7.51 (d, $J$ = 8.2 Hz, 1H, ArH), 2.48 (s, 3H, -$CH_3$); $^{13}$C NMR (101 MHz, DMSO-$d^6$) δ: 158.09, 155.78, 151.07, 140.54, 135.49, 133.66, 131.53, 127.77, 126.46, 123.78, 121.88, 118.38, 118.11, 116.14, 21.31; IR (KBr) ν: 2824 (-C—H, -N—H), 1347-1600 (-C=N) $cm^{-1}$.

7-methyl-N-(5-phenyl-1,3,4-thiadiazol-2-yl)-10,10a-dihydroacridin-9(8aH)-imine (**6g**): Yellow solid, Yield 56%; m.p. 294–296 °C; ESI-MS m/z: 369 ($[M + H]^+$); $^1$H NMR (400 MHz, DMSO-$d^6$) δ: 8.17 (d, $J$ = 8.5 Hz, 1H, ArH), 8.07 (s, 1H, ArH), 7.97–7.85 (m, 2H, ArH), 7.80 (d, $J$ = 6.9 Hz, 4H, ArH), 7.59-7.49 (m, 3H, ArH), 7.48–7.34 (m, 1H, ArH), 2.45 (s, 3H, -$CH_3$); $^{13}$C NMR (101 MHz, DMSO-$d^6$) δ: 160.12, 157.49, 140.60, 138.98, 137.32, 135.07, 133.97, 131.26, 130.45, 129.82, 127.17, 127.15, 125.56, 124.06, 119.50, 119.21, 118.10, 117.00, 21.44; IR (KBr) ν: 2792 (C—H, N—H), 1366-1627 (C=N) $cm^{-1}$.

7-methyl-N-(5-(4-methoxyphenyl)-1,3,4-thiadiazol-2-yl)-10,10a-dihydroacridin-9(8aH)-imine (**6h**): Orange-yellow solid, Yield 84%; m.p. >300 °C; ESI-MS m/z: 399 ($[M + H]^+$); $^1$H NMR (400 MHz, DMSO-$d^6$) δ: 8.33 (d, $J$ = 8.7 Hz, 1H, ArH), 8.18 (s, 1H, ArH), 8.10–8.01 (m, 3H, ArH), 7.98 (d, $J$ = 8.8 Hz, 2H, ArH), 7.91–7.77 (m, 1H, ArH), 7.68 (dd, $J$ = 8.6, 2.3 Hz, 1H, ArH), 7.62 (dd, $J$ = 14.4, 7.9 Hz, 1H, ArH), 7.08 (d, $J$ = 8.7 Hz, 1H, ArH), 2.48 (s, 3H, -$CH_3$); $^{13}$C NMR (101 MHz, DMSO-$d^6$) δ: 162.17, 159.02, 140.47, 139.22, 137.04, 136.13, 135.63,

128.91, 127.11, 127.06, 126.92, 126.85, 125.57, 125.06, 120.71, 120.03, 119.59, 118.29, 117.35, 113.02, 56.20, 21.52; IR (KBr) ν: 3090, 2920, 2830 (-C—H, -N—H), 1486-1628 (-C=N) $cm^{-11}$.

7-methyl-N-(5-(4-nitrophenyl)-1,3,4-thiadiazol-2-yl)-10,10a-dihydroacridin-9(8aH)-imine (**6i**): Purple-red solid, Yield 53%; m.p. 224–226 °C; ESI-MS m/z: 414 ($[M + H]^+$); $^1H$ NMR (400 MHz, DMSO-$d^6$) δ: 12.17 (s, 1H, -NH), 8.35 (d, $J$ = 8.9 Hz, 2H, ArH), 8.15 (d, $J$ = 8.9 Hz, 2H, ArH), 7.96 (d, $J$ = 5.1 Hz, 2H, ArH), 7.72 (t, $J$ = 7.0 Hz, 1H, ArH), 7.65–7.56 (m, 2H, ArH), 7.54 (d, $J$ = 8.5 Hz, 1H, ArH), 7.18 (t, $J$ = 7.6 Hz, 1H, ArH), 2.35 (s, 3H, $-CH_3$); $^{13}C$ NMR (100 MHz, DMSO-$d^6$) δ: 165.41, 148.81, 143.72, 140.52, 139.40, 138.81, 136.22, 135.69, 135.63, 127.93, 126.99, 125.65, 125.02, 120.12, 119.72, 118.15, 117.34, 113.14, 21.54; IR (KBr) ν: 2918 (-C—H, -N—H), 1340-1627 (-C=N) $cm^{-1}$.

7-methyl-N-(5-(4-chlorophenyl)-1,3,4-thiadiazol-2-yl)-10,10a-dihydroacridin-9(8aH)-imine (**6j**): Orange-yellow solid, Yield 67%, m.p. 258–260 °C; ESI-MS m/z: 403 ($[M + H]^+$); $^1H$ NMR (400 MHz, DMSO-$d^6$) δ: 8.23 (d, $J$ = 8.6 Hz, 1H, ArH), 8.12 (s, 1H, ArH), 8.04–7.93 (m, 2H, ArH), 7.88 (s, 2H, ArH), 7.81 (d, $J$ = 8.5 Hz, 2H, ArH), 7.58 (d, $J$ = 8.5 Hz, 2H, ArH), 7.51 (t, $J$ = 7.5 Hz, 1H, ArH), 2.49 (s, 3H, $-CH_3$); $^{13}C$ NMR (101 MHz, DMSO-$d^6$) δ: 160.71, 146.36, 140.56, 139.09, 137.87, 135.87, 135.50, 129.89, 129.14, 128.58, 127.04, 125.34, 124.71, 119.72, 119.37, 118.16, 117.13, 21.47; IR (KBr) ν: 2795 (-C—H, -N—H), 1366–1627 (-C=N) $cm^{-1}$.

#### 3.1.6. Preparation of Single Crystal Compounds **5b** and **6d** and Their X-ray Single Crystal Diffraction Method

A single crystal of **5b** and **6d** suitable for X-ray diffraction study was cultivated from 95% ethyl alcohol and N, N-dimethylformamide respectively, by a slow evaporation method at room temperature. All measurements were performed with Mo Kα radiation (λ = 0.7107 Å) on a Brucker SMART 1000 CCD X diffractometer (Billerica, MA, USA). The structure was solved by direct methods with SHELXS-97 [25] and refined by SHELXL-97 [26]. All non-hydrogen atoms were refined with anisotropic thermal parameters. The final full-matrix least-squares refinement of **5b** gave R = 0.0914, $\omega = 1/[s^2(Fo^2) + (0.0431\ p)^2 + 0.2721\ p]$ where $p = (Fo^2 + 2Fc^2)/3$, S = 1.043, $(\Delta/\sigma)_{max}$ = 0.237 and $(\Delta/\sigma)_{min} = -0.267$ e/Å$^3$. In addition, the final full-matrix least-squares refinement of **6d** gave R = 0.0914, $\omega = (1/[s^2(Fo^2) + (0.0650\ p)^2 + 0.0224\ p]$ where $p = (Fo^2 + 2Fc^2)/3$, S = 1.028, $(\Delta/\sigma)_{max}$ = 0.236 and $(\Delta/\sigma)_{min} = -0.197$ e/Å$^3$.

### *3.2. In-Vivo Antitumor Activity*

#### 3.2.1. Antiproliferative Activity

Test samples, including compounds **4–6** and commercial classical anticancer drugs (5-FU and cis-platinum), were screened for their anti-cancer activity against HFF human foreskin fibroblast cells, MGC-803 human gastric cancer cells, BEL-7404 human hepatocellular carcinoma cells, NCI-H460 human large cell carcinoma cells, and T24 human bladder carcinoma cells using the 3-(4,5-dimethylthiazol-2-yl)-2,5-diphenyltetrazoliumbromide (MTT) assay method cited in the literature [27]. The initial concentration of all the test samples was 100 µg/mL, which was serially diluted in complete medium with ten-fold dilutions to give six concentrations per compound. Their cytotoxicity was determined in 96-well flat bottomed microtiter plates. All the test samples were tested in triplicate. The results were expressed as the drug concentration that inhibited cell growth by 50% as compared to the controls ($IC_{50}$). The $IC_{50}$ values were calculated from regression lines obtained from the percent cell growth inhibition plotted as a function of the logarithm of the dose.

#### 3.2.2. Apoptosis and Cell Cycle Analysis

The apoptosis assay and the cell cycle analysis were carried out by cytometry (FACS-Verse, BD, Piscataway, NJ, USA) at an excitation wavelength of 488 nm according to the method described in the literature with slight modifications [28]. The cells were seeded at

$2 \times 10^6$/well and washed by cold PBS. The buffer solutions were prepared using 0.1 M pH 7.4 Hepes/NaOH, 1.4 M NaCl, and 25 mM $CaCl_2$.

3.2.3. Topo I Inhibitory Activity

Topo I and pBR322 were obtained commercially from Takara Bio Inc. (Shiga, Japan). And the enzyme inhibitory activity was determined by our previous methods [19].

3.2.4. Anti-Angiogenic Effect Using the Zebrafish Model

(a) Zebrafish toxicity assay

The zebrafish embryos were collected at 6 hpf and randomly divided into naive control (embryos maintained in distilled water), vehicle control (embryos treated with 2% DMSO), and drug groups. Stock solutions of all drugs were prepared in 2% DMSO as a solubilizing agent and diluted to three concentrations (0.5 mg/mL, 1 mg/mL, and 2 mg/mL). Each group had 20 embryos per test concentration. The zebrafish embryos were maintained in an incubator at 28 °C and read at 72 hpf for their mortality and teratogenicity (including non-hatching, egg condensation, spinal curvature, pericardial enlargement, etc.). Each compound was evaluated in three independent biological experiments.

(b) Angiogenesis assay

The 24 hpf zebrafish embryos were dechorionated with a 1 $gL^{-1}$ pronase treatment and maintained in distilled water in 12-well cell culture plates (each well contained 20 embryos). A negative control group containing zebrafish embryos in distilled water and a vehicle treatment group that was treated with 2% dimethyl sulfoxide (DMSO) were prepared. The anti-angiogenic compounds were diluted to 1 mg/mL. After incubating at 28 °C for 72 h, the embryos were immersed in 4% paraformaldehyde and dehydrated by gradient ethanol. Subsequently, the embryos were balanced in NTMT buffer (5 M NaCl + 1 M Tris [pH 9.0–9.5] + 1 M $MgCl_2$+ 10% Tween), and nitrotetrazolium blue chloride (NBT) and p-toluidine salt (BCIP) staining were performed. To evaluate the effect of compounds on the angiogenesis of zebrafish embryos, the growth of embryonic sub-intestinal veins (SIVs) at 72 hpf was observed using an IX71 Olympus microscope (Hamburger, Germany). The length of the SIVs was calculated using the image J 1.8.0 software (Bethesda, MD, USA).

## 4. Conclusions

A new series of acridine-triazole and acridine-thiadiazole derivatives were synthesized and characterized by spectral studies. All the synthesized compounds were evaluated for their in vitro cytotoxic activities against HFF, MGC-803, BEL-7404, NCI-H460, and T24 by the MTT assay method. Most of the compounds were sensitive to MGC-803 and T24 cell lines. Compared to all the prepared compounds, **4a**, **5d** and **6h** exhibited the best anticancer activity against MGC-803 cell lines, and compounds **4h**, **5h** and **6h** showed the most excellent antitumor activity against T24. Preliminary studies of antitumor mechanisms revealed that the representative compounds (**5d** and **6h** or **4h** and **6h**) could suppress cell proliferation by inducing apoptosis in the Q3 period of MGC-803 or T24 cell lines. Compound **5d** might inhibit the growth of tumor cells by arresting cells in the G2 phase, while compound **4h** had a great effect on the S phase. In the zebrafish experiment, compound **5d** displayed a superior antiangiogenic effect and lower toxicity than other compounds. Therefore, compound **5d** has the potential to be an antitumor drug with high efficiency and low toxicity.

**Author Contributions:** Contributed to the synthesis of compounds, L.H., X.L. and H.X.; performed pharmacological activity testing and docking studies of the target products, R.C., Y.N., Z.J. and Y.L.; wrote the manuscript, L.H. and R.C.; supervision of the project and design of experiment, L.H., R.C. and R.L.; contributed to improvising the manuscript, provided intellectual inputs, and edited the language of the manuscript, Y.J. and L.W. All authors have read and agreed to the published version of the manuscript.

**Funding:** This work acknowledges the support of the National Natural Science Foundation of China (22067001) and the Inheritance and innovation team of Guangxi Traditional Chinese Medicine (2022B005). We also thank the Guangxi Key Laboratory of Zhuang and Yao Ethnic Medicine [(2013) No. 20], the Collaborative Innovation Center of Zhuang and Yao Ethnic Medicine [(2014) No. 32], the Guangxi Key Discipline Zhuang Pharmacology ((2013) No. 16), the First-class Discipline in Guangxi of traditional Chinese Pharmacology (Direction of Ethnic Medicine) ((2018) No. 12) and Guangxi Joint Graduate education Base (Guangxi Degree (2021) No. 6).

**Institutional Review Board Statement:** There are no any ethical issues with animals.

**Informed Consent Statement:** These studies were not involving humans.

**Data Availability Statement:** We have presented all of our main data in the form of tables and figures. CCDC 2214949 contain supplementary crystallographic data for compound **5b** and CCDC 2214923 contain supplementary crystallographic data for compound **6d**. These datas can be obtained free of charge via http://www.ccdc.cam.ac.uk/conts/retrieving.html (accessed on 20 December 2022) or the Cambridge Crystallographic Data Centre, 12 Union Road, Cambridge CB2 1EZ, UK; Fax: +44-1223-336-033; or e-mail: deposit@ccdc.cam.ac.uk.

**Conflicts of Interest:** The authors declare no conflict of interest.

## References

1. Cheng, G.X.; Gao, Y. Progress in the study of the mechanism of action of anti-tumor drugs. *Jilin Med.* **2009**, *30*, 3080–3083.
2. Martins, C.; Gunaratnam, M.; Stuart, J.; Makwana, V.; Greciano, O.; Reszka, A.P.; Kelland, L.R.; Neidle, S. Structure-based design of benzylamino-acridine compounds as G-quadruplex DNA telomere targeting agents. *Bioorg. Med. Chem. Lett.* **2007**, *17*, 2293–2298. [CrossRef]
3. Kamal, A.; Srinivas, O.; Ramulu, P.; Ramesh, G.; Kumar, P.P. Synthesis of C8-linked pyrrolo[2,1-c][1,4]benzodiazepine-acridone/acridine hybrids as potential DNA-binding agents. *Bioorg. Med. Chem. Lett.* **2004**, *14*, 4107–4111. [CrossRef]
4. Demeunynck, M.; Charmantray, F.; Martelli, A. Interest of acridine derivatives in the anticancer chemotherapy. *Curr. Pharm. Des.* **2001**, *7*, 1703–1724.
5. Graham, J.A.; Gordon, W.R.; Bruce, C.B.; William, A.D. Potential antitumor agents. 50. in vivo solid-tumor activity of derivatives of N-[2-(Dimethylamino) ethy]acridine-4-carboxamide. *J. Med. Chem.* **1987**, *30*, 664–669.
6. Wainwright, M. Acridine-a neglected antibacterial chromophore. *J. Antimicrob. Chemother.* **2001**, *47*, 1–13. [CrossRef]
7. Hamulakova, S.; Janovec, L.; Soukup, O.; Jun, D.; Kuca, K. Synthesis, in vitro acetylcholinesterase inhibitory activity and molecular docking of new acridine-coumarin hybrids. *Int. J. Biol. Macromol.* **2017**, *104*, 333–338. [CrossRef]
8. Denny, W.A. Acridine Derivatives as Chemotherapeutic Agents. *Curr. Med. Chem.* **2002**, *9*, 1655–1665. [CrossRef]
9. Lee, Y.C.; Chen, Y.J.; Huang, C.H.; Chang, L.S. Amsacrine-induced apoptosis of human leukemia U937 cells is mediated by the inhibition of AKT- and ERK-induced stabilization of MCL1. *Apoptosis* **2017**, *22*, 406–420. [CrossRef]
10. Mi, J.L.; Wu, J.; Zhou, C.H. Research progress of triazole antitumor drugs. *West China J. Pharm. Sci.* **2008**, *23*, 84–86.
11. Kumar, D.; Kumar, N.M.; Chang, K.H.; Shah, K. Synthesis and anticancer activity of 5-(3-indolyl)-1,3,4-thiadiazoles. *Eur. J. Med. Chem.* **2010**, *45*, 4664–4668. [CrossRef]
12. Joubert, J.P.; Smit, F.J.; du Plessis, L.; Smith, P.J.; N'Da, D.D. Synthesis and in vitro biological evaluation of aminoacridines and artemisinin-acridine hybrids. *Eur. J. Pharm. Sci.* **2014**, *56*, 16–27. [CrossRef]
13. Malachowska-Ugarte, M.; Cholewinski, G.; Dzierzbicka, K.; Trzonkowski, P. Synthesis and biological activity of novel mycophenolic acid conjugates containing nitro-acridine/acridone derivatives. *Eur. J. Med. Chem.* **2012**, *54*, 197–201. [CrossRef]
14. Barros, F.W.A.; Silva, T.G.; da Rocha Pitta, M.G.; Bezerra, D.P.; Costa-Lotufo, L.V.; de Moraes, M.O.; Pessoa, C.; de Moura, M.A.F.B.; de Abreu, F.C.; de Lima, M.D.C.A.; et al. Synthesis and cytotoxic activity of new acridine-thiazolidine derivatives. *Bioorg. Med. Chem.* **2012**, *20*, 3533–3539. [CrossRef]
15. Chen, C.-H.; Lin, Y.-W.; Zhang, X.; Chou, T.-C.; Tsai, T.-J.; Kapuriya, N.; Kakadiya, R.; Su, T.-L. Synthesis and in vitro cytotoxicity of 9-anilinoacridines bearing N-mustard residue on both anilino and acridine rings. *Eur. J. Med. Chem.* **2009**, *44*, 3056–3059. [CrossRef]
16. Tomascikova, J.; Imrich, J.; Danihel, I.; Bohm, S.; Kristian, P.; Pisarcikova, J.; Sabol, M.; Klika, K.D. Regioselectivity and Tautomerism of Novel Five-Membered Ring Nitrogen Heterocycles Formed via Cyclocondensation of Acylthiosemicarbazides. *Molecules* **2008**, *13*, 501–518. [CrossRef]
17. Pommier, Y. Diversity of DNA topoisomerases I and inhibitors. *Biochimie* **1998**, *80*, 255–270. [CrossRef]
18. Ismail, M.M.F.; Rateb, H.S.; Hussein, M.M.M. Synthesis and docking studies of novel benzopyran-2-ones with anticancer activity. *Eur. J. Med. Chem.* **2010**, *45*, 3950–3959. [CrossRef]
19. Chen, R.; Huo, L.N.; Jaiswal, Y.; Wei, J.H.; Li, D.P.; Zhong, J.; Williams, L.; Xia, X.; Liang, Y. Synthesis and Evaluation of Anticancer Activity of New 4-Acyloxy Derivatives of Robustic Acid. *Int. J. Mol. Sci.* **2019**, *20*, 5336. [CrossRef]
20. Zhang, J.; Gao, B.L.; Zhang, W.C.; Qian, Z.J.; Xiang, Y. Monitoring antiangiogenesis of bevacizumab in zebrafish. *Drug Des. Dev. Ther.* **2018**, *12*, 2423–2430. [CrossRef]

21. Howe, K.; Clark, M.D.; Torroja, C.F.; Torrance, J.; Berthelot, C.; Muffato, M.; Collins, J.E.; Humphray, S.; McLaren, K.; Matthews, L.; et al. The zebrafish reference genome sequence and its relationship to the human genome. *Nature* **2013**, *496*, 498–503. [CrossRef] [PubMed]
22. Gore, A.V.; Monzo, K.; Cha, Y.R.; Pan, W.; Weinstein, B.M. Vascular development in the zebrafish. *CSH Perspect. Med.* **2012**, *2*, a006684. [CrossRef] [PubMed]
23. Schuermann, A.; Helker, C.S.M.; Herzog, W. Angiogenesis in zebrafish. *Semin. Cell Dev. Biol.* **2014**, *31*, 106–114. [CrossRef]
24. Huo, L.N.; Chen, R.; Liao, Y.F.; Liu, H.G.; Li, P.Y.; Lu, R.M.; Zhong, Z.G. Synthesis, crystal structure and biological evaluation of acridine-1, 2, 3-triazole derivatives. *Chin. J. Struct. Chem.* **2016**, *35*, 698–704.
25. Bruker. *APEXII Software*, Version 6.3.1; Bruker AXS Inc.: Madison, WI, USA, 2004.
26. Sheldrick, G.M. *SHELXS-97 and SHELXL-97, Program for X-ray Crystal Structure Refinement*; University of Göttingen: Göttingen, Germany, 1997.
27. Su, W.; Zhang, B.Q.; Peng, B.H.; Tang, Z.F.; Li, P.Y. Synthesis and Anticancer Activity of a New Di-nuclear Ruthenium Arene Compound with Thiosemicarbazones. *Chin. J. Struct. Chem.* **2020**, *39*, 1112–1118.
28. Huang, R.Z.; Wang, C.Y.; Li, J.F.; Yao, G.Y.; Pan, Y.M.; Ye, M.Y.; Wang, H.S.; Zhang, Y. Synthesis, antiproliferative and apoptosis-inducing effects of novel asiatic acid derivatives containing α-aminophosphonates. *RSC Adv.* **2016**, *6*, 62890–62906. [CrossRef]

International Journal of
*Molecular Sciences*

MDPI

*Article*

# BPR0C261, An Analogous of Microtubule Disrupting Agent D-24851 Enhances the Radiosensitivity of Human Non-Small Cell Lung Cancer Cells via p53-Dependent and p53-Independent Pathways

**Jyh-Der Leu [1,2,†], Shih-Ting Lin [3,†], Chiung-Tong Chen [4], C.-Allen Chang [3] and Yi-Jang Lee [3,5,*]**

1 Division of Radiation Oncology, Taipei City Hospital RenAi Branch, Taipei 106, Taiwan
2 Institute of Neuroscience, National Chengchi University, Taipei 116, Taiwan
3 Department of Biomedical Imaging and Radiological Sciences, National Yang Ming Chiao Tung University, Taipei 11221, Taiwan
4 Institute of Biotechnology and Pharmaceutical Research, National Health Research Institutes, Zhunan 350, Taiwan
5 Cancer Progression Research Center, National Yang Ming Chiao Tung University, Taipei 11221, Taiwan
* Correspondence: yjlee2@nycu.edu.tw; Tel.: +886-2-28267189; Fax: +886-2-28201095
† These authors contributed equally to this work.

**Citation:** Leu, J.-D.; Lin, S.-T.; Chen, C.-T.; Chang, C.-A.; Lee, Y.-J. BPR0C261, An Analogous of Microtubule Disrupting Agent D-24851 Enhances the Radiosensitivity of Human Non-Small Cell Lung Cancer Cells via p53-Dependent and p53-Independent Pathways. *Int. J. Mol. Sci.* **2022**, *23*, 14083. https://doi.org/10.3390/ijms232214083

Academic Editor: Laura Paleari

Received: 24 August 2022
Accepted: 23 September 2022
Published: 15 November 2022

**Publisher's Note:** MDPI stays neutral with regard to jurisdictional claims in published maps and institutional affiliations.



**Abstract:** (1) Destabilization of microtubule dynamics is a primary strategy to inhibit fast growing tumor cells. The low cytotoxic derivative of microtubule inhibitor D-24851, named BPR0C261 exhibits antitumor activity via oral administration. In this study, we investigated if BPR0C261 could modulate the radiation response of human non-small cell lung cancer (NSCLC) cells with or without p53 expression. (2) Different doses of BPR0C261 was used to treat human NSCLC A549 (p53+/+) cells and H1299 (p53−/−) cells. The cytotoxicity, radiosensitivity, cell cycle distribution, DNA damage, and protein expression were evaluated using an MTT assay, a colony formation assay, flow cytometry, a comet assay, and an immunoblotting analysis, respectively. (3) BPR0C261 showed a dose-dependent cytotoxicity on A549 cells and H1299 cells with $IC_{50}$ at 0.38 μM and 0.86 μM, respectively. BPR0C261 also induced maximum $G_2$/M phase arrest and apoptosis in both cell lines after 24 h of treatment with a dose-dependent manner. The colony formation analysis demonstrated that a combination of low concentration of BPR0C261 and X-rays caused a synergistic radiosensitizing effect on NSCLC cells. Additionally, we found that a low concentration of BPR0C261 was sufficient to induce DNA damage in these cells, and it increased the level of DNA damage induced by a fractionation radiation dose (2 Gy) of conventional radiotherapy. Furthermore, the p53 protein level of A549 cell line was upregulated by BPR0C261. On the other hand, the expression of PTEN tumor suppressor was found to be upregulated in H1299 cells but not in A549 cells under the same treatment. Although radiation could not induce PTEN in H1299 cells, a combination of low concentration of BPR0C261 and radiation could reverse this situation. (4) BPR0C261 exhibits specific anticancer effects on NSCLC cells by the enhancement of DNA damage and radiosensitivity with p53-dependent and p53-independent/PTEN-dependent manners. The combination of radiation and BPR0C261 may provide an important strategy for the improvement of radiotherapeutic treatment.

**Keywords:** NSCLC; BPR0C261; radiosensitivity; DNA damage; microtubule inhibitor; p53; PTEN

## 1. Introduction

Human non-small cell lung cancer (NSCLC) is the primary type of lung cancer globally. In the United States, 236,740 new cases and 350 deaths per day are reported for lung cancer, which is still the leading cause of cancer death in 2022 [1]. The good news is that the incidence and the mortality of NSCLC are decreasing because of the increasing knowledge on potential risk factors [2]. According to the stage of NSCLC, the treatments range from pure surgery, chemo-radiotherapy, immunotherapy, and targeted drug therapy [3]. The adjuvant chemo-radiotherapy is the most traditional and common method for NSCLC therapy in different stages [4,5]. However, effective therapeutic approaches for NSCLC remain to be developed.

Hundreds of *N*-heterocyclic indolyl glyoxylamides have been synthesized and evaluated for their anticancer activity, including cancer cells of murine leukemia, human gastric, breast, and uterus sources [6]. These compounds are structurally analogous to N-(pyridin-4-yl)-[1-(4-chlorbenzyl)-indol-3-yl]-glyoxyl-amid (D-24851) that functions as a microtubule inhibitor containing oral antitumor activity in vivo, as well as multidrug-resistance tumor cells [7]. D-24851 has been reported to induce apoptosis in malignant glioma cells without active p53, a well-known tumor suppressor protein [8]. Interestingly, two compounds (BPR0C123 and BPR0C259) of *N*-heterocyclic indolyl glyoxylamides exhibited different cytotoxicity on human NSCLC cell lines, but also induced p53-independent apoptosis and radiosensitivity to different levels [9]. BPR0C261 is identified as the most cytotoxic molecules against a broad spectrum of mammalian cancer cells in this series of compounds [6]. The antitumor actions of BPR0C261 include antimitosis and anti-angiogenesis in vivo [10]. Because BPR0C261 can also be administrated orally and extend the lifespan of tumor-bearing mice, it is interesting to be considered in clinical applications [10]. Whether BPR0C261 also induces cytotoxicity and radiosensitivity in NSCLC with or without p53 tumor suppressor gene is still unclear.

The homozygous mutation of the p53 tumor suppressor gene is engaged in 50–60% of human cancers [11]. The biochemical role of the p53 gene is to encode a transcription factor and transactivate downstream genes for apoptosis, DNA repair, senescence, and cell cycle arrest [12]. A recent report focused on the crosstalk between p53 and immunity and demonstrated that mutant p53 could suppress innate immunity to promote tumorigenesis [13]. In addition, p53 can transcriptionally suppress the expression of vascular endothelial growth factor (VEGF) and induce the production of arrestin to inhibit angiogenesis in human tumor [14,15]. Although p53 is known to be important for the induction of apoptosis in cancer cells exposed to genotoxicity, the p53-independent pathways are also frequently reported to mediate drug-induced apoptosis [16–18]. For instance, the PTEN (phosphatase and tensin homolog) tumor suppressor has been reported to be involved in p53-independnet apoptosis [19]. PTEN is a negative regulator of the phosphatidylinositol -3 kinase (PI3K)/Akt/mTOR pathway that phosphorylates phosphatidylinositol (4,5)-trisphosphate ($PIP_2$) to $PIP_3$ to trigger a series of signal transduction and promote cell survival, cell cycle progression, cell growth, and angiogenesis [20,21]. Interestingly, the promoter of the PTEN gene harbors a p53-binding element that can be directly regulated by wild-type p53, although PTEN is constitutively expressed through a p53-independnet element [22]. PTEN mutations are mainly discovered in endometrial carcinomas and glioblastomas and are associated with 13.5% of human cancers [23,24]. PTEN can inhibit the cell cycle progression and promote apoptosis in NSCLC cells [25]. Moreover, a combination of radiation and paclitaxel has been reported to trigger a PTEN–PI3K–Akt–Bax signaling cascade in NSCLC xenograft tumors and suppress tumor growth in the absence of functional p53 [26]. However, the effect of BPR0C261 on the expression of p53 and/or PTEN tumor suppressors in NSCLC cells remains to be investigated.

In this study, we combine BPR0C261 and radiation for the treatment of NSCLC cells and compare the results with the individual effect of both methods. The cell cycle, DNA damage, radiosensitivity, and cell apoptosis are measured and evaluated. We also demonstrate that p53 expression is significant in NSCLC cells containing wild-type p53, while PTEN becomes dominant in those with deleted p53. The association of tumor suppressors with the BPR0C261-enhanced radiosensitivity is discussed.

## 2. Results

### 2.1. Effects of BPR0C261 on NSCLC Cells

First, NSCLC cells were treated with increased concentrations of BPR0C261 and the drug effects on cells were detected by morphological change and the MTT assay. Starting from 0.01 μM BPR0C261, an increase of round-up cells could be detected in A549 cells and H1299 cells exposed to increased concentrations of the drug for 12 h (Figure 1A). We also measured the $IC_{50}$ of BPR0C261 by a dose-dependent experiment and showed that A549 cells were more sensitive to the drug than H1299 cells (Figure 1B). The $IC_{50}$ of BPR0C261 on A549 cells and H1299 cells were 0.38 μM and 0.86 μM, respectively.

### 2.2. Effects of BPR0C261 on the Redistribution of Cell Cycle

We next examined the cell cycle of A549 cells and H1299 cells treated with BPR0C261. Through the flow cytometry, the percentage of cells at each phase after BPR0C261 treatment was indicated from 6 h to 24 h. The results showed that BPR0C261 caused significant $G_2$/M phase arrest in both A549 cells and H1299 using individual $IC_{50}$ (Figure 2A,B). Notably, A549 cells harbored wild-type p53 while H1299 cells contained null-p53, and this was associated with different patterns of cell cycle progression treated with BPR0C261 for 3 h to 6 h. That is, the $G_2$/M phase arrest of H1299 cells happened three hours earlier than that of A549 cells treated with BPR0C261. Nevertheless, the highest G2/M phase arrest was still detected in both cell lines after 12 h of treatment. The percentage of each phase of the cell cycle was also quantified in both cell lines treated with the corresponding $IC_{50}$ of BPR0C261 (Figure 2C,D).

### 2.3. Effects of High Concentration of BPR0C261 on Induction of Sub-G1 Population in NSCLC Cells

We next examined cell cycle redistribution including the percentage of sub-G1 phase of NSCLC cells treated with BPR0C261 at 0.1 M and 1 M that were below and over $IC_{50}$, respectively. Compared to 1 M BPR0C261, 0.1 M BPR0C261 did not induce significant percentage of sub-G1 phase in A549 cells after 24 h of treatment, although both concentrations still induced a significant G2/M phase arrest after 12 h of treatment (Figure 3A). A similar phenomenon was also detected in H1299 cells treated with BPR0C261, but a low concentration of BPR0C261 appeared to induce significant sub-G1 phase compared to untreated control (Figure 3B). The results of DNA histograms of A549 cells and H1299 cells treated with different concentrations of BPR0C261 were quantified (Figure 3C,D).

### 2.4. The Combination Treatment of Radiation and BPR0C261 Decreased the Survival Fraction in NSCLC Cells

The surviving fraction of A549 and H1299 were evaluated after a treatment with BPR0C261 or radiation alone and then the combination of the two treatments using the colony formation assay. The concentrations of BPR0C261 used were 0.1 μM and 1 μM, which are lower or higher than the $IC_{50}$, respectively. Compared to the radiation treatment alone, BPR0C261 enhanced the radiosensitivity of both A549 cells and H1299 cells (Figure 4 and Supplementary Material S1). Intuitively, the higher the concentration of BPR0C261, the lower the cell survival rate. Using the method of Valeriote and Carpentier (see Materials and Methods), the combination of BPR0C261 and radiation showed a synergistic effect on NSCLC cells compared to the treatment with radiation alone (Tables 1 and 2).

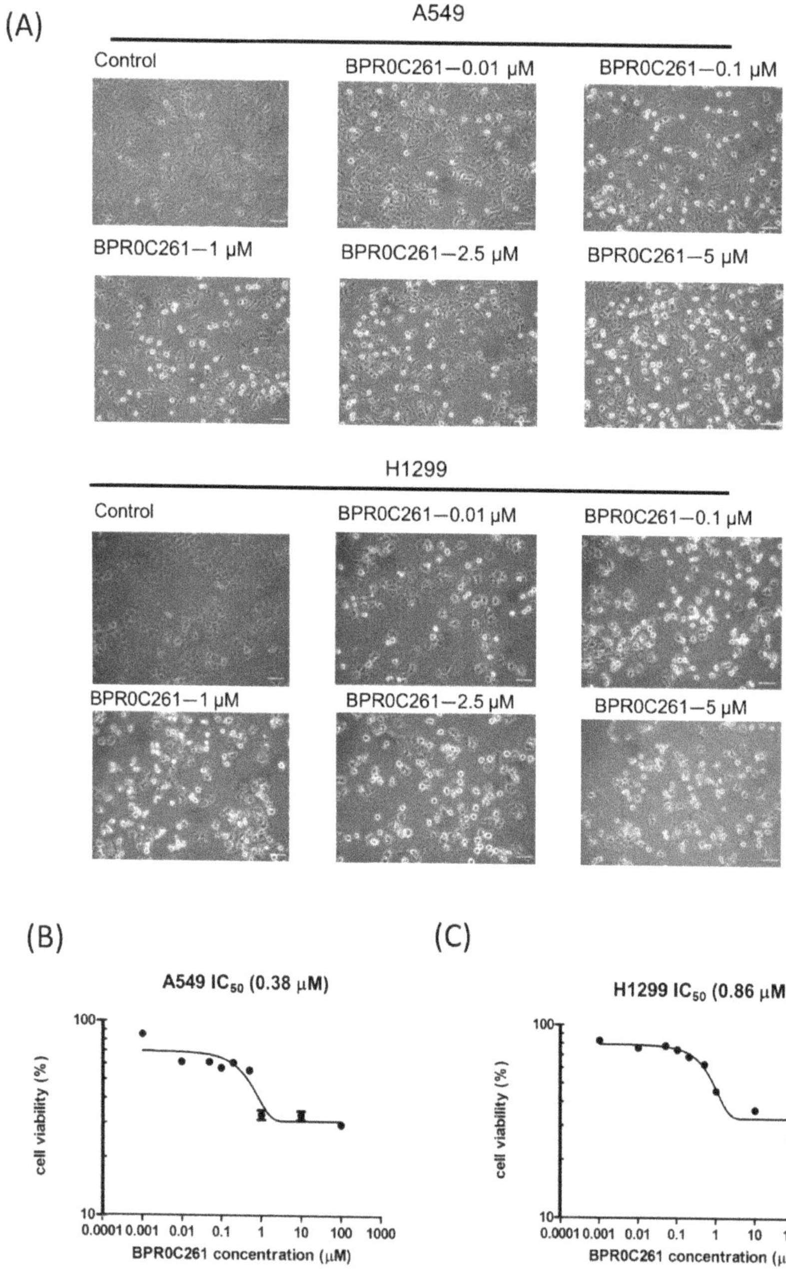


**Figure 1.** Cytotoxic effects of BPR0C261 on NSCLC cells. (**A**) Microscopic visualization of cell morphological changes by BPR0C261. Magnification: 40×; scale bar: 100 m. (**B**,**C**) The MTT assay for estimation of $IC_{50}$ of BPR0C261 on A549 cells and H1299 cells after 24 h of treatment, respectively.

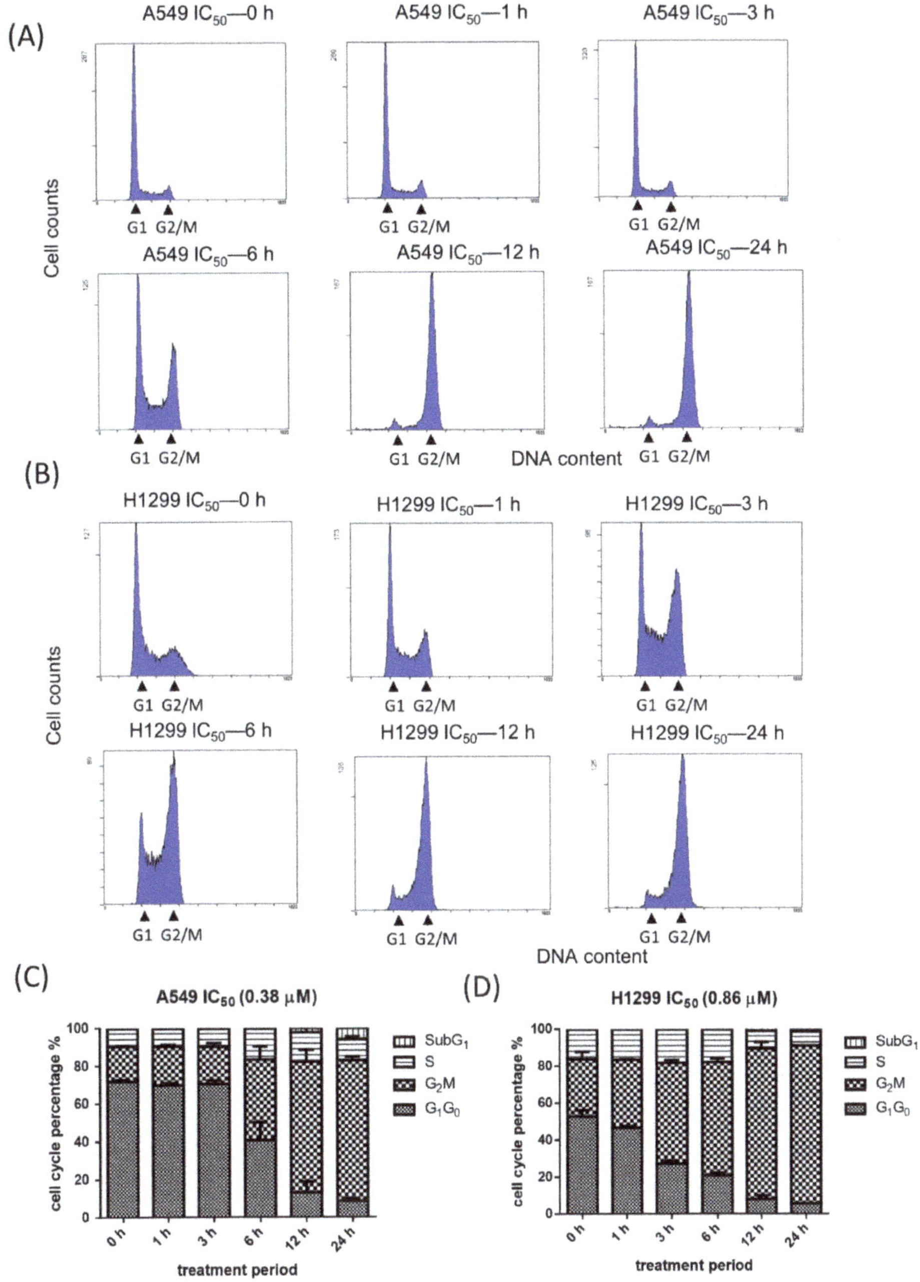


**Figure 2.** Analysis of cell cycle distribution by flow cytometry after NSCLC cells were treated with BPR0C261. (**A**) A549 cells and (**B**) H1299 cells were treated with their corresponding $IC_{50}$ of BPR0C261 with a time-dependent cell cycle redistribution. (**C**,**D**) Quantification of percentage of each cell cycle phase in A549 cells and H1299 cells, respectively. Each datum represents the mean of three independent experiments ± S.D.

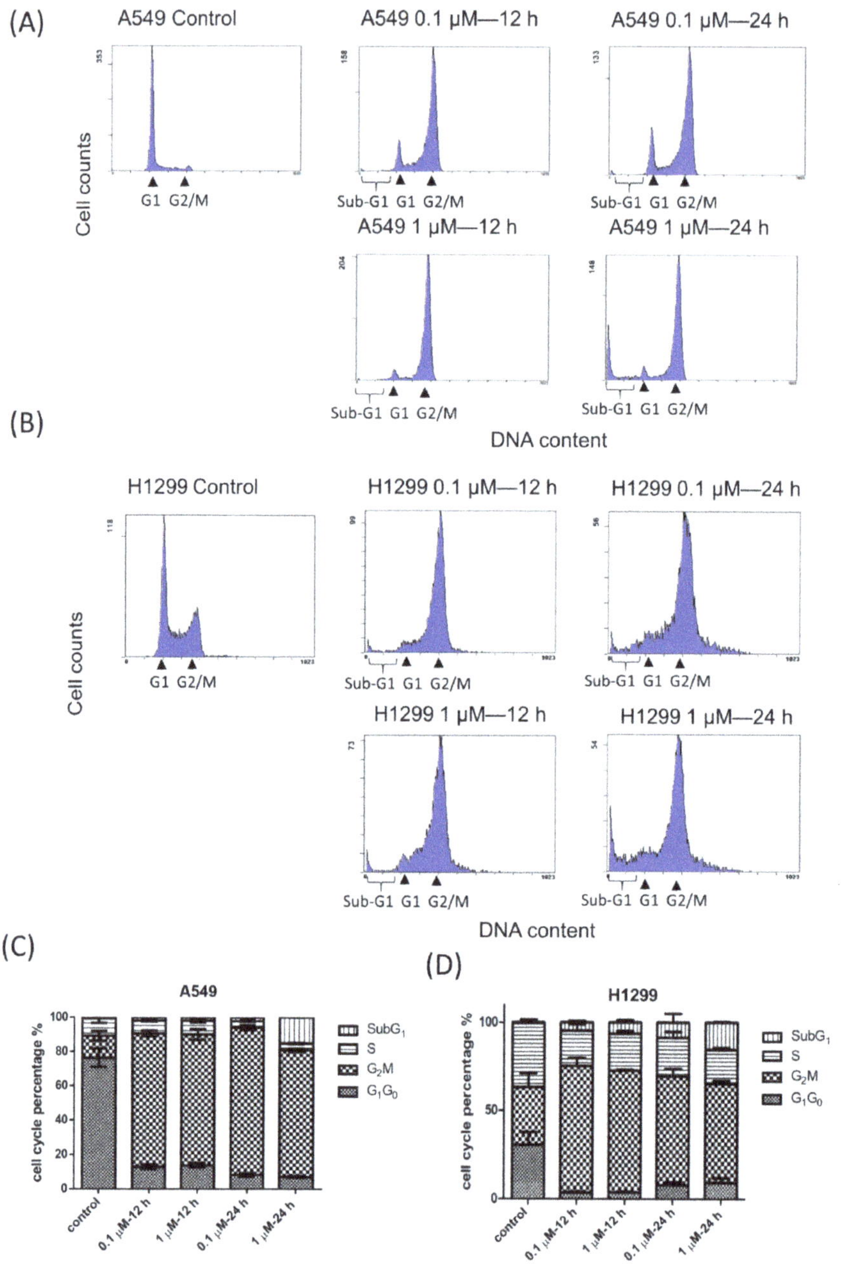


**Figure 3.** Comparison of sub-G1 phase of NSCLC cells treated with high and low concentration of BPR0C261. The sub-G1 phase of (**A**) A549 cells and (**B**) H1299 cells treated with 0.1 M and 1 M of BPR0C261 for 12 h and 24 h. (**C**,**D**) Quantification of percentage of sub-G1 phase in A549 cells and H1299 cells treated with BPR0C261, respectively. Each datum represents the mean of three independent experiments ± S.D.

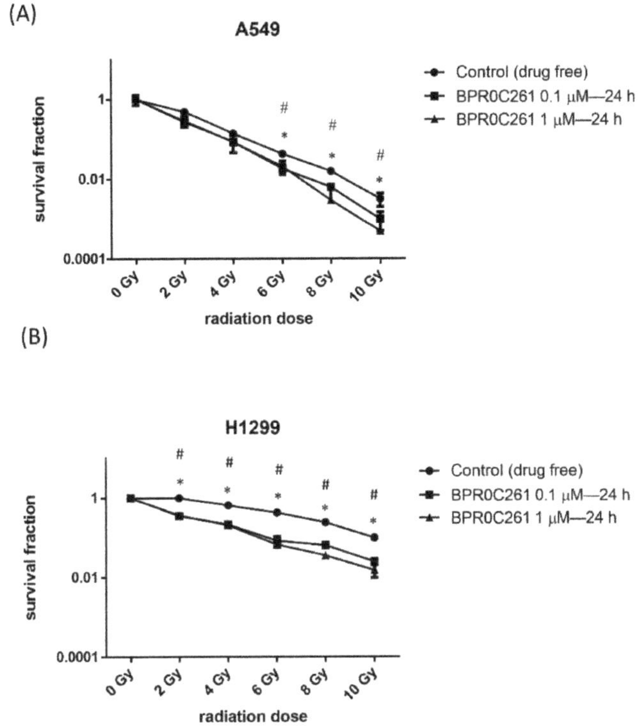


**Figure 4.** Effect of BPR0C261 on radiosensitivity of NSCLC cells. The survival fraction of (**A**) A549 cells and (**B**) H1299 cells exposed to different doses of X-rays after treatment with BPR0C261. *: $p < 0.05$ for 0.1 μM BPR0C261 treatment compared to radiation alone. #: $p < 0.05$ for 1 μM BPR0C261 treatment compared to radiation alone.

**Table 1.** Effect of BPR0C261 combining different doses of X-rays on A549 cells.

| The $SF_D$ of BPR0C261 0.1 μM = 0.368 | | | |
|---|---|---|---|
| **Radiation (Gy)** | $SF_R$ | $SF_{R+D}$ | $SF_R \times SF_D$ |
| 2 Gy | 0.4824 | 0.0822 (synergism) | 0.1774 |
| 4 Gy | 0.1394 | 0.0264 (synergism) | 0.0513 |
| 6 Gy | 0.0433 | 0.0056 (synergism) | 0.0159 |
| 8 Gy | 0.0159 | 0.0020 (synergism) | 0.0058 |
| 10 Gy | 0.0032 | 0.0003 (synergism) | 0.0012 |
| **The $SF_D$ of BPR0C261 1 μM = 0.305** | | | |
| **Radiation (Gy)** | $SF_R$ | $SF_{R+D}$ | $SF_R \times SF_D$ |
| 2 Gy | 0.4824 | 0.1057 (synergism) | 0.1471 |
| 4 Gy | 0.1394 | 0.0313 (synergism) | 0.0425 |
| 6 Gy | 0.0433 | 0.0076 (synergism) | 0.0132 |
| 8 Gy | 0.0159 | 0.0011 (synergism) | 0.0048 |
| 10 Gy | 0.0032 | 0.0002 (synergism) | 0.0010 |

**Table 2.** Effect of BPR0C261 combining different doses of X-rays on H1299 cells.

| | The $SF_D$ of BPR0C261 0.1 μM = 0.504 | | |
|---|---|---|---|
| **Radiation (Gy)** | **$SF_R$** | **$SF_{R+D}$** | **$SF_R \times SF_D$** |
| 2 Gy | 0.9917 | 0.3326 (synergism) | 0.4998 |
| 4 Gy | 0.6482 | 0.2123 (synergism) | 0.3267 |
| 6 Gy | 0.4202 | 0.0826 (synergism) | 0.2118 |
| 8 Gy | 0.2375 | 0.0640 (synergism) | 0.1197 |
| 10 Gy | 0.0983 | 0.0249 (synergism) | 0.0495 |
| | **The $SF_D$ of BPR0C261 1 μM = 0.412** | | |
| **Radiation (Gy)** | **$SF_R$** | **$SF_{R+D}$** | **$SF_R \times SF_D$** |
| 2 Gy | 0.9917 | 0.3555 (synergism) | 0.4085 |
| 4 Gy | 0.6482 | 0.2038 (synergism) | 0.2670 |
| 6 Gy | 0.4202 | 0.0665 (synergism) | 0.1731 |
| 8 Gy | 0.2375 | 0.0358 (synergism) | 0.0978 |
| 10 Gy | 0.0983 | 0.0152 (synergism) | 0.0405 |

### 2.5. Effects of BPR0C261 and Radiation on Induction of DNA Damage in NSCLC Cells

To determine the DNA damage in NSCLC cells treated with BPR0C261 and/or radiation, single-cell gel electrophoresis (comet assay) was exploited for the analysis. The quantification of DNA damage by the comet assay was accomplished by calculating the tail moment. The radiation doses ranging from 2 Gy to 10 Gy showed a dose-dependent increase of DNA damage in A549 cells and H1299 cells (Figure 5A). Surprisingly, the same effect was detected in cells treated with BPR0C261 from 0.01 M to 1 M for 24 h (Figure 5B). We further treated cells with 0.1 M BPR0C261 for 24 h followed by 2 Gy of X-rays exposure. The results showed that this combination induced higher tail moments than the single treatment with BPR0C261 or X-rays on both NSCLC cell lines (Figure 5C). These data indicated that a low concentration of BPR0C261 enhanced the efficacy of a therapeutic radiation dose (2 Gy) for inducing DNA damage.

### 2.6. The Expression of p53 and PTEN in NSCLC Cells Treated with BPR0C261 and Radiation

As BPR0C261 induced cell cycle arrest and DNA damage in NSCLC cells with or without p53, we next examined the expression of p53 and PTEN mentioned above after cells were treated with BPR0C261 and/or radiation. We first irradiated A549 cells (p53+/+) and H1299 cells (p53-null) from 2 Gy to 10 Gy and showed that p53 was significantly upregulated in A549 cells after radiation exposure, and PTEN was also upregulated (Figure 6A). On the other hand, PTEN expression was downregulated and phosphorylated Akt (p-Akt) was upregulated in p53-null H1299 cells after irradiation over 8 Gy (Figure 6A). We next treated NSCLC cells with BPR0C261 from 0.01 M to 5 M for 24 h. The results showed that BPR0C261 could efficiently upregulate p53 but not PTEN, and it could suppress the expression of Akt protein and p-Akt in A549 cells (Figure 6B). In H1299 cells, PTEN could be upregulated by BPR0C261 but Akt and p-Akt were not affected (Figure 6B). The drug treatment or radiation alone did not dramatically reduce the ratios of p-Akt/Akt. Furthermore, a combination of 0.1 M BPR0C261 and incremental doses of X-rays showed that p53 and PTEN were upregulated in A549 cells and H1299 cells, respectively (Figure 6C). The expression of Akt protein was not significantly affected by the cotreatment of BPR0C261 and radiation. Compared to untreated control, however, the p-Akt levels of A549 cells were more affected than those of H1299 cells after cotreatment of BPR0C261 and radiation.

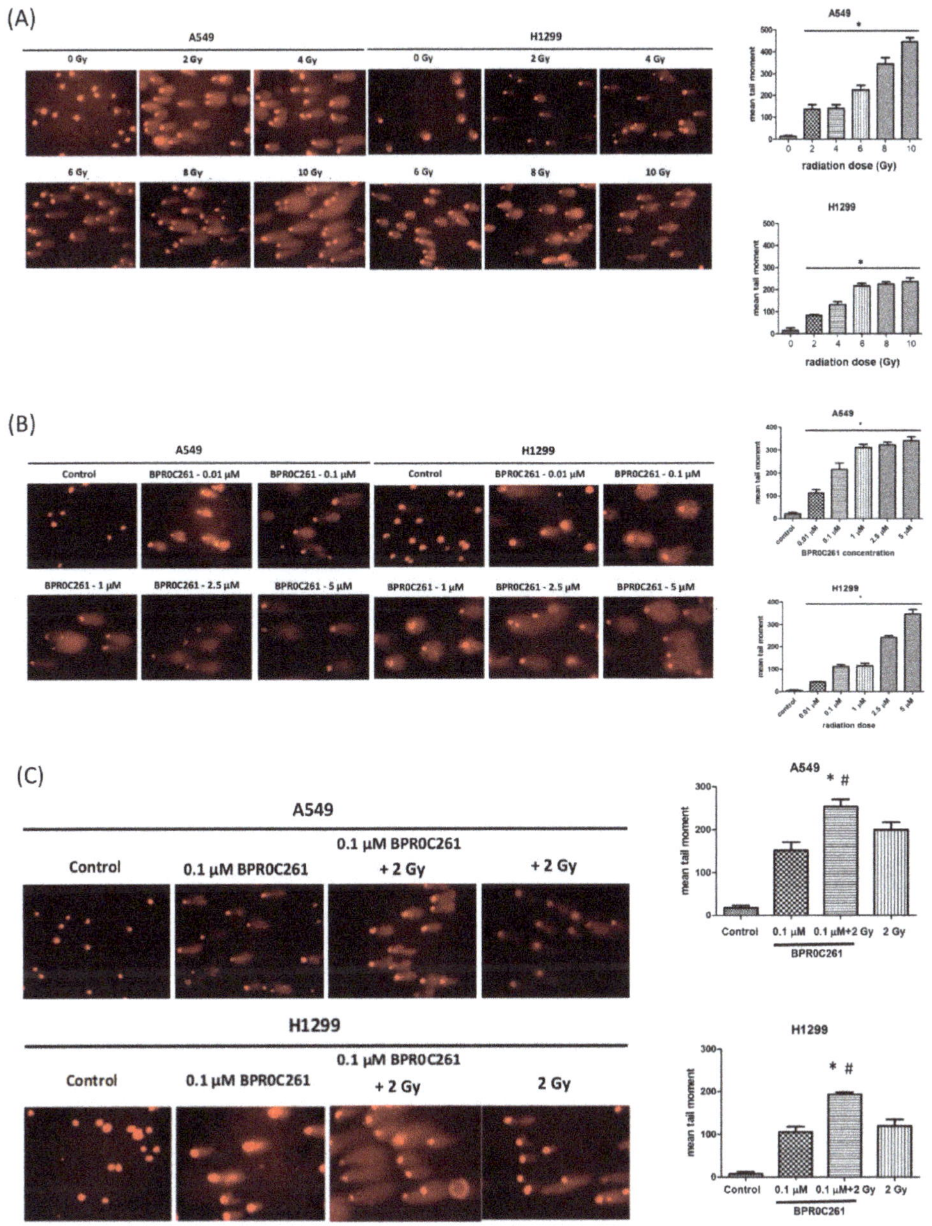


**Figure 5.** Analysis of DNA damage in NSCLC cells using the comet assay. The images of DNA damage in A549 cells and H1299 cells exposed to (**A**) different doses of X-rays and (**B**) increased concentrations of BPR0C261. The level of DNA damage was quantified by measuring the tail moments. (**C**) Comparison of DNA damage induced by a combination of low concentration of BPR0C261 and 2 Gy X-rays, and individual treatment alone. The microscopic magnification for visualization of comets were 20×. *: $p < 0.05$ compared with BPR0C261 alone. #: $p < 0.05$ compared with radiation alone.

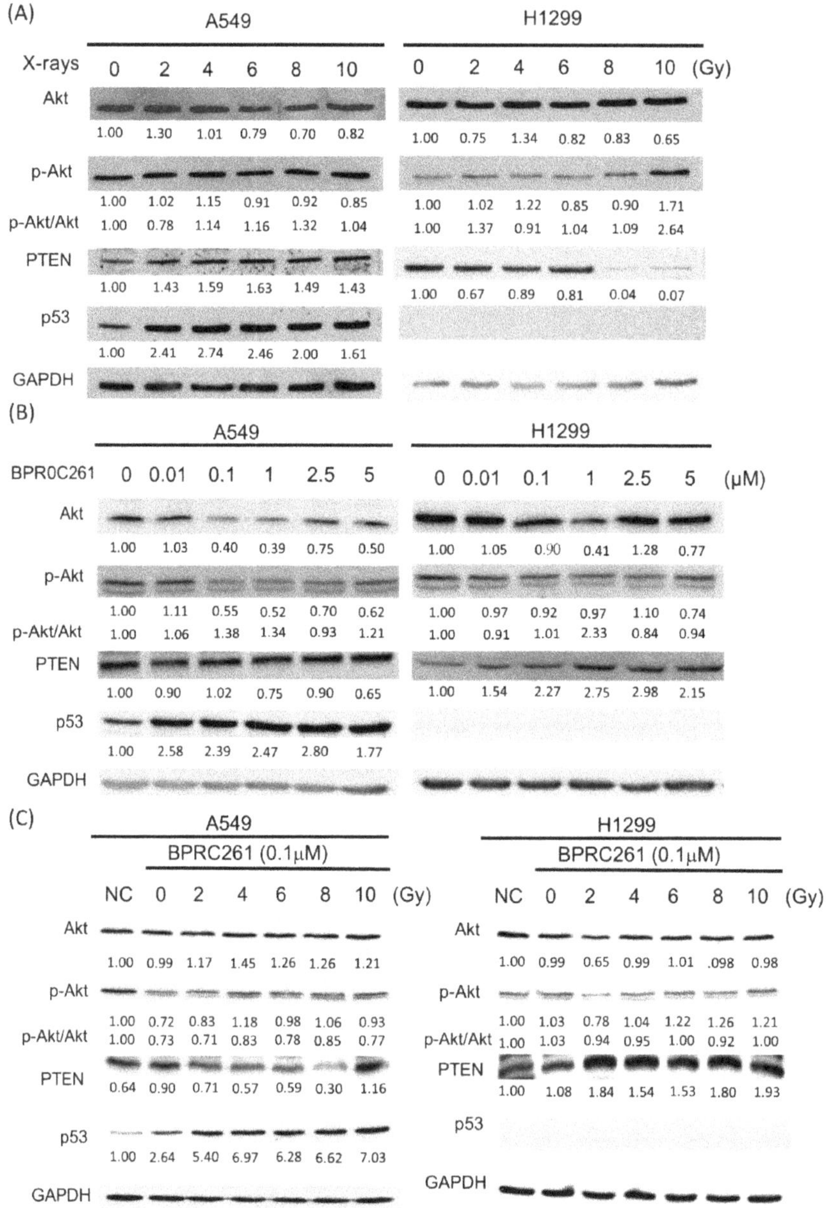

**Figure 6.** Expression of p53 and PTEN in NSCLC cells treated with BPR0C261 and radiation. (**A**) Western blot analysis of Akt, p-Akt, PTEN, and p53 in A549 cells and H1299 cells exposed to different doses of X-rays. (**B**) Same as (**A**) but treated with different concentrations of BPR0C261. (**C**) A combination of BPR0C261 and X-rays on the expression of proteins described above. NC is the negative control without treatment. The blots of Akt, pAkt, PTEN, and p53 were normalized to that of GAPDH, and then compared to untreated control or NC to determine the ratio of each treatment. The results are from three independent experiments.

## 3. Discussion

It has been reported that human NSCLC cells are more resistant to ionizing radiation than small cell lung cancer (SCLC) cells in vitro [27]. Additionally, NSCLC accounts for 80–85% of human lung cancer [28]. Therefore, the development of effective and biocompatible radiosensitizers is still ongoing for the treatment of NSCLC.

In a previous study, we demonstrated that BPR0C123 and BPR0C259 could induce $G_2/M$ phase arrest in A549 and H1299 cells. However, the significant enhancement of radiosensitivity by these two compounds was only found in NSCLC cells exposed to 10 Gy -rays [9]. Because 2 Gy is routinely used for fractionation radiotherapy in clinics, it is expected to have a radiosensitizer that can raise the effect at low-dose radiation. In this study, we showed that BPR0C261, a well-studied D-24851 derivative [10], could enhance the radiosensitivity at lower radiation doses, especially the p53-null NSCLC cells. Moreover, a low concentration (0.1 M) and high concentration (1 M) of BPR0C261 induced similar radiosensitive effects. This is important because 0.1 M BPR0C261 induced very low cytotoxicity, which is a required criterion for an ideal radiosensitizer [29]. Notably, according to the pattern of survival curves shown in NSCLC cells treated with BPR0C261, it seemed that H1299 cells exhibited stronger responses to X-rays than A549 cells. Because the intrinsic radiosensitivity of A549 cells is higher than that of H1299 cells, this is most likely due to the presence of wild-type p53 that contributes to the radiosensitivity of lung cancer [30].

BPR0C261 is designed to target microtubules and cause antimitotic effect in cancer cells [10]. Thus, it is no doubt that cells treated with this drug will lead to G2/M phase arrest, although the level may be different. The rationale of BPR0C261-induced radiosensitivity is mainly based on the fact that the $G_2/M$ phase is the most sensitive phase to radiation [31]. BPR0C261 induced $G_2/M$ arrest in A549 and H1299 cells, but H1299 cells exhibited an earlier accumulation of $G_2/M$ phase compared to A549 cells. This phenomenon may also be associated with the p53 gene. When cells are exposed to genotoxic agents, p53-dependent G1 phase arrest will be triggered via the transactivation of the p21 gene, a cyclin-dependent kinase inhibitor (CKI) [32,33]. It is speculated that p53 would be activated by BPR0C261 to induce the G1 phase arrest prior to the $G_2/M$ phase arrest in A549 cells. On the other hand, the p53-null H1299 cells entered the G2/M phase at an earlier time point after cells were treated with BPR0C261. Although p53 is also demonstrated to be important for G2/M phase arrest in normal human fibroblasts [34], BPR0C261-induced G2/M phase arrest does not depend on the p53 status. Therefore, it may further explain the p53-independent induction of radiosensitivity by this compound.

It has been reported that fractional DNA content is a characteristic of apoptosis that can be recognized in a DNA histogram, namely a sub-G1 or hypodiploid subpopulation detected by flow cytometry [35,36]. Because the sub-G1 phase is believed to be composed of apoptotic cells and necrotic cells, it would be interesting to use other apoptotic and necrotic specific biomarkers to evaluate the types of cell death induced by BPR0C261 in the future [37–39]. BPR0C261 has been reported to induce apoptosis by observing the increase of DNA fragmentation and positive TUNEL signals in gastric MKN-45 cells [10]. The analysis of the sub-G1 subpopulation in NSCLC cells suggests that current data agree with previous report that BPR0C261 may also induce apoptosis in this cancer type. Although p53 is important for inducing apoptosis [40], BPR0C261 could increase the percentage of the sub-G1 phase in both p53-positive and p53-negative NSCLC cells. Therefore, other cell death mechanisms should also be involved in BPR0C261-mediated cytotoxicity. Notably, the induction of the sub-G1 phase in NSCLC cells was more robust using a high concentration of BPR0C261. However, it could induce similar levels of G2/M phase arrest at low and high concentrations, suggesting that a low concentration of BPR0C261 may be biocompatible for combining with other therapeutic methods.

The synergistic effect is another important criterion to evaluate an ideal radiosensitizer. For instance, tirapazamine conjugated to gold nanoparticles exhibited a synergistic radiosensitizing effect on human hepatoma HepG2 cells [41]. Codrug-loaded nanoparticles has also been reported to improve the synergistic therapeutic efficacy of chemoradiotherapy [42]. It is interesting to combine this compound with nanoparticles which are usually used for drug carriers and tracking, to passively target tumor in vivo through the enhanced permeability and retention (EPR) effect [43,44]. In the present study, BPR0C261 also performed synergistic radiosensitizing effects on p53-wild-type A549 cells and p53-null H1299 cells based on the method of Valeriote and Carpentier [45,46]. Drug-induced PTEN in p53-null NSCLC cells may be important for this phenomenon as PTEN mutation has recently been reported to decrease radiosensitivity in NSCLC cells [47]. An improvement of the synergistic radiosensitizing effect may be also expected by combining BPR0C261 and nanoparticles for in vivo treatment in the future.

Radiation is known to induce DNA damage, but whether D-24851-related derivatives could influence the integrity of DNA has been little studied. Although a low concentration of BPR0C261 (0.1 M) did not cause significant cytotoxicity, the comet assay showed that this treatment could increase the tail moment in this assay. Actually, the use of 0.01 M BPR0C261 was already sufficient to induce DNA damage. The induction of DNA damage by BPRC261 in both NSCLC cell lines was dose-dependent. Because of this effect, a low concentration of BPR0C261 also increased the clinical dose (2 Gy) radiation-induced DNA damage. The impairment of microtubule dynamics by paclitaxel, a microtubule stabilizer, has been reported to induce DNA damage [48]. To the best of our knowledge, this is the first report showing that BPR0C261, a microtubule inhibitor containing antitumor activity also induces DNA damage. It may explain the role of BPR0C261 as a potent radiosensitizer, and the underlying mechanism would be interesting to be further investigated.

The molecular mechanisms disturbed by BPR0C261 to influence the radiosensitivity and DNA damage have not been fully investigated. We started from two common tumor suppressor genes p53 and PTEN and showed that they could account for the p53-dependent and the p53-independent pathways raised by BPR0C261. However, current data only demonstrated the upregulation of the p53 protein in A549 cells treated with radiation and/or BPR0C261. The activation and function of p53 are mainly determined by post-translational modifications [49]. It would be important to investigate the mechanisms of p53 activation by BPR0C261-mediated radiation responses in the future. As mentioned, we have demonstrated that another two BPR0C derivatives exhibited p53-independent apoptosis [9]. However, the molecules responsible for this phenomenon remain unknown. In this study, PTEN may represent the tumor suppressor gene in p53-independent pathway after cells were treated by BPR0C261, suggesting that PTEN may also play a role in other BPR0C series compounds. Because somatic PTEN mutation is less found in NSCLC [50,51], the induction of PTEN by BPR0C261 may be an important mechanism for this compound to be applied in various lung cancers. We also found that the level of PTEN in p53-null H1299 cells were suppressed by a high dose of radiation (>8 Gy). Because PTEN is a direct downstream target gene transactivated by p53 [22], it would be interesting to further investigate if the maintenance of the PTEN level at a high dose radiation is associated with p53 status. The counterpart of PTEN, named Akt oncoprotein, was also found to be activated by a high dose of radiation. This is consistent with previous reports that radiation can activate the Akt pathway in different cancer types [52,53]. Although the combination of BPR0C261 and radiation induced, respectively, p53 and PTEN in A549 cells and H1299 cells, the expression and activity of Akt was not induced in both cell types compared to untreated control. However, whether the activity of Akt and related signaling are associated with BPR0C261-modulated radiation responses would be interesting to further investigate in the future.

Taken together, BPR0C261 not only decreased cell viability but also induced G2/M phase arrest and DNA damage with a dose-dependent manner in NSCLC cells. BPR0C261 is regarded as a potent radiosensitizer because it showed a synergistic radiosensitizing effect when combining with conventional dose (2 Gy) for fractionation radiotherapy. The underlying mechanism of BPR0C261-induced radiosensitivity is associated with p53 and PTEN molecules in p53-dependent and p53-independent pathways. Current data suggest that BPR0C261 may modulate the effect of radiation on the treatment of lung cancer cells in vitro. To evaluate the radiosensitizing role of BPR0C261 in potent clinical application, an animal study using a xenograft tumor model should be designed soon to confirm the radiosensitizing efficacy of BPR0C261 in vivo.

## 4. Materials and Methods

### 4.1. Cell Culture

Human A549 epithelial adenocarcinoma cells and human lymph node invaded H1299 non-small cell lung carcinoma cells were cultured in Dulbecco's Modified Eagle Medium (DMEM medium) supplemented with 10% fetal bovine serum (FBS), 1% penicillin–streptomycin solution (P/S), and 1% L-glutamine. Cells were incubated at 37 °C (5% $CO_2$ in air) and subcultured by trypsinization every 48 h.

### 4.2. Reagent

BPR0C261 with the chemical formula N-heterocyclic indolyl glyoxylamide N-(3-methyl-5-isothiazolyl)-2-1[(3-methyl-5-isoxazolyl-)methyl]-1H-3-indoyl-2-oxoacetamide was synthesized in the institute of Biotechnology and Pharmaceutical Research, National Health and Research Institute, Taiwan [10]. The stock solution (1 mM) of BPR0C261 was dissolved in dimethyl sulfoxide (DMSO). An appropriate volume of DMEM medium was added to dilute the stock solution for specific experiments.

### 4.3. Radiation Source

Radiation was delivered by the cabinet digital X-ray machine (RS 2000 Biological Research X-ray Irradiator; Rad Source Technologies, Inc., Suwanee, GA, USA) operating at 160 kVp and 25 mA. Cells were suspended in a T-25 flask and exposed to radiation at a dose rate of 38.37 mGy per second.

### 4.4. Cell Viability Assay

For cell viability assays, $4.5 \times 10^3$ cells were seeded in each well of a 96-well plate. Different concentrations of BPR0C261 were used to treat cells for 24 h. Subsequently, 0.5 mM 3-(4,5-Dimethylthiazol-2-yl)-2,5-diphenyltetrazolium bromide (MTT) dissolved in serum free DMEM was added (100 μL/well) into each well and incubated at 37 °C for 3 h. DMSO was then used to dissolve the insoluble purple formazan crystals. The absorbance of this solution was quantified at the wavelength of 570 nm by a Tecan's Sunrise absorbance microplate reader (TECAN Group Ltd., Männedorf, Switzerland). The cell viability was determined by the following formula: (the optical density (OD value) of the experiments group/the OD value of control group) × 100%, and the control group was normalized to 100%. The $IC_{50}$ was calculated using Prism statistical software (Ver.5, GraphPad Software, San Diego, CA, USA). The change of cell viability was also visualized by a bright field microscope (Leica DM IRB, Wetzlar, Germany). The images were captured by a camera (Powershot A260, Canon U.S.A. Inc., New York, NY, USA).

### 4.5. Flow Cytometry Analysis

The cell cycle analysis were executed by following a previous report with slight modification [54]. In brief, A549 cells and H1299 cells ($1 \times 10^6$ each) were seeded in 10 cm dishes and exposed to radiation or drug. The cells were then rinsed, trypsinized, and collected by centrifugation at 600 rpm for 5 min. To fix the cells, 70% precooled ethanol was added to cell pellets and kept at 4 °C overnight. The cells were then collected and treated with RNase A for 30 min at room temperature. After centrifugation, the cells were resuspended in 20 μg/mL of propidium iodide (PI) and sieved through a 37 m mesh. The DNA histogram was analyzed using a Beckman Coulter Cytomic FC500 flow cytometer and its bundled software (Beckman Coulter, Inc., Bream CA, USA). The sub-G1, G1, S, and G2/M phases were individually gated for quantification.

## 4.6. Colony Formation Assay

The radiosensitivity was determined by calculating the survival fractions of radiation doses. After radiation exposure as mentioned above, the cells were counted and seeded into 6 cm culture dishes and incubated at 37 °C for 14 days. The colonies were stained with 1.25% crystal violet in 100% alcohol. A colony should contain at least 50 cells visualized under a bright field microscope. The formula for the plating efficiency (PE) was: PE = (number of colonies/number of cell seeded) × 100%. Survival fraction (SF) = PE of irradiated cells/PE of unirradiated cells. The survival fractions of drug-combining radiation was measured and assayed according to the method of Valeriote and Carpentier [45,46]. In Tables 1 and 2, $SF_R$ and $SF_D$ are the survival fractions of cells treated by radiation and BPR0C261, respectively. $SF_{R+D}$ is the survival fraction of cells treated with radiation and BPR0C261 at different concentrations. The value of $SF_{R+D}$ was compared to the corresponding $SF_R \times SF_D$, and the combination treatment effect was considered synergistic when $SF_{R+D} < SF_R \times SF_D$.

## 4.7. Single-Cell Gel Electrophoresis Assay (Comet Assay)

Irradiated cells ($1 \times 10^4$) were mixed with 75 L 0.75% low melting agarose (LAMDA Biotech Inc., St. Louis, MO, USA). It was loaded onto a slide covered by a spread of 1% normal agarose gel. After solidification, another 1% normal agarose gel was used to spread and cover on the low-melting agarose mixed cells. The prepared slide was immersed in the lysis solution (1% Triton X-100, 2.5 M NaCl, 0.1 M ethylenediaminetetraacetic acid (EDTA) and 10 mM Tris, pH = 10) at 4 °C for 1 h. Cells in the slide were subjected to electrophoresis in the alkaline electrophoresis buffer (300 mM NaOH, 1 mM EDTA and the pH value was adjusted to higher than 12) for 40 min. The condition of electrophoresis was 1 V/cm with a constant 300 mA. The gels were washed three times by the neutralization buffer (0.4 mM Tris, pH 7.5). Cells were then stained using 20 µg/mL of ethidium bromide (EtBr) as a nucleic acid staining reagent and 25 µL of 20 µg/mL EtBr was used and visualized under a fluorescence microscope (Leica DM IRB, Wetslar, Germany). The images were acquired using a digital camera (Powershot A260, Canon U.S.A. Inc.) and quantified by CometScore 2.0 software (Ver. 2.0, The TriTek Corporation, Fairfax, VA, USA). More than 50 cells were counted in each group.

## 4.8. Western Blot Analysis

The Western blot analysis was performed by following a previous report with slight modification [55]. In brief, 30–50 µg of protein lysates were run on the sodium dodecyl sulfate-polyacrylamide gel electrophoresis (SDS-PAGE), and electrotransferred to a nitrocellulose membrane for the detection of specific proteins by primary antibodies. The primary antibodies used in this study included anti-p53 antibody (GTX70214; GeneTex, Inc., Irvine, CA, USA), anti-PTEN antibody (Cat#9552; Cell Signaling Technology®, Danvers, MA, USA), anti-Akt antibody (sc-7126), phosphor-specific anti-Akt antibody, (sc-7985, Santa Cruz Biotechnology, Inc., Dallas, TX, USA), and anti-glyceraldehyde-3-phosphate dehydrogenase (GAPDH) antibody (MA5-15738; Thermo Fisher Scientific, Waltham, MA, USA). Enhanced chemiluminescent (ECL) agents (Bio-Rad Laboratories, Hercules, CA, USA) were used for imaging the membrane that was acquired by ImageQuant LAS4000 (GE Healthcare, Buckinghamshire, UK). The blots were quantified using the densitometry involved in ImageJ software (National Institutes of Health, Bethesda, MD, USA).

## 4.9. Statistical Analysis

Each datum is presented as means ± S.D. The statistical analysis was determined using a *t*-test. The statistical significance was defined as $p < 0.05$. Tables were drawn using GraphPad Prism statistical software (Ver.5, GraphPad Software, San Diego, CA, USA).

**Supplementary Materials:** The supporting information can be downloaded at: https://www.mdpi.com/article/10.3390/ijms232214083/s1.

**Author Contributions:** J.-D.L., C.-A.C. and Y.-J.L. were responsible for the conceptualization and funding acquisition. S.-T.L. was responsible for conducting the experiments and for the methodology. C.-T.C. provided the resources and acted as a consultant for the drug functions. Y.-J.L. wrote the manuscript and validated and visualized the data. All authors have read and agreed to the published version of the manuscript.

**Funding:** This study was funded by grants of Taipei City Hospital, RenAi Branch (TPECH-111-12), the Ministry of Science and Technology of Taiwan (109-2314-B-010-021-MY3), and the Cancer Progression Research Center, National Yang Ming Chiao Tung University from The Featured Areas Research Center Program within the framework of the Higher Education Sprout Project by the Ministry of Education (MOE) in Taiwan (111W31405 and 111W31101).

**Institutional Review Board Statement:** Not applicable.

**Informed Consent Statement:** Not applicable.

**Data Availability Statement:** The data presented in this study are available in Supplementary Materials.

**Acknowledgments:** We thank Abigail Lin for editing part of the English writing. We thank Fang-Yu Lee's assistance in data collection. We also thank Ya-Huei Wang for administrative works of this research.

**Conflicts of Interest:** The authors declare no conflict of interest.

## References

1. Siegel, R.L.; Miller, K.D.; Fuchs, H.E.; Jemal, A. Cancer statistics, 2022. *CA Cancer J. Clin.* **2022**, *72*, 7–33. [CrossRef]
2. Wong, M.C.S.; Lao, X.Q.; Ho, K.F.; Goggins, W.B.; Tse, S.L.A. Incidence and mortality of lung cancer: Global trends and association with socioeconomic status. *Sci. Rep.* **2017**, *7*, 14300. [CrossRef] [PubMed]
3. Zappa, C.; Mousa, S.A. Non-small cell lung cancer: Current treatment and future advances. *Transl. Lung Cancer Res.* **2016**, *5*, 288–300. [CrossRef] [PubMed]
4. Conibear, J.; AstraZeneca, U.K.L. Rationale for concurrent chemoradiotherapy for patients with stage III non-small-cell lung cancer. *Br. J. Cancer* **2020**, *123* (Suppl. S1), 10–17. [CrossRef] [PubMed]
5. Ohe, Y. Chemoradiotherapy for lung cancer: Current status and perspectives. *Int. J. Clin. Oncol.* **2004**, *9*, 435–443. [CrossRef]
6. Li, W.T.; Hwang, D.R.; Chen, C.P.; Shen, C.W.; Huang, C.L.; Chen, T.W.; Lin, C.H.; Chang, Y.L.; Chang, Y.Y.; Lo, Y.K.; et al. Synthesis and biological evaluation of N-heterocyclic indolyl glyoxylamides as orally active anticancer agents. *J. Med. Chem.* **2003**, *46*, 1706–1715. [CrossRef]
7. Bacher, G.; Nickel, B.; Emig, P.; Vanhoefer, U.; Seeber, S.; Shandra, A.; Klenner, T.; Beckers, T. D-24851, a novel synthetic microtubule inhibitor, exerts curative antitumoral activity in vivo, shows efficacy toward multidrug-resistant tumor cells, and lacks neurotoxicity. *Cancer Res.* **2001**, *61*, 392–399.
8. Ito, H.; Kanzawa, T.; Kondo, S.; Kondo, Y. Microtubule inhibitor D-24851 induces p53-independent apoptotic cell death in malignant glioma cells through Bcl-2 phosphorylation and Bax translocation. *Int. J. Oncol.* **2005**, *26*, 589–596. [CrossRef]
9. Huang, T.H.; Chiu, S.J.; Chiang, P.H.; Chiou, S.H.; Li, W.T.; Chen, C.T.; Chang, C.A.; Chen, J.C.; Lee, Y.J. Antiproliferative effects of N-heterocyclic indolyl glyoxylamide derivatives on human lung cancer cells. *Anticancer Res.* **2011**, *31*, 3407–3415.
10. Hu, C.B.; Chen, C.P.; Yeh, T.K.; Song, J.S.; Chang, C.Y.; Chuu, J.J.; Tung, F.F.; Ho, P.Y.; Chen, T.W.; Lin, C.H.; et al. BPR0C261 is a novel orally active antitumor agent with antimitotic and anti-angiogenic activities. *Cancer Sci.* **2011**, *102*, 182–191. [CrossRef]
11. Baugh, E.H.; Ke, H.; Levine, A.J.; Bonneau, R.A.; Chan, C.S. Why are there hotspot mutations in the TP53 gene in human cancers? *Cell. Death Differ.* **2018**, *25*, 154–160. [CrossRef] [PubMed]
12. Gilbert, S.F. *Developmental Biology*, 9th ed.; Sinauer Associates: Sunderland, MA, USA, 2010; 80p, ISBN 9780878933846.
13. Ghosh, M.; Saha, S.; Bettke, J.; Nagar, R.; Parrales, A.; Iwakuma, T.; van der Velden, A.W.M.; Martinez, L.A. Mutant p53 suppresses innate immune signaling to promote tumorigenesis. *Cancer Cell.* **2021**, *39*, 494–508.e5. [CrossRef] [PubMed]
14. Assadian, S.; El-Assaad, W.; Wang, X.Q.; Gannon, P.O.; Barres, V.; Latour, M.; Mes-Masson, A.M.; Saad, F.; Sado, Y.; Dostie, J.; et al. p53 inhibits angiogenesis by inducing the production of Arresten. *Cancer Res.* **2012**, *72*, 1270–1279. [CrossRef] [PubMed]
15. Zhang, L.; Yu, D.; Hu, M.; Xiong, S.; Lang, A.; Ellis, L.M.; Pollock, R.E. Wild-type p53 suppresses angiogenesis in human leiomyosarcoma and synovial sarcoma by transcriptional suppression of vascular endothelial growth factor expression. *Cancer Res.* **2000**, *60*, 3655–3661.
16. Lanni, J.S.; Lowe, S.W.; Licitra, E.J.; Liu, J.O.; Jacks, T. p53-independent apoptosis induced by paclitaxel through an indirect mechanism. *Proc. Natl. Acad. Sci. USA* **1997**, *94*, 9679–9683. [CrossRef]
17. Miyake, N.; Chikumi, H.; Takata, M.; Nakamoto, M.; Igishi, T.; Shimizu, E. Rapamycin induces p53-independent apoptosis through the mitochondrial pathway in non-small cell lung cancer cells. *Oncol. Rep.* **2012**, *28*, 848–854. [CrossRef]
18. Anaya-Eugenio, G.D.; Tan, C.Y.; Rakotondraibe, L.H.; Carcache de Blanco, E.C. Tumor suppressor p53 independent apoptosis in HT-29 cells by auransterol from Penicillium aurantiacobrunneum. *Biomed Pharmacother.* **2020**, *127*, 110124. [CrossRef]

19. Ma, X.; Choudhury, S.N.; Hua, X.; Dai, Z.; Li, Y. Interaction of the oncogenic miR-21 microRNA and the p53 tumor suppressor pathway. *Carcinogenesis* **2013**, *34*, 1216–1223. [CrossRef]
20. Yehia, L.; Ngeow, J.; Eng, C. PTEN-opathies: From biological insights to evidence-based precision medicine. *J. Clin. Investig.* **2019**, *129*, 452–464. [CrossRef]
21. Hemmings, B.A.; Restuccia, D.F. The PI3K-PKB/Akt pathway. *Cold Spring Harb. Perspect. Biol.* **2015**, *7*, a026609. [CrossRef]
22. Stambolic, V.; MacPherson, D.; Sas, D.; Lin, Y.; Snow, B.; Jang, Y.; Benchimol, S.; Mak, T.W. Regulation of PTEN transcription by p53. *Mol. Cell.* **2001**, *8*, 317–325. [CrossRef]
23. Masson, G.R.; Williams, R.L. Structural Mechanisms of PTEN Regulation. *Cold Spring Harb Perspect Med* **2020**, *10*, a036152. [CrossRef] [PubMed]
24. Bonneau, D.; Longy, M. Mutations of the human PTEN gene. *Hum. Mutat.* **2000**, *16*, 109–122. [CrossRef]
25. Liu, L.; Huang, L.; He, J.; Cai, S.; Weng, Y.; Huang, S.; Ma, S. PTEN inhibits non-small cell lung cancer cell growth by promoting G0/G1 arrest and cell apoptosis. *Oncol. Lett.* **2019**, *17*, 1333–1340. [CrossRef]
26. Li, G.; Zhao, J.; Peng, X.; Liang, J.; Deng, X.; Chen, Y. Radiation/paclitaxel treatment of p53-abnormal non-small cell lung cancer xenograft tumor and associated mechanism. *Cancer Biother. Radiopharm.* **2012**, *27*, 227–233. [CrossRef] [PubMed]
27. Carmichael, J.; Degraff, W.G.; Gamson, J.; Russo, D.; Gazdar, A.F.; Levitt, M.L.; Minna, J.D.; Mitchell, J.B. Radiation sensitivity of human lung cancer cell lines. *Eur. J. Cancer Clin. Oncol.* **1989**, *25*, 527–534. [CrossRef]
28. Molina, J.R.; Yang, P.; Cassivi, S.D.; Schild, S.E.; Adjei, A.A. Non-small cell lung cancer: Epidemiology, risk factors, treatment, and survivorship. *Mayo. Clin. Proc.* **2008**, *83*, 584–594. [CrossRef]
29. Zdrowowicz, M.; Datta, M.; Rychlowski, M.; Rak, J. Radiosensitization of PC3 Prostate Cancer Cells by 5-Thiocyanato-2′-deoxyuridine. *Cancers* **2022**, *14*, 2035. [CrossRef]
30. Cheng, G.; Kong, D.; Hou, X.; Liang, B.; He, M.; Liang, N.; Ma, S.; Liu, X. The tumor suppressor, p53, contributes to radiosensitivity of lung cancer cells by regulating autophagy and apoptosis. *Cancer Biother. Radiopharm.* **2013**, *28*, 153–159. [CrossRef]
31. Liu, C.; Nie, J.; Wang, R.; Mao, W. The Cell Cycle G2/M Block Is an Indicator of Cellular Radiosensitivity. *Dose Response* **2019**, *17*, 1559325819891008. [CrossRef]
32. Di Leonardo, A.; Linke, S.P.; Clarkin, K.; Wahl, G.M. DNA damage triggers a prolonged p53-dependent G1 arrest and long-term induction of Cip1 in normal human fibroblasts. *Genes. Dev.* **1994**, *8*, 2540–2551. [CrossRef] [PubMed]
33. el-Deiry, W.S.; Harper, J.W.; O'Connor, P.M.; Velculescu, V.E.; Canman, C.E.; Jackman, J.; Pietenpol, J.A.; Burrell, M.; Hill, D.E.; Wang, Y.; et al. WAF1/CIP1 is induced in p53-mediated G1 arrest and apoptosis. *Cancer Res.* **1994**, *54*, 1169–1174.
34. Agarwal, M.L.; Agarwal, A.; Taylor, W.R.; Stark, G.R. p53 controls both the G2/M and the G1 cell cycle checkpoints and mediates reversible growth arrest in human fibroblasts. *Proc. Natl. Acad. Sci. USA* **1995**, *92*, 8493–8497. [CrossRef] [PubMed]
35. Umansky, S.R.; Korol, B.A.; Nelipovich, P.A. In vivo DNA degradation in thymocytes of gamma-irradiated or hydrocortisone-treated rats. *Biochim. Biophys. Acta* **1981**, *655*, 9–17. [CrossRef]
36. Gong, J.; Traganos, F.; Darzynkiewicz, Z. A selective procedure for DNA extraction from apoptotic cells applicable for gel electrophoresis and flow cytometry. *Anal. Biochem.* **1994**, *218*, 314–319. [CrossRef] [PubMed]
37. Yu, S.F.; Chen, T.M.; Chen, Y.H. Apoptosis and necrosis are involved in the toxicity of Sauropus androgynus in an in vitro study. *J. Formos. Med. Assoc.* **2007**, *106*, 537–547. [CrossRef]
38. Darzynkiewicz, Z.; Juan, G.; Li, X.; Gorczyca, W.; Murakami, T.; Traganos, F. Cytometry in cell necrobiology: Analysis of apoptosis and accidental cell death (necrosis). *Cytometry* **1997**, *27*, 1–20. [CrossRef]
39. Darzynkiewicz, Z.; Bedner, E.; Traganos, F. Difficulties and pitfalls in analysis of apoptosis. *Methods Cell Biol* **2001**, *63*, 527–546.
40. Aubrey, B.J.; Kelly, G.L.; Janic, A.; Herold, M.J.; Strasser, A. How does p53 induce apoptosis and how does this relate to p53-mediated tumour suppression? *Cell. Death Differ.* **2018**, *25*, 104–113. [CrossRef]
41. Liu, X.; Liu, Y.; Zhang, P.; Jin, X.; Zheng, X.; Ye, F.; Chen, W.; Li, Q. The synergistic radiosensitizing effect of tirapazamine-conjugated gold nanoparticles on human hepatoma HepG2 cells under X-ray irradiation. *Int. J. Nanomed.* **2016**, *11*, 3517–3531. [CrossRef]
42. Nosrati, H.; Charmi, J.; Abhari, F.; Attari, E.; Bochani, S.; Johari, B.; Rezaeejam, H.; Kheiri Manjili, H.; Davaran, S.; Danafar, H. Improved synergic therapeutic effects of chemoradiation therapy with the aid of a co-drug-loaded nano-radiosensitizer under conventional-dose X-ray irradiation. *Biomater Sci.* **2020**, *8*, 4275–4286. [CrossRef] [PubMed]
43. Kang, H.; Rho, S.; Stiles, W.R.; Hu, S.; Baek, Y.; Hwang, D.W.; Kashiwagi, S.; Kim, M.S.; Choi, H.S. Size-Dependent EPR Effect of Polymeric Nanoparticles on Tumor Targeting. *Adv. Healthc. Mater.* **2020**, *9*, e1901223. [CrossRef] [PubMed]
44. Gao, Q.; Zhang, J.; Gao, J.; Zhang, Z.; Zhu, H.; Wang, D. Gold Nanoparticles in Cancer Theranostics. *Front Bioeng. Biotechnol.* **2021**, *9*, 647905. [CrossRef] [PubMed]
45. Valeriote, F.; Lin, H. Synergistic interaction of anticancer agents: A cellular perspective. *Cancer Chemother. Rep.* **1975**, *59*, 895–900. [PubMed]
46. Carpentier, Y.; Demange, L.; Loirette, M.; Hivet, J.; Desoize, B. Chronology of combined chemotherapy (5FU) and radiotherapy. I. In vitro study. *Anticancer Res.* **1993**, *13*, 2177–2180.
47. Fischer, T.; Hartmann, O.; Reissland, M.; Prieto-Garcia, C.; Klann, K.; Pahor, N.; Schulein-Volk, C.; Baluapuri, A.; Polat, B.; Abazari, A.; et al. PTEN mutant non-small cell lung cancer require ATM to suppress pro-apoptotic signalling and evade radiotherapy. *Cell. Biosci.* **2022**, *12*, 50. [CrossRef]

48. Branham, M.T.; Nadin, S.B.; Vargas-Roig, L.M.; Ciocca, D.R. DNA damage induced by paclitaxel and DNA repair capability of peripheral blood lymphocytes as evaluated by the alkaline comet assay. *Mutat. Res.* **2004**, *560*, 11–17. [CrossRef]
49. Appella, E.; Anderson, C.W. Post-translational modifications and activation of p53 by genotoxic stresses. *Eur. J. Biochem.* **2001**, *268*, 2764–2772. [CrossRef]
50. Jin, G.; Kim, M.J.; Jeon, H.S.; Choi, J.E.; Kim, D.S.; Lee, E.B.; Cha, S.I.; Yoon, G.S.; Kim, C.H.; Jung, T.H.; et al. PTEN mutations and relationship to EGFR, ERBB2, KRAS, and TP53 mutations in non-small cell lung cancers. *Lung Cancer* **2010**, *69*, 279–283. [CrossRef]
51. Gkountakos, A.; Sartori, G.; Falcone, I.; Piro, G.; Ciuffreda, L.; Carbone, C.; Tortora, G.; Scarpa, A.; Bria, E.; Milella, M.; et al. PTEN in Lung Cancer: Dealing with the Problem, Building on New Knowledge and Turning the Game Around. *Cancers* **2019**, *11*, 1141. [CrossRef]
52. Park, J.H.; Kim, Y.H.; Shim, S.; Kim, A.; Jang, H.; Lee, S.J.; Park, S.; Seo, S.; Jang, W.I.; Lee, S.B.; et al. Radiation-Activated PI3K/AKT Pathway Promotes the Induction of Cancer Stem-Like Cells via the Upregulation of SOX2 in Colorectal Cancer. *Cells* **2021**, *10*, 135. [CrossRef] [PubMed]
53. Li, H.F.; Kim, J.S.; Waldman, T. Radiation-induced Akt activation modulates radioresistance in human glioblastoma cells. *Radiat. Oncol.* **2009**, *4*, 43. [CrossRef] [PubMed]
54. Wang, C.Y.; Chang, C.Y.; Wang, C.Y.; Liu, K.; Kang, C.Y.; Lee, Y.J.; Chen, W.R. N-Dihydrogalactochitosan Potentiates the Radiosensitivity of Liver Metastatic Tumor Cells Originated from Murine Breast Tumors. *Int. J. Mol. Sci.* **2019**, *20*, 5581. [CrossRef] [PubMed]
55. Lee, Y.J.; Sheu, T.J.; Keng, P.C. Enhancement of radiosensitivity in H1299 cancer cells by actin-associated protein cofilin. *Biochem Biophys. Res. Commun.* **2005**, *335*, 286–291. [CrossRef] [PubMed]

International Journal of *Molecular Sciences*

MDPI

*Review*

# Statins as Repurposed Drugs in Gynecological Cancer: A Review

**Kai-Hung Wang [1], Chin-Hung Liu [2] and Dah-Ching Ding [3,4,*]**

1 Department of Medical Research, Hualien Tzu Chi Hospital, Buddhist Tzu Chi Medical Foundation, Hualien 970, Taiwan
2 Department of Pharmacology, School of Medicine, Tzu Chi University, Hualien 970, Taiwan
3 Department of Obstetrics and Gynecology, Hualien Tzu Chi Hospital, Buddhist Tzu Chi Foundation, and Tzu Chi University, Hualien 970, Taiwan
4 Institute of Medical Sciences, College of Medicine, Tzu Chi University, Hualien 970, Taiwan
* Correspondence: dah1003@tzuchi.com.tw; Tel.: +886-38561825-13381

**Abstract:** Discovering new drugs is an expensive and time-consuming process, including target identification, bioavailability, pharmacokinetic (PK) tests, pharmacodynamic (PD) tests, toxicity profiles, recommended dosage test, and observation of the side effects, etc. Repurposed drugs could bypass some steps, starting from phase II trials, and shorten the processes. Statins, also known as HMG-CoA inhibitors (HMGCR), are commonly used to manage and prevent various cardiovascular diseases and have been shown to improve the morbidity and mortality of patients. In addition to the inhibitory effects on the production of cholesterol, the beneficial effects of statins on the prognosis and risk of various cancers are also shown. Statins not only inhibited cell proliferation, metastasis, and chemoresistance but affected the tumor microenvironment (TME). Thus, statins have great potential to be repurposed in oncology. Hence, we review the meta-analysis, cohort, and case-control studies of statins in gynecological cancers, and elucidate how statins regulate cell proliferation, apoptosis, tumor growth, and metastasis. Although the results in gynecological cancers remain controversial and the effects of different statins in different histotypes of gynecological cancers and TME are needed to elucidate further, statins are excellent candidates and worthy of being repurposed drugs in treating gynecological cancers.

**Keywords:** statins; repurposed drugs; gynecological cancer; endometrial cancer; ovarian cancer; cervical cancer

**Citation:** Wang, K.-H.; Liu, C.-H.; Ding, D.-C. Statins as Repurposed Drugs in Gynecological Cancer: A Review. *Int. J. Mol. Sci.* **2022**, *23*, 13937. https://doi.org/10.3390/ijms232213937

Academic Editor: Laura Paleari

Received: 30 October 2022
Accepted: 10 November 2022
Published: 11 November 2022



## 1. Introduction

Gynecological cancer is any cancer that starts in a woman's reproductive organs, including cervical, endometrial, ovarian, vaginal, and vulvar cancer. The treatments generally include surgery, radiation therapy, chemotherapy, target therapy, and immunotherapy. Combination therapy is a trend worldwide. However, discovering new drugs or targets is always the mission against cancers. There is an established setting for new drug discovery from pre-clinical results, in vitro and in vivo, to human studies, phase I and II trials, and a phase III randomized controlled trial (RCT). It is expensive and takes over 10 years in all processes [1]. Thus, if the existing drugs could be repurposed, it can dramatically reduce costs and save time, benefiting patients who suffer from these malignant and lethal diseases.

Statins, as 3-hydroxy-3-methyl-glutaryl-CoA (HMG-CoA) reductase competitive inhibitors (HMGCR), are commonly used as lipid-lowering drugs, preventing cardiovascular diseases. However, the anti-cancer properties of statins have been investigated in recent decades, showing better prognoses in various types of cancer through various mechanisms [2,3]. The evidence of the anti-cancer effects of statins in gynecological cancers is sparse and controversial, thus, we review and assess the potential of statins as repurposed drugs in gynecological cancers.

## 2. Cervical Cancer and Human Papillomavirus (HPV)

Cervical cancer is the fourth most common cancer in women and the fourth highest mortality rate worldwide. Cervical cancer diagnoses for 6.6% of all cancer types with a mortality rate of 7.5% in 2018 [4]. For diagnosis of cervical cancer from cytologic examination, the precancerous stage includes low-grade squamous intra-epithelial lesion (LSIL or mild dysplasia) and high-grade squamous intra-epithelial lesion [HSIL or moderate dysplasia, severe dysplasia, and carcinoma in situ (CIS)], and the cancer types include squamous cell carcinoma (SCC), and adenocarcinoma. The diagnosis of cervical cancer from histologic examination includes cervical intraepithelial neoplasia 1 (CIN1), CIN2, and CIN3 and cancer lesions. LSIL is relatively equal to CIN1, while HSIL is relatively equal to CIN2 and CIN3 [5].

Human papillomavirus (HPV) has been defined as a carcinogen, especially the high-risk types, and the persistence of HPV infection was a necessary etiological cause of cervical cancer [6]. High-risk HPV (HR-HPV) types include HPV16, 18, 31, 33, 35, 39, 45, 51, 52, 56, 58, 59, 66, and 68 [7]. Inoculation of HPV vaccines showed long-term efficacy and could prevent cervical cancer [8]. The ideal age for the administration of the HPV vaccines is 10 to 13 years. In low-resource settings, the simple and inexpensive way is to start with visual inspections only or in combination with HPV tests. In high-resource situations, it starts with cytologic tests (pap smear test) and HPV tests to screen cervical cancer patients [9].

## 3. Endometrial Cancer and Its Risk Factors

Endometrial cancer (EC) is the sixth most common cancer in females and the second most commonly diagnosed cancer of female reproductive organs. Around 417,000 new cases were detected, and 97,000 women died worldwide from the disease in 2020 [10] [11]. There are two main types of ECs that were characterized. Type I ECs, around ~80%, are mostly well differentiated with endometrioid histology and show a high level of estrogen receptor (ER). Type II ECs are poorly differentiated with serous or clear cell histology and show a high recurrence rate (80%~90%) within 3 years, representing a poor prognosis [12]. In addition, ECs can be low-grade (grades 1 and 2) tumors which are generally associated with a better prognosis, or high-grade carcinomas (grade 3) carrying an intermediate prognosis [13]. The risk factors for EC include high body mass index (BMI: kg/m$^2$), often with other components of metabolic syndrome (e.g., hypertension, diabetes), nulliparity or infertility, early menarche, and late menopause.

The relative risk (RR) for developing EC with metabolic syndrome was 1.89 [95% confidence interval (CI) 1.34 to 2.67, $p < 0.001$] among the components of metabolic syndrome. Obesity (BMI > 30) was associated with the greatest increase in RR of 2.21 ($p < 0.001$). Other components of the metabolic syndrome linked to endometrial cancer include hypertension, with a RR of 1.81 ($p < 0.05$). [14] Type II Diabetes mellitus (DM) showed an independent risk factor for EC, with an approximate doubling of risk [Odds ratio (OR) was 2.18, 95% CI 1.40 to 3.41] [15]. Among the causes of infertility, polycystic ovarian syndrome (PCOS) showed an increase in risk (OR = 2.79, 95% CI 1.31 to 5.95) [16]. Both early menarche (RR was 2.4 for women <12 vs. ≥15 years) [17], and late menopause (RR = 1.8 for women ≥55 versus <50 years) [18] are associated with increased risk for EC.

## 4. Ovarian Cancer and Its Risk Factors

Ovarian cancer is the leading disease of death in females diagnosed with gynecological cancers. In the meantime, it is women's fifth most frequent cause of death. There are approximately 21,750 new ovarian cancer cases in the US, comprising 1.2% of all cancer cases. The estimated number of deaths related to ovarian cancer was 13,940 in 2020 [19]. Among the ovarian cancer subtypes, type II high-grade serous carcinoma (HGSC) is the most prevalent and lethal, representing more than 70% of ovarian cancer. Type I tumor includes low-grade serous, endometrioid, clear-cell, and mucinous carcinomas, presenting at an early stage and carrying a good prognosis except for clear-cell [20]. HGSCs arise from serous tubal intraepithelial carcinoma (STIC) in the fimbriae of the tube, undergoing malignant transformation and metastasizing to the nearby ovaries and peritoneum [21,22].

The risks of ovarian cancer were increasing in postmenopausal women and those with a family history of breast or ovarian cancer. At the same time, pregnancy, lactation, and oral contraceptive pills reduced the risks [23]. Moreover, obesity was an independent prognostic factor in addition to advanced tumor staging and positive ascites cytology. The hazard ratio (HR) of overall survival (OS) was 1.871, 95% CI 1.005 to 3.486 in all ovarian cancer patients, and the HR was elevated to 2.405, 95% CI 1.335 to 4.333 in pT3 stage patients [24].

## 5. Statins, HMG-CoA Reductase Inhibitor (HMGCR) and the Role in the Tumor Microenvironment (TME)

### 5.1. Statins, Lipid-Lowering Drugs

Statins are traditionally applied in cardiovascular diseases to reduce cholesterol [25] and could be divided into two groups: type-I derivatives (from fermentation, including mevastatin, lovastatin, pravastatin, and simvastatin), and type-II drugs (from the synthetic origin, including fluvastatin, atorvastatin, cerivastatin, pitavastatin, and rosuvastatin) [26,27]. The main role of statin in the mevalonate pathway is inhibiting HMG-CoA reductase (HMGCR), resulting in the depletion of LDL cholesterol [28] (Figure 1). The statins were used between 10–80 mg, and the metabolic pathway of statins was major through CYP3A4 [29] (Table 1). However, recent studies suggested that statins could have anti-tumor effects (Figures 1 and 2), from meta-analysis and bench, in vitro and in vivo.

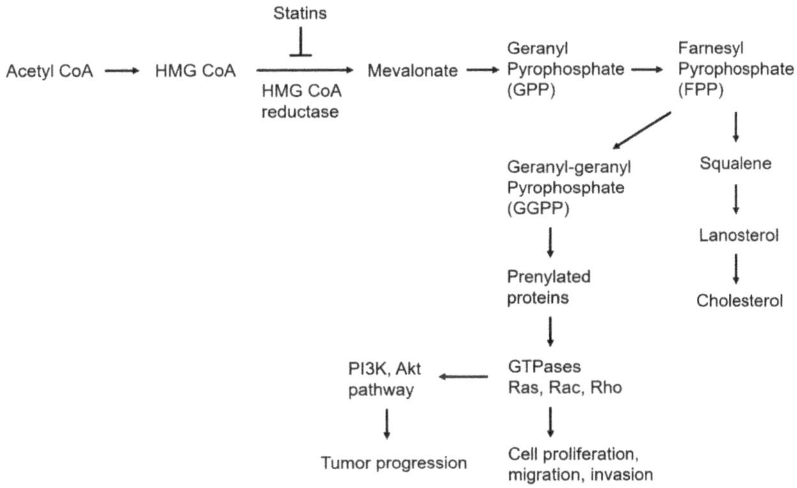


**Figure 1.** The mevalonate pathway and the role of statin in regulating tumor progression and biosynthesis of cholesterol.

**Table 1.** The doses and metabolic pathway of statins.

| Drugs | Property | High Dose | Moderate Dose | Low Dose | Metabolism Pathway |
|---|---|---|---|---|---|
| Atorvastatin | lipophilic | 40–80 mg | 10–20 mg | | CYP3A4 |
| Fluvastatin | lipophilic | | 80 mg | 20–40 mg | CYP2C9 |
| Lovastatin | lipophilic | | 40 mg | 20 mg | CYP3A4 |
| Pravastatin | hydrophilic | | 40–80 mg | 10–20 mg | Sulfation |
| Rosuvastatin | hydrophilic | 20–40 mg | 5–10 mg | | CYP2C9 |
| Simvastatin | lipophilic | | 20–40 mg | 10 mg | CYP3A4 |

CYP3A4: cytochrome P450, subfamily IIIA, polypeptide 4. CYP2C9: cytochrome P450, subfamily IIC, polypeptide 9. High dose: reduce LDL ≥ 50%, moderate dose: reduce LDL 30–49%, low dose: reduce LDL < 30%.

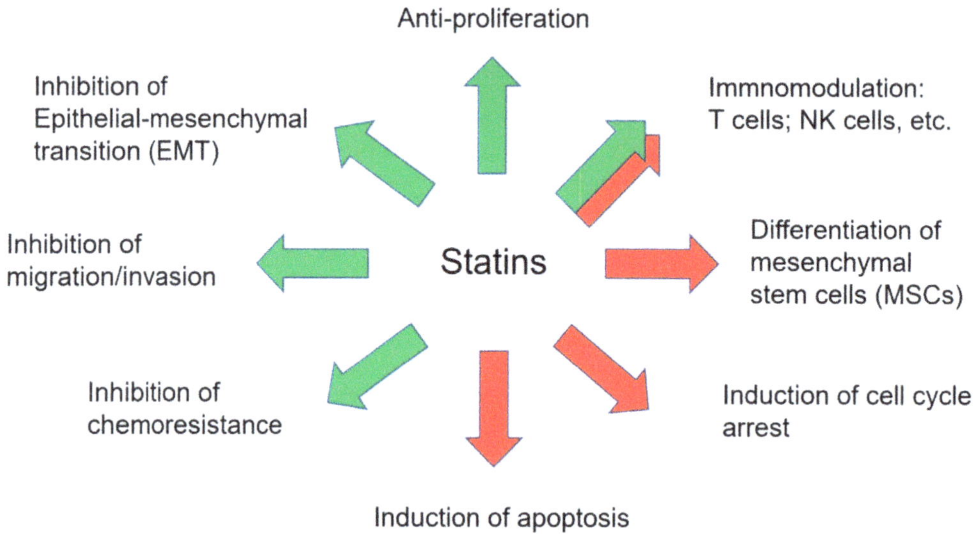


**Figure 2.** The effects of statins on anti-tumor progression and tumor microenvironment (TME). TME includes immune cells and MSCs. Green arrows represent inhibitory effects, and red arrows represent promoting effects.

### 5.2. Statins in Cancer

The most investigated statin in cancer is simvastatin. In general, the role of statins was tumor suppressor. Statins could induce cancer cell apoptosis through traditional caspases cascade and inhibit cell proliferation, migration, invasion, epithelial-mesenchymal transition (EMT), and chemoresistances in various types of cancer (Figure 2), including breast, lung, pancreas, and liver cancer [30]. Statins induced apoptosis of cancer cells through NFκB and the canonical caspase pathway and reduced proliferation through MEK1/2, ERK1/2, and JNK pathways [31]. Statins also induced cell cycle arrest of cancer cells by activating AMPK and increasing p21 and p27 expression [32]. Simvastatin suppresses the invasion of cancer cells by decreasing Pituitary Tumor-Transforming Gene 1 (PTTG1) [33]. Furthermore, statins could also regulate epigenetic machinery resulting in cell cycle arrest. DNMTs could be the targets of statins and the downstream p16 protein [34] and p21 [35]. In conclusion, statins showed anti-tumor progression in various cancers (Figure 2).

### 5.3. Statins in Immune Cells

Mostly, statins showed anti-inflammatory effects and enhanced the number of regulatory T cells (Treg) [36], which may result in the suppression of the Th1 immune response [37]. In addition, statin treatment reduced the Th17 population [38]. Treg obtained immunosuppressive effects on immunotherapy. However, a high dose of Atorvastatin could reduce the in vitro function of conventional T and regulatory T (Treg) cells [39]. Furthermore, statins were associated with better clinical outcomes in patients treated with PD-1 inhibitors [40]. Statins plus Th1 cytokines or dendritic cells (DC)-based immunotherapy could suppress breast tumor growth [41]. Statins could stimulate immunogenicity and promote an anti-melanoma immune response [42]. These data showed the conflicting roles of statins in immunotherapy. Thus, the roles of statins in immune cells and immunotherapy are needed to be elucidated.

### 5.4. Statins in MSCs

Statins had several effects on mesenchymal stem cells (MSCs). Statins could enhance the osteogenic differentiation, angiogenic potential, migration, homing, survival, and proliferation of MSCs [43], which may have improved therapeutic outcomes in regenerative medicine. The evidence of statins in regulating cancer-associated MSCs (CaMSCs) is limited. Simvastatin could decrease CCL3 expression from cancer cells and ICAM-1, VCAM1, IL-6, and CCL2 expression from CaMSCs, disrupting the crosstalk of the cancer cells and tumor microenvironments (TME) and inhibiting tumor progression [44]. Therefore, the roles of statins in the TME—not only in immune cells but also MSCs, especially CaMSCs— need further investigation.

## 6. Statins as Potential Anti-Cancer Agents in Gynecological Cancers

### *6.1. Meta-Analysis in EC*

Statin use was associated with lower risks of EC (RR = 0.81, 95% CI 0.70 to 0.94, $p = 0.001$) but not with mortalities (HR = 0.71, 95% CI 0.64 to 0.80, $p = 0.144$) [45]. In another study, it was shown that statin use could increase overall survival (OS) (HR = 0.80, 95% CI 0.66 to 0.95) [46]. However, not all studies suggested positive results. It was shown that statin use did not reduce the risk of EC (RR = 0.88, 95% CI 0.78 to 1.00, $p = 0.05$), even in the long-term statin user (>5 years) (RR = 0.79, 95% CI 0.58 to 1.08, $p = 0.14$) [47]. There was also no protective effect on EC in another study (RR = 0.94, 95% CI 0.82 to 1.07) [48] (Table 2).

**Table 2.** Clinical studies of statins in endometrial cancer.

| Study Type | Findings in Statin Use Group | Results | References |
|---|---|---|---|
| Meta-analysis | Decrease risks and mortality of endometrial cancer. | Risk: RR = 0.81, 95% CI 0.70 to 0.94; OS: HR = 0.71, 95% CI 0.64 to 0.80. | [45] |
| Meta-analysis | Decrease mortality of endometrial cancer. | Mortality: HR = 0.80, 95% CI, 0.66 to 0.95. | [46] |
| Meta-analysis | No protective effect on endometrial cancer. | Risk: RR = 0.88, 95% CI 0.78 to 1.00; long-term use (>5 years) RR = 0.79, 95% CI 0.58 to 1.08. | [47] |
| Meta-analysis | No protective effect on endometrial cancer. | Risk: RR = 0.94, 95% CI 0.82 to 1.07. | [48] |
| Cohort study | No protective effect on endometrial cancer. | HR = 0.67, 95% CI: 0.39 to 1.17. | [49] |
| Cohort study | No protective effect on survival in endometrial cancer. | Mortality: Type I HR 0.92, 95% CI 0.70 to 1.2; type II HR = 0.92, 95%CI 0.65 to 1.29. | [50] |
| Cohort study | No protective effect on endometrial cancer. | Risk: HR = 0.83, 95% CI 0.64 to 1.08. | [51] |
| Cohort study | No protective effect on endometrial cancer. | Recurrence-free survival (82% vs. 83%); disease-specific survival (86% vs. 84%); and OS (77% vs. 75%). | [52] |
| Cohort study | Decrease mortality of endometrial cancer. | Mortality: HR = 0.41, 95% CI 0.20 to 0.82. | [53] |
| Cohort study | Decrease mortality of endometrial cancer. | Mortality: HR = 0.80, 95% CI, 0.74 to 0.88. | [3] |
| Cohort study | Improve DSS of endometrial cancer, especially concurrent use with aspirin. | DSS: HR 0.63, 95% CI 0.40 to 0.99; concurrent use with aspirin HR 0.25, 95% CI 0.09 to 0.70. | [54] |
| Cohort study | Improve OS and PFS of hyperlipidaemic high-grade endometrial cancer. | Mortality: HR = 0.42, 95% CI, 0.20 to 0.87; PFS: HR = 0.47, 95% CI 0.23 to 0.95. | [55] |
| Cohort study | Decrease mortality of endometrial cancer. | Mortality: continuing user HR = 0.70, 95% CI 0.53 to 0.92; new users HR = 0.43, 95% CI 0.29 to 0.65. | [56] |
| Cohort study | Decrease mortality of type I endometrial cancer and statin new user. | Mortality: type I HR = 0.87, 95% CI 0.76 to 1.00; hydrophilic statins HR = 0.84, 95% CI 0.68 to 1.03; new user HR = 0.75, 95% CI 0.59 to 0.95. | [57] |
| Cohort study | Decrease risks of endometrial cancer. | Risk: HR = 0.74, 95% CI 0.59 to 0.94. | [58] |

CI: confidence interval. RR: relative risk. OR: odds ratio. OS: overall survival. HR: hazard ratio. DSS: disease-specific survival. PFS: progression-free survival.

### 6.2. Cohort Studies in EC

The results of statins use in EC are controversial, including no protective effects on risks (HR = 0.67; 95% CI: 0.39–1.17) [49], OS for type I (HR = 0.92, 95% CI 0.70 to 1.2) and type II (HR = 0.92, 95% CI 0.65 to 1.29, $p$ = 0.62) EC patients [50]. There was no significant association in post-diagnostic use of statins (new users) (adjusted HR 0.83, 95% CI 0.64 to 1.08) [51] and no difference between statin users and nonusers in 5-year recurrence-free survival (82% vs. 83%; $p$ = 0.508), disease-specific survival (86% vs. 84%; $p$ = 0.549), or overall survival (77% vs. 75%; $p$ = 0.901) [52] (Table 2).

In contrast, statin use decreased the mortalities in several studies, including OS (HR = 0.41, 95% CI 0.20 to 0.82) [53], OS (HR = 0.80; 95% CI 0.74–0.88) [3], disease-specific survival (DSS) (HR = 0.63, 95% CI 0.40 to 0.99), DSS in concurrent statin and aspirin user (HR = 0.25, 95% CI 0.09 to 0.70) [54], OS in hyperlipidemic patients (HR = 0.42; 95% CI 0.20 to 0.87; $p$ = 0.02), PFS (HR = 0.47; 95% CI 0.23 to 0.95; $p$ = 0.04) [55], OS in continuing (pre- and postdiagnosis) users (HR = 0.70, 95% CI 0.53 to 0.92), new (postdiagnosis only) users (HR = 0.43, 95% CI 0.29 to 0.65) [56]. Furthermore, statin use decreased EC-specific mortality in type I cancers (HR = 0.87; 95% CI 0.76 to 1.00), for hydrophilic statins (HR = 0.84; 95% CI 0.68 to 1.03) and the new user (HR = 0.75; 95% CI 0.59 to 0.95) [57]. In addition, the risk of EC for statin use was decreased (HR = 0.74, 95% CI 0.59 to 0.94) [58] (Table 2).

In summary, the effects of statins in treating EC are still controversial. However, large results suggested that statins may be potent drugs to decrease the risks and mortalities of EC, and are worth performing clinical trials.

## 7. Meta-Analysis in Ovarian Cancer

Statin use was not significantly associated with the risks (RR = 0.92, 95% CI 0.85 to 1.00) but decreased the mortality (HR = 0.78, 95% CI 0.73 to 0.83) of ovarian cancer [45]. Another study showed that statin use did not reduce the risk of ovarian cancer (RR = 0.88, 95% CI 0.76 to 1.03, $p$ = 0.12). Furthermore, no association was found between long-term statin use (>5 years) and the risk of ovarian cancer (RR = 0.73, 95% CI 0.51 to 1.04, $p$ = 0.08) [47]. It was shown that the risks were not significantly associated with statin type (lipophilic RR = 0.88, 95% CI 0.69 to 1.12; hydrophilic RR = 1.06, 95% CI 0.72 to 1.57), histotypes of ovarian cancer (serous: RR: 0.95, 95% CI 0.69 to 1.30; clear cells: RR = 1.17, 95% CI 0.74 to 1.86), and long-term user (RR = 0.77, 95% CI 0.54 to 1.10) [59] (Table 3).

**Table 3.** Clinical studies of statins in ovarian cancer.

| Study Type | Findings in Statin Use Group | Results | References |
|---|---|---|---|
| Meta-analysis | No association with risks but decreased mortality of ovarian cancer. | Risk: RR = 0.92, 95% CI 0.85 to 1.00; OS: HR = 0.78, 95% CI 0.73 to 0.83. | [45] |
| Meta-analysis | No protective effect on ovarian cancer. | Risk: RR = 0.88, 95% CI 0.76 to 1.03; long-term use (>5 years) RR = 0.73, 95% CI 0.51 to 1.04. | [47] |
| Meta-analysis | No protective effect on ovarian cancer, regardless of the statin type, tumor histotypes: serous and clear cells, and long-term user. | Risk: lipophilic RR = 0.88, 95% CI 0.69 to 1.12; hydrophilic RR = 1.06, 95% CI 0.72 to 1.57), histotypes of cancer (serous: RR: 0.95, 95% CI 0.69 to 1.30; clear cells: RR = 1.17, 95% CI 0.74 to 1.86), and long-term user (RR = 0.77, 95% CI 0.54 to 1.10). | [59] |
| Meta-analysis | No association with risks but decreased mortality in ovarian cancer. | Risk: RR = 0.88, 95% CI 0.75 to 1.03; OS: RR = 0.76, 95% CI 0.67 to 0.86. | [60] |
| Meta-analysis | Decrease risks of ovarian cancer, especially in long-term use group. | Risk: RR = 0.79, 95% CI 0.64 to 0.98; long-term use (>5 years) RR = 0.48, 95% CI 0.28 to 0.80. | [61] |
| Meta-analysis | Improve OS and decrease cancer-specific mortality in ovarian cancer. | Mortality: HR = 0.74, 95%CI 0.63 to 0.87; cancer-specific mortality (HR = 0.87, 95% CI 0.80 to 0.95. | [62] |

**Table 3.** *Cont.*

| Study Type | Findings in Statin Use Group | Results | References |
|---|---|---|---|
| Meta-analysis | Decrease mortality of ovarian cancer. | Mortality: HR = 0.76, 95% CI: 0.68–0.85. | [63] |
| Meta-analysis | * Decrease ovarian cancer risks in genetically proxied HMG-CoA reductase inhibition population as well as in BRCA1/2 carrier. | Risk: OR = 0.60, 95% CI 0.43 to 0.83; BRCA1/2 carrier HR = 0.69, 95% CI 0.51 to 0.93. | [64] |
| Cohort study | No protective effect on ovarian cancer. | Risks: HR = 0.69, 95% CI 0.32–1.49. | [49] |
| Case-control study | No protective effect on ovarian cancer. | Risks: OR = 0.98, 95% CI 0.87 to 1.10. | [65] |
| Case-control study | No protective effect on ovarian cancer. | Risks: HR = 0.99, 95% CI 0.78 to 1.25. | [66] |
| Cohort study | Increase the risk of ovarian cancer, especially in pravastatin user. | Risks: HR = 1.30, 95% CI 1.04 to 1.62; pravastatin HR = 1.89, 95% CI 1.24 to 2.88. | [58] |
| Cohort study | No protective effect on ovarian cancer. | Mortality: HR = 0.57, 95% CI 0.21–1.51 | [67] |
| Cohort study | No protective effect on ovarian cancer | Mortality: HR = 0.90, 95% CI 0.78 to 1.04. | [68] |
| Cohort study | No protective effect on ovarian cancer, neither in lipophilic nor hydrophilic statins. | Mortality: HR = 0.90, 95% CI 0.70 to 1.15; lipophilic statins HR = 0.82, 95% CI 0.61 to 1.11; hydrophilic statins HR = 1.04, 95% CI 0.72 to 1.49. | [69] |
| Cohort study | No protective effect on ovarian cancer with hyperlipidemia, but the mortality was decreased in non-serous-papillary subtypes. | Mortality: hyperlipidemia HR = 0.80, 95% CI 0.50 to 1.29; non-serous-papillary subtypes HR = 0.23, 95% CI 0.05 to 0.96. | [70] |
| Cohort study | Decrease mortality of ovarian cancer | Mortality: HR = 0.45, 95% CI 0.23 to 0.88. | [71] |
| Cohort study | Decrease mortality of ovarian cancer | Mortality: HR = 0.47, 95% CI 0.26 to 0.85. | [72] |
| Cohort study | Decrease mortality of ovarian cancer | Mortality: HR = 0.81, 95% CI 0.72 to 0.90. | [73] |
| Cohort study | Decrease mortality of ovarian cancer | Mortality: HR = 0.74, 95% CI 0.61 to 0.91. | [74] |
| Cohort study | Decreases ovarian cancer mortality, both in serous and non-serous types. | Mortality: HR = 0.66, 95% CI 0.55 to 0.81; serous type HR = 0.69, 95% CI 0.54 to 0.87; non-serous type HR = 0.63, 95% CI 0.44 to 0.90. | [75] |
| Cohort study | Decrease mortality in all patients and in those who were serous type. | Mortality: HR = 0.76, 95% CI 0.64 to 0.89; serous type HR = 0.80, 95%CI 0.67 to 0.96. | [76] |

* Genetically proxied HMG-CoA reductase inhibition population contained single nucleotide polymorphism (SNP). CI: confidence interval. RR: relative risk. OS: overall survival. HR: hazard ratio.

Similar to previous studies, statin use was not associated with the risk (RR = 0.88, 95% CI 0.75 to 1.03) but could significantly decrease mortality (RR = 0.76, 95% CI 0.67 to 0.86) of ovarian cancer [60]. Another study showed that statin use decreased the risks (RR = 0.79, 95% CI, 0.64 to 0.98) of ovarian cancer, especially in long-term users (>5 years) (RR = 0.48, 95% CI 0.28 to 0.80) [61]. Post-diagnostic statin use could decrease OS (HR = 0.74, 95% CI 0.63 to 0.87) and cancer-specific mortality (HR = 0.87, 95% CI 0.80 to 0.95) [62]. This could be seen in another study, showing improved OS in statin users (HR: 0.76, 95% CI: 0.68–0.85) [63]. Intriguingly, genetically proxied HMG-CoA reductase inhibition equivalent to 1-mmol/L (38.7-mg/dL) reduction in LDL cholesterol, significantly decreased the risk of ovarian cancer (OR = 0.60, 95% CI 0.43 to 0.83) as well as in BRCA1/2 mutation carriers, (HR = 0.69, 95% CI 0.51 to 0.93). [64] (Table 3).

*Cohort Studies and Case-Control Studies in Ovarian Cancer*

There was no association between the risk of ovarian cancer and statin user, HR = 0.69, 95% CI 0.32–1.49 [49], OR = 0.98, 95% CI 0.87 to 1.10 [65], and HR = 0.99, 95% CI 0.78 to 1.25) [66]. Moreover, the risk was even higher (HR = 1.30, 95% CI 1.04–1.62), which was largely attributed to the effect of the hydrophilic statin, especially pravastatin (HR = 1.89, 95% CI 1.24–2.88) [58]. Statin use was not associated with mortalities of ovarian cancer, HR = 0.57, 95% CI 0.21–1.51 [67], HR = 0.90, 95% CI 0.78 to 1.04 [68], and HR = 0.90, 95% CI 0.70 to 1.15, including lipophilic statin use (HR = 0.82, 95% CI 0.61 to 1.11) and hydrophilic statins (HR = 1.04, 95% CI 0.72 to 1.49) [69], and even in the patients with hyperlipidemia

(HR = 0.80, 95% CI 0.50 to 1.29) [70]. However, the mortalities were significantly decreased in non-serous-papillary subtypes (HR = 0.23, 95% CI 0.05 to 0.96) [70] (Table 3).

On the contrary, statin use decreased the mortalities of ovarian cancer, HR = 0.45, 95% CI 0.23 to 0.88 [71], HR = 0.47, 95% CI 0.26 to 0.85 [72], HR = 0.81, 95% CI 0.72 to 0.90 [73], HR = 0.74, 95% CI 0.61 to 0.91 [74], and HR = 0.66, 95% CI 0.55 to 0.81 [75], both in serous (HR = 0.69, 95% CI 0.54 to 0.87) and non-serous (HR = 0.63, 95% CI 0.44 to 0.90) histologies [75]. It was also shown that statin use decreased mortality in another study, HR = 0.76, 95% CI 0.64 to 0.89 for all patients and HR = 0.80, 95%CI 0.67 to 0.96 for patients with serous types [76] (Table 3).

Because ovarian cancer has different histotypes, statin use did not show significant differences in risks in serous and clear cell types [59], but the mortality decreased [75,76]. The results of statin use in ovarian cancer patients remained controversial. Thus, additional studies are needed to elucidate the effects of different statins on different histotypes of ovarian cancer.

## 8. Cohort Studies in Other Gynecological Cancers

The HR association between the risk of cervical cancer and statin use was 0.83, 95% CI of 0.67 to 0.99. Statin use was associated with decreased total gynecological cancer mortality, (HR = 0.70, 95% CI 0.50 to 0.98) [77]. The statin use group had a better prognosis compared with the non-user (progression-free survival: HR = 0.062, 95% CI 0.008 to 0.517; overall survival: HR = 0.098, 95% CI 0.041–0.459) in cervical cancer patients [78] (Table 4). The effects of statin use against cervical cancer and vulvar cancer are not conclusive due to too few studies and case numbers [61]. In conclusion, statin use may obtain protective effects on cervical cancer, but the evidence is too few.

**Table 4.** Clinical studies of statins in other gynecological cancer.

| Study Type | Findings in Statin Use Group | Results | References |
|---|---|---|---|
| Cohort study | Decrease risks of cervical cancer; decrease mortality in total gynecological cancer. | Risk: HR = 0.83 (95% CI 0.67 to 0.99; total gynecological cancer HR = 0.70, 95% CI 0.50 to 0.98. | [77] |
| Cohort study | Decrease mortality of cervical cancer | Progression-free survival: HR = 0.062, 95% CI 0.008 to 0.517; overall survival: HR = 0.098, 95% CI 0.041–0.459. | [78] |

CI: confidence interval. HR: hazard ratio.

## 9. The Mechanisms of the Anti-Tumor Effects of Statins on Gynecological Cancer

Simvastatin exhibits anti-metastatic and anti-tumorigenic effects in ECC-1 and Ishikawa EC cells through mitogen-activated protein kinase (MAPK) but not the Akt/mTOR pathway [79]. The drug for diabetes, metformin, combined with simvastatin, synergistically inhibited cell growth and induced Bim expression and apoptosis in RL95-2, HEC1B, and Ishikawa EC cells. The combination treatment of metformin and simvastatin upregulated phosphorylated AMPK (pAMPK) and downregulated downstream phosphorylated S6 (pS6), suggesting the mTOR pathway may regulate these anti-proliferative effects [80]. Lipophilic (lovastatin and simvastatin) but not hydrophilic (pravastatin) statins induced apoptosis in ovarian cancer cell lines A2780 and UCI 101; endometrial cancer cell line, Ishikawa; and cervical cancer cell line, HeLa, which all expressed high levels of HMG-CoA reductase [81] (Table 5).

**Table 5.** The preclinical studies of statin in gynecological cancers.

| Treatment | Experiments | Cell Lines | Effects of Statins | Pathway/Mechanism | References |
|---|---|---|---|---|---|
| Simvastatin | in vitro | ECC-1 and Ishikawa | Anti-proliferative and anti-metastatic effects. | MAPK pathway. | [79] |
| Simvastatin + metformin | in vitro | RL95-2, HEC-1B, and Ishikawa | Induce apoptosis; synergized with metformin. | Bim, AMPK/S6. | [80] |
| Lovastatin and simvastatin | in vitro | A2780, UCI 101, Ishikawa, and HeLa | Induce apoptosis. | | [81] |
| Lovastatin and Pravastatin | in vitro and in vivo | SKOV3 | Anti-metastatic effects, reduce peritoneal dissemination. | RhoA. | [82] |
| Lovastatin and atorvastatin | in vitro | Hey 1B and Ovcar-3 | Induce apoptosis. | JNK/Rac1/Cdc42. | [83] |
| Lovastatin + doxorubicin | in vitro | A2780 | Induced apoptosis; synergized with doxorubicin. | | [84] |
| Lovastatin | in vitro and in vivo | SKOV3 and OVCAR5, mogp-TAg mice | Anti-tumor growth and induce autophagy. | | [85] |
| Simvastatin | in vitro and in vivo | RMG-1 and TOV-21G | Induce apoptosis and anti-tumor growth. | Osteopontin (OPN). | [86] |
| Simvastatin, atorvastatin, rosuvastatin, lovastatin, fluvastatin, pravastatin | in vitro | A2780, Igrov-1, SKOV-3, Ovcar-4, Ovcar-5 and Ovcar-8 | Induce apoptosis; both activate and block the autophagy. Lipophilic statins were more potent than hydrophilic statins. | Rab7/p62/LC3-II. | [87] |
| simvastatin | in vitro and in vivo | SKOV3, OVCAR3, and ID8 | Induce apoptosis and inhibit tumor growth. | | [88] |
| simvastatin | in vitro and in vivo | Hey, SKOV3, and KpB mice | Anti-metastatic and anti-tumorigenic effects. | MAPK and AKT/mTOR. | [89] |
| Atorvastatin, fluvastatin, simvastatin | in vitro | CaSki, HeLa, and ViBo | Induce apoptosis. | | [90] |
| simvastatin + paclitaxel | in vitro and in vivo | SiHa, C33A, HeLa, and ViBo | Induce apoptosis and inhibit tumor growth; synergized with paclitaxel. | Raf, ERK1/2, Akt, mTOR, and prenylated Ras. | [91] |
| Atorvastatin | in vitro and in vivo | SiHa and Caski | Induce apoptosis and autophagy and inhibit tumor growth. | AMPK, Akt/mTOR. | [92] |

Lovastatin and Pravastatin decreased metastasis through RhoA signaling in vitro and in vivo of SKOV3 ovarian cancer cells [82]. In addition, lovastatin and atorvastatin induced apoptosis in Hey 1B and Ovcar-3 ovarian cancer cells and suppressed anchorage-independent growth of these cells through the JNK/Rac1/Cdc42 pathway [83]. Lovastatin synergizes with doxorubicin to induce apoptosis by a novel mevalonate-independent mechanism [84]. In the mogp-TAg mice model, the promoter of oviduct glycoprotein-1 was used to drive the expression of SV40 T-antigen, and serous tubal intraepithelial carcinomas (STICs) were developed in gynecologic tissues. Lovastatin significantly reduced the development of STICs in mogp-TAg mice and inhibited ovarian tumor growth in the mouse xenograft model. Furthermore, lovastatin induced autophagy in ovarian cancer cells in vitro [85]. Simvastatin inhibited the proliferation of ovarian clear cell RMG-1 and TOV21-G in vitro and tumor growth in vivo [86]. All statins except pravastatin demonstrated single-agent activity against monolayers (IC50 = 1–35 μM) and spheroids (IC50 = 1–13 μM) of ovarian cancer cells. Furthermore, simvastatin could activate and block autophagy through the Rab7/p62/LC3-II pathway, and the lipophilic statins, simvastatin, and fluvastatin were more potent than hydrophilic statins [87]. In the ID8 syngeneic mice model, simvastatin induced apoptosis and inhibit tumor growth of ovarian cancer [88]. In a K18-gT121+/− p53fl/fl Brca1fl/fl (KpB) mouse model, a unique serous ovarian cancer mouse model specifically and somatically deletes Brca1 and p53 and inactivates the retinoblastoma (Rb) proteins; simvastatin reduced the orthotropic xenograft tumor. In vitro studies showed that simvastatin obtained anti-metastatic and anti-tumorigenic effects through MAPK and AKT/mTOR pathways [89] (Table 5).

Atorvastatin, fluvastatin, and simvastatin induced apoptosis in cervical cancer cells, CaSki, HeLa, and ViBo (established from a biopsy derived from a cervical tumor) [90]. Moreover, simvastatin reduced the phosphorylation of Raf, ERK1/2, Akt, and mTOR and prenylated Ras, resulting in the induction of apoptosis and inhibition of cervical cancer tumor growth. A combination of simvastatin and paclitaxel abolished tumor growth in vivo [91]. In addition, apoptosis and autophagy were induced by atorvastatin through the AMPK/Akt/mTOR pathway. The xenograft tumor was reduced when treated with atorvastatin [92] (Table 5).

In summary, statins showed great potential to reduce cell proliferation and tumor growth of gynecological cancer in vitro and in in vivo. Akt/mTOR is the most important pathway in regulating cell proliferation, and a combination of statins with chemo drugs could synergize the anti-tumorigenic effects. Based on this foundation, statins may be a candidate repurposed drug for gynecological cancers.

## 10. Conclusions and Perspective

The mevalonate pathway and lipid metabolism are linked to the key regulators, influencing gene expression, chromatin remodeling, cellular differentiation, stress response, and tumor microenvironment that collectively enhance tumor development [93]. Statins obtain pleiotropic roles to decrease tumor progression through mevalonate-dependent and independent pathways. Statins reduced the prenylated small GTPase and other signaling pathways, such as Akt/mTOR, to induce apoptosis and autophagy and inhibit cell proliferation and metastasis, resulting in anti-tumor development.

This review of statin use in gynecological cancers showed positive and negative results. Some studies cannot avoid confounders, including multiple comorbidities, lifestyle factors, health-related behaviors, stage and grade of disease, and other medications. The study or clinical trials of different statins (e.g., lipophilic or hydrophilic) on different histotypes of cancer (e.g., serous type or non-serous type; type I or type II) and in combination with chemo drugs are required to validate since there are only 3 trials on EC, 6 trials on ovarian cancer, compared to breast cancer which has 52 trials.

If statins are to be applied clinically to gynecological cancers, they may be used as a single agent. We advocate that using statins in combination with other drugs is more potent. In addition, the identification of the response prediction markers is just undergoing. In ovarian cancer cells, *VDAC1* and *LDLRAP1* were positively and negatively correlated with the response to statins, respectively [94].

The value of statins as therapeutic drugs against gynecological cancer is inestimable because the repurposing of inexpensive, commonly used, and FDA-approved medications to exploit their anti-cancer effects yields the development of cost-effective approaches to cancer therapy. Most important, it can directly benefit the patients, life-saving or prolonging.

**Author Contributions:** Conceptualization, K.-H.W., C.-H.L. and D.-C.D.; methodology, K.-H.W.; formal analysis, K.-H.W.; investigation, K.-H.W.; resources, K.-H.W.; data curation, K.-H.W.; writing—original draft preparation, K.-H.W. and D.-C.D.; writing—review and editing, D.-C.D. All authors have read and agreed to the published version of the manuscript.

**Funding:** This research was funded by Hualien Tzu Chi Hospital (TCRD 108-57, 108-49, 109-62, TCRD 111-071, TCRD 111-080) and the Buddhist Tzu Chi Medical Foundation (TCMF-EP 111-01, TCMF-CP 111-05).

**Institutional Review Board Statement:** Not applicable.

**Informed Consent Statement:** Not applicable.

**Data Availability Statement:** Not applicable.

**Conflicts of Interest:** The authors declare no conflict of interest.

## References


1. Prasad, V.; Mailankody, S. Research and Development Spending to Bring a Single Cancer Drug to Market and Revenues After Approval. *JAMA Intern. Med.* **2017**, *177*, 1569–1575. [CrossRef] [PubMed]
2. Matusewicz, L.; Meissner, J.; Toporkiewicz, M.; Sikorski, A.F. The Effect of Statins on Cancer Cells—Review. *Tumour Biol.* **2015**, *36*, 4889–4904. [CrossRef] [PubMed]
3. Wang, A.; Aragaki, A.K.; Tang, J.Y.; Kurian, A.W.; Manson, J.E.; Chlebowski, R.T.; Simon, M.; Desai, P.; Wassertheil-Smoller, S.; Liu, S.; et al. Statin Use and All-Cancer Survival: Prospective Results from the Women's Health Initiative. *Br. J. Cancer* **2016**, *115*, 129–135. [CrossRef] [PubMed]
4. Bray, F.; Ferlay, J.; Soerjomataram, I.; Siegel, R.L.; Torre, L.A.; Jemal, A. Global Cancer Statistics 2018: GLOBOCAN Estimates of Incidence and Mortality Worldwide for 36 Cancers in 185 Countries. *CA Cancer J. Clin.* **2018**, *68*, 394–424. [CrossRef] [PubMed]
5. Alrajjal, A.; Pansare, V.; Choudhury, M.S.R.; Khan, M.Y.A.; Shidham, V.B. Squamous Intraepithelial Lesions (SIL: LSIL, HSIL, ASCUS, ASC-H, LSIL-H) of Uterine Cervix and Bethesda System. *Cytojournal* **2021**, *18*, 16. [CrossRef]
6. Bouvard, V.; Baan, R.; Straif, K.; Grosse, Y.; Secretan, B.; El Ghissassi, F.; Benbrahim-Tallaa, L.; Guha, N.; Freeman, C.; Galichet, L.; et al. A Review of Human Carcinogens—Part B: Biological Agents. *Lancet Oncol.* **2009**, *10*, 321–322. [CrossRef]
7. Clifford, G.M.; de Vuyst, H.; Tenet, V.; Plummer, M.; Tully, S.; Franceschi, S. Effect of HIV Infection on Human Papillomavirus Types Causing Invasive Cervical Cancer in Africa. *J. Acquir. Immune Defic. Syndr.* **2016**, *73*, 332–339. [CrossRef]
8. GlaxoSmithKline Vaccine HPV-007 Study Group; Romanowski, B.; de Borba, P.C.; Naud, P.S.; Roteli-Martins, C.M.; De Carvalho, N.S.; Teixeira, J.C.; Aoki, F.; Ramjattan, B.; Shier, R.M.; et al. Sustained Efficacy and Immunogenicity of the Human Papillomavirus (HPV)-16/18 AS04-Adjuvanted Vaccine: Analysis of a Randomised Placebo-Controlled Trial up to 6.4 years. *Lancet* **2009**, *374*, 1975–1985.
9. Ngan, H.Y.S.; Garland, S.M.; Bhatla, N.; Pagliusi, S.R.; Chan, K.K.L.; Cheung, A.N.Y.; Chu, T.-Y.; Domingo, E.J.; Qiao, Y.L.; Park, J.S.; et al. Asia Oceania Guidelines for the Implementation of Programs for Cervical Cancer Prevention and Control. *J. Cancer Epidemiol.* **2011**, *2011*, 794861. [CrossRef]
10. Brüggmann, D.; Ouassou, K.; Klingelhöfer, D.; Bohlmann, M.K.; Jaque, J.; Groneberg, D.A. Endometrial Cancer: Mapping the Global Landscape of Research. *J. Transl. Med.* **2020**, *18*, 386. [CrossRef]
11. Sung, H.; Ferlay, J.; Siegel, R.L.; Laversanne, M.; Soerjomataram, I.; Jemal, A.; Bray, F. Global Cancer Statistics 2020: GLOBOCAN Estimates of Incidence and Mortality Worldwide for 36 Cancers in 185 Countries. *CA Cancer J. Clin.* **2021**, *71*, 209–249. [CrossRef] [PubMed]
12. Connor, E.V.; Rose, P.G. Management Strategies for Recurrent Endometrial Cancer. *Expert Rev. Anticancer Ther.* **2018**, *18*, 873–885. [CrossRef] [PubMed]
13. Soslow, R.A.; Tornos, C.; Park, K.J.; Malpica, A.; Matias-Guiu, X.; Oliva, E.; Parkash, V.; Carlson, J.; McCluggage, W.G.; Gilks, C.B. Endometrial Carcinoma Diagnosis: Use of FIGO Grading and Genomic Subcategories in Clinical Practice: Recommendations of the International Society of Gynecological Pathologists. *Int. J. Gynecol. Pathol.* **2019**, *38* (Suppl. S1), S64–S74. [CrossRef] [PubMed]
14. Esposito, K.; Chiodini, P.; Capuano, A.; Bellastella, G.; Maiorino, M.I.; Giugliano, D. Metabolic Syndrome and Endometrial Cancer: A Meta-Analysis. *Endocrine* **2014**, *45*, 28–36. [CrossRef]
15. Rosato, V.; Zucchetto, A.; Bosetti, C.; Dal Maso, L.; Montella, M.; Pelucchi, C.; Negri, E.; Franceschi, S.; La Vecchia, C. Metabolic Syndrome and Endometrial Cancer Risk. *Ann. Oncol.* **2011**, *22*, 884–889. [CrossRef]
16. Barry, J.A.; Azizia, M.M.; Hardiman, P.J. Risk of Endometrial, Ovarian and Breast Cancer in Women with Polycystic Ovary Syndrome: A Systematic Review and Meta-Analysis. *Hum. Reprod. Update* **2014**, *20*, 748–758. [CrossRef] [PubMed]
17. Brinton, L.A.; Berman, M.L.; Mortel, R.; Twiggs, L.B.; Barrett, R.J.; Wilbanks, G.D.; Lannom, L.; Hoover, R.N. Reproductive, Menstrual, and Medical Risk Factors for Endometrial Cancer: Results from a Case-Control Study. *Am. J. Obstet. Gynecol.* **1992**, *167*, 1317–1325. [CrossRef]
18. Zucchetto, A.; Serraino, D.; Polesel, J.; Negri, E.; De Paoli, A.; Dal Maso, L.; Montella, M.; La Vecchia, C.; Franceschi, S.; Talamini, R. Hormone-Related Factors and Gynecological Conditions in Relation to Endometrial Cancer Risk. *Eur. J. Cancer Prev.* **2009**, *18*, 316–321. [CrossRef]
19. Siegel, R.L.; Miller, K.D.; Jemal, A. Cancer Statistics, 2020. *CA Cancer J. Clin.* **2020**, *70*, 7–30. [CrossRef]
20. Koshiyama, M.; Matsumura, N.; Konishi, I. Recent Concepts of Ovarian Carcinogenesis: Type I and Type II. *Biomed Res. Int.* **2014**, *2014*, 934261. [CrossRef]
21. Lee, Y.; Miron, A.; Drapkin, R.; Nucci, M.R.; Medeiros, F.; Saleemuddin, A.; Garber, J.; Birch, C.; Mou, H.; Gordon, R.W.; et al. A Candidate Precursor to Serous Carcinoma That Originates in the Distal Fallopian Tube. *J. Pathol.* **2007**, *211*, 26–35. [CrossRef] [PubMed]
22. Kim, J.; Coffey, D.M.; Creighton, C.J.; Yu, Z.; Hawkins, S.M.; Matzuk, M.M. High-Grade Serous Ovarian Cancer Arises from Fallopian Tube in a Mouse Model. *Proc. Natl. Acad. Sci. USA* **2012**, *109*, 3921–3926. [CrossRef] [PubMed]
23. Momenimovahed, Z.; Tiznobaik, A.; Taheri, S.; Salehiniya, H. Ovarian Cancer in the World: Epidemiology and Risk Factors. *Int. J. Women's Health* **2019**, *11*, 287–299. [CrossRef] [PubMed]

24. Iyoshi, S.; Sumi, A.; Yoshihara, M.; Kitami, K.; Mogi, K.; Uno, K.; Fujimoto, H.; Miyamoto, E.; Tano, S.; Yoshikawa, N.; et al. Obesity Contributes to the Stealth Peritoneal Dissemination of Ovarian Cancer: A Multi-Institutional Retrospective Cohort Study. *Obesity* **2022**, *30*, 1599–1607. [CrossRef]
25. Goldstein, J.L.; Brown, M.S. A Century of Cholesterol and Coronaries: From Plaques to Genes to Statins. *Cell* **2015**, *161*, 161–172. [CrossRef]
26. Oryan, A.; Kamali, A.; Moshiri, A. Potential Mechanisms and Applications of Statins on Osteogenesis: Current Modalities, Conflicts and Future Directions. *J. Control. Release* **2015**, *215*, 12–24. [CrossRef]
27. Endo, A. The Origin of the Statins. *Atheroscler. Suppl.* **2004**, *5*, 125–130. [CrossRef]
28. Schonewille, M.; de Boer, J.F.; Mele, L.; Wolters, H.; Bloks, V.W.; Wolters, J.C.; Kuivenhoven, J.A.; Tietge, U.J.F.; Brufau, G.; Groen, A.K. Statins Increase Hepatic Cholesterol Synthesis and Stimulate Fecal Cholesterol Elimination in Mice. *J. Lipid Res.* **2016**, *57*, 1455–1464. [CrossRef]
29. Abd, T.T.; Jacobson, T.A. Statin-Induced Myopathy: A Review and Update. *Expert Opin. Drug Saf.* **2011**, *10*, 373–387. [CrossRef]
30. Di Bello, E.; Zwergel, C.; Mai, A.; Valente, S. The Innovative Potential of Statins in Cancer: New Targets for New Therapies. *Front. Chem.* **2020**, *8*, 516. [CrossRef]
31. Campbell, M.J.; Esserman, L.J.; Zhou, Y.; Shoemaker, M.; Lobo, M.; Borman, E.; Baehner, F.; Kumar, A.S.; Adduci, K.; Marx, C.; et al. Breast Cancer Growth Prevention by Statins. *Cancer Res.* **2006**, *66*, 8707–8714. [CrossRef] [PubMed]
32. Wang, S.-T.; Ho, H.J.; Lin, J.-T.; Shieh, J.-J.; Wu, C.-Y. Simvastatin-Induced Cell Cycle Arrest through Inhibition of STAT3/SKP2 Axis and Activation of AMPK to Promote p27 and p21 Accumulation in Hepatocellular Carcinoma Cells. *Cell Death Dis.* **2017**, *8*, e2626. [CrossRef] [PubMed]
33. Yin, L.; He, Z.; Yi, B.; Xue, L.; Sun, J. Simvastatin Suppresses Human Breast Cancer Cell Invasion by Decreasing the Expression of Pituitary Tumor-Transforming Gene 1. *Front. Pharmacol.* **2020**, *11*, 574068. [CrossRef] [PubMed]
34. Zhu, B.; Gong, Y.; Yan, G.; Wang, D.; Wang, Q.; Qiao, Y.; Hou, J.; Liu, B.; Tang, C. Atorvastatin Treatment Modulates p16 Promoter Methylation to Regulate p16 Expression. *FEBS J.* **2017**, *284*, 1868–1881. [CrossRef]
35. Dongoran, R.A.; Wang, K.-H.; Lin, T.-J.; Yuan, T.-C.; Liu, C.-H. Anti-Proliferative Effect of Statins Is Mediated by DNMT1 Inhibition and p21 Expression in OSCC Cells. *Cancers* **2020**, *12*, 2084. [CrossRef]
36. Forero-Peña, D.A.; Gutierrez, F.R.S. Statins as Modulators of Regulatory T-Cell Biology. *Mediat. Inflamm.* **2013**, *2013*, 167086. [CrossRef]
37. Shahbaz, S.K.; Sadeghi, M.; Koushki, K.; Penson, P.E.; Sahebkar, A. Regulatory T Cells: Possible Mediators for the Anti-Inflammatory Action of Statins. *Pharmacol. Res.* **2019**, *149*, 104469. [CrossRef]
38. Ulivieri, C.; Baldari, C.T. Statins: From Cholesterol-Lowering Drugs to Novel Immunomodulators for the Treatment of Th17-Mediated Autoimmune Diseases. *Pharmacol. Res.* **2014**, *88*, 41–52. [CrossRef]
39. Rodríguez-Perea, A.L.; Rojas, M.; Velilla-Hernández, P.A. High Concentrations of Atorvastatin Reduce in-Vitro Function of Conventional T and Regulatory T Cells. *Clin. Exp. Immunol.* **2019**, *196*, 237–248. [CrossRef]
40. Cantini, L.; Pecci, F.; Hurkmans, D.P.; Belderbos, R.A.; Lanese, A.; Copparoni, C.; Aerts, S.; Cornelissen, R.; Dumoulin, D.W.; Fiordoliva, I.; et al. High-Intensity Statins Are Associated with Improved Clinical Activity of PD-1 Inhibitors in Malignant Pleural Mesothelioma and Advanced Non-Small Cell Lung Cancer Patients. *Eur. J. Cancer* **2021**, *144*, 41–48. [CrossRef]
41. Oechsle, C.M.; Showalter, L.E.; Novak, C.M.; Czerniecki, B.J.; Koski, G.K. Statin Drugs Plus Th1 Cytokines Potentiate Apoptosis and Ras Delocalization in Human Breast Cancer Lines and Combine with Dendritic Cell-Based Immunotherapy to Suppress Tumor Growth in a Mouse Model of HER-2pos Disease. *Vaccines* **2020**, *8*, 72. [CrossRef] [PubMed]
42. Sarrabayrouse, G.; Pich, C.; Teiti, I.; Tilkin-Mariame, A.F. Regulatory Properties of Statins and Rho Gtpases Prenylation Inhibitiors to Stimulate Melanoma Immunogenicity and Promote Anti-Melanoma Immune Response. *Int. J. Cancer* **2017**, *140*, 747–755. [CrossRef] [PubMed]
43. Gorabi, A.M.; Kiaie, N.; Pirro, M.; Bianconi, V.; Jamialahmadi, T.; Sahebkar, A. Effects of Statins on the Biological Features of Mesenchymal Stem Cells and Therapeutic Implications. *Heart Fail. Rev.* **2021**, *26*, 1259–1272. [CrossRef] [PubMed]
44. Galland, S.; Martin, P.; Fregni, G.; Letovanec, I.; Stamenkovic, I. Attenuation of the pro-Inflammatory Signature of Lung Cancer-Derived Mesenchymal Stromal Cells by Statins. *Cancer Lett.* **2020**, *484*, 50–64. [CrossRef]
45. Chen, Y.; Han, L.; Zheng, A. Association between Statin Use and the Risk, Prognosis of Gynecologic Cancer: A Meta-Analysis. *Eur. J. Obstet. Gynecol. Reprod. Biol.* **2022**, *268*, 74–81. [CrossRef]
46. Li, J.; Liu, R.; Sun, Z.; Tang, S.; Wang, L.; Liu, C.; Zhao, W.; Yao, Y.; Sun, C. The Association between Statin Use and Endometrial Cancer Survival Outcome: A Meta-Analysis. *Medicine* **2018**, *97*, e13264. [CrossRef]
47. Wang, Y.; Ren, F.; Song, Z.; Chen, P.; Liu, S.; Ouyang, L. Statin Use and the Risk of Ovarian and Endometrial Cancers: A Meta-Analysis. *BMC Cancer* **2019**, *19*, 730. [CrossRef]
48. Yang, J.; Zhu, Q.; Liu, Q.; Wang, Y.; Xie, W.; Hu, L. Statin Use and Endometrial Cancer Risk: A Meta-Analysis. *Oncotarget* **2017**, *8*, 62425–62434. [CrossRef]
49. Yu, O.; Boudreau, D.M.; Buist, D.S.M.; Miglioretti, D.L. Statin Use and Female Reproductive Organ Cancer Risk in a Large Population-Based Setting. *Cancer Causes Control* **2009**, *20*, 609–616. [CrossRef]
50. Yoon, L.S.; Goodman, M.T.; Rimel, B.J.; Jeon, C.Y. Statin Use and Survival in Elderly Patients with Endometrial Cancer. *Gynecol. Oncol.* **2015**, *137*, 252–257. [CrossRef]

51. Sanni, O.B.; Mc Menamin, Ú.C.; Cardwell, C.R.; Sharp, L.; Murray, L.J.; Coleman, H.G. Commonly Used Medications and Endometrial Cancer Survival: A Population-Based Cohort Study. *Br. J. Cancer* **2017**, *117*, 432–438. [CrossRef] [PubMed]
52. Segev, Y.; Gemer, O.; Helpman, L.; Hag-Yahia, N.; Eitan, R.; Raban, O.; Vaknin, Z.; Ben-Arie, A.; Amit, A.; Levy, T.; et al. An Israeli Gynecologic Oncology Group Study of Statin Use and Endometrial Cancer Prognosis. *Int J. Gynecol. Obstet.* **2020**, *148*, 79–86. [CrossRef] [PubMed]
53. Arima, R.; Marttila, M.; Hautakoski, A.; Arffman, M.; Sund, R.; Ilanne-Parikka, P.; Kangaskokko, J.; Urpilainen, E.; Läärä, E.; Hinkula, M.; et al. Antidiabetic Medication, Statins and the Risk and Prognosis of Non-Endometrioid Endometrial Cancer in Women with Type 2 Diabetes. *Anticancer Res.* **2018**, *38*, 4169–4178. [CrossRef]
54. Nevadunsky, N.S.; Van Arsdale, A.; Strickler, H.D.; Spoozak, L.A.; Moadel, A.; Kaur, G.; Girda, E.; Goldberg, G.L.; Einstein, M.H. Association Between Statin Use and Endometrial Cancer Survival. *Obstet. Gynecol.* **2015**, *126*, 144–150. [CrossRef] [PubMed]
55. Feng, C.H.; Miller, C.M.; Tenney, M.E.; Lee, N.K.; Yamada, S.D.; Hasan, Y. Statin Use Significantly Improves Overall Survival in High-Grade Endometrial Cancer. *Int. J. Gynecol. Cancer* **2016**, *26*, 1642–1649. [CrossRef]
56. Sperling, C.D.; Verdoodt, F.; Kjaer Hansen, M.; Dehlendorff, C.; Friis, S.; Kjaer, S.K. Statin Use and Mortality among Endometrial Cancer Patients: A Danish Nationwide Cohort Study. *Int. J. Cancer* **2018**, *143*, 2668–2676. [CrossRef]
57. Feng, J.-L.; Dixon-Suen, S.C.; Jordan, S.J.; Webb, P.M. Is There Sufficient Evidence to Recommend Women Diagnosed with Endometrial Cancer Take a Statin: Results from an Australian Record-Linkage Study. *Gynecol. Oncol.* **2021**, *161*, 858–863. [CrossRef]
58. Desai, P.; Wallace, R.; Anderson, M.L.; Howard, B.V.; Ray, R.M.; Wu, C.; Safford, M.; Martin, L.W.; Rohan, T.; Manson, J.E.; et al. An Analysis of the Association between Statin Use and Risk of Endometrial and Ovarian Cancers in the Women's Health Initiative. *Gynecol. Oncol.* **2018**, *148*, 540–546. [CrossRef]
59. Irvin, S.; Clarke, M.A.; Trabert, B.; Wentzensen, N. Systematic Review and Meta-Analysis of Studies Assessing the Relationship between Statin Use and Risk of Ovarian Cancer. *Cancer Causes Control* **2020**, *31*, 869–879. [CrossRef]
60. Mohammadian-Hafshejani, A.; Sherwin, C.M.T.; Heidari-Soureshjani, S. Do Statins Play Any Role in Reducing the Incidence and Mortality of Ovarian Cancer? A Systematic Review and Meta-Analysis. *J. Prev. Med. Hyg.* **2020**, *61*, E331–E339.
61. Liu, Y.; Qin, A.; Li, T.; Qin, X.; Li, S. Effect of Statin on Risk of Gynecologic Cancers: A Meta-Analysis of Observational Studies and Randomized Controlled Trials. *Gynecol. Oncol.* **2014**, *133*, 647–655. [CrossRef] [PubMed]
62. Li, X.; Zhou, J. Impact of Postdiagnostic Statin Use on Ovarian Cancer Mortality: A Systematic Review and Meta-Analysis of Observational Studies. *Br. J. Clin. Pharmacol.* **2018**, *84*, 1109–1120. [CrossRef] [PubMed]
63. Majidi, A.; Na, R.; Dixon-Suen, S.; Jordan, S.J.; Webb, P.M. Common Medications and Survival in Women with Ovarian Cancer: A Systematic Review and Meta-Analysis. *Gynecol. Oncol.* **2020**, *157*, 678–685. [CrossRef] [PubMed]
64. Yarmolinsky, J.; Bull, C.J.; Vincent, E.E.; Robinson, J.; Walther, A.; Smith, G.D.; Lewis, S.J.; Relton, C.L.; Martin, R.M. Association Between Genetically Proxied Inhibition of HMG-CoA Reductase and Epithelial Ovarian Cancer. *JAMA* **2020**, *323*, 646–655. [CrossRef] [PubMed]
65. Baandrup, L.; Dehlendorff, C.; Friis, S.; Olsen, J.H.; Kjær, S.K. Statin Use and Risk for Ovarian Cancer: A Danish Nationwide Case-Control Study. *Br. J. Cancer* **2015**, *112*, 157–161. [CrossRef]
66. Urpilainen, E.; Marttila, M.; Hautakoski, A.; Arffman, M.; Sund, R.; Ilanne-Parikka, P.; Arima, R.; Kangaskokko, J.; Puistola, U.; Läärä, E.; et al. The Role of Metformin and Statins in the Incidence of Epithelial Ovarian Cancer in Type 2 Diabetes: A Cohort and Nested Case-Control Study. *BJOG-Int. J. Obstet. Gynaecol.* **2018**, *125*, 1001–1008. [CrossRef] [PubMed]
67. Chen, H.-Y.; Wang, Q.; Xu, Q.-H.; Yan, L.; Gao, X.-F.; Lu, Y.-H.; Wang, L. Statin as a Combined Therapy for Advanced-Stage Ovarian Cancer: A Propensity Score Matched Analysis. *Biomed Res. Int.* **2016**, *2016*, 9125238. [CrossRef]
68. Verdoodt, F.; Kjaer Hansen, M.; Kjaer, S.K.; Pottegård, A.; Friis, S.; Dehlendorff, C. Statin Use and Mortality among Ovarian Cancer Patients: A Population-Based Cohort Study. *Int. J. Cancer* **2017**, *141*, 279–286. [CrossRef]
69. Majidi, A.; Na, R.; Jordan, S.J.; De Fazio, A.; Webb, P.M.; OPAL Study Group. Statin Use and Survival Following a Diagnosis of Ovarian Cancer: A Prospective Observational Study. *Int. J. Cancer* **2021**, *148*, 1608–1615. [CrossRef]
70. Habis, M.; Wroblewski, K.; Bradaric, M.; Ismail, N.; Yamada, S.D.; Litchfield, L.; Lengyel, E.; Romero, I.L. Statin Therapy Is Associated with Improved Survival in Patients with Non-Serous-Papillary Epithelial Ovarian Cancer: A Retrospective Cohort Analysis. *PLoS ONE* **2014**, *9*, e104521. [CrossRef]
71. Elmore, R.G.; Ioffe, Y.; Scoles, D.R.; Karlan, B.Y.; Li, A.J. Impact of Statin Therapy on Survival in Epithelial Ovarian Cancer. *Gynecol. Oncol.* **2008**, *111*, 102–105. [CrossRef] [PubMed]
72. Lavie, O.; Pinchev, M.; Rennert, H.S.; Segev, Y.; Rennert, G. The Effect of Statins on Risk and Survival of Gynecological Malignancies. *Gynecol. Oncol.* **2013**, *130*, 615–619. [CrossRef] [PubMed]
73. Couttenier, A.; Lacroix, O.; Vaes, E.; Cardwell, C.R.; De Schutter, H.; Robert, A. Statin Use Is Associated with Improved Survival in Ovarian Cancer: A Retrospective Population-Based Study. *PLoS ONE* **2017**, *12*, e0189233. [CrossRef] [PubMed]
74. Harding, B.N.; Delaney, J.A.; Urban, R.R.; Weiss, N.S. Use of Statin Medications Following Diagnosis in Relation to Survival among Women with Ovarian Cancer. Cancer Epidemiol. *Cancer Epidemiol. Biomark. Prev.* **2019**, *28*, 1127–1133. [CrossRef] [PubMed]
75. Vogel, T.J.; Goodman, M.T.; Li, A.J.; Jeon, C.Y. Statin Treatment Is Associated with Survival in a Nationally Representative Population of Elderly Women with Epithelial Ovarian Cancer. *Gynecol. Oncol.* **2017**, *146*, 340–345. [CrossRef] [PubMed]

76. Hanley, G.E.; Kaur, P.; Berchuck, A.; Chase, A.; Grout, B.; Deurloo, C.M.; Pike, M.; Richardson, J.; Terry, K.L.; Webb, P.M.; et al. Cardiovascular Medications and Survival in People with Ovarian Cancer: A Population-Based Cohort Study from British Columbia, Canada. *Gynecol. Oncol.* **2021**, *162*, 461–468. [CrossRef] [PubMed]
77. Kim, D.-S.; Ahn, H.S.; Kim, H.J. Statin Use and Incidence and Mortality of Breast and Gynecology Cancer: A Cohort Study Using the National Health Insurance Claims Database. *Int. J. Cancer* **2022**, *150*, 1156–1165. [CrossRef] [PubMed]
78. Song, M.-K.; Shin, B.-S.; Ha, C.-S.; Park, W.-Y. Would Lipophilic Statin Therapy as a Prognostic Factor Improve Survival in Patients With Uterine Cervical Cancer? *Int. J. Gynecol. Cancer* **2017**, *27*, 1431–1437. [CrossRef]
79. Schointuch, M.N.; Gilliam, T.P.; Stine, J.E.; Han, X.; Zhou, C.; Gehrig, P.A.; Kim, K.; Bae-Jump, V.L. Simvastatin, an HMG-CoA Reductase Inhibitor, Exhibits Anti-Metastatic and Anti-Tumorigenic Effects in Endometrial Cancer. *Gynecol. Oncol.* **2014**, *134*, 346–355. [CrossRef]
80. Kim, J.S.; Turbov, J.; Rosales, R.; Thaete, L.G.; Rodriguez, G.C. Combination Simvastatin and Metformin Synergistically Inhibits Endometrial Cancer Cell Growth. *Gynecol. Oncol.* **2019**, *154*, 432–440. [CrossRef]
81. Kato, S.; Smalley, S.; Sadarangani, A.; Chen-Lin, K.; Oliva, B.; Brañes, J.; Carvajal, J.; Gejman, R.; Owen, G.I.; Cuello, M. Lipophilic but Not Hydrophilic Statins Selectively Induce Cell Death in Gynaecological Cancers Expressing High Levels of HMGCoA Reductase. *J. Cell. Mol. Med.* **2010**, *14*, 1180–1193. [PubMed]
82. Horiuchi, A.; Kikuchi, N.; Osada, R.; Wang, C.; Hayashi, A.; Nikaido, T.; Konishi, I. Overexpression of RhoA Enhances Peritoneal Dissemination: RhoA Suppression with Lovastatin May Be Useful for Ovarian Cancer. *Cancer Sci.* **2008**, *99*, 2532–2539. [CrossRef] [PubMed]
83. Liu, H.; Liang, S.-L.; Kumar, S.; Weyman, C.M.; Liu, W.; Zhou, A. Statins Induce Apoptosis in Ovarian Cancer Cells through Activation of JNK and Enhancement of Bim Expression. *Cancer Chemother. Pharmacol.* **2009**, *63*, 997–1005. [CrossRef] [PubMed]
84. Martirosyan, A.; Clendening, J.W.; Goard, C.A.; Penn, L.Z. Lovastatin Induces Apoptosis of Ovarian Cancer Cells and Synergizes with Doxorubicin: Potential Therapeutic Relevance. *BMC Cancer* **2010**, *10*, 103. [CrossRef] [PubMed]
85. Kobayashi, Y.; Kashima, H.; Wu, R.-C.; Jung, J.-G.; Kuan, J.-C.; Gu, J.; Xuan, J.; Sokoll, L.; Visvanathan, K.; Shih, I.-M.; et al. Mevalonate Pathway Antagonist Suppresses Formation of Serous Tubal Intraepithelial Carcinoma and Ovarian Carcinoma in Mouse Models. *Clin. Cancer Res.* **2015**, *21*, 4652–4662. [CrossRef]
86. Matsuura, M.; Suzuki, T.; Suzuki, M.; Tanaka, R.; Ito, E.; Saito, T. Statin-Mediated Reduction of Osteopontin Expression Induces Apoptosis and Cell Growth Arrest in Ovarian Clear Cell Carcinoma. *Oncol. Rep.* **2011**, *25*, 41–47. [CrossRef]
87. Robinson, E.; Nandi, M.; Wilkinson, L.L.; Arrowsmith, D.M.; Curtis, A.D.M.; Richardson, A. Preclinical Evaluation of Statins as a Treatment for Ovarian Cancer. *Gynecol. Oncol.* **2013**, *129*, 417–424. [CrossRef]
88. Greenaway, J.B.; Virtanen, C.; Osz, K.; Revay, T.; Hardy, D.; Shepherd, T.; DiMattia, G.; Petrik, J. Ovarian Tumour Growth Is Characterized by Mevalonate Pathway Gene Signature in an Orthotopic, Syngeneic Model of Epithelial Ovarian Cancer. *Oncotarget* **2016**, *7*, 47343–47365. [CrossRef]
89. Stine, J.E.; Guo, H.; Sheng, X.; Han, X.; Schointuch, M.N.; Gilliam, T.P.; Gehrig, P.A.; Zhou, C.; Bae-Jump, V.L. The HMG-CoA Reductase Inhibitor, Simvastatin, Exhibits Anti-Metastatic and Anti-Tumorigenic Effects in Ovarian Cancer. *Oncotarget* **2016**, *7*, 946–960. [CrossRef]
90. Crescencio, M.E.; Rodríguez, E.; Páez, A.; Masso, F.A.; Montaño, L.F.; López-Marure, R. Statins Inhibit the Proliferation and Induce Cell Death of Human Papilloma Virus Positive and Negative Cervical Cancer Cells. *Int. J. Biomed. Sci.* **2009**, *5*, 411–420.
91. Pan, Q.; Xu, J.; Ma, L. Simvastatin Enhances Chemotherapy in Cervical Cancer via Inhibition of Multiple Prenylation-Dependent GTPases-Regulated Pathways. *Fundam. Clin. Pharmacol.* **2020**, *34*, 32–40. [CrossRef] [PubMed]
92. Sheng, B.; Song, Y.; Zhang, J.; Li, R.; Wang, Z.; Zhu, X. Atorvastatin Suppresses the Progression of Cervical Cancer via Regulation of Autophagy. *Am. J. Transl. Res.* **2020**, *12*, 5252–5268. [PubMed]
93. Li, Z.; Kang, Y. Lipid Metabolism Fuels Cancer's Spread. *Cell Metab.* **2017**, *25*, 228–230. [CrossRef] [PubMed]
94. Kobayashi, Y.; Takeda, T.; Kunitomi, H.; Chiwaki, F.; Komatsu, M.; Nagai, S.; Nogami, Y.; Tsuji, K.; Masuda, K.; Ogiwara, H.; et al. Response Predictive Markers and Synergistic Agents for Drug Repositioning of Statins in Ovarian Cancer. *Pharmaceuticals* **2022**, *15*, 124. [CrossRef] [PubMed]

International Journal of
*Molecular Sciences*

MDPI

*Review*

# Emerging Roles of the α-Catenin Family Member α-Catulin in Development, Homeostasis and Cancer Progression

**Mateusz Gielata, Kamila Karpińska, Tomasz Pieczonka and Agnieszka Kobielak ***

Laboratory of the Molecular Biology of Cancer, Centre of New Technologies, University of Warsaw, 02-097 Warsaw, Poland
* Correspondence: a.kobielak@cent.uw.edu.pl; Tel.: +48-22-55-43-735

**Abstract:** α-catulin, together with vinculin and the α-catenins, belongs to the vinculin family of proteins, best known for their actin-filament binding properties and crucial roles in cell-cell and cell-substrate adhesion. In the past few years, an array of binding partners for α-catulin have surfaced, which has shed new light on the possible functions of this protein. Despite all this information, the molecular basis of how α-catulin acts in cells and controls a wide variety of signals during morphogenesis, tissue homeostasis, and cancer progression remains elusive. This review aims to highlight recent discoveries on how α-catulin is involved in a broad range of diverse biological processes with an emphasis on cancer progression.

**Keywords:** α-catulin; CTNNAL1; catenin; invasion; epithelial-mesenchymal transition; EMT; vascular mimicry

**Citation:** Gielata, M.; Karpińska, K.; Pieczonka, T.; Kobielak, A. Emerging Roles of the α-Catenin Family Member α-Catulin in Development, Homeostasis and Cancer Progression. *Int. J. Mol. Sci.* **2022**, 23, 11962. https://doi.org/10.3390/ijms231911962

Academic Editor: Laura Paleari

Received: 14 September 2022
Accepted: 6 October 2022
Published: 8 October 2022



## 1. Introduction

Homeostasis in healthy tissues strongly depends on cadherin- and integrin-mediated, cell-to-cell and cell-to-extracellular matrix (ECM) adhesion, respectively [1]. Both types of adhesion are crucial for maintaining tissue architecture and sensing and responding to changes in their environments. Cadherins are transmembrane glycoproteins that mediate calcium-dependent cell-cell adhesion. Through their homophilic binding interactions, cadherins play a role in cell-sorting mechanisms, conferring adhesion specificities on cells. The regulated expression of cadherins also controls cell polarity and tissue morphology. Classical cadherins are located at adherens junctions and are characterized by five homologous repeats at the extracellular domain. In contrast, the intracellular classical cadherin cytoplasmic domain (CCD) binds armadillo family proteins β-catenin (Ctnnb1) and p120ctn (Ctnnd1). The interaction with β-catenin links cadherins to α-catenin and the actin cytoskeleton, whereas p120ctn is involved in cadherin turnover. By regulating contact formation and stability, cadherins play a crucial role in tissue morphogenesis and homeostasis [1].

The adhesion of cells to the extracellular matrix (ECM) is mainly mediated by integrins, which undergo a conformational change upon activation to recruit structural and signaling molecules. Thus, integrins not only mechanically couple the cytoskeleton to the ECM but also transmit molecular signaling cascades to regulate cellular functions in response to extracellular cues [1].

During tissue morphogenesis, wound healing or pathological alterations in diseases like cancer, the ability of cells to rapidly and reversibly change adhesive properties is a key feature. This cell plasticity is driven by the programs of the epithelial–mesenchymal transition (EMT) and mesenchymal–epithelial transition (MET), both of which play essential roles during normal embryogenesis and tissue homeostasis [2]. However, the aberrant activation of these processes can also drive different stages of cancer progression, including invasion, cell dissemination, metastatic colonization, and secondary tumor outgrowth [3]. EMT enables physically connected epithelial cells to disassociate their characteristic classical cadherin/catenin cell-cell contacts, lose their apical-basolateral polarity, and increase expression and activity of integrins displaying leading-edge asymmetry to become motile and mesenchymal-like and capable of degrading the basement membrane. The events occurring during EMT include the downregulation of cytokeratins and E-cadherin, epithelial-specific markers, and an increase in mesenchymal markers, such as fibronectin, N-cadherin, and vimentin. Transcription factors, including Snail1/Snail, Snail2/Slug, Twist, and ZEB1, are well known to be involved in the orchestration of EMT. Cell–cell adhesion and cell–ECM sites contain overlapping functional constituents containing common and distinct proteins. The crosstalk between these adhesion sites is crucial to coordinate cell migration with dynamic interactions between cells. Because both integrins and cadherins associate with the cytoskeleton and many common signaling molecules, it is likely that the cell–ECM and cell–cell adhesion processes mediated by these two types of receptors act in a coordinated manner in regulating cellular functions [4]. Changes in expression or mutations of these proteins, especially cadherins, catenins, and integrins, are frequently associated with diseases ranging from developmental defects to carcinogenesis and metastasis [5–7]. It is well established that two significant hallmarks of cancer, namely loss of cell-to-cell adhesion and anchorage-independent growth, are both dependent on cell adhesion molecules. Vinculin and α-catenin are two related proteins that play crucial roles in those processes [4,8]. However, the function of their recently characterized homolog α-catulin is still poorly understood. α-catulin is a protein whose name is composed of "α-cat", which comes from α-catenin, and "ulin", which comes from vinculin as it is a homolog of α-catenin protein belonging to the vinculin superfamily. Despite the sequence homology and shared superfamily with α-catenin, α-catulin's localization and functions appear to differ. Multiple reports describe α-catulin as an important factor contributing to cancer cell migration and invasion; however, the exact molecular mechanism leading to this phenomenon remains unclear. A growing number of reported α-catulin-interacting partners and new connections imply even more complex regulatory functions for this protein. This review aims to highlight recent discoveries emphasizing how α-catulin is involved in the coordination of a network of signals and actin cytoskeleton regulation.

## 2. α-Catulin—A Member of the α-Catenin Family

Whereas all other catenins (β-catenin, plakoglobin and p120 catenin) share relatively high sequence homology and belong to the Armadillo family of proteins, vinculin and α-catenins differ in sequence and structural organization and form the vinculin family [9] together with the recently characterized homolog α-catulin [10]. Although it is well known that α-catenin is necessary for cadherin-catenin-mediated cell-cell adhesion, and vinculin is important for integrin-mediated cell-ECM adhesion and cell-cell adhesion, the function of α-catulin is still not well understood. α-catulin (catenin alpha like 1) protein is encoded by the *CTNNAL1* gene located on chromosome 9 in loci q31–32 (Figure 1B) positions 108,942,569–109,013,522 in a minus-strand orientation, resulting in a base length of 70,954 (https://www.genecards.org/cgi-bin/carddisp.pl?gene=CTNNAL1 (accessed on 5 September 2022)). The protein is 734 amino acids long and weighs 81,896 Da (https://www.uniprot.org/uniprotkb/Q9UBT7/entry (accessed on 5 September 2022)). However, two other alternative splicing isoforms have been described, one with substitution in positions 714–734 [10] and another with missing aa in positions 397–480 [11]. There is no 3D structure for α-catulin that has been deposited in the PDB file. The only available structure is the one predicted by AlphaFold (https://alphafold.ebi.ac.uk/ (accessed on 5 September 2022)) which still has a poor structural prognosis in some locations (Figure 1C). mRNA of α-catulin is widely expressed in the human body. It has been reported to be expressed in the thymus, prostate, testes, ovary, small intestine, colon [12], skeletal muscle, lung, heart, and placenta [10]. The human protein atlas also confirms that α-catulin protein is widely found in the human body, interestingly having the highest score in endocrine tissues, female and muscle tissues (proteinatlas.org). In 2002, Park et al. demonstrated that when using the Blast tool and analyzing the *CTNNAL1* sequence, they reported having high similarity to α-catenin. The BESTFIT similarity alignment of α-catulin with its closest human homolog α-catenin showed 27% identity, and alignment with vinculin showed almost 20% identity (Figure 1A). They also showed that α-catulin is characterized by an extra 16 N-terminal amino acids not present in mammalian α-catenins. α-catulin and α-catenin homology is represented by two blocks; the first homology sequence is between α-catulin residues 18–524 and αN-catenin positions 2–504. The following sequence is a region of 110 amino acids present in α-catenin that is omitted in α-catulin. The second homologous block extends α-catulin residues 525–734 [12]. α-catulin also shows high sequence similarity with vinculin, hence being categorized as a part of the vinculin superfamily of proteins. The homology with vinculin, however, is lower, reaching 21%. The similarity is essentially high in the N-terminal domain of the protein, shown to have putative binding sites for β-catenin, talin, and α-actinin [10]. Amphipathic helices in the C-terminal region corresponding to α-catenin contain potential binding sites for the actin cytoskeleton. This region also contains potential binding sites for ZO-1, the protein important for tight junctions [10,12], another type of intercellular adhesion complex that forms the border between the apical and basolateral cell surface domains in polarized epithelia and controls paracellular permeability. Despite the sequence homology between α-catulin and α-catenin, their subcellular localization pattern is different as shown by Park et al. α-catulin localized to both the membrane-rich (pellet) fraction and the soluble (cytosolic) fraction, whereas α-catenin was found to localize almost exclusively to the membrane-rich fraction. They confirmed those results with two different experiments, one with high-speed fractionation into cytosolic and membrane-rich fractions followed by Western blotting, and the second with Myc-tagged α-catulin (pcDNA Myc:α-catulin) and indirect immunofluorescence. Despite the above-mentioned characteristics of α-catulin, it is still very poorly characterized.

A

| | **α-catulin** | **α-catenin** | **Vinculin** |
|---|---|---|---|
| **α-catulin** | 100 | 27.09 | 19.15 |
| **α-catenin** | 27.09 | 100 | 25.84 |
| **Vinculin** | 19.15 | 25.84 | 100 |

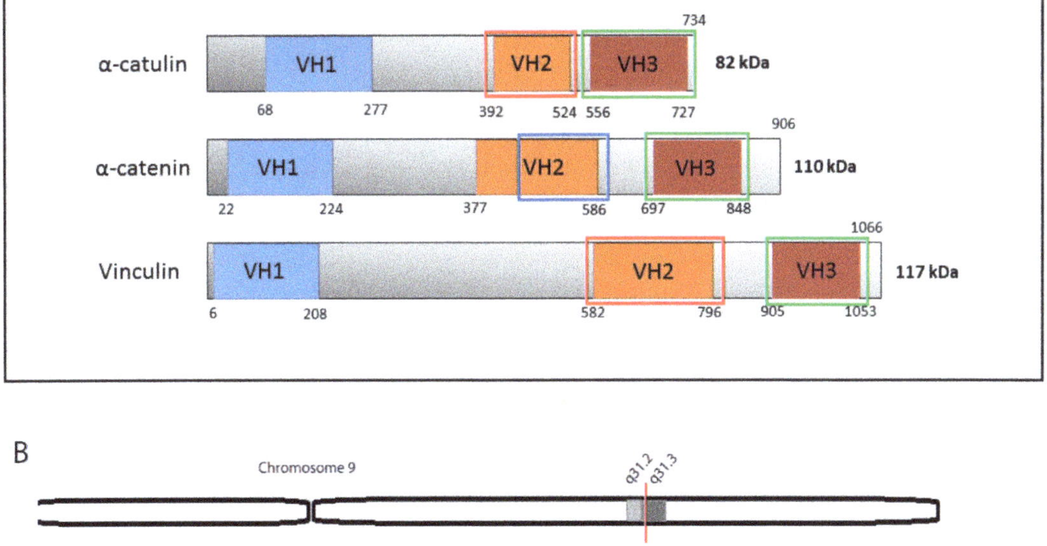


**Figure 1.** *Cont.*

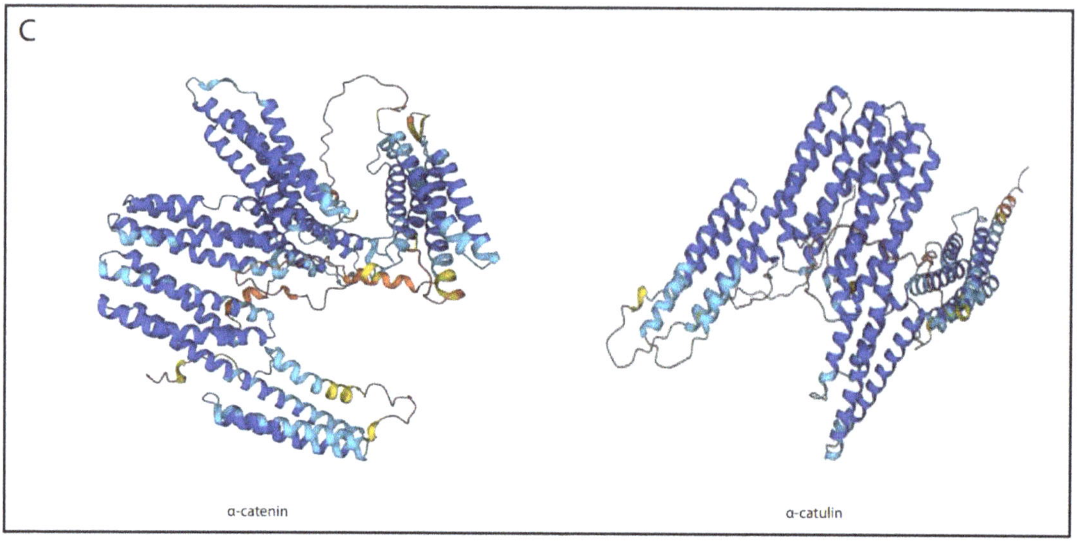


**Figure 1.** Structural features of α-catulin. (**A**) Table represents amino acid sequence similarities (%) between α-catulin, α-catenin and vinculin. α-catulin shares 27.09% homology with α-catenin and 19.15% with vinculin. (**B**) Schematic representation of α-catulin (CTNNAL1) gene on chromosome 9 locus 31.3. (**C**) Scheme shows the predicted 3D structure of α-catulin and α-catenin protein by AlphaFold.

## 3. Binding Partners of α-Catulin

One of the first described interacting partners of α-catulin is Lbc Rho guanine nucleotide exchange factor. Rho guanine nucleotide exchange factor (GEF) functions for the RhoA small GTPase protein [13]. RhoA is inactive when bound to the GDP, but when acted on by the Rho GEFs, GDP can be released, and GTP might be attached, leading to the activation of RhoA. Furthermore, active RhoA can bind to and activate distant effectors or enzymes. Interestingly, in this particular case, RhoA is a major regulator of the cell actin cytoskeleton [14]. One of the GEFs specific for Rho is a DH domain containing Lbc oncogenic product GEF [15,16]. All Lbc Rho GEF forms possess common C-terminal regions following DH domain cassette [17]. Park et al. showed a direct association between Lbc Rho GEF and α-catulin using three independent systems: yeast two-hybrid interaction, direct binding in vitro, and complex formation in mammalian cells. The required site of interaction within the Lbc C-terminal region was mapped to the ~253-residue IDR (intrinsically disordered region). They also determined that the α-catulin site required for the interaction lies in the N-terminal residues 34–524. Coexpression of α-catulin and wt-Lbc led to increased GTP-Rho formation in cooperative action. This implies that α-catulin is an upstream regulator of Rho. Overall, the authors conclude that α-catulin acts as a scaffold protein for Lbc Rho GEF and facilitates Lbc-induced Rho signals [12,17].

α-Catulin has also been shown to interact with the dystrophin complex through direct interaction with dystrobrevin in *C. elegans*. This interaction is conserved and also present in mouse skeletal muscles [18]. Dystrophin has been known as a cause of Duchenne muscular dystrophy, yet dystrophin usually functions in protein complexes known as dystrophin-associated protein complex (DAPC) [19]. It had been previously shown that mutations in the CTNNAL1 gene lead to the interruption of DAPC localization near dense bodies [20]. In the above-mentioned publication, the reciprocal action of α-catulin with dystrobrevin was validated by co-immunoprecipitation as well as by mass spectrometry and yeast two-hybrid screen. The authors observed an increase in α-catulin expression levels in the skeletal muscle of dystrophin-deficient mice, where dystrophin-associated protein complex is disassembled, and the link between the costamere and the sarcolemma is absent. To bind α-catulin, dystrobrevin requires a C-terminus as well as an α-helix H2 proximal to the C-terminal region [18]. Similar results have been obtained in other studies. Lyssand et al. showed that the C-terminal domain of dystrobrevin recruits α-catulin to the $\alpha_{1D}$-AR signalosome. Adrenergic receptors (ARs) and G protein-coupled receptors (GPCR) are important regulators of cardiovascular system function. Their function revolves around increasing blood pressure and promoting vascular remodeling. [21]. Sequence analysis revealed that, similar to α-catenin, α-catulin has a putative binding domain for β-catenin; therefore, a group led by Deniz Toksoz took a closer look into this interaction, mapping it to the N-terminal 163 amino acids of the protein [21,22]. When performing co-immunoprecipitation, they noticed that α-catulin indeed co-precipitates with β-catenin, but the amount of α-catulin associated with β-catenin appeared to be smaller than that of α-catenin associated with β-catenin. Given that endogenously in cells, the pool of β-catenin is naturally bound to α-catenin, these results were not surprising. α-catulin might associate with a different fraction of β-catenin than α-catenin does. There might be other pools of β-catenin, such as tyrosine phosphorylated β-catenin, in which protein interactions are altered [21,23–25]. Here, the authors additionally proposed the antiproliferative role of α-catulin, as it attenuates cyclin D1 transcription, leading to decreased cyclin D1 protein levels. They also observed that expression of α-catulin had a negative impact on cancer cell colony formation ability, leading to the statement that α-catulin modulates endogenous growth signaling pathways [21,22]. As β-catenin functions at the adherens junctions and also acts in the nucleus after stabilization of a pool of β-catenin in response to the upstream Wnt signals, it is crucial to further investigate the catulin-β-catenin interaction. Another α-catulin interacting protein was reported in the publication by Wiesner et al. in 2008. It was shown that α-catulin can modulate the NF-κB pathway by binding to IKK-β [21,26]. The NF-κB pathway plays a pivotal role in a variety of biological processes like innate and adaptive immune responses, tissue differentiation and apoptosis [21,27,28]. The targets of NF-κB include its own inhibitors IκBα and IκBβ [21,29]. Different extracellular stimuli activate the IκB kinases IKK-α and -β, which phosphorylate IκBα, which results in the degradation of IκBα and translocation of NF-κB to the nucleus [21,30]. Wiesner et al. provided evidence that α-catulin binds to IKK-β by immunoprecipitation. Moreover, they limited the interaction site in α-catulin to its C-terminal 87 amino acids. As α-catulin binds to IKK-β in the C terminus and to Lbc Rho GEF in the N-terminus, the authors claim that it may allow simultaneous stimulation of both pathways, being a bridge between those two. As there is evidence that both the NF-κB and RhoA signaling pathways play multiple roles in tumorigenesis, cell migration, invasion, and escape from apoptosis, α-catulin, as a linker of those two pathways, might serve as a crucial clinical target [21,26,31]. The interaction of α-catulin with Lbc, dystrophin complex and other proteins and resulting pathways activation have been represented in Figure 2.

Finally, α-catulin has been described as an interactor protein of human NEK1 protein kinase. NEK kinases are involved in regulating diverse cellular processes like the cell cycle, mitosis, cilia formation, and the DNA damage response and the etiology of polycystic kidney disease (PKD). α-catulin has been described as one of the 11 new binding proteins of NEK1. Moreover, it has been proven to interact with both regulatory and kinase domains (NRD and NKD) [32]. Interestingly, aberrant expression of NEKs appears to be involved in the initiation, maintenance, progression and metastasis of cancer and is associated with a poor prognosis [33]. A better understanding of NEK1 kinase interaction with α-catulin may lead to more successful clinical trials of NEK inhibitors.

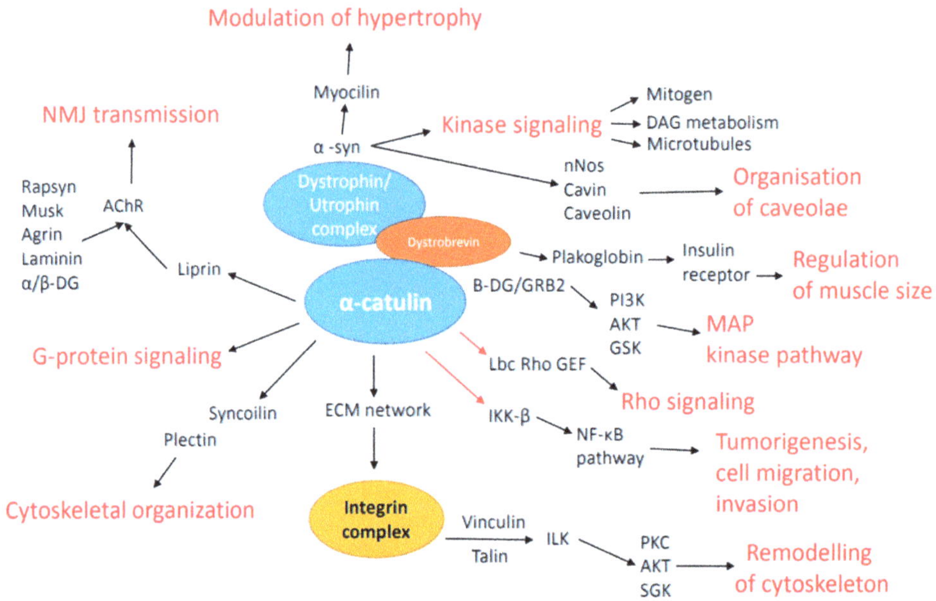


**Figure 2.** Overview of the function of α-catulin in dystrophin complex. In blue circles, α-catulin and dystrophin complex are shown. The interaction occurs via dystrobrevin, highlighted in orange. Shown are distinct interactors of the complex as well as direct interactors of α-catulin. Enlarged is also integrin complex, having interactions indirectly via ECM and impacting cytoskeleton remodeling. Highlighted in red are key pathways and functions resulting from either interaction of the complex or α-catulin directly.

## 4. α-Catulin and Its Function during Development

The plethora of interacting proteins indicates that α-catulin may play essential roles in various vital regulatory processes. Thus far, the important role of α-catulin has been shown in the process of neurulation during mouse development [34], where cell-cell and cell-ECM interactions are constantly under remodeling to enable proper architecture and function of forming tissues. The actin-cytoskeleton and actomyosin contractility integrated at the cell-cell and cell-ECM adhesions cooperatively are crucial to shape the cells and tissues [35–37]. The adherens junctions are required for the transmission of force across an epithelium, and the actomyosin cortex, which spans the apical surface of an epithelium, transitions between elongation and active state of actin nucleation while still attached to the adherens junction, allowing for apical constriction, which is crucial, for example, during neurulation [37–39]. It is important that actomyosin machinery is located at the right place and time to generate the required force to pull the neural folds together [40]. Interestingly, α-catulin was shown to participate in the apical actomyosin network regulation by serving as a scaffold protein that may be important for properly directing Rho family GTPase signaling during neurulation. α-catulin-deficient mice show neural tube (NT) closure defects. They are embryonically lethal with massive disorganization of their neuroepithelium, extra bending, absence of apically localized actin filaments, nestin and phosphorylated myosin, and inappropriate basement membrane assembly due to very low expression of its components: laminin and fibronectin. The neuroepithelium of α-catulin deficient mice lack apically localized actin filaments and P-Mlc, which typically correlate with proper Rho-dependent cell constriction. In vitro studies performed in a three-dimensional model of MDCK cells showed that α-catulin is localized specifically at the apical parts of cells membranes and is important for proper cell polarization, organization of actomyosin cytoskeleton, stabilization of intercellular junction as well as distribution of active Rho A. Taken together, data collected both from in vivo mouse model and in vitro 3D studies indicated a pivotal role of alpha-catulin protein in neurulation during embryonic development, as it can act as a scaffold for RhoA in apical parts of cells, which results in correct spatial activation of downstream myosin to influence actin-myosin dynamics and the stability of cell-cell junctions, which allows generating the appropriate tension needed for the apical constriction of cells and proper bending of the neural plate [34].

## 5. Role of α-Catulin in Homeostasis

In the last decade, numerous studies have also demonstrated the importance of α-catulin in the maintenance of tissue homeostasis. A-catulin was reported to play potential functions in hematopoietic stem cells (HSCs), bronchial epithelium, muscles and intestine [18,20,41–45]. In hematopoietic stem cells, α-catulin is expressed only in a specific population of 0.02% of bone marrow hematopoietic cells. Generation of a mouse model with green fluorescent protein (GFP) knocked-in into the α-catulin locus allowed to show that α-catulin together with c-kit marks the population of cells that possess the long-term multilineage reconstitution ability of bone marrow after irradiation [41]. In addition, the distribution of α-catulin$^+$ c-kit1$^+$ cells indicates that HSCs are more common in the central marrow than near the bone surface [41,42]. Even though α-catulin proved to be a great marker for HSC visualization in the bone, the exact function of this protein in those cells was not established.

Furthermore, high expression of α-catulin was also detected in bronchial epithelium under ozone-stressed conditions. Results from this study suggest that elevated α-catulin expression may be a protective response aimed at maintaining airway epithelial integrity [43].

Moreover, in neuromuscular junctions, dystrobrevin utilizes α-catulin for proper neurotransmitter receptor (AChR) clustering on myotubes, indicating its important role in a synaptic machinery organization [44]. As an anchor protein that locates potassium channels and neurotransmitter receptors in specific nanodomains, α-catulin plays a key role in the physiological processes related to the neurosecretion as well as excitation of neurons and muscles. Dysfunction of this important protein may be linked to muscular and neurological disorders [20,44]. It has also been reported that α-catulin ortholog is a critical cytoskeletal regulator in C.elegans, crucial for the proper localization of calcium-dependent potassium channels in both neurons and muscles. In muscles, α-catulin, via the dystrophin complex, binds the calcium-dependent potassium channels near L-type calcium channels. In turn, in neurons, α-catulin controls the localization of the potassium channels independently of the dystrophin complex [20,21,46]. The interaction with dystrophin complex seems to be the best characterized so far for α-catulin.

Recent studies performed on Chinese patients with Hirschsprung disease revealed that α-catulin can be attributed to genetic factors or gene-gene interaction networks responsible for enteric neuronal dysfunction [45]. Interestingly, catulin expression was observed in the enteric innervation of newborn mice [34].

## 6. α-Catulin in Cancer Invasion and Metastasis

Even though α-catulin is overall a very poorly described protein, its participation in cancerogenesis and influence on cancer cell invasion and metastasis has been reported and researched in many papers. Both structural and signaling functions of α-catulin may play a role in cancer progression. As mentioned above, α-catulin in the N-terminal region contains binding sites for β-catenin, talin, α-actinin, and the actin cytoskeleton. This suggests that it may function as a cytoskeletal linker protein that is able to modulate cell migration [8]. Cell migration is a process that plays a pivotal role in carcinogenesis and participates in the metastasizing of cancer cells. Metastasis is a complex phenomenon that occurs in all types of cancers and is responsible for death [47]. It is based on the fact that cancer cells escape from the primary tumor, migrate, enter the lumen of blood and lymphatic vessels and reach distant organs, where they can repopulate the tumor mass [48,49]. Cancer cells need to acquire a mesenchymal phenotype in the process called epithelial-to-mesenchymal transition (EMT) [50]. EMT is a phenomenon where cells downregulate proteins involved in apical cell-cell contact and adherence junction formation, such as E-cadherin and α-catenin, and start upregulating proteins specific for mesenchymal features of the cell, such as N-cadherin and vimentin, which results in the enhanced motility of the cells. This switch between relatively stable cell-cell contacts and an increase in motility is crucial for cancer invasion [51]. It has been observed that when α-catenin, a cell-cell junction protein, is conditionally lost in the epithelium, cells begin to demonstrate increased proliferation rates, migrative properties, and the squamous cell carcinoma (SCC) phenotype [52]. Using microarray analysis to compare mouse α-catenin cKO keratinocytes, which failed to form cell–cell junctions, and WT epithelial cells, it was observed that α-catulin is highly upregulated in the cells with increased motility and mesenchymal phenotypes [52]. This data suggested the participation of α-catulin in cancer progression and was further investigated by our group in a model of human head and neck squamous cell carcinoma (hHNSCC), which is a very aggressive tumor type and accounts for more than 450,000 malignancies diagnosed each year. Despite new treatment options, patients are still faced with a very high rate of recurrence and metastatic disease, with a 5-year survival rate of only 50% [53–55]. It was shown that α-catulin is upregulated in the metastatic cells in the xenotransplant in vivo model and also in vitro in the hSCC (human squamous cell carcinoma) cell line after EMT induction. Moreover, α-catulin is highly expressed at the invasion front and in migrating, metastatic streams of cells in human samples of HNSCC and in higher grades of tumor samples when compared with normal mucosa epithelium [56]. Most importantly, ablation of α-catulin in hSCC cells decreased the ability of these cells to migrate and invade in vitro and decreased their metastatic potential in vivo [56]. Given that the expression of α-catulin not only correlates with tumor grade, but also appears to be involved in the regulation of the invasive character of the HNSCC cells, it suggests that α-catulin may represent a novel yet critical mediator of oral cancer progression. As this type of cancer usually spreads locally, utilizing collective migration, α-catulin could be important for spatiotemporal fine-tuning of Rho GTPases within a group of cancer cells to control divergent cell-cell and cell-ECM adhesion as well as cytoskeletal functions to achieve cellular coordination and mechanocoupling. This is one of the options that will require further testing to better understand the role of catulin in the process of HNSCC invasion. As α-catulin expression and function correlated with the early onset of hSCC cell invasion, our group used the human α-catulin promoter fragment driving GFP expression to develop a reporter system. This unique system, for the first time, allowed us to isolate in vivo a small population of invasive cells at the human tumor invasion front [57]. After verifying the reporter system, we showed that cells highly expressing GFP driven from α-catulin expression localize at the invasion front in a spheroid model of hSCC cells. Additionally, this system marked the cells with higher migratory, invasive, and tumorigenic potential in vitro in the 3D model. Cells highly expressing α-catulin were also observed in a small population of invasive cells at the tumor front in the in vivo model of head and neck squamous cell carcinoma. Expression of GFP under α-catulin promoter correlated with the loss of an epithelial marker, E-cadherin expression, indicative of ongoing EMT. The reporter system allowed for isolation and transcriptional characterization of those highly invasive cells, providing a list of deregulated genes that are involved in cellular movement, ILK and integrin signalling, as well as axonal guidance signalling [57]. This functional genomic study of the purified population of invasive cells revealed enrichment in genes involved in cellular movement and invasion, providing a molecular profile of HNSCC invasive cells. Interestingly, this profile overlapped partially with the expression of signature genes related to partial EMT available from single-cell analysis of human HNSCC specimens [58]. This comparison strengthens the idea that α-catulin in this type of cancer might be important for spatiotemporal regulation of Rho GTPases within a group of cancer cells to control dynamic plasticity and crosstalk between cadherin-mediated cell-cell contact and integrin-dependent cell-ECM adhesion, which is crucial during collective invasion and migration. Further research on catulin revealed that its role in cancer progression is not only limited to HNSCC specifically, which utilizes collective invasion for local spread. It was recently published that α-catulin is also expressed in human breast cancer samples and triple-negative breast cancer cell lines, and its expression correlates with tumor progression [59]. Breast cancer is now the most common cancer worldwide [60], and the worst outcome is presented by triple-negative breast cancer [61]. Knockdown of α-catulin in triple-negative human breast cancer cell lines MDA-MB-231 and HCC1806 revealed a decrease in the invasion capability of those cells in 3D spheroid model assays [59]. The use of a catulin-GFP-promoter-based reporter system in a 3D spheroid model of triple-negative breast cancer cell lines showed that the most invading cells co-express α-catulin and known EMT marker vimentin. Transcriptional profiling of GFP-positive cells isolated from tumors that formed after injection of a catulin-GFP triple-negative breast cancer cell line disclosed the list of deregulated genes involved in cellular movement and invasion and, interestingly, migration of endothelial cells [59]. Top pathways deregulated in the α-catulin GFP + cells involved epithelial adherens junction signaling and remodeling of epithelial adherens junctions. Special attention was paid to genes involved in the vasculature, as it was observed that tumor areas enriched in GFP+ cells presented visible dense vasculature. Surprisingly, some cells highly expressing GFP co-expressed MCAM (CD146), an endothelial marker but also a cellular surface receptor of different ligands, are actively involved in signaling in the numerous physiological and pathological processes involving metastases of different cancer types. Cells highly expressing GFP and co-expressing MCAM formed vasculogenic structures resembling vessels. This suggests that α-catulin marks highly invasive breast cancer cells that are characterized by increased plasticity and might participate in the process of vascular mimicry, allowing cancer cells to metastasize [59]. In addition, ablation of α-catulin in the in vivo model resulted in decreased tumor size and decreased stemness potential of cancer cells with lowered expression of CD44, which is known to be enriched in breast cancer (BC) stem cells [59]. These data implicate that α-catulin might play an important role in cancer type-specific tumor-microenvironment interplay. Moreover, it may be involved in the inflection of adhesive properties of tumor cells. The possible mechanism of increased α-catulin expression in invasive cancer cells might be explained by the research performed by Cassandri et al. [62]. They showed that zinc-finger protein 750 (ZNF750) is a negative regulator of the migration and invasion of breast cancer cells. It functions as a repressor of a prometastatic transcriptional program. This transcriptional program was shown to express genes that are involved in focal adhesion and extracellular matrix interactions with an emphasis on CTNNAL1. They showed that the expression of CTNNAL1 and LAMB3 contradictorily correlated with ZNF750 expression in a breast cancer model. ZNF750 recruits epigenetic modifiers KDM1A and HDAC1 to the promoter region of the α-catulin gene, which affects histone marks and trans activates these genomic sites. Additionally, they also showed gene expression analysis in cancer patient datasets that indicated ZNF750 and its targets to be negative prognostic factors in breast cancer [62]. In 2011, Fan et al. published a paper confirming the previously described participation of α-catulin in tumorigenesis. They showed that α-catulin expression is elevated in oral cancer cells versus normal cells. They also found that the knockdown of α-catulin resulted in the accumulation of cell populations in S and G2/M cell-cycle phases with decreased cyclin A and cyclin B1 expression. α-catulin knockdown induced cellular senescence as the major phenotype of cell death in two oral cancer cell lines, OC2 and A549. In patients, α-catulin expression correlated with tumor size, whereas α-catulin knockdown supressed tumorigenicity in xenograft models. Knockdown of α-catulin in cancer cells bearing either wild-type or mutant p53 was sufficient to trigger DNA damage response and eventually induce cellular senescence in vitro [63]. In addition to structural functions and regulation of actin cytoskeleton during cancer invasion and migration, α-catulin enhances cancer metastasis by influencing signaling pathways. Liang et al. showed that α-catulin expression correlates with the cell invasiveness potential in vitro and metastatic potential in vivo. It occurs via an ILK/NF-κB/integrin network where α-catulin directly interacts with ILK, which in turn activates the ILK/Akt/NF-κB signaling pathway and upregulates fibronectin and integrin $\alpha_v\beta_3$. α-catulin as an integrin signaling adaptor might play a pivotal role in regulating integrin-mediated cellular functions via binding to ILK [64]. Later, the same scientific group focused on the participation of α-catulin in cancer stemness and EMT. They found that cells overexpressing α-catulin have genes such as *FGF2*, *BMI1*, *ALDH1A3*, *POU5F1* and *NANOG* upregulated. Additionally, high expression of α-catulin was required to maintain stemness in a lung cancer model, and klf5 was indicated as a new interacting protein that plays an important role in stemness maintenance by cooperating with α-catulin to enhance the transcription of *POU5F1* and *NANOG*. Knockdown of klf5 in cells overexpressing α-catulin abolished the sphere formation capacity. α-catulin not only interacts with klf5 but also protects this protein by blocking the WWP1-mediated proteasomal degradation of KLF5 [65]. The participation of α-catulin in cancer cell migration and invasion has also been proven in a melanoma cancer model. Kreiseder et al. showed that α-catulin is highly expressed in melanoma cells, resulting in reduced E-cadherin and increased N-cadherin expression. Upregulation of α-catulin promotes expression of EMT markers Snail/Slug and Zeb1/2; in addition, α-catulin regulated PTEN and RKIP, inhibitors of the NF-κB pathway. They also found MCAM, plakoglobin, and occludin to be altered in α-catulin-deficient cells. Their results further confirmed that α-catulin is not only responsible for the downregulation of E-cadherin but is also required for melanoma invasion by the upregulation of MMP 2 and 9 and the activation of ROCK/Rho [66]. They further studied the role of α-catulin and, in 2015, published a paper showing that α-catulin is responsible for the chemoresistance of melanoma cells to cisplatin. This reduction in cisplatin-mediated apoptosis of melanoma cancer cells is due to the fact that α-catulin is responsible for NF-κB, AP-1 activation and ERK phosphorylation, and, in the case of knockdown of α-catulin, the cisplatin-mediated apoptosis was shown to be enhanced [67].

## 7. Conclusions

α-catulin, together with vinculin and α-catenins, belongs to the vinculin family of proteins, best known for their actin-filament binding properties and crucial roles in cell-cell and cell-substrate adhesion; however, despite sequence homology, α-catulin seems to have independent roles. α-catulin has been shown to be important in inflammation, apoptotic resistance, cytoskeletal reorganization, senescence resistance, cancer progression, and EMT. Multiple binding proteins of α-catulin revealed in recent years suggest a molecular hub function, integrating a cytoskeleton with a number of signaling pathways. Unfortunately, the molecular mechanisms of α-catulin action are still poorly characterized and need further investigation, especially in the field of cancer progression.

Increased α-catulin expression was observed in the invading front of squamous cell carcinoma, and its depletion led to decreased invasion and metastasis in a xenograft transplant mouse model. α-catulin was also reported to be upregulated in a highly invasive non-small cell lung cancer cell line, as well as in breast and prostate cancer. Despite multiple reports describing α-catulin as an important factor contributing to cancer cell migration and invasion, the exact molecular mechanism leading to this phenotype remains unclear.

As α-catulin depletion was shown to have a strong effect on RhoA signaling and the actomyosin cytoskeleton arrangement, it will be crucial to further investigate the role of α-catulin in spatial RhoA distribution during cell migration. Further experiments encompassing mass spectrometry are currently underway in our laboratory in order to identify potential α-catulin interaction partners contributing to the front-rear stabilization of migrating cancer cells.

A question also remains about potential α-catulin and cadherin interactions. The role of α-catulin in the process of EMT and the switch between relatively stable cell-cell contacts and an increase in motility, which is crucial for cancer invasion, is of great interest. Catulin expression was reported to be upregulated when α-catenin, a cell-cell junction protein, was conditionally lost in the epithelium, which was accompanied by an increased proliferation rate and migrative properties. Therefore, further investigations at the structural, cellular, and functional levels are also needed to understand the exact sequence of molecular interactions and conformational changes operating between the cadherin/β-catenin/α-catenin complex and α-catulin and F-actin and tension-dependent remodeling of cell-cell adhesion.

An analysis of α-catulin dynamics using high-resolution live imaging should help us to map α-catulin's localisation and interactions in time and space. These could bring us closer to solving how α-catulin orchestrates adhesion and the actin cytoskeleton.

As α-catulin is broadly expressed and plays multiple physiological functions both during development and adult life, direct therapeutic strategies towards silencing its gene may not be applicable. On the other hand, targeted disruption of signaling pathways originating or ending at α-catulin may be a more promising therapeutic target.

**Author Contributions:** M.G. contributed: literature review and manuscript writing; K.K. contributed: literature review and manuscript writing; T.P. contributed: literature review and manuscript writing; A.K. contributed: literature review and manuscript writing. All authors have read and agreed to the published version of the manuscript.

**Funding:** This work was funded by the Polish National Science Center (NCN) grants: 2015/17/B/NZ5/02551; 2017/25/B/NZ5/01597 and 2020/37/B/NZ5/03950 to A. Kobielak.

**Institutional Review Board Statement:** Not applicable.

**Informed Consent Statement:** Not applicable.

**Conflicts of Interest:** The authors declare no conflict of interest.

## References


1. Weber, G.F.; Bjerke, M.A.; DeSimone, D.W. Integrins and cadherins join forces to form adhesive networks. *J. Cell Sci.* **2011**, *124 Pt 8*, 1183–1193. [CrossRef] [PubMed]
2. Nieto, M.A.; Huang RY, J.; Jackson, R.A.; Thiery, J.P. Emt: 2016. *Cell* **2016**, *166*, 21–45. [CrossRef] [PubMed]
3. Lu, W.; Kang, Y. Epithelial-Mesenchymal Plasticity in Cancer Progression and Metastasis. *Dev. Cell* **2019**, *49*, 361–374. [CrossRef] [PubMed]
4. Bachir, A.I.; Horwitz, A.R.; Nelson, W.J.; Bianchini, J.M. Actin-Based Adhesion Modules Mediate Cell Interactions with the Extracellular Matrix and Neighboring Cells. *Cold Spring Harb. Perspect. Biol.* **2017**, *9*, a023234. [CrossRef]
5. Cano, A.; Pérez-Moreno, M.A.; Rodrigo, I.; Locascio, A.; Blanco, M.J.; Del Barrio, M.G.; Portillo, F.; Nieto, M.A. The transcription factor snail controls epithelial-mesenchymal transitions by repressing E-cadherin expression. *Nat. Cell Biol.* **2000**, *2*, 76–83. [CrossRef]
6. Yang, J.; Weinberg, R.A. Epithelial-mesenchymal transition: At the crossroads of development and tumor metastasis. *Dev. Cell* **2008**, *14*, 818–829. [CrossRef]
7. Nieto, M.A. The ins and outs of the epithelial to mesenchymal transition in health and disease. *Annu. Rev. Cell Dev. Biol.* **2011**, *27*, 347–376. [CrossRef]
8. Kobielak, A.; Fuchs, E. alpha-catenin: At. the junction of intercellular adhesion and actin dynamics. *Nat. Rev. Mol. Cell Biol.* **2004**, *5*, 614–625. [CrossRef]
9. McCrea, P.D.; Gu, D. The catenin family at a glance. *J. Cell Sci.* **2010**, *123 Pt 5*, 637–642. [CrossRef]
10. Janssens, B.; Staes, K.; van Roy, F. Human alpha-catulin, a novel alpha-catenin-like molecule with conserved genomic structure, but deviating alternative splicing. *Biochim. Biophys Acta* **1999**, *1447*, 341–347. [CrossRef]
11. Bechtel, S.; Rosenfelder, H.; Duda, A.; Schmidt, C.P.; Ernst, U.; Wellenreuther, R.; Mehrle, A.; Schuster, C.; Bahr, A.; Blöcker, H.; et al. The full-ORF clone resource of the German cDNA Consortium. *BMC Genom.* **2007**, *8*, 399. [CrossRef] [PubMed]
12. Park, B.; Nguyen, N.T.; Dutt, P.; Merdek, K.D.; Bashar, M.; Sterpetti, P.; Toksoz, D. Association of Lbc Rho guanine nucleotide exchange factor with alpha-catenin-related protein, alpha-catulin/CTNNAL1, supports serum response factor activation. *J. Biol. Chem.* **2002**, *277*, 45361–45370. [CrossRef] [PubMed]
13. Hart, M.J.; Sharma, S.; Elmasry, N.; Qiu, R.-G.; McCabe, P.; Polakis, P.; Bollag, G. Identification of a novel guanine nucleotide exchange factor for the Rho GTPase. *J. Biol. Chem.* **1996**, *271*, 25452–25458. [CrossRef] [PubMed]
14. Thumkeo, D.; Watanabe, S.; Narumiya, S. Physiological roles of Rho and Rho effectors in mammals. *Eur. J. Cell Biol.* **2013**, *92*, 303–315. [CrossRef]
15. Toksoz, D.; Williams, D.A. Novel human oncogene lbc detected by transfection with distinct homology regions to signal transduction products. *Oncogene* **1994**, *9*, 621–628.
16. Zheng, Y.; Olson, M.F.; Hall, A.; Cerione, R.A.; Toksoz, D. Direct involvement of the small GTP-binding protein Rho in lbc oncogene function. *J. Biol. Chem.* **1995**, *270*, 9031–9034. [CrossRef]

17. Sterpetti, P.; Hack, A.A.; Bashar, M.P.; Park, B.; Cheng, S.D.; Knoll, J.H.; Toksoz, D. Activation of the Lbc Rho exchange factor proto-oncogene by truncation of an extended C terminus that regulates transformation and targeting. *Mol. Cell Biol.* **1999**, *19*, 1334–1345. [CrossRef]
18. Oh, H.J.; Abraham, L.S.; van Hengel, J.; Stove, C.; Proszynski, T.J.; Gevaert, K.; DiMario, J.X.; Sanes, J.R.; van Roy, F.; Kim, H. Interaction of alpha-catulin with dystrobrevin contributes to integrity of dystrophin complex in muscle. *J. Biol. Chem.* **2012**, *287*, 21717–21728. [CrossRef]
19. Durbeej, M.; Campbell, K.P. Muscular dystrophies involving the dystrophin-glycoprotein complex: An overview of current mouse models. *Curr. Opin. Genet. Dev.* **2002**, *12*, 349–361. [CrossRef]
20. Abraham, L.S.; Oh, H.J.; Sancar, F.; Richmond, J.E.; Kim, H. An Alpha-Catulin Homologue Controls Neuromuscular Function through Localization of the Dystrophin Complex and BK Channels in Caenorhabditis elegans. *PLoS Genet.* **2010**, *6*, e1001077. [CrossRef]
21. Lyssand, J.S.; Whiting, J.L.; Lee, K.-S.; Kastl, R.; Wacker, J.L.; Bruchas, M.R.; Miyatake, M.; Langeberg, L.K.; Chavkin, C.; Scott, J.D.; et al. alpha-Dystrobrevin-1 recruits alpha-catulin to the alpha(1D)-adrenergic receptor/dystrophin-associated protein complex signalosome. *Proc. Natl. Acad. Sci. USA* **2010**, *107*, 21854–21859. [CrossRef] [PubMed]
22. Merdek, K.D.; Nguyen, N.T.; Toksoz, D. Distinct activities of the alpha-catenin family, alpha-catulin and alpha-catenin, on beta-catenin-mediated signaling. *Mol. Cell Biol.* **2004**, *24*, 2410–2422. [CrossRef]
23. Bullions, L.C.; A Notterman, D.; Chung, L.S.; Levine, A.J. Expression of wild-type alpha-catenin protein in cells with a mutant alpha-catenin gene restores both growth regulation and tumor suppressor activities. *Mol. Cell Biol.* **1997**, *17*, 4501–4508. [CrossRef] [PubMed]
24. Obama, H.; Ozawa, M. Identification of the domain of alpha-catenin involved in its association with beta-catenin and plakoglobin (gamma-catenin). *J. Biol. Chem.* **1997**, *272*, 11017–11020. [CrossRef] [PubMed]
25. Kim, K.; Lee, K.Y. Tyrosine phosphorylation translocates beta-catenin from cell–>cell interface to the cytoplasm, but does not significantly enhance the LEF-1-dependent transactivating function. *Cell Biol. Int.* **2001**, *25*, 421–427. [CrossRef] [PubMed]
26. Wiesner, C.; Winsauer, G.; Resch, U.; Hoeth, M.; Schmid, J.A.; van Hengel, J.; van Roy, F.; Binder, B.R.; de Martin, R. Alpha-catulin, a Rho signalling component, can regulate NF-kappaB through binding to IKK-beta, and confers resistance to apoptosis. *Oncogene* **2008**, *27*, 2159–2169. [CrossRef] [PubMed]
27. Baeuerle, P.A.; Baltimore, D. NF-kappa B: Ten years after. *Cell* **1996**, *87*, 13–20. [CrossRef]
28. Baldwin, A.S., Jr. Series introduction: The transcription factor NF-kappaB and human disease. *J. Clin. Investig.* **2001**, *107*, 3–6. [CrossRef]
29. Stehlik, C.; De Martin, R.; Kumabashiri, I.; Schmid, J.A.; Binder, B.R.; Lipp, J. Nuclear factor (NF)-kappaB-regulated X-chromosome-linked iap gene expression protects endothelial cells from tumor necrosis factor alpha-induced apoptosis. *J. Exp. Med.* **1998**, *188*, 211–216. [CrossRef]
30. Zandi, E.; Rothwarf, D.M.; Delhase, M.; Hayakawa, M.; Karin, M. The IkappaB kinase complex (IKK) contains two kinase subunits, IKKalpha and IKKbeta, necessary for IkappaB phosphorylation and NF-kappaB activation. *Cell* **1997**, *91*, 243–252. [CrossRef]
31. Basseres, D.S.; Baldwin, A.S. Nuclear factor-kappaB and inhibitor of kappaB kinase pathways in oncogenic initiation and progression. *Oncogene* **2006**, *25*, 6817–6830. [CrossRef] [PubMed]
32. Surpili, M.J.; Delben, T.M.; Kobarg, J. Identification of proteins that interact with the central coiled-coil region of the human protein kinase NEK1. *Biochemistry* **2003**, *42*, 15369–15376. [CrossRef] [PubMed]
33. Melo-Hanchuk, T.D.; Martins, M.B.; Cunha, L.L.; Soares, F.A.; Ward, L.S.; Vassallo, J.; Kobarg, J. Expression of the NEK family in normal and cancer tissue: An immunohistochemical study. *BMC Cancer* **2020**, *20*, 23. [CrossRef]
34. Karpińska, K.; Cao, C.; Yamamoto, V.; Gielata, M.; Kobielak, A. Alpha-Catulin, a New Player in a Rho Dependent Apical Constriction That Contributes to the Mouse Neural Tube Closure. *Front. Cell Dev. Biol.* **2020**, *8*, 154. [CrossRef]
35. Takeichi, M.; Nakagawa, S. Cadherin-dependent cell-cell adhesion. *Curr. Protoc. Cell Biol.* 2001. [CrossRef] [PubMed]
36. Maitre, J.L.; Heisenberg, C.P. Three functions of cadherins in cell adhesion. *Curr. Biol.* **2013**, *23*, R626–R633. [CrossRef] [PubMed]
37. Martin, A.C.; Gelbart, M.; Fernandez-Gonzalez, R.; Kaschube, M.; Wieschaus, E.F. Integration of contractile forces during tissue invagination. *J. Cell Biol.* **2010**, *188*, 735–749. [CrossRef] [PubMed]
38. Arnold, T.R.; Stephenson, R.E.; Miller, A.L. Rho GTPases and actomyosin: Partners in regulating epithelial cell-cell junction structure and function. *Exp. Cell Res.* **2017**, *358*, 20–30. [CrossRef]
39. Jodoin, J.N.; Coravos, J.S.; Chanet, S.; Vasquez, C.G.; Tworoger, M.; Kingston, E.R.; Perkins, L.A.; Perrimon, N.; Martin, A.C. Stable Force Balance between Epithelial Cells Arises from F-Actin Turnover. *Dev. Cell* **2015**, *35*, 685–697. [CrossRef]
40. Sawyer, J.M.; Harrell, J.R.; Shemer, G.; Sullivan-Brown, J.; Roh-Johnson, M.; Goldstein, B. Apical constriction: A cell shape change that can drive morphogenesis. *Dev. Biol.* **2010**, *341*, 5–19. [CrossRef]
41. Acar, M.; Kocherlakota, K.S.; Murphy, M.M.; Peyer, J.G.; Oguro, H.; Inra, C.N.; Christabel, J.; Zhao, Z.; Luby-Phelps, K.; Morrison, S.J. Deep imaging of bone marrow shows non-dividing stem cells are mainly perisinusoidal. *Nature* **2015**, *526*, 126. [CrossRef] [PubMed]
42. Kokkaliaris, K.D.; Kunz, L.; Cabezas-Wallscheid, N.; Christodoulou, C.; Renders, S.; Camargo, F.; Trumpp, A.; Scadden, D.T.; Schroeder, T. Adult blood stem cell localization reflects the abundance of reported bone marrow niche cell types and their combinations. *Blood* **2020**, *136*, 2296–2307. [CrossRef] [PubMed]

43. Xiang, Y.; Qin, X.-Q.; Liu, H.-J.; Tan, Y.-R.; Liu, C.; Liu, C.-X. Identification of Transcription Factors Regulating CTNNAL1 Expression in Human Bronchial Epithelial Cells. *PLoS ONE* **2012**, *7*, e31158. [CrossRef] [PubMed]
44. Gingras, J.; Gawor, M.; Bernadzki, K.M.; Grady, R.M.; Hallock, P.; Glass, D.J.; Proszynski, T.J. alpha-Dystrobrevin-1 recruits Grb2 and alpha-catulin to organize neurotransmitter receptors at the neuromuscular junction. *J. Cell Sci.* **2016**, *129*, 898–911.
45. Wang, Y.; Jiang, Q.; Cai, H.; Xu, Z.; Wu, W.; Gu, B.; Li, L.; Cai, W. Genetic variants in RET, ARHGEF3 and CTNNAL1, and relevant interaction networks, contribute to the risk of Hirschsprung disease. *Aging-Us* **2020**, *12*, 4379–4393. [CrossRef]
46. Oh, K.H.; Abraham, L.S.; Gegg, C.; Silvestri, C.; Huang, Y.C.; Alkema, M.J.; Kim, H. Presynaptic BK channel localization is dependent on the hierarchical organization of alpha-catulin and dystrobrevin and fine-tuned by CaV2 calcium channels. *BMC Neurosci.* **2015**, *16*, 26. [CrossRef]
47. Wu, M.-H.; Hong, T.-M.; Cheng, H.-W.; Pan, S.-H.; Liang, Y.-R.; Hong, H.-C.; Chiang, W.-F.; Wong, T.-Y.; Shieh, D.-B.; Shiau, A.-L.; et al. Galectin-1-Mediated Tumor Invasion and Metastasis, Up-Regulated Matrix Metalloproteinase Expression, and Reorganized Actin Cytoskeletons. *Mol. Cancer Res.* **2009**, *7*, 311–318. [CrossRef]
48. Deryugina, E.I.; Quigley, J.P. Matrix metalloproteinases and tumor metastasis. *Cancer Metastasis Rev.* **2006**, *25*, 9–34. [CrossRef]
49. Fidler, I.J. Timeline-The pathogenesis of cancer metastasis: The 'seed and soil' hypothesis revisited. *Nat. Rev. Cancer* **2003**, *3*, 453–458. [CrossRef]
50. Thierauf, J.; Veit, J.A.; Hess, J. Epithelial-to-Mesenchymal Transition in the Pathogenesis and Therapy of Head and Neck Cancer. *Cancers* **2017**, *9*, 76. [CrossRef]
51. Christofori, G. New signals from the invasive front. *Nature* **2006**, *441*, 444–450. [CrossRef] [PubMed]
52. Kobielak, A.; Fuchs, E. Links between alpha-catenin, NF-kappa B, and squamous cell carcinoma in skin. *Proc. Natl. Acad. Sci. USA* **2006**, *103*, 2322–2327. [CrossRef] [PubMed]
53. Kamangar, F.; Dores, G.M.; Anderson, W.F. Patterns of cancer incidence, mortality, and prevalence across five continents: Defining priorities to reduce cancer disparities in different geographic regions of the world. *J. Clin. Oncol.* **2006**, *24*, 2137–2150. [CrossRef] [PubMed]
54. Winn, D.M.; Diehl, S.R.; Horowitz, A.M.; Gutkind, S.; Sandberg, A.L.; Kleinman, D.V. Scientific progress in understanding oral and pharyngeal cancers. *J. Am. Dent. Assoc.* **1998**, *129*, 713–718. [CrossRef]
55. Liebertz, D.J.; Lechner, M.G.; Masood, R.; Sinha, U.K.; Han, J.; Puri, R.K.; Correa, A.J.; Epstein, A.L. Establishment and Characterization of a Novel Head and Neck Squamous Cell Carcinoma Cell Line USC-HN1. *Head Neck Oncol.* **2010**, *2*, 1–14. [CrossRef]
56. Cao, C.; Chen, Y.; Masood, R.; Sinha, U.K.; Kobielak, A. alpha-Catulin Marks the Invasion Front of Squamous Cell Carcinoma and Is Important for Tumor Cell Metastasis. *Mol. Cancer Res.* **2012**, *10*, 892–903. [CrossRef]
57. Karpińska, K.; Gielata, M.; Gwiazdowska, A.; Boryń, Ł.; Kobielak, A. Catulin Based Reporter System to Track and Characterize the Population of Invasive Cancer Cells in the Head and Neck Squamous Cell Carcinoma. *Int. J. Mol. Sci.* **2022**, *23*, 140. [CrossRef]
58. Puram, S.V.; Tirosh, I.; Parikh, A.S.; Patel, A.P.; Yizhak, K.; Gillespie, S.; Rodman, C.; Luo, C.L.; Mroz, E.A.; Emerick, K.S.; et al. Single-Cell Transcriptomic Analysis of Primary and Metastatic Tumor Ecosystems in Head and Neck Cancer. *Cell* **2017**, *171*, 1611–1624.e24. [CrossRef]
59. Gielata, M.; Karpińska, K.; Gwiazdowska, A.; Boryń Kobielak, A. Catulin reporter marks a heterogeneous population of invasive breast cancer cells with some demonstrating plasticity and participating in vascular mimicry. *Sci. Rep.* **2022**, *12*, 12673. [CrossRef]
60. Boyle, P.; Howell, A. The globalisation of breast cancer INTRODUCTION. *Breast Cancer Res.* **2010**, *12*. [CrossRef] [PubMed]
61. Kumar, P.; Aggarwal, R. An overview of triple-negative breast cancer. *Arch. Gynecol. Obs.* **2016**, *293*, 247–269. [CrossRef] [PubMed]
62. Cassandri, M.; Butera, A.; Amelio, I.; Lena, A.M.; Montanaro, M.; Mauriello, A.; Anemona, L.; Candi, E.; Knight, R.A.; Agostini, M.; et al. ZNF750 represses breast cancer invasion via epigenetic control of prometastatic genes. *Oncogene* **2020**, *39*, 4331–4343. [CrossRef] [PubMed]
63. Fan, L.-C.; Chiang, W.-F.; Liang, C.-H.; Tsai, Y.-T.; Wong, T.-Y.; Chen, K.-C.; Hong, T.-M.; Chen, Y.-L. alpha-Catulin knockdown induces senescence in cancer cells. *Oncogene* **2011**, *30*, 2610–2621. [CrossRef] [PubMed]
64. Liang, C.H.; Chiu, S.Y.; Hsu, I.; Wu, Y.Y.; Tsai, Y.T.; Ke, J.Y.; Hong, T.M. alpha-Catulin drives metastasis by activating ILK and driving an alphavbeta3 integrin signaling axis. *Cancer Res.* **2013**, *73*, 428–438. [CrossRef] [PubMed]
65. Tung, C.-H.; Huang, M.-F.; Liang, C.-H.; Wu, Y.-Y.; Wu, J.-E.; Hsu, C.-L.; Chen, Y.-L.; Hong, T.-M. alpha-Catulin promotes cancer stemness by antagonizing WWP1-mediated KLF5 degradation in lung cancer. *Theranostics* **2022**, *12*, 1173–1186. [CrossRef] [PubMed]
66. Kreiseder, B.; Orel, L.; Bujnow, C.; Buschek, S.; Pflueger, M.; Schuett, W.; Hundsberger, H.; de Martin, R.; Wiesner, C. a-Catulin downregulates E-cadherin and promotes melanoma progression and invasion. *Int. J. Cancer* **2013**, *132*, 521–530. [CrossRef]
67. Kreiseder, B.; Holper-Schichl, Y.M.; Muellauer, B.; Jacobi, N.; Pretsch, A.; Schmid, J.A.; Wiesner, C. Alpha-catulin contributes to drug-resistance of melanoma by activating NF-kappaB and AP-1. *PLoS ONE* **2015**, *10*, e0119402. [CrossRef]

International Journal of
*Molecular Sciences*

MDPI

*Article*

# Dysregulation of miRISC Regulatory Network Promotes Hepatocellular Carcinoma by Targeting PI3K/Akt Signaling Pathway

**Maheshkumar Kannan [1], Sridharan Jayamohan [1], Rajesh Kannan Moorthy [1], Siva Chander Chabattula [2], Mathan Ganeshan [3] and Antony Joseph Velanganni Arockiam [1,*]**

1 Molecular Oncology Laboratory, Department of Biochemistry, School of Life Sciences, Bharathidasan University, Tiruchirappalli 620024, India
2 Department of Biotechnology, Bhupat and Jyoti Mehta School of Biosciences, Indian Institute of Technology Madras, Chennai 600036, India
3 Molecular Cancer Biology Laboratory, Department of Biomedical Science, Bharathidasan University, Tiruchirappalli 620024, India
* Correspondence: ajvelanganni@gmail.com; Tel.: +91-0431-2407071 (ext. 570)

**Abstract:** Hepatocellular carcinoma (HCC) remains the third leading malignancy worldwide, causing high mortality in adults and children. The neuropathology-associated gene AEG-1 functions as a scaffold protein to correctly assemble the RNA-induced silencing complex (RISC) and optimize or increase its activity. The overexpression of oncogenic miRNAs periodically degrades the target tumor suppressor genes. Oncogenic miR-221 plays a seminal role in the carcinogenesis of HCC. Hence, the exact molecular and biological functions of the oncogene clusters miR-221/AEG-1 axis have not yet been examined widely in HCC. Here, we explored the expression of both miR-221 and AEG-1 and their target/associate genes by qRT-PCR and western blot. In addition, the role of the miR-221/AEG-1 axis was studied in the HCC by flow cytometry analysis. The expression level of the AEG-1 did not change in the miR-221 mimic, and miR-221-transfected HCC cells, on the other hand, decreased the miR-221 expression in AEG-1 siRNA-transfected HCC cells. The miR-221/AEG-1 axis silencing induces apoptosis and G2/M phase arrest and inhibits cellular proliferation and angiogenesis by upregulating p57, p53, RB, and PTEN and downregulating LSF, LC3A, Bcl-2, OPN, MMP9, PI3K, and Akt in HCC cells.

**Keywords:** HCC; miR-221; astrocyte elevated gene-1; angiogenesis; cell proliferation; regulatory genes

**Citation:** Kannan, M.; Jayamohan, S.; Moorthy, R.K.; Chabattula, S.C.; Ganeshan, M.; Arockiam, A.J.V. Dysregulation of miRISC Regulatory Network Promotes Hepatocellular Carcinoma by Targeting PI3K/Akt Signaling Pathway. *Int. J. Mol. Sci.* **2022**, *23*, 11300. https://doi.org/10.3390/ijms231911300

Academic Editor: Laura Paleari

Received: 31 August 2022
Accepted: 19 September 2022
Published: 25 September 2022



## 1. Introduction

Hepatocellular carcinoma (HCC) is the most common cancer and the third leading cancer-related death with a poor prediction [1]. HCC is the fastest rising cancers, following others, and has the highest incidence in developing countries, such as India [2]. Hepatitis B virus (HBV) and hepatitis C virus (HCV) are major risk factors in developing human HCC, with approximately 50–80% of cases from HBV and 10 to 25% from HCV infections, respectively [3]. Non-alcoholic fatty liver disease (NAFLD) has a role in developing HCC caused by the accumulation of fat exceeding 5% of liver weight in the absence of alcohol. NAFLD is a non-viral risk factor for developing HCC worldwide that majorly causes liver damage that leads to HCC from liver cirrhosis [4,5]. Surgery is considered the primary treatment for HCC, and recent studies have reported that chemotherapy is the most suitable treatment for HCC [6]. Nevertheless, the overall survival time of drug-treated patients is merely 2–5 years [7]. The molecular studies and targeted therapies for the regulatory protein networks of the apoptosis, cell cycle, and angiogenesis progression could be promising for treating HCC.

MicroRNAs (miRNAs) are a class of short-term endogenous non-coding RNAs, having 18–25 nucleotides in length. miRNAs act as oncogenes or tumor suppressors that dysregulate their target gene expression at the post-transcription level [8,9]. miRNAs are considered key mediators in various biological processes, including cell proliferation, apoptosis, and angiogenesis by dysregulation of specific target mRNAs [10,11] or the signaling pathways, such as PTEN/PI3K/Akt [12]. It is targeted by several miRNAs, particularly miR-221/222 [13], miR-543 [14], miR-146b [15], and miR-181a/b-1 [16] through binding in the 3′ Untranslated Region (UTR). Furthermore, miRNAs are involved in the NAFLD progression in various cell types and are considered a potential biomarkers [4]. Recent studies suggest that both tumor suppressors and oncogenic miRNAs correlate with several human diseases and play a critical role in cancer initiation and progression, especially the miR-221, which acts as an oncogene and induces carcinogenic activities in various human cancers, including HCC [17–19]. However, miR-221 specificity and sensitivity in HCC are not yet completely explored.

Astrocyte elevated gene -1 (AEG-1), also called Metadherin (*MTDH*) or protein Lysine Rich CECAM1 (LYRIC), has been reported as a human immunodeficiency virus (HIV)-1-inducible gene in human fetal astrocytes [20,21]. AEG-1 is overexpressed frequently and acts as an oncogene in several cancers, including HCC, but the clinical level studies of AEG-1 in the tumor initiation and progression in HCC are still unclear. AEG-1 plays a crucial role in cancer metastasis by regulating cell invasion and migration, apoptosis, angiogenesis, and chemo-resistance. The AEG-1 oncogene promotes the cells as aggressive cancers from the normal condition by dysregulating the corresponding proteins through the signaling pathways, such as PI3K/Akt [22,23]. During the RISC Complex formation, AEG-1 works as a scaffold protein and activates the miRNA as Onco-miR, which leads to the activation or degradation of specific regulatory genes [24].

Several studies have proven that the ectopic expression of miR-221 regulates the HCC by individual or cluster [25,26]. Meanwhile, several tumor suppressor miRNAs regulate AEG-1, and emerging evidence revealed that the lack of targeting miRNAs' expression during carcinogenesis leads to the activation of AEG-1 [25,27]. However, numerous studies analyzed the specific miRNA/oncogene and their regulatory network, especially miR-221-targeted genes (PHF2 [25], C1QTNF1-AS1 [26]), and AEG-1-targeted miRNAs (miR-375 [27,28] and miR-195 [29]) in HCC. Hence, the regulation of both oncogene–onco-miR cluster such as AEG-1/miR-221 axis on the HCC regulatory network is still limited. In this study, we aimed to investigate the miR-221/AEG-1 axis regulations and molecular mechanisms for the potential targets in HCC.

## 2. Results

### 2.1. Clinical Significance of AEG-1 and miR-221 in Hepatocellular Carcinoma Patients. - MiR-221 and AEG-1 Were Highly Expressed in Hepatocellular Carcinoma Patients

TIMER 2.0 (http://timer.cistrome.org/) (accessed on 13 August 2022) analysis showed that AEG-1 was upregulated in BRCA (breast invasive carcinoma), CHOL (cholangiocarcinoma), COAD (colon adenocarcinoma), ESCA (esophageal carcinoma), HNSC (head and neck cancer), KIRC (kidney renal clear cell carcinoma), LIHC (liver hepatocellular carcinoma), LUAD (lung adenocarcinoma), LUSC (lung squamous cell carcinoma), READ (rectum adenocarcinoma), STAD (stomach adenocarcinoma), THCA (thyroid carcinoma), and UCEC (uterine corpus endometrial carcinoma) (Figure 1A). AEG-1 was 15% amplified in LIHC (Figure 1B) (www.cbioportal.org) (accessed on 13 August 2022). Meanwhile, we investigated the expressions of AEG-1 and miR-221 in LIHC from TCGA data using the UALCAN and (http://starbase.sysu.edu.cn) (accessed on 13 August 2022) portal. Findings showed that AEG-1 and miR-221 were highly expressed in LIHC compared to normal tissues (Figure 1C,D). Next, we observed that low expressions of AEG-1 and miR-221 had high rates of OS (overall survival), RFS (relapse-free survival), PFS (progression-free survival), and DSS (disease-specific survival) in the Kaplan–Meier plotter (Figure 1E,F). The correlation of PTEN, Akt, PI3K, and AEG-1 mRNA expression was performed using GEPIA 2 plotter. The results showed that PTEN, Akt, and PI3K mRNA expression was significantly correlated with AEG-1 mRNA expression (Figure 1G). Altogether, these results suggest that the low expression of AEG-1 and miR-221 was associated with a better prognosis in hepatocellular carcinoma patients.

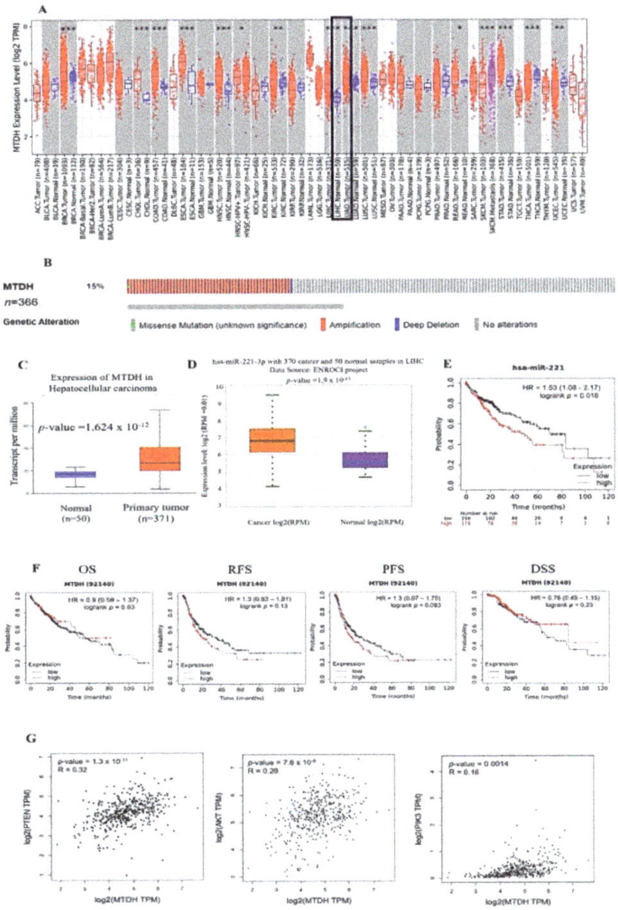


**Figure 1.** High expression of AEG-1 and miR-221 with poorer prognosis in liver cancer patients. (**A**). The mRNA expression of AEG-1 showed across all TCGA tumors, T (tumor—red) and N (normal—blue), Wilcoxon test used for statistical significance analysis (* $p$-value < 0.05; ** $p$-value < 0.01; *** $p$-value < 0.001) by using TIMER 2.0. (**B**) AEG-1 was amplified (red) in 15% of LIHC (liver hepatocellular carcinoma) (www.cbioportal.org) (accessed on 13 August 2022). (**C**) The AEG-1 mRNA expression was examined in normal (blue) ($n$ = 50) and tumor (red) tissues ($n$ = 371) from the cases of LIHC in the TCGA database using UALCAN (accessed on 13 August 2022). (**D**) miR-221 expression was examined in normal (blue) ($n$ = 50) and tumor (orange) tissues ($n$ = 370) from the cases of LIHC using the publicly available TCGA database using (http://starbase.sysu.edu.cn) (accessed on 13 August 2022) portal. (**E**) The Kaplan –Meier plot shows that a lower expression of miR-221 is associated with better overall survival of hepatocellular carcinoma patients. (**F**) Kaplan–Meier survival curves of liver cancer patients showed AEG-1 expression on overall survival (OS), relapse-free survival (RFS), progression-free survival (PFS), and disease-specific survival (DSS); the cutoff value of AEG-1 expression was set as its median value (accessed on 13 August 2022). (**G**) Pearson's correlation between AEG-1 and PTEN, Akt, and PI3K expression in LIHC patients using the TCGA database using GEPIA 2 (Gene Expression Profiling Interactive Analysis) (accessed on 13 August 2022).

### 2.2. Ectopic Expression of miR-221/AEG-1 in HCC Cell Line Panel

We investigated the miR-221 and AEG-1 expressions in normal human liver epithelial cells (THLE-2) and HCC cell line panels (HepG2, Huh7, and Hep3B) by qRT-PCR. The results revealed that the miR-221 and AEG-1 expressions were upregulated in all HCC cell lines compared to human liver epithelial cells (Figure 2A,B). Next, we analyzed the expression of miR-221 in the miR-221 mimic- (Figure 2C), miR-221 inhibitor- (Figure 2D), and AEG-1 siRNA- (Figure 2E) transfected groups. The miR-221 inhibitor- and AEG-1 siRNA-transfected groups showed decreased miR-221 expression, and the miR-221 mimic-transfected group showed ectopic expression of miR-221. Moreover, we examined the role of miR-221 mimic, miR-221 inhibitor, and AEG-1 siRNA on AEG-1. The results showed that the miR-221 mimic and miR-221 inhibitor did not alter the AEG-1 mRNA levels (Figure 2F) and AEG-1 siRNA silenced the AEG-1 mRNA levels in HCC cells (Figure 2G).

### 2.3. miR-221/AEG-1 Axis Regulates Apoptosis, Cell Cycle, Angiogenesis, and Autophagy Mechanism by the Activation of Regulatory Genes in HCC Cells In Vitro

Next, we analyzed the AEG-1 and miR-221 roles in their regulatory network, which regulates cell cycle, apoptosis, angiogenesis, and autophagy in miR-221 mimic-, miR-221 inhibitor-, AEG-1 siRNA-, and their corresponding controls-transfected HCC panel by qRT-PCR. We observed the decreased expression of LSF, MMP9, OPN, Bcl2, LC3A, and PI3K/Akt and increased the PTEN, p57, p53, and RB expressions (Figures 3 and 4) in the AEG-1 siRNA- and miR-221 inhibitor-transfected groups in HCC cells. The outcomes revealed that AEG-1 and miR-221 could play an essential role in the HCC regulatory networks.

### 2.4. Knockdown of miR-221 and AEG-1 Inhibits Invasion, Migration, and Cellular Proliferation In Vitro

Next, we examined the effects of AEG-1 and miR-221 in the invasion, migration, and cellular proliferation activity by transwell assays, wound healing, and cell viability assays (MTT) in miR-221 mimic/inhibitor-, AEG-1 siRNA-, with their corresponding control-transfected HCC cells in in vitro. The results confirmed that the silencing of miR-221 and AEG-1 effectively inhibits HCC cell invasion (Figure 5A), migration (Figure 5B), and cellular proliferation (Figure 5C) (Supplementary Figure S1A–C) compared with their controls. These results confirmed that onco-miR-221 and AEG-1 oncogene are possible regulators of cell proliferation and migration in HCC cells.

### 2.5. Downregulation of miR-221 and AEG-1 Promotes Apoptosis and Cell Cycle Arrest in HCC Cells In Vitro

Furthermore, we performed flow cytometry analysis to confirm the miR-221 and AEG-1 regulation of cell cycle and apoptosis in HCC cells. HCC cells were treated with miR-221 mimic/inhibitor and AEG-1 siRNA and performed cell cycle analysis by PI staining and apoptosis assay by Alexa Fluor-conjugated Annexin V-FITC/PI dual-staining. The apoptosis (Figure 6A) and cell cycle (Figure 6B) results showed that the percentage of the apoptotic cells increased in G0-G1 and induced cell cycle arrest in sub-G1 and G2/M compared with miR-221 mimic and their corresponding control groups. These results proved that miR-221 and AEG-1 were involved and could control cell cycle regulation and apoptosis.

### 2.6. Downregulation of miR-221 and AEG-1 Inhibits Angiogenesis and Enhances Apoptosis and Cell Cycle Arrest by Modulating Regulatory Proteins In Vitro

We analyzed the AEG-1 protein expression HCC panel and confirmed that the relative protein expression of AEG-1 was overexpressed significantly in HepG2, Huh-7, and Hep3B compared to THLE-2 cells (Figure 7A). Following, the HCC cells were transfected with miR-221 mimic, miR-221 inhibitor, and AEG-1 siRNA, with corresponding controls to analyze the relative AEG-1 expression. The results showed that miR-221 mimic, miR-221 inhibitor, and their corresponding controls did not alter AEG-1 expression and significantly decreased in the AEG-1 siRNA-transfected group when compared to the control (Figure 7B).

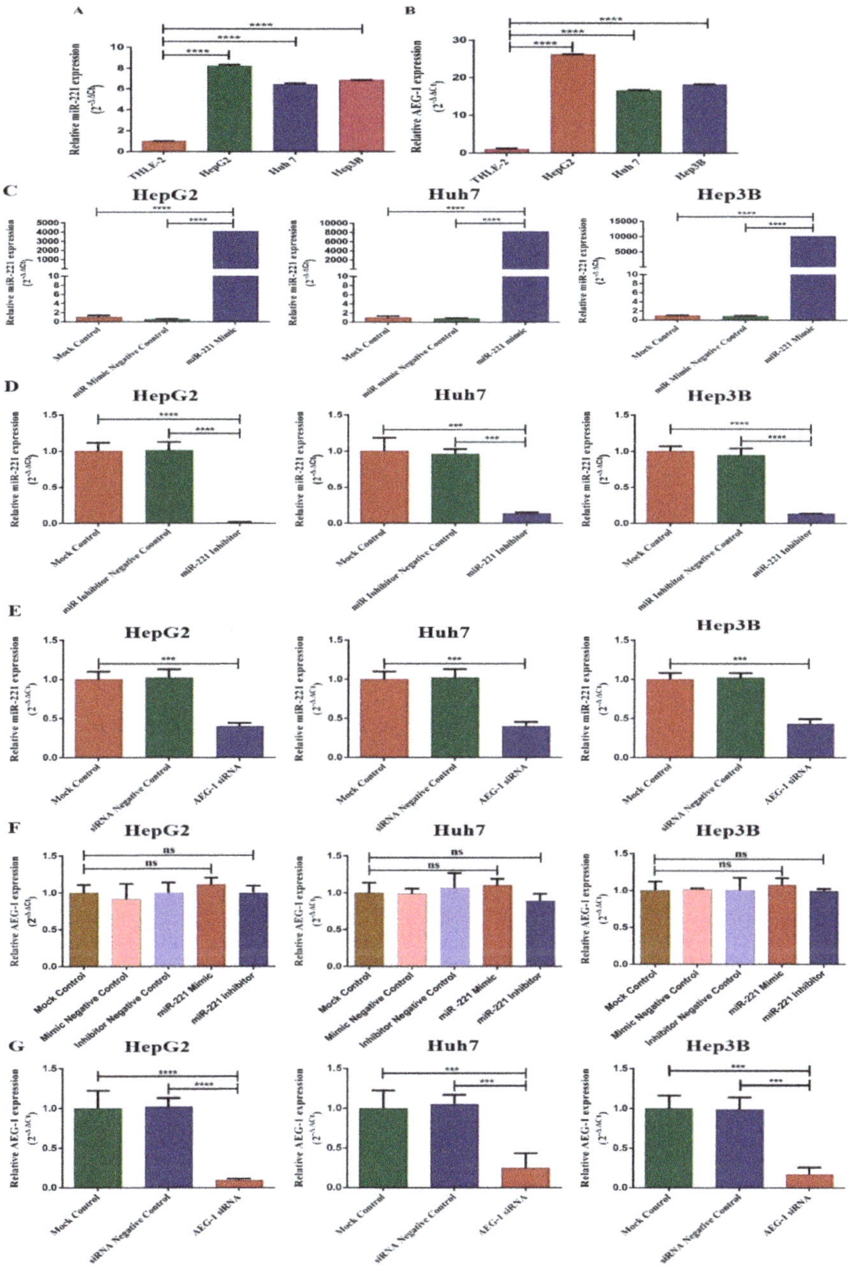


**Figure 2.** The relative miR-221 and AEG-1 mRNA levels in HCC cell lines. The miR-221 (**A**) and AEG-1 (**B**) relative mRNA levels analyzed in HCC cell line panels, the relative miR-221 expression in miR-221 mimic- (**C**), miR-221 inhibitor- (**D**), and AEG-1 siRNA- (**E**) transfected HCC cells. The relative mRNA levels of AEG-1 in miR-221 mimic/inhibitor- (**F**) and AEG-1 siRNA-transfected groups in HCC cells (**G**). The RNU6 and GAPDH were used as internal controls. Error bars presented as mean ± s.d and *p*-values represented as *** $p < 0.001$, **** $p < 0.0001$. ns represented as non-significance.

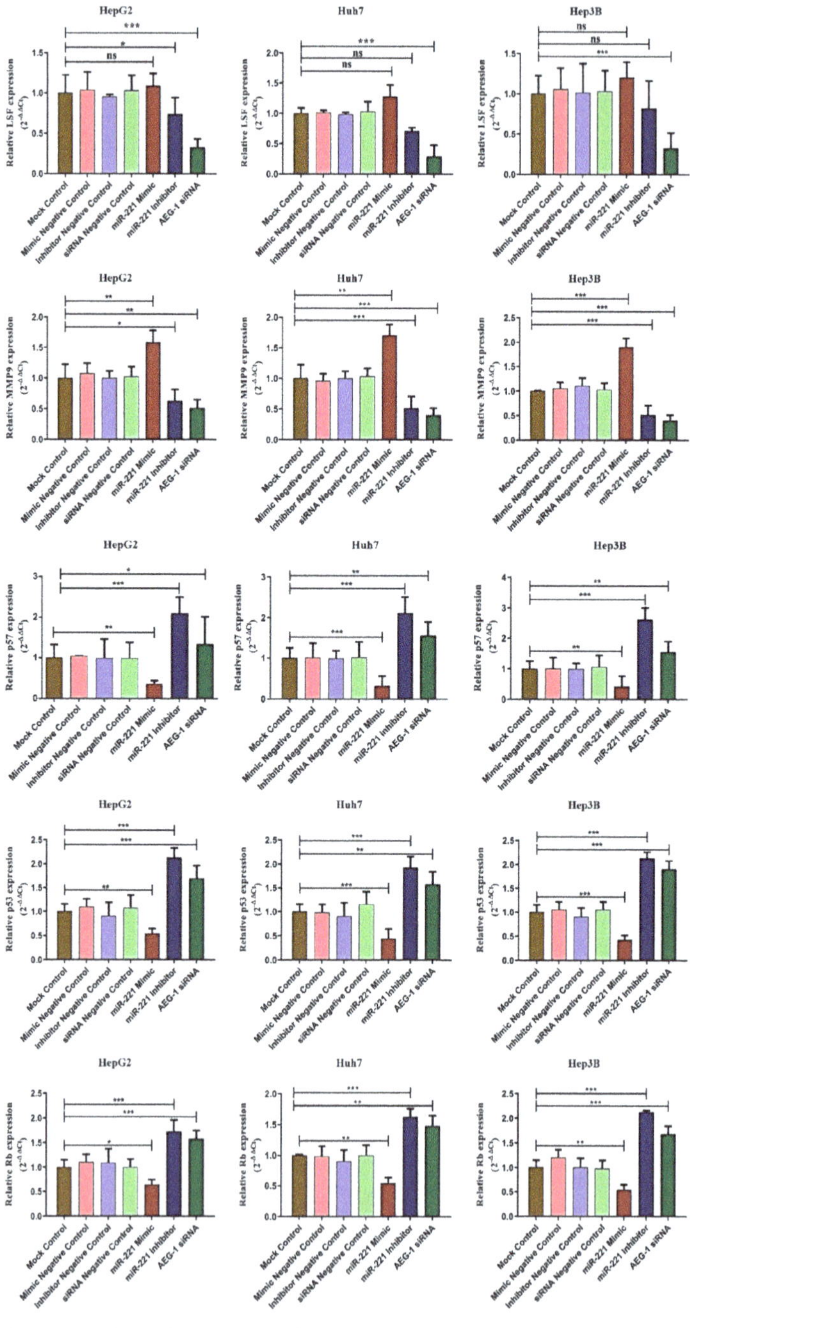


**Figure 3.** AEG-1 and miR-221 regulate angiogenesis and cell cycle regulatory mRNA expressions in the HCC cell lines. The regulatory mRNAs expressions which regulate angiogenesis (LSF and MMP9) and cell cycle (p57, p53, and RB) were analyzed in miR-221 mimic-, miR-221 inhibitor-, AEG-1 siRNA-, and their corresponding control-transfected HepG2, Huh7, and Hep3B cells by using qRT-PCR. The GAPDH was used as an internal control. Error bars presented as mean ± s.d and $p$-values represented as * $p < 0.05$, ** $p < 0.01$, and *** $p < 0.001$. ns represented as non-significance.

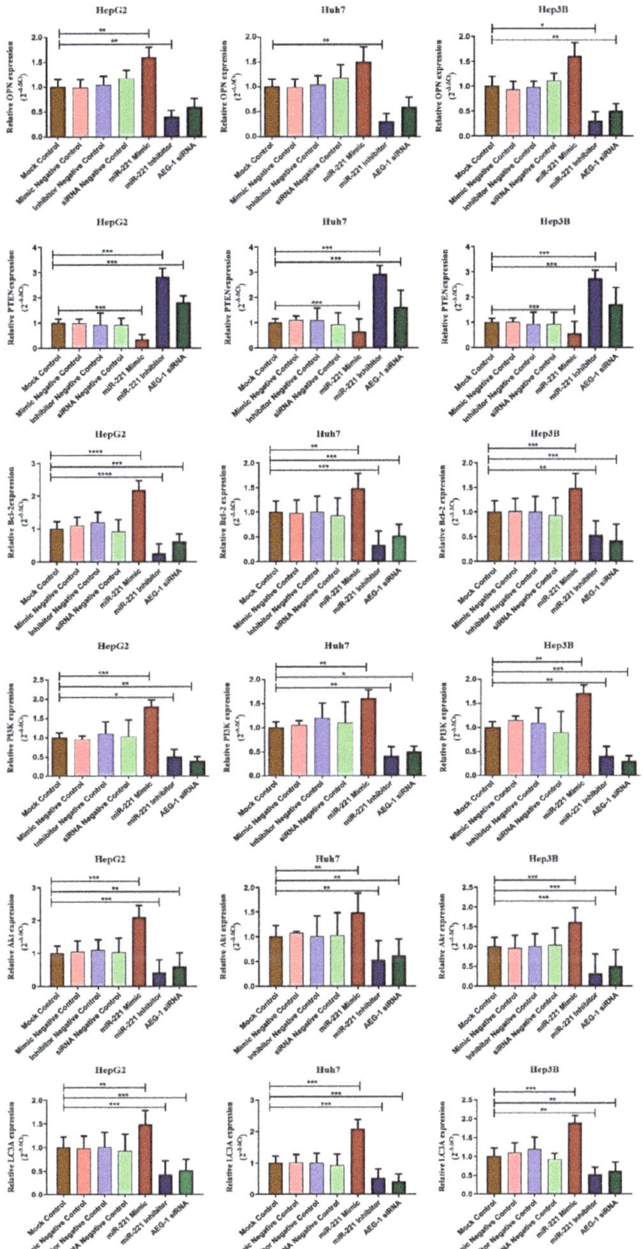

**Figure 4.** AEG-1/miR-221 regulates the apoptosis and autophagy regulatory mRNAs in the HCC cell lines. The apoptosis and autophagy regulatory mRNAs (OPN, Bcl-2, PTEN, LC3A, PI3K, and Akt) expressions were analyzed in miR-221 mimic-, miR-221 inhibitor, and AEG-1 siRNA-transfected HCC cells by using qRT-PCR, and GAPDH used as an internal control. Error bars presented as mean ± s.d. and *p*-value represented as * $p < 0.05$, ** $p < 0.01$, *** $p < 0.001$, **** $p < 0.0001$ compared with corresponding controls.

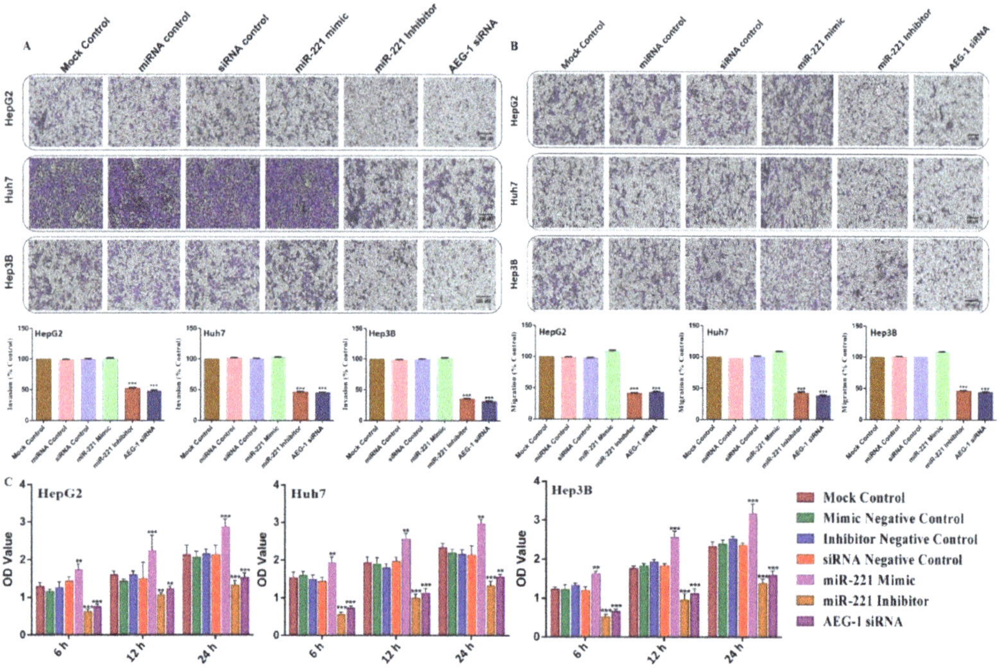

**Figure 5.** Silencing of AEG-1 and miR-221 inhibits HCC cell migration and proliferation in vitro. The effect of the miR-221/AEG-1 measured on HCC cell migration and cell proliferation in the miR-221 mimic-, miR-221 inhibitor-, AEG-1 siRNA-, and their control-transfected HepG2, Huh7, and Hep3B cells by invasion (**A**), migration (**B**), and MTT assay (**C**) in vitro. Images analyzed using Image J (NIH) (scale bar, 100 μm). Error bars presented as mean ± s.d and $p$-value represented as ** $p < 0.01$, *** $p < 0.001$ compared to the corresponding controls.

Furthermore, we investigated the impact of miR-221 and AEG-1 knockdown on regulatory proteins, which regulate the apoptosis, cell cycle, angiogenesis, and autophagy in HepG2 (Figure 7C), Huh7 (Figure 7D), and Hep3B (Figure 7E) cells. The results showed that the knockdown of miR-221 and AEG-1 significantly inhibited LSF and MMP9 and upregulated p57, p53, and RB protein levels when compared to the control. Moreover, we demonstrated the apoptosis and autophagy regulatory protein expression in miR-221 inhibitor- and AEG-1 siRNA-transfected groups of HepG2 (Figure 8A), Huh7 (Figure 8B), and Hep3B (Figure 8C) cells. The results indicated that the silencing of miR-221 and AEG-1 significantly increased PTEN protein levels and decreased OPN, Bcl-2, LC3A/B, PI3K, and p-Akt protein levels when compared to miR-221 mimic and control groups. Thus, we identified that miR-221 and AEG-1 are possible oncogenes in HCC.

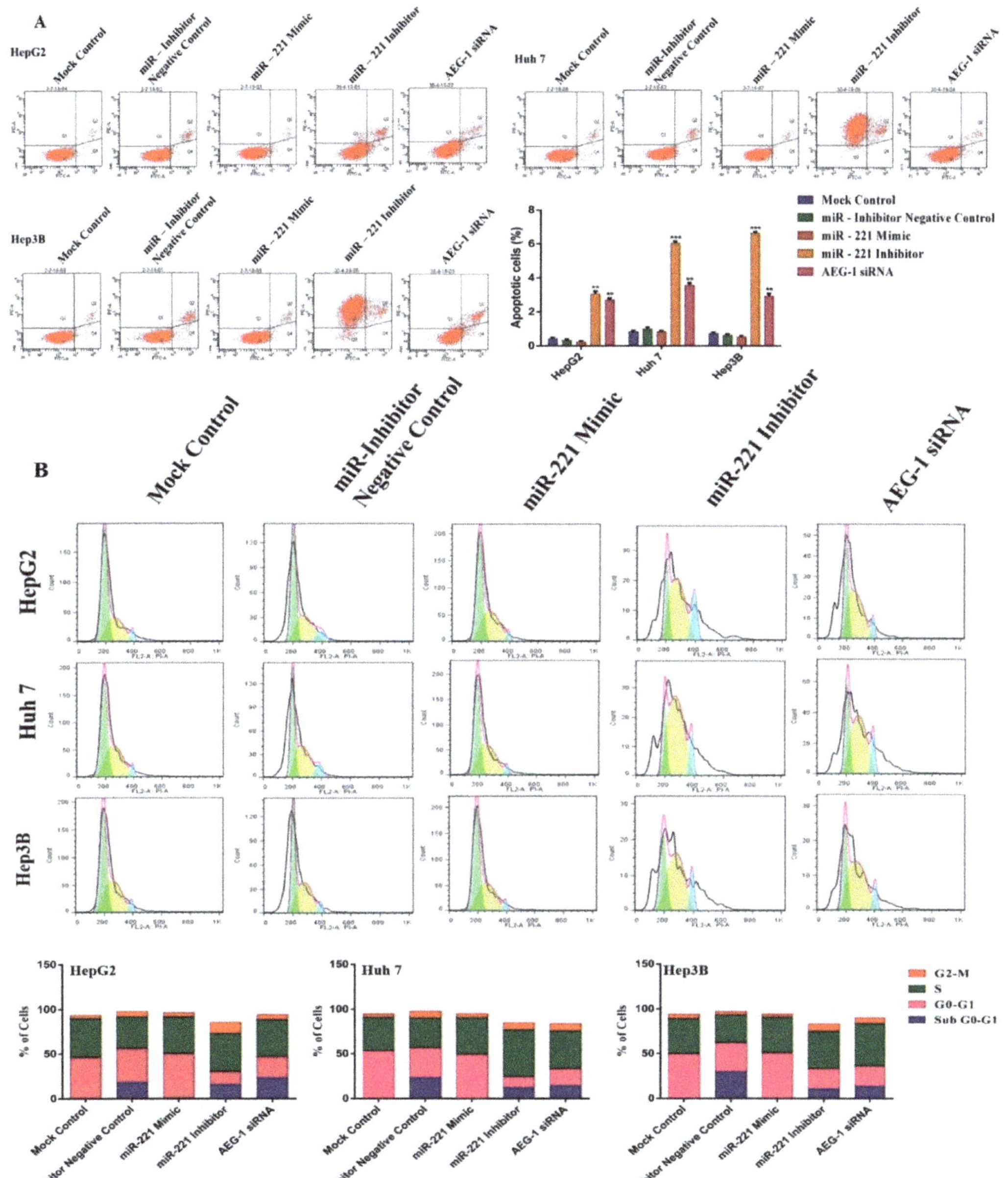


**Figure 6.** The effects of AEG-1/miR-221 on HCC cell cycle regulation and apoptosis. The effects of AEG-1/miR-221 on apoptosis (**A**) and cell cycle (**B**) analyzed in the miR-221 mimic-, and miR-221 inhibitor-, AEG-1 siRNA-, and their corresponding control-transfected HCC cells in vitro by flow cytometry analysis. Error bars presented as the mean ± s.d. and $p$-values represented as ** $p < 0.01$, *** $p < 0.001$ compared to the corresponding controls.

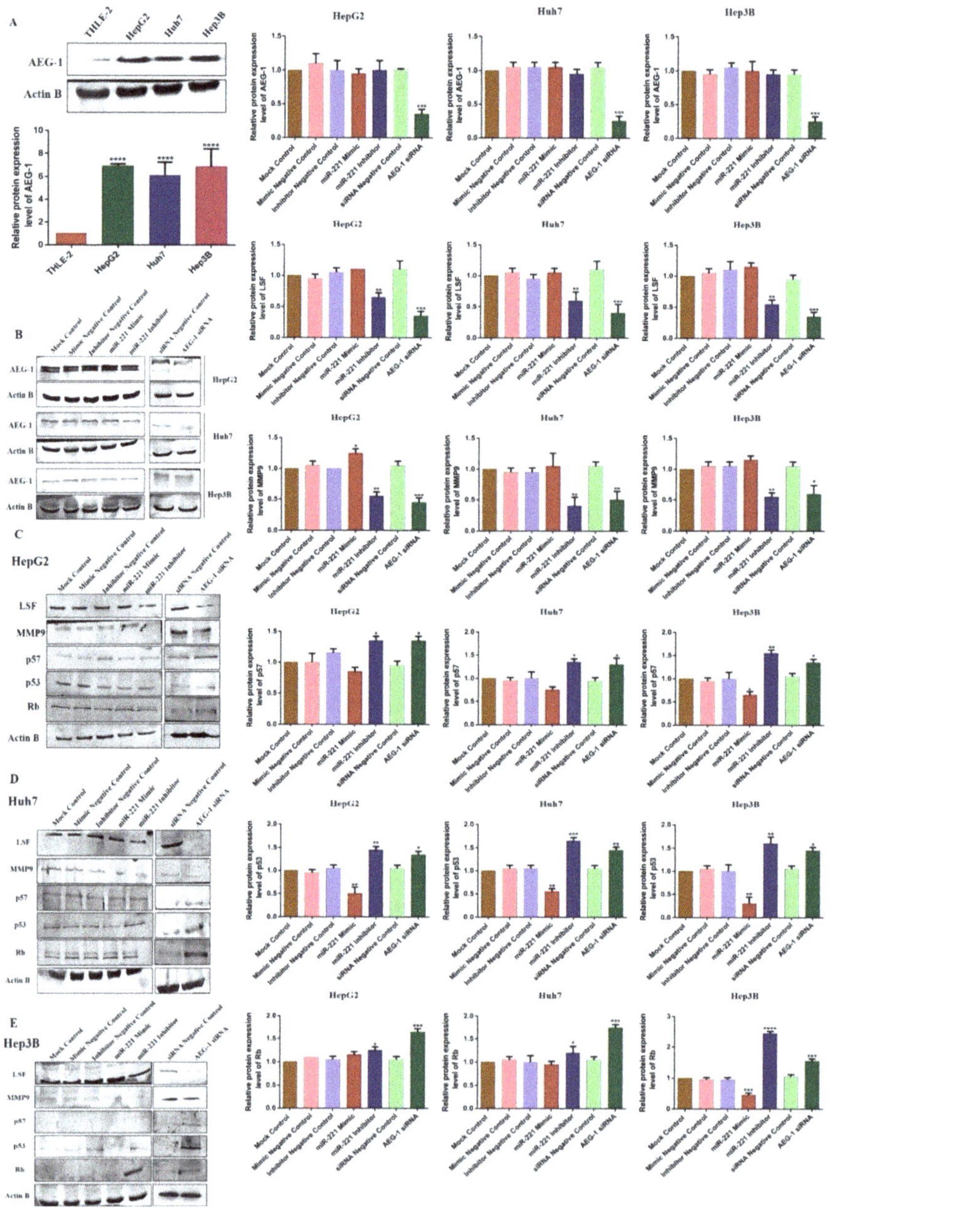


**Figure 7.** The effect of AEG-1/miR-221 was analyzed on cell cycle and angiogenesis regulatory proteins in HCC cell lines. The relative AEG-1 protein expression in THLE-2, HepG2, Huh7, and Hep3B cells (**A**) and miR-221 mimic-, miR-221 inhibitor-, and AEG-1 siRNA-transfected HCC cells (**B**). The regulatory protein expressions which regulate angiogenesis (LSF and MMP 9) and cell cycle (p57, p53, and RB) were analyzed in miR-221 mimic-, miR-221 inhibitor-, AEG-1 siRNA-, and their controls-transfected HepG2 (**C**), Huh 7 (**D**), and Hep3B (**E**) cells by western blot. *p*-Values presented as * $p < 0.05$, ** $p < 0.01$,*** $p < 0.001$, **** $p < 0.0001$ and error bars presented as the mean ± s.d. ns—(non-significant) compared to the controls.

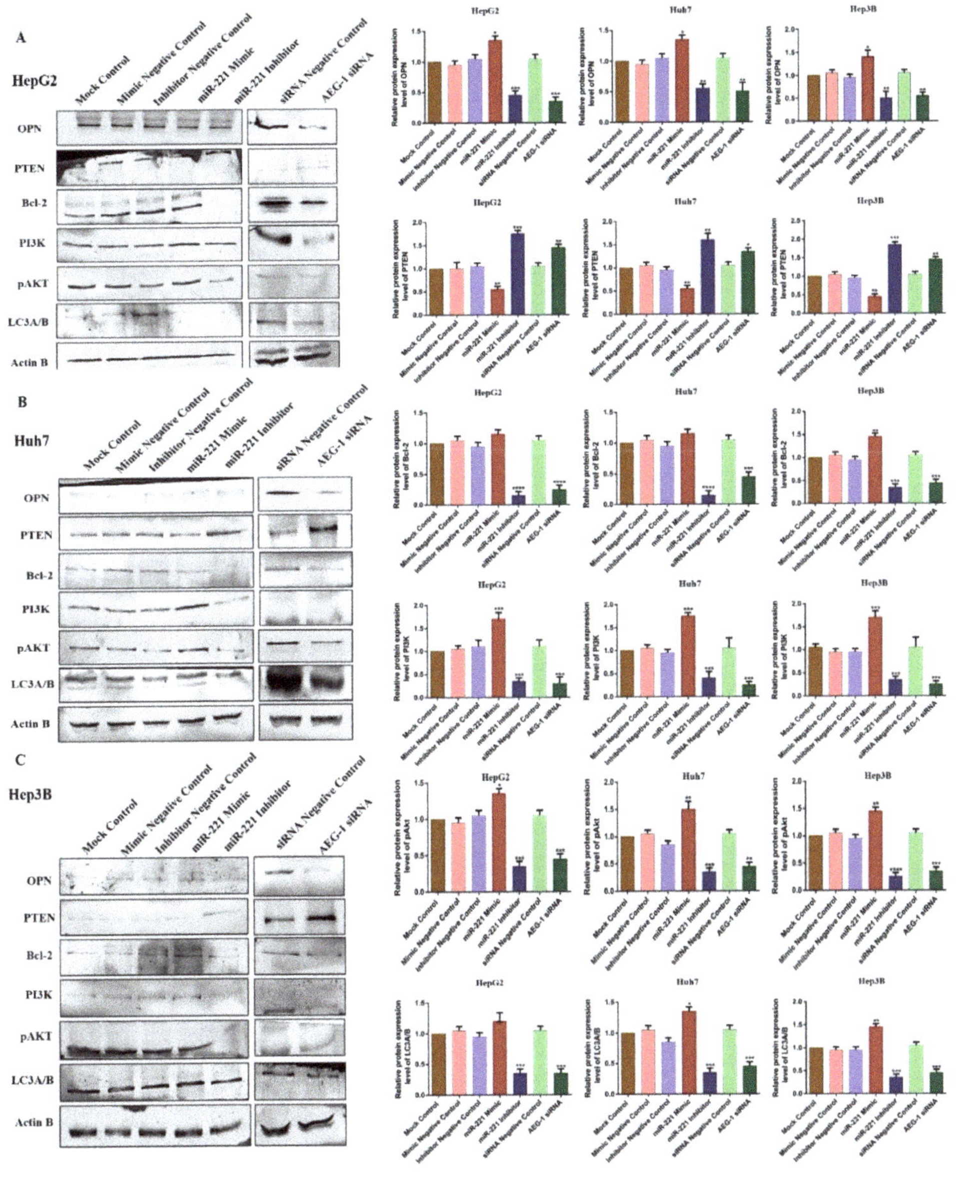


**Figure 8.** Ectopic expression of AEG-1/miR-221 dysregulates apoptosis and autophagy regulatory protein levels in HCC cell lines. The relative expressions of apoptosis (OPN and Bcl-2), autophagy (LC3A/B), PTEN, and PI3K/Akt proteins were analyzed by western blotting in the mock control, miR-221 mimic, miR-221 inhibitor, AEG-1 siRNA, and corresponding controls transfected HepG2 (**A**), Huh7 (**B**), and Hep3B (**C**) cells using β-actin as an internal control. Error bars are represented as mean ± s.d. and * $p < 0.05$, ** $p < 0.01$, *** $p < 0.001$, **** $p < 0.0001$ compared to the control group. ns: non-significance.

## 3. Discussion

Recently, cancer has been the main challenge globally with limited treatment. Chemotherapy is considered a potential treatment that improves the survival of the patients. However, early cancer detection and prevention remain challenged because it is the only disease affecting the normal cell directly without any injury. In recent days, targeted therapies have been considered an advanced treatment for early detection and cancer metastasis [30]. HCC is one of the aberrant cancers in growing countries such as India. HBV, HCV, and NAFLD are the common risk factors for causing HCC [31], which stimulates cell proliferation by regulating several oncogenes or oncogenic miRNAs, especially AEG-1 [32] and miR-221 [33]. Astrocyte-elevated genes are one of the targets of the tumor-associated antigen (TAA), which also is a component of RISC that leads to regulating the miRNAs [20,21,34]. TAA is a class of tumor antigens that are overexpressed in cancer cells and observed in a lower expression in normal cells. TAA promotes tumorigenesis by regulating cell proliferation, chemoresistance, metastasis, and apoptosis in cancers like HCC [28]. These findings revealed that AEG-1 is frequently overexpressed in human cancers and associated with several hallmarks of HCC. We affirmed the previous findings that AEG-1 is overexpressed in HCC cells, which indicates that AEG-1 is a therapeutic hallmark for HCC.

The advanced evidence showed the miRNAs that contributes in several human pathological process, including cancers. The oncogenic or tumor suppressor miRNAs dysregulate the specific regulatory mRNAs, which are mainly involved in the malignant activities including metastasis, survival, proliferation, chemoresistance, and apoptosis in various cancers, including HCC. miR-221 is one of the onco-miRNAs that is significantly overexpressed in cancers and regulates cell proliferation and metastasis. Our findings revealed the aberrant expression of miR-221 and its regulation in HCC. These findings prove that miR-221 plays a crucial role in liver carcinogenesis.

Earlier studies confirmed that AEG-1 was targeted by some tumor suppressor miRNAs, especially miR-375 and miR-195. Also, it showed lower expression of these tumor suppressor miRNAs during carcinogenesis. The ectopic expression of miR-375 and miR-195 inhibited cell proliferation, angiogenesis, and cancer metastasis by targeting and inhibiting the AEG-1 expression in HCC [8,27–29] and cervical cancers [11]. However, the ectopic expression of onco-miR-221 enhanced cellular proliferation and metastasis in HCC through the dysregulation of their targeted oncogenes and tumor suppressor mRNAs [24–26]. Moreover, miR-221 enhanced the angiogenesis activity in HCC by upregulation of SND1 oncogene [35]. These findings confirmed that AEG-1 and miR-221 play major oncogenic roles and regulate human carcinogenesis, including HCC. The silencing of AEG-1 and miR-221 inhibits angiogenesis activity and cell proliferation in several cancers, including HCC. Therefore, we figured that the silencing of AEG-1 and miR-221 inhibits angiogenesis and cellular proliferation by AEG-1 siRNA and anti-miR-221 in HCC cells. The miRNAs play a vital role in cell cycle regulation and apoptosis and induce cell cycle regulation by targeting the regulatory mRNA during carcinogenesis. The ectopic expression of tumor suppressor miRNAs miR-875-5p and miR-375 enhanced G0 phase apoptosis by G1 Phase arrest. It confirmed that the upregulation of miR-875-5p [36] and miR-375 [9] induce apoptosis and cell cycle arrest through the silencing of AEG-1 in cervical cancer and HCC. Our results confirmed that the silencing of AEG-1 and miR-221 induces apoptosis in sub-G0-G1 and G2-M phase arrest and inhibits angiogenesis and cellular proliferation in HCC cells.

Previous studies revealed the onco-miR–oncogene cluster regulations and proved that the upregulation of onco-miR-26a and the CENTG1 oncogene enhances cell proliferation and metastasis in Glioma cancer cells [37]. However, the molecular mechanisms of the onco-miR/oncogene cluster and their correlations are still unclear and limited in human cancers, including HCC, especially the miR-221 and AEG-1 cluster. Therefore, we focused on the onco-miR–oncogene regulation and correlations in HCC, especially the miR-221 and AEG-1 axis. Also, we analyzed the miR-221 and AEG-1 expressions, correlations, and effects of this cluster on their regulatory network proteins, which involve apoptosis, cell proliferation, invasion, autophagy, and angiogenesis in HCC.

Several findings showed that oncogenes or onco-miRNAs target several oncogenes and tumor suppressor regulatory genes directly, or association through pathways. One of these oncogenes is the LSF/TFCP2 (transcription factor LSF) involved in the angiogenesis regulations and is overexpressed in human cancers, including HCC [38]. The LSF targets some downstream signal genes and different regulatory mRNAs, including OPN [39] and MMP9 [40]. Osteopontin (OPN) is a phospho-glycoprotein involved in tumor metastasis and cell death. The matrix metalloproteinase 9 (MMP9) is also responsible for cell growth and cancer metastasis. It helps tumor growth by enhancing angiogenesis regulation during carcinogenesis. The overexpression of LSF enhances the angiogenesis activity, cell invasion, and migration in HCC cells. In addition, it was previously identified that the transcription factor LSF is a direct downstream target of AEG-1, and the LSF mRNA levels significantly increased during the overexpression of AEG-1 in HCC cells [41].

Interestingly, miR-221 regulates OPN in osteoblasts and induces the OPN protein levels during carcinogenesis [42]. On the other hand, the AEG-1 targets MMP2/9 in thyroid cancer cells and enhances cell invasion and migration by inducing the MMP2/9 [43]. Moreover, the upregulation of miR-221/222 regulates and induces the MMP-9 protein expression in pancreatic cancers [44]. Our outcomes confirmed that the LSF, OPN, MMP-9 mRNA, and protein levels decreased while silencing the miR-221 and AEG-1 in HCC cells. The miRNAs or oncogenes regulate their downstream target proteins through direct targets or signaling pathways. The signaling pathways playing critical roles and promoting carcinogenesis include tumor initiation and progression. NF-κB is an essential transcription factor, and several pathological and physiological processes regulate this pathway in human cancers, including cancer development and metastasis. The NF-kB is an aberrant expression in tumors by the miRNAs or some oncogenes through different molecular mechanisms [45], including AEG-1 [46] and miR-221 [47].

miRNAs play a crucial role in cell proliferation that controls or promotes cell differentiation by dysregulating the cell cycle or apoptosis regulatory proteins such as PTEN, p27, and p57 in human cancers. The PTEN is a tumor suppressor protein that helps prevent cell proliferation and induces cell cycle arrest to promote apoptosis in normal conditions (48,49). The cyclin-dependent kinase inhibitor 1B (P27) and cyclin-dependent kinase inhibitor 1C (p57) work as inhibitors of the cyclin/cyclin-dependent kinase (CDK) complexes in the different phases of the cycle to prevent cell differentiation and induce the cell cycle arrest. The ectopic expression of miR-221 regulates PTEN, inhibiting its expression in lung cancer [48] and inhibiting the p57 and p27 protein levels in HCC. During knockdown of the miRNA-221, the p57 and p27 expression level increases, resulting in induced cell death in HCC cells [49]. On the other hand, some findings showed that the AEG-1 oncogene targets PTEN and p57 and regulates their protein level in HCC [24]. Our results also prove that the PTEN, p27, and p57 expressions significantly increase while silencing the AEG-1 and miR-221 in HCC cells.

The activation of the NF-kB signaling pathway alters several regulatory proteins, including PTEN. The overexpression of NF-kB inhibits the transcription level of PTEN by the activation of anti-apoptotic protein Bcl-2 and promotes cell proliferation [50]. NF-kB inhibits cell death and promotes cell proliferation in prostate cancer by the overexpression of Bcl-2 [51]. Some previous studies confirmed that AEG-1 regulates Bcl-2 and significantly downregulates the Bcl-2 protein levels while silencing the AEG-1 in HCC [52]. Moreover, the miR-221 regulates Bcl-2 and BAX proteins, enhances the BAX protein level, and inhibits Bcl-2 while silencing the miR-221 in bladder cancer [53]. The PI3K/Akt is another essential signaling pathway that involves and regulates cancer activation and metastasis. The upregulation of PI3K/Akt induces MDM2-mediated p53 degradation by silencing PTEN protein levels in gastric cancer [54].

In cervical cancer, retinoblastoma (RB1) is a tumor suppressor protein regulated by PTEN. PTEN inhibits the PI-3 Kinase-mediated RB phosphorylation in cervical cancer by deregulating phosphatidylinositol 3,4,5 triphosphate Phosphorylation [55]. Moreover, RB1 regulates NF-kB and inhibits NF-kB transcriptional activity in prostate cancer [56]. Autophagy is a mechanism considered essential for cancer cell survival and cell death. Most of the signaling pathways and regulatory genes are involved in the autophagy mechanism and regulate their function, including Bcl-2 [57], p53 [58], PTEN, PI3K/Akt [59], and NF-kB [60] in human diseases, including cancer. In this way, onco-miR-221 and AEG-1 clusters may regulate the apoptosis, cell cycle, angiogenesis, and autophagy regulatory proteins PI3K, Akt, p53, p57, Bcl-2, RB1, OPN, MMP9, and LSF in HCC. We obtained similar results in AEG-1 siRNA- and miR-221 inhibitor-transfected HCC cells.

Based on the earlier studies, both AEG-1 and miR-221 regulate several cancers, including HCC, by direct targeting or through signaling pathways. Hence, the interaction between the AEG-1 and miR-221 in HCC is not yet known. We identified a part of this, such as AEG-1 and miR-221 regulating cooperatively in HCC tumorigenesis.

## 4. Materials and Methods

### 4.1. Cell Lines, Cell Culture

Human HCC cell lines (HepG2, Huh7, and Hep3B) were obtained from the National Centre for Cell Science (NCCS), Pune, India, and cultured in DMEM with 10% FBS (fetal bovine serum) and 1% PS (penicillin streptomycin) (Invitrogen, Carlsbad, CA, USA). The normal liver epithelial cell line THLE-2 (ATCC-CRL-2706) was purchased from the American Type Culture Collection (ATCC, Manassas, VA, USA) and cultured in Bronchial Epithelial Cell Growth Medium (BMEM) with 0.08% phosphoethanolamine (Sigma Aldrich St. Louis, MO, USA), 10% FBS, and 0.06% epidermal growth factor (EGF), human recombinant (Corning, NY, USA). The cells were cultured in a 37 °C humidified chamber with 5% $CO_2$.

### 4.2. miRNA and siRNA Transfection

HCC cells were cultured in a 6-well plate ($1 \times 10^6$) and transfected with miR-221 mimic, miR-221 inhibitor, and their control miRNAs (5 μM), small interfering RNA (siRNA) negative control, and AEG-1 siRNA (sense: 5′-GACACUGGAGAUGCUAAUAUU-3′, antisense: 5′-UAUUAGCAUCUCCAGUGUCUU-3′) (2 nM) as per the manufacture protocols using Lipofectamine 2000 and RNAi MAX reagents (Invitrogen, USA) in Opti-MEM medium (Thermo Fisher, Waltham, MA, USA). The mock control group was treated with transfection reagent alone in Opti-MEM (Without miRNAs and siRNAs). The transfection complex was prepared, and the cells were added directly to the transfection complex and incubated at 37 °C for 48 h. The transfection complex was removed in 6–8 h incubation and replaced with the fresh culture medium. The cells were collected and analyzed for the effects of miRNA and siRNA on the treated groups after 48 h transfection by quantitative real-time PCR (qRT-PCR) (Life Technologies, Burlington, Ontario, Canada).

### 4.3. Quantitative Real-Time PCR

Total RNA was isolated from the RNAi-transfected HCC cells using TRIZOL (Invitrogen, CA, USA). The 2 μg of total RNA was collected and randomly primed to synthesize the complementary DNA (cDNA) using the SuperScript First Strand cDNA Synthesis kit (K1622) (Thermo Fisher Scientific, Waltham, MA, USA) as per the manufacturer's protocol. Following, we performed the qRT-PCR using Step One plus qRT-PCR system (Life Technologies, Burlington, ON, Canada) with SYBR™ Green Master Mix—Real-Time PCR Master Mix (Applied Biosystems, Waltham, MA, USA) to analyze the miR-221, AEG-1, LSF, MMP9, p57, p53, RB, OPN, PTEN, Bcl-2, PI3K, Akt, and LC3A mRNA expressions. The primers were purchased from Eurofins Genomics (Louisville, KY, USA) and listed in Supplementary Table S1. The U6 snRNA and GAPDH were internal controls for miR-221 and AEG-1. The data was collected, and the specificity of the primers was verified by melt curve analysis. The fold changes of miR-221 and mRNAs were calculated using the $2^{-\Delta\Delta Ct}$ method.

### 4.4. Transwell Migration and Invasion Assay

The transwell chamber (Corning) with the Matrigel-coated insert (2 mg/mL) was used for the invasion assay, and the Matrigel-free inserts were used for the migration assay. For invasion assay, cells were transfected with miR-inhibitor, miR-221 mimic, AEG-1 siRNA, and their corresponding controls for 48 h. Following the incubation, cells were collected and seeded (1 × $10^5$) in the upper chamber of the 8 μm transwell inserts with a serum-free DMEM. FBS (10%) containing DMEM was filled in the lower chamber and incubated for 48 h at 37 °C. After, the upper chamber was removed from the plate and fixed with methanol for 20 min at RT and stained with crystal violet (0.1 mg/mL). Following the staining, the non-invading cells were removed from the surface of the Matrigel, a cotton swab was used for removal, and the invading cells on the bottom of the upper chamber were imaged and counted using a light microscope with 20× magnification (Carl Zeiss, North York, ON, Canada). The same method was performed on the Matrigel-free membrane inserts.

### 4.5. Annexin V-FITC/PI Double Staining of Cell Apoptosis

The Annexin dual staining assay was performed to detect the apoptosis activity in HCC cells using Annexin V-FITC and propidium iodide (PI) (Invitrogen, Carlsbad, CA, USA). The cells were collected after 48 h incubation and washed twice with ice-cold phosphate buffer saline (PBS), and then the cells were resuspended in 100 μL of binding buffer containing Annexin V-FITC (5 μL) and PI (1 μL) (1 mg/mL) and kept in the dark (RT) for 15 min. Following the incubation, 400 μL of the binding buffer was added to each tube, and the cells were mixed well and analyzed within an hour using BD Accuri C6 Flow Cytometer (BD Biosciences, Franklin Lakes, NJ, USA). Flowjo™ software (v.10.0.6) (Becton, Dickinson and Company, Ashland, OR, USA) was used for data analysis.

### 4.6. Propidium Iodide Staining of Cell Cycle Analysis

The flow cytometry analysis was performed to analyze the cell cycle using propidium iodide staining. After 48 h transfection, cells were collected and washed with ice-cold PBS twice and fixed with 70% ice-cold ethanol at −20 °C for 24 h. Following the incubation, cells were stained with 250 μL of propidium iodide (PI) containing RNAs staining solution (50 mg mL/1 mg) (MP Biomedicals, Santa Ana, CA, USA) and incubated at 4 °C in the dark for 30 min. Following the incubation, samples were mixed well and analyzed using FACS Calibur flow cytometry (BD FACS). FlowJo™ Software v.10.0.6 (Becton, Dickinson and Company, Ashland, OR, USA) was used for data analysis.

### 4.7. Wound-Healing Assay

The HCC cells were seeded in 12-well plates (1 × $10^5$) and transfected with RNAi. After the 48-h incubation, a wound was created on the surface of the monolayer using a 200 μL tip, and images were captured at different time intervals (0, 12, and 24 h) using Floid Cell Imaging Station (Life Technologies, Burlington, ON, Canada). The wound gap and width were measured by using ImageJ (v.1.46r) (National Institutes of Health, Bethesda, ML, USA).

### 4.8. MTT Assay

After 48 h of RNAi transfection, the HCC cells were collected and seeded in 96-well plates (1 × $10^4$). The cells treated with MTT ((3-(4,5-dimethylthiazol-2-yl)-2,5-diphenyltetrazolium bromide) (0.5 mg/mL) and incubated for 4 h at 37 °C. Following the incubation, the medium was removed and 150 uL of DMSO was added (Sigma-Aldrich, Burlington, MA, USA) to dissolve the crystals of MTT formazan and the plates were gently mixed for 5 min. The plate read at 570 nm using a microplate reader (Bio-Rad, Hercules, CA, USA) at different time intervals (0, 12, and 24 h).

### 4.9. Western Blotting Analysis

HCC cells were treated with the miR-221 inhibitor, miR-221 mimic, AEG-1 siRNA, and the corresponding controls. Following the 48 h, the cell lysate was collected and homogenized using RIPA (Radio Immunoprecipitation Assay) lysis buffer (Santa Cruz Biotechnology, Dallas, TX, USA). Lowry's method was performed to estimate the protein.

The 50 μg of total lysate was mixed with a 4 × SDS loading buffer at a 1:3 ratio, and samples were heated at 95 °C for 9 min. The samples were loaded and separated on SDS-polyacrylamide gels (12%) (100 V, 2.30 h) and then transferred onto a 0.2 μm nitrocellulose membrane (150 V, 1.30 h) (Bio-Rad, Hercules, CA, USA) by using Towbin buffer. The membrane was incubated with 5% BSA or skimmed milk powder (5% skimmed milk powder in 100 mL TBST (20 mM Tris, 136 mM NaCl, 0.1% Tween 20, (pH 7.6)) at RT to block the non-specific antibody binding. After 1 h incubation, the membranes were incubated with primary antibodies against AEG-1 (ab124789), Late SV40 factor (LSF) (ab80445), MMP9 (ab76003), LC3A/B (ab128025), Osteopontin (OPN) (ab91655), PTEN (ab32199), p57, Kip2 (ab75974), p53 (ab1101), RB (ab24), β-actin (ab6276), Bcl-2 (sc-7382), PI3K (sc-1637), and p-Akt (sc-514032) overnight at 4 °C. The membranes were washed with TBST and TBS 3 times (5 min/each) and incubated with alkaline phosphatase (ALP)-conjugated anti-mouse (ab97020) or anti-rabbit (ab6722) secondary antibodies for 2 h at room temperature. The BCIP/NBT solution (Merk, Kenilworth, NJ, USA) was used to detect the bands, and the densitometry of the band was analyzed using ImageJ (v.1.46r) (National Institutes of Health, Bethesda, ML, USA).

### 4.10. Statistical Analysis

GraphPad Prism (v 7.0) (GraphPad Software, San Diego, CA, USA) was used for statistical analyses, and one-way ANOVA was used for comparison. $p < 0.05$ was considered statistically significant, and all the data were performed and presented in at least three separate experiments.

## 5. Conclusions

We investigated and proved that miR-221 and AEG-1 were overexpressed in HCC and enhanced the HCC tumorigenesis by the regulatory network and signaling pathways. Recent studies considered targeted gene silencing by oncogenic or tumor suppressor miRNAs for diagnosis and therapy. We showed both onco-miR-221/AEG-1 oncogene axis regulations and their role in HCC. Our results indicated that the miR-221 expression level decreased in AEG-1 siRNA-transfected group, but the miR-221 did not alter the AEG-1 expression level in HCC cells. Our results showed that the miR-221 interaction with AEG-1 through RISC activation of miR-221 may be associated with AEG-1 during RISC complex formation and regulation by optimizing or enhancing the miR-221 activity. This complex leads to the dysregulation of their target/associated genes in HCC tumorigenesis.

Moreover, the silencing of both miR-221 and AEG-1 significantly increased the p53, p57, RB, and PTEN expression levels and decreased BCL2, LSF, MMP9, LC3A, PI3K, and p-Akt mRNA and protein levels in HCC cells. These findings revealed that the AEG-1/miR-221 plays a crucial role during HCC tumorigenesis by direct targeting or through PTEN/PI3K/Akt signaling pathways. Thus, the AEG-1 and miR-221 describe a novel therapeutic strategy to treat HCC. However, the exact interaction between miR-221 and AEG-1 in RISC and their molecular mechanism remains unclear. Further studies on the relationship between miR-221 and AEG-1 are essential to confirm their regulation in vivo. Further, our future study will explore the miR-221 and AEG-1 effects on RISC in NAFLD or liver cirrhosis-associated HCC.

**Supplementary Materials:** The following supporting information can be downloaded at: https://www.mdpi.com/article/10.3390/ijms231911300/s1.

**Author Contributions:** M.K., S.J. and A.J.V.A. performed all experiments; M.K., S.J., R.K.M., M.G. and A.J.V.A. helped with collection and data analysis; S.C.C. contributed to flow cytometry experiments; S.J. and A.J.V.A. critically edited the manuscript; M.K. and A.J.V.A. described the study and drafted the manuscript. All authors have read and agreed to the published version of the manuscript.

**Funding:** This research work was supported by the Department of Biotechnology (DBT), India through Pilot Project on Cancer (6242-P49/RGCB/PMD/DBT/AAJV/2015) to AA.

**Data Availability Statement:** Data not presented in full in the body of this paper can be obtained by writing to the corresponding author.

**Acknowledgments:** We wish to thank DBT for providing financial assistance to Maheshkumar Kannan through Junior Research Fellowship (JRF). We also thank the DST-FIST program for the instrument facility.

**Conflicts of Interest:** The authors declare no possible conflict of interest.

## References

1. Siegel, R.L.; Miller, K.D.; Jemal, A. Cancer statistics, 2019. *CA A Cancer J. Clin.* **2019**, *69*, 7–34. [CrossRef] [PubMed]
2. Bray, F.; Ferlay, J.; Soerjomataram, I.; Siegel, R.L.; Torre, L.A.; Jemal, A. Global cancer statistics 2018: GLOBOCAN estimates of incidence and mor tality worldwide for 36 cancers in 185 countries. *CA Cancer J. Clin.* **2018**, *68*, 394–424. [CrossRef] [PubMed]
3. Di-Bisceglie, A.M. Hepatitis B and hepatocellular carcinoma. *Hepatology* **2009**, *49*, 56–60. [CrossRef] [PubMed]
4. Dongiovanni, P.; Meroni, M.; Longo, M.; Fargion, S.; Fracanzani, A.L. miRNA Signature in NAFLD: A Turning Point for a Non-Invasive Diagnosis. *Int. J. Mol. Sci.* **2018**, *19*, 3966. [CrossRef] [PubMed]
5. Agrawal, S.; Duseja, A. Nonalcoholic Fatty Liver Disease–The Clinician's Perspective. *Trop. Gastroenterol.* **2014**, *35*, 212–221. [CrossRef]
6. Jindal, A.; Thadi, A.; Shailubhai, K. Hepatocellular carcinoma: Etiology, current and future drugs. *J. Clin. Exp. Hepatol.* **2019**, *9*, 221–232. [CrossRef]
7. Forner, A.; Reig, M.; Bruix, J. Hepatocellular carcinoma. *Lancet* **2018**, *391*, 1301–1314. [CrossRef]
8. Wang, R.; Zhao, N.; Li, S.; Fang, J.H.; Chen, M.X.; Yang, J.; Jia, W.H.; Yuan, Y.; Zhuang, S.M. MicroRNA-195 suppresses angiogenesis and metastasis of hepatocellular carcinoma by inhibiting the expression of VEGF, VAV2, and CDC42. *Hepatology* **2013**, *58*, 642–653. [CrossRef]
9. Luo, B.; Kang, N.; Chen, Y.; Liu, L.; Zhang, Y. Oncogene miR-106a promotes proliferation and metastasis of prostate cancer cells by directly targeting PTEN in vivo and in vitro. *Minerva Med.* **2018**, *109*, 24–30. [CrossRef]
10. Xue, Y.; Xu, W.; Zhao, W.; Wang, W.; Zhang, D.; Wu, P. miR-381 inhibited breast cancer cells proliferation, epithelial-to-mesenchymal transition and metastasis by targeting CXCR4. *Biomed. Pharmacother.* **2017**, *86*, 426–433. [CrossRef]
11. Jayamohan, S.; Kannan, M.; Moorthi, R.K.; Rajasekaran, N.; Jung, H.S.; Shin, Y.K.; Arockiam, A.J.V. Dysregulation of miR-375/AEG-1 axis by Human Papillomavirus 16/18-E6/E7 Promotes Cellular Proliferation, Migration and Invasion in Cervical Cancer. *Front. Oncol.* **2019**, *9*, 847. [CrossRef] [PubMed]
12. Slattery, M.L.; Mullany, L.E.; Sakoda, L.C.; Wolff, R.K.; Stevens, J.R.; Samowitz, W.S.; Herrick, J.S. The PI3K/AKT signaling pathway: Associations of miRNAs with dysregulated gene expression in colorectal cancer. *Mol. Carcinog.* **2018**, *57*, 243–261. [CrossRef] [PubMed]
13. Amini-Farsani, Z.; Sangtarash, M.H.; Shamsara, M.; Teimori, H. MiR-221/222 promote chemoresistance to cisplatin in ovarian cancer cells by targeting PTEN/PI3K/AKT signaling pathway. *Cytotechnology* **2018**, *70*, 203–213. [CrossRef]
14. Zheng, H.; Liu, J.; Tycksen, E.; Nunley, R.; McAlinden, A. MicroRNA-181a/b-1 over-expression enhances osteogenesis by modulating PTEN/PI3K/AKT signaling and mitochondrial metabolism. *Bone* **2019**, *123*, 92–102. [CrossRef] [PubMed]
15. Ramírez-Moya, J.; Wert-Lamas, L.; Santisteban, P. MicroRNA-146b promotes PI3K/AKT pathway hyperactivation and thyroid cancer progression by targeting PTEN. *Oncogene* **2018**, *37*, 3369–3383. [CrossRef] [PubMed]
16. Liu, G.; Zhou, J.; Dong, M. Down-regulation of miR-543 expression increases the sensitivity of colorectal cancer cells to 5-Fluorouracil through the PTEN/PI3K/AKT pathway. *Biosci. Rep.* **2019**, *39*, 3. [CrossRef]
17. Calin, G.A.; Croce, C.M. MicroRNA signatures in human cancers. *Nat. Rev. Cancer* **2006**, *6*, 857–866. [CrossRef] [PubMed]
18. Xie, D.; Yuan, P.; Wang, D.; Jin, H.; Chen, H. Expression and prognostic significance of miR-375 and miR-221 in liver cancer. *Oncol. Lett.* **2017**, *14*, 2305–2309. [CrossRef] [PubMed]
19. Pineau, P.; Volinia, S.; McJunkin, K.; Marchio, A.; Battiston, C.; Terris, B.; Mazzaferro, V.; Lowe, S.W.; Croce, C.M.; Dejean, A. miR-221 overexpression contributes to liver tumorigenesis. *Proc. Natl. Acad. Sci. USA* **2010**, *107*, 264–269. [CrossRef]
20. Britt, D.E.; Yang, D.F.; Yang, D.Q.; Flanagan, D.; Callanan, H.; Lim, Y.P.; Lin, S.H.; Hixson, D.C. Identification of a novel protein, LYRIC, localized to tight junctions of polarized epithelial cells. *Exp. Cell. Res.* **2004**, *300*, 134–148. [CrossRef]
21. Kang, D.C.; Su, Z.Z.; Sarkar, D.; Emdad, L.; Volsky, D.J.; Fisher, P.B. Cloning and characterization of HIV-1- inducible astrocyte elevated gene-1, AEG-1. *Gene* **2005**, *353*, 8–15. [CrossRef] [PubMed]
22. Reghupaty, S.C.; Sarkar, D. Current Status of Gene Therapy in Hepatocellular Carcinoma. *Cancers* **2019**, *11*, 1265. [CrossRef]
23. Hu, B.; Emdad, L.; Bacolod, M.D.; Kegelman, T.P.; Shen, X.N.; Alzubi, M.A.; Das, S.K.; Sarkar, D.; Fisher, P.B. Astrocyte elevated gene-1 interacts with Akt isoform 2 to control glioma growth, survival, and pathogenesis. *Cancer Res.* **2014**, *74*, 7321–7332. [CrossRef]
24. Yoo, B.K.; Santhekadur, P.K.; Gredler, R.; Chen, D.; Emdad, L.; Bhutia, S.; Pannell, L.; Fisher, P.B.; Sarkar, D. Increased RNA-induced silencing complex (RISC) activity contributes to hepatocellular carcinoma. *Hepatology* **2011**, *53*, 1538–1548. [CrossRef]
25. Fu, Y.; Liu, M.; Li, F.; Qian, L.; Zhang, P.; Fengwei, L.; Cheng, W.; Hou, R. MiR-221 Promotes Hepatocellular Carcinoma Cells Migration via Targeting PHF2. *Biomed. Res. Int.* **2019**, *2019*, 11. [CrossRef] [PubMed]

26. Li, H.; Zhang, B.; Ding, M.; Lu, S.; Zhou, H.; Sun, D.; Wu, G.; Gan, X. C1QTNF1-AS1 regulates the occurrence and development of hepatocellular carcinoma by regulating miR-221-3p/SOCS3. *Hepatol. Int.* **2019**, *13*, 277–292. [CrossRef] [PubMed]
27. Ke, Q.H.; Chen, H.Y.; He, Z.L.; Lv, Z.; Xu, X.F.; Qian, Y.G.; Zheng, S.S. Silencing of microRNA-375 affects immune function in mice with liver failure by upregulating astrocyte elevated gene-1 through reducing apoptosis of Kupffer cells. *J. Cell. Biochem.* **2019**, *120*, 253–263. [CrossRef]
28. He, X.X.; Chang, Y.; Meng, F.Y.; Wang, M.Y.; Xie, Q.H.; Tang, F.; Li, P.Y.; Song, Y.H.; Lin, J.S. MicroRNA-375 targets AEG-1 in hepatocellular carcinoma and suppresses liver cancer cell growth in vitro and in vivo. *Oncogene* **2012**, *31*, 3357–3369. [CrossRef] [PubMed]
29. Yan, J.J.; Chang, Y.; Zhang, Y.N.; Lin, J.S.; He, X.X.; Huang, H.J. miR-195 inhibits cell proliferation via targeting AEG-1 in hepatocellular carcinoma. *Oncol. Lett.* **2017**, *13*, 3118–3126. [CrossRef]
30. Yarchoan, M.; Johnson, B.A.; Lutz, E.R.; Laheru, D.A.; Jaffee, E.M. Targeting neoantigens to augment antitumor immunity. *Nat. Rev. Cancer* **2017**, *17*, 209. [CrossRef]
31. Liu, S.; Koh, S.S.; Lee, C.G. Hepatitis B Virus X Protein and Hepatocarcinogenesis. *Int. J. Mol. Sci.* **2016**, *17*, 940. [CrossRef]
32. Zhao, J.; Wang, W.; Huang, Y.; Wu, J.; Chen, M.; Cui, P.; Zhang, W.; Zhang, Y. HBx elevates oncoprotein AEG-1 expression to promote cell migration by downregulating miR-375 and miR-136 in malignant hepatocytes. *DNA Cell Biol.* **2014**, *33*, 715–722. [CrossRef]
33. Chen, J.J.; Tang, Y.S.; Huang, S.F.; Ai, J.G.; Wang, H.X.; Zhang, L.P. Hbx protein-induced upregulation of microRNA-221 promotes aberrant proliferation in HBv related hepatocellular carcinoma by targeting estrogen receptor-α. *Oncol. Rep.* **2015**, *33*, 792–798. [CrossRef]
34. Brown, D.M.; Ruoslahti, E. Metadherin, a cell surface protein in breast tumors that mediates lung metastasis. *Cancer Cell* **2004**, *5*, 365–374. [CrossRef]
35. Santhekadur, P.K.; Das, S.K.; Gredler, R.; Chen, D.; Srivastava, J.; Robertson, C.; Baldwin, A.S.; Fisher, P.B.; Sarkar, D. Multifunction protein staphylococcal nuclease domain containing 1 (SND1) promotes tumor angiogenesis in human hepatocellular carcinoma through novel pathway that involves nuclear factor kappaB and miR-221. *J. Biol. Chem.* **2012**, *287*, 13952–13958. [CrossRef]
36. Caixia, H.; Shichang, C.; Zheng, J.; Yin, T.; Junsheng, L.; Long, J.; Zhang, W.; Wang, X.; Sheng, S.; Cong, L.; et al. MiR-875-5p inhibits hepatocellular carcinoma cell proliferation and migration by repressing astrocyte elevated gene-1 (AEG-1) expression. *Transl. Cancer Res.* **2018**, *7*, 158–169.
37. Kim, H.; Huang, W.; Jiang, X.; Pennicooke, B.; Park, P.J.; Johnson, M.D. Integrative genome analysis reveals an oncomir/oncogene cluster regulating glioblastoma survivorship. *Proc. Natl. Acad. Sci. USA* **2009**, *107*, 2183–2188. [CrossRef]
38. Yoo, B.K.; Emdad, L.; Gredler, R.; Fuller, C.; Dumur, C.I.; Jones, K.H.; Jackson-Cook, C.; Su, Z.Z.; Chen, D.; Sarkar, D.; et al. Transcription factor Late SV40 Factor (LSF) functions as an oncogene in hepatocellular carcinoma. *Proc. Natl. Acad. Sci. USA* **2010**, *107*, 8357–8362. [CrossRef]
39. Yoo, B.K.; Gredler, R.; Chen, D.; Santhekadur, P.K.; Fisher, P.B.; Sarkar, D. c-Met activation through a novel pathway involving osteopontin mediates oncogenesis by the transcription factor LSF. *J. Hepatol.* **2011**, *55*, 1317–1324. [CrossRef]
40. Santhekadur, P.K.; Gredler, R.; Chen, D.; Siddiq, A.; Shen, X.N.; Das, S.K.; Emdad, L.; Fisher, P.B.; Sarkar, D. Late SV40 factor (LSF) enhances angiogenesis by transcriptionally up-regulating matrix metalloproteinase-9 (MMP-9). *J. Biol. Chem.* **2012**, *287*, 3425–3432. [CrossRef]
41. Yoo, B.K.; Emdad, L.; Su, Z.Z.; Villanueva, A.; Chiang, D.Y.; Mukhopadhyay, N.D.; Mills, A.S.; Waxman, S.; Fisher, R.A.; Sarkar, D.; et al. Astrocyte elevated gene-1 regulates hepatocellular carcinoma development and progression. *J. Clin. Investig.* **2009**, *119*, 465–477. [CrossRef] [PubMed]
42. Zheng, X.; Dai, J.; Zhang, H.; Ge, Z. MicroRNA-221 promotes cell proliferation, migration, and differentiation by regulation of ZFPM2 in osteoblasts. *Braz. J. Med. Biol. Res.* **2018**, *51*, e7574. [CrossRef] [PubMed]
43. Huang, L.L.; Wang, Z.; Cao, C.J.; Ke, Z.F.; Wang, F.; Wang, R.; Luo, C.Q.; Lu, X.; Wang, L.T. AEG-1 associates with metastasis in papillary thyroid cancer through upregulation of MMP2/9. *Int. J. Oncol.* **2017**, *51*, 812–822. [CrossRef]
44. Xu, Q.; Li, P.; Chen, X.; Zong, L.; Jiang, Z.; Nan, L.; Lei, J.; Duan, W.; Zhang, D.; Wang, Z.; et al. miR-221/222 induces pancreatic cancer progression through the regulation of matrix metalloproteinases. *Oncotarget* **2015**, *6*, 14153–14164. [CrossRef]
45. Wu, J.; Ding, J.; Yang, J.; Guo, X.; Zheng, Y. MicroRNA Roles in the Nuclear Factor Kappa B Signaling Pathway in Cancer. *Front. Immunol.* **2018**, *9*, 546. [CrossRef] [PubMed]
46. Zhao, D.; Zhuang, N.; Ding, Y.; Kang, Y.; Shi, L. miR-221 activates the NF-kB pathway by targeting A20. *Biochm. Biophys. Res. Comm.* **2016**, *472*, 11–18. [CrossRef]
47. Emdad, L.; Sarkar, D.; Su, Z.; Randolph, A.; Boukerche, H.; Fisher, P.B. Activation of the Nuclear Factor KB Pathway by Astrocyte Elevated Gene-1: Implications for Tumor Progression and Metastasis. *Cancer Res.* **2006**, *66*, 3. [CrossRef] [PubMed]
48. Wang, N.; Zhu, C.; Xu, Y.; Qian, W.; Zheng, M. Negative Regulation of PTEN by MicroRNA-221 and Its Association with Drug Resistance and Cellular Senescence in Lung Cancer Cells. *Biomed. Res. Int.* **2018**, *2018*, 7908950. [CrossRef]
49. Fornari, F.; Gramantieri, L.; Ferracin, M.; Veronese, A.; Sabbioni, S.; Calin, G.A.; Grazi, G.L.; Giovannini, C.; Croce, C.M.; Negrini, M.; et al. MiR-221 controls CDKN1C/p57 and CDKN1B/p27 expression in human hepatocellular carcinoma. *Oncogene* **2008**, *27*, 5651–5661. [CrossRef] [PubMed]
50. Zhang, W.; Kone, B.C. NF-kappaB inhibits transcription of the H(+)-K(+)-ATPase alpha(2)-subunit gene: Role of histone deacetylases. *Am. J. Physiol. Renal Physiol.* **2002**, *283*, 904–911. [CrossRef]

51. Catz, S.D.; Johnson, J.L. Transcriptional regulation of bcl-2 by nuclear factor kappa B and its significance in prostate cancer. *Oncogene* **2001**, *20*, 7342–7351. [CrossRef] [PubMed]
52. Zhou, Z.; Deng, H.; Yan, W.; Luo, M.; Tu, W.; Xia, Y.; He, J.; Han, P.; Fu, Y.; Tian, D. AEG-1 promotes anoikis resistance and orientation chemotaxis in hepatocellular carcinoma cells. *PLoS ONE* **2014**, *9*, e100372.
53. Fu, B.; Wang, Y.; Zhang, X.; Lang, B.; Zhou, X.; Xu, X.; Zeng, T.; Liu, W.; Zhang, X.; Wang, G.; et al. MiR-221-induced PUMA silencing mediates immune evasion of bladder cancer cells. *Int. J. Oncol.* **2015**, *46*, 1169–1180. [CrossRef] [PubMed]
54. Wang, S.Q.; Wang, C.; Chang, L.M.; Zhou, K.R.; Wang, J.W.; Ke, Y.; Yang, D.X.; Shi, H.G.; Wang, R.; Liu, H.M.; et al. Geridonin and paclitaxel act synergistically to inhibit the proliferation of gastric cancer cells through ROS-mediated regulation of the PTEN/PI3K/Akt pathway. *Oncotarget* **2016**, *7*, 72990–73002. [CrossRef]
55. Paramio, J.M.; Navarro, M.; Segrelles, C.; Gómez-Casero, E.; Jorcano, J.L. PTEN tumour suppressor is linked to the cell cycle control through the retinoblastoma protein. *Oncogene* **1999**, *18*, 7462–7468. [CrossRef]
56. Jin, X.; Ding, D.; Yan, Y.; Li, H.; Wang, B.; Ma, L.; Ye, Z.; Ma, T.; Wu, Q.; Huang, H.; et al. Phosphorylated RB Promotes Cancer Immunity by Inhibiting NF-κB Activation and PD-L1 Expression. *Mol. Cell* **2019**, *73*, 22–35. [CrossRef]
57. Fernandez, A.F.; Sebti, S.; Wei, Y.; Zou, Z.; Shi, M.; McMillan, K.L.; He, C.; Ting, T.; Liu, Y.; Levine, B.; et al. Disruption of the beclin 1/Bcl-2 autophagy regulatory complex promotes longevity in mice. *Nature* **2018**, *558*, 136–140. [CrossRef]
58. Maiuri, M.C.; Galluzzi, L.; Morselli, E.; Kepp, O.; Malik, S.A.; Kroemer, G. Autophagy regulation by p53. *Curr. Opin. Cell. Biol.* **2010**, *22*, 181–185. [CrossRef]
59. Jeyamohan, S.; Moorthy, R.K.; Kannan, M.K.; Arockiam, A.J. Parthenolide induces apoptosis and autophagy through the suppression of PI3K/Akt signaling pathway in cervical cancer. *Biotech. Lett.* **2016**, *38*, 1251–1260. [CrossRef]
60. Peng, X.; Wang, Y.; Li, H.; Fan, J.; Shen, J.; Yu, X.; Zhou, Y.; Mao, H. ATG5-mediated autophagy suppresses NF-κB signaling to limit epithelial inflammatory response to kidney injury. *Cell Death Dis.* **2019**, *10*, 253. [CrossRef]

International Journal of
*Molecular Sciences*

MDPI

*Review*

# Zinc Fingers and Homeobox Family in Cancer: A Double-Edged Sword

**Yonghua Bao [1], Haifeng Zhang [1], Zhixue Han [1], Yongchen Guo [2,]* and Wancai Yang [3,]***

1 Department of Pathology, Mudanjiang Medical University, Mudanjiang 157011, China
2 Department of Immunology, Mudanjiang Medical University, Mudanjiang 157011, China
3 Comprehensive Cancer Center, The Ohio State University, Columbus, OH 43210, USA
* Correspondence: guoyongchen@mdjmu.edu.cn (Y.G.); wancai.yang@osumc.edu (W.Y.)

**Abstract:** The zinc fingers and homeobox (ZHX) family includes ZHX1, ZHX2, and ZHX3, and their proteins have similar unique structures, containing two C2H2-type zinc finger motifs and four or five HOX-like homeodomains. The members of the ZHX family can form homodimers or heterodimers with each other or with a subunit of nuclear factor Y. Previous studies have suggested that ZHXs can function as positive or negative transcriptional regulators. Recent studies have further revealed their biological functions and underlying mechanisms in cancers. This review summarized the advances of ZHX-mediated functions, including tumor-suppressive and oncogenic functions in cancer formation and progression, the molecular mechanisms, and regulatory functions, such as cancer cell proliferation, migration, invasion, and metastasis. Moreover, the differential expression levels and their association with good or poor outcomes in patients with various malignancies and differential responses to chemotherapy exert opposite functions of oncogene or tumor suppressors. Therefore, the ZHXs act as a double-edged sword in cancers.

**Keywords:** zinc fingers; homeoboxes; ZHX1; ZHX2; ZHX3; cancer

**Citation:** Bao, Y.; Zhang, H.; Han, Z.; Guo, Y.; Yang, W. Zinc Fingers and Homeobox Family in Cancer: A Double-Edged Sword. *Int. J. Mol. Sci.* **2022**, 23, 11167. https://doi.org/10.3390/ijms231911167

Academic Editor: Laura Paleari

Received: 8 August 2022
Accepted: 19 September 2022
Published: 22 September 2022

**Publisher's Note:** MDPI stays neutral with regard to jurisdictional claims in published maps and institutional affiliations.



## 1. Introduction

The zinc fingers and homeobox (ZHX) family have three members, including ZHX1–3 [1–4]. They share similar gene and protein structures [5]. ZHX family members contain a unique protein structure, two C2-H2 (Cys-Xaa2-Cys-Xaa12-His-Xaa4-His) zinc finger motifs and four or five HOX-like homeodomains (HDs).

The humansZHX1 and ZHX2 are located on chromosome 8 [1,4], and ZHX3 is located on chromosome 20 [6]. ZHX1–3 can constitute homodimers or heterodimers and can interact with the alpha subunit of nuclear factor Y (NFYA) to repress transcription [1–3,6–8].

### 1.1. ZHX1 Gene and Biological Functions

ZHX1 was initially identified from a murine bone marrow stromal cell [9]. Human ZHX1 was identified by screening a human liver cDNA library for NFY-interacting protein isolation [4]. The ZHX1 gene of humans is located in Chr8q24.13, spanning approximately 27 kb. ZHX1 has eight transcripts, including two main ones, 4.5 kilobases (kb) and 5 kb, respectively. These two transcripts are expressed ubiquitously, although the latter is more abundant in most tissues [4].

Human ZHX1 contains 873 amino acid residues and is structurally composed of two C2-H2 motifs and five homeodomains [4]. This is why it is classified as a zinc finger of homeodomain transcription factor. ZHX1 forms homodimers or heterodimers through the homeodomain 1 region (amino acids 272–432). ZHX1 functions as a transcriptional repressor via an acidic region (amino acids 831–873). Achieving complete inhibitory activity requires dimerization [7].

ZHX1 proteins of humans and mice possess 91% amino acid similarity [8]. The ZHX1 gene of mice is located in chromosome 15, approximately 29 kb in length [2]. ZHX1 mRNA is widely expressed in adult mice, with a high expression in the brain and lower levels in the liver and kidney [9]. It has been reported that the mouse ZHX1 gene expression could be induced by IL-2 in CTLL-2 cells [10].

Besides interacting with NFYA, ZHX1 binds DNA methyltransferase 3B [11] and the transcriptional corepressor BS69 to induce repression activity [12].

### *1.2. ZHX2 and Biological Functions*

Human ZHX2 was cloned as a novel ubiquitous transcription factor and a ZHX1-interacting protein from a yeast two-hybrid screen on a size-fractionated brain cDNA library [1]. The ZHX2 gene is located on Chr8q24.13, spanning approximately 193 kb. This gene has two transcripts (splice variants) coding 837aa and 166aa. ZHX2 mRNA is expressed in various tissues, with the highest levels in the ovaries, and sequentially lower levels in the lung, heart, kidney, brain, and liver. The pancreas, spleen, testis, and skeletal muscle have intermediate expression levels. A genome-wide association study has demonstrated that individuals with the ZHX2 polymorphism G779A have a strong response to the smallpox vaccine [13].

The mouse ZHX2 gene encodes an 836aa protein, having 87% amino acid identify with the human ZHX2 protein [14]. ZHX2 contains two C2-H2 zinc finger motifs and five HDs. This region, called the P domain, is rich in proline and resides between HD1 and HD2. ZHX2 forms a homodimer with itself or a heterodimer with ZHX1 at the region containing HD1. In addition, ZHX2 can interact with NFYA at the region between HD1 and HD2. Further studies have revealed that nuclear ZHX2, not cytoplasmic ZHX2, acts as a transcriptional repressor. The repressor domain resides in a region between the HD1 and P domains [1,3].

ZHX2 was previously thought to suppress gene transcription. However, in vitro and in vivo studies showed that ZHX2 positively regulates major urinary proteins' (Mups) gene expression [15].

### *1.3. ZHX3 and Biological Functions*

The human ZHX3 gene is located on Chr20q12, spanning approximately 139 kb, and has 21 transcripts (splice variants). ZHX3 has 956 amino acid residues and, like ZHX1, contains two C2-H2 motifs and five HDs [6]. ZHX3 forms a heterodimer with ZHX1 and interacts with NFYA. The pleiotropic HD1 region is responsible for the dimerization with ZHX1, interaction with NFYA, and repressor function. In addition, two nuclear localization signals were mapped to the N-terminus at zinc fingers 1 and the HD2 region. Moreover, protein–protein interaction assays showed that ZHX3 formed a heterodimer with ZHX2 via a region containing HD1. In mice, there are three ZHX3 transcripts of 9.5 kb, 6.5 kb, and 4.4 kb in length, which are ubiquitously and differentially expressed. Functional studies have revealed that nuclear ZHX3 is a ubiquitous transcriptional repressor and functions as a dimer [8,16].

Accumulating evidence has demonstrated that ZHX family members play important roles in cell development, differentiation, and various cancers [17–22]. The aberrant expression and dysfunction of ZHXs are associated with the occurrence and progression of various diseases. These include hematological, neurological, and glomerular diseases, as well as carcinogenesis and its progression [22]. The expression levels are also linked to the outcomes of several malignancies. Interestingly, the biological functions of ZHXs in cancers are not consistent. For example, ZHXs act as tumor suppressors or oncogenes in different types of cancers. The expression levels of ZHXs are also associated with poor or good outcomes in patients with various malignancies, and act as a "double-edged sword". This review focuses on the clinical significance of the oncogenic or tumor-suppressive functions of ZHX family members in cancers.

## 2. Oncogenic or Tumor-Suppressive Functions of ZHXs in Cancers

### 2.1. ZHX1 and Cancer

ZHX1 has recently been involved in the tumorigenesis of various tumors. The regulation of ZHX1 expression, its clinical significance, and role in cancer is controversial. ZHX1 acts as an oncogene or tumor suppressor in different types of cancer (Table 1). ZHX1 contributes to cancer cell proliferation, mobility, migration, and invasion (Figure 1). In glioblastoma [23], the messenger RNA and protein expression of ZHX1 are increased compared with normal brain tissue, and ZHX1 overexpression was associated with short overall survival. In vitro, the knockdown of ZHX1 expression suppressed cell proliferation, mobility, migration, and invasion in glioblastoma. However, increasing the expression of ZHX1 enhanced malignant features, which are linked to the regulation of TWIST and SNAIL, both of which are transcriptional factors for the epithelial–mesenchymal transition (EMT) and metastasis.

It has been reported that long non-coding RNA (lncRNA), microRNA (miRNA), and ZHX1 participate in tumor initiation and progression. In vitro and in vivo studies have shown that LncRNA MALAT1 facilitates glioblastoma cell proliferation and progression by reducing miR-199a to promote ZHX1 expression. The LncRNA MALAT1-elevated expression is associated with the poor prognosis of glioblastoma patients [24]. Another study on glioma also showed that ZHX1 is an oncogene, and it was identified as a target gene of miR-23b-3p. LncRNA SNHG17 can absorb miR-23b-3p like a sponge to elevate ZHX1expression; therefore, increased SNHG17 and ZHX1 expressions and the reduced miR-23b-3p expression were found in glioma tissues. Oncogenic SNHG17 improves cell proliferation, migration, and invasion in glioma [25]. In addition, lncRNA LINC01140 promote the development of glioma by downregulating miR-199a-3p expression and indirectly upregulating ZHX1 expression [26]. These findings strongly suggest that targeting the lncRNA/miRNA/ZHX1 pathway might provide a new therapeutic strategy for tumors of the nervous system.

Another study has demonstrated the oncogenic roles of lncRNA DLG1-AS1/miR-107/ZHX1 axis in cervical cancer [27]. lncRNA DLG1-AS1 was remarkably overexpressed in cervical cancer tissues, and the cervical cancer patients with a high DLG1-AS1 expression have worse outcomes. Manipulating gene expression in vitro showed that the knockdown of the DLG1-AS1 gene inhibited the proliferation of cervical cancer cells. Mechanical studies showed that the inhibition of miR-107 abrogated DLG1-AS1-mediated ZHX1 expression via competitive binding between ZHX1 and miR-107. Vice versa, increasing the expression of miR-107 rescued the role of DLG1-AS1 in cervical cancer cell proliferation, indicating that the oncogenic effects of DLG-AS in cervical cancer are miR-107/ZHX1-dependent. ZHX1 also acts as an oncogene in cholangiocarcinoma [28]. Immunohistochemical staining analysis revealed that ZHX1 is amplified and overexpressed in cholangiocarcinoma tissues. In vitro studies showed that overexpressing ZHX1 facilitated cholangiocarcinoma cell proliferation, migration, and invasion. Conversely, ZHX1 knockdown using specific small interfering RNAs decreased malignant characteristics. Mechanistic studies showed that ZHX1-mediated tumor promotion might be partially associated with early growth response 1 (EGR1).

**Table 1.** Studies on ZHX1 expression, function, clinical significance, and molecular mechanism.

| ZHX Family Member | Expression Levels | Functions | Cancer Types | Clinical Significance | Associated Effects | Involved Molecules | References |
|---|---|---|---|---|---|---|---|
| ZHX1 | Increased | Oncogene | Glioblastoma | ZHX1 overexpression was associated with short overall survival. | Facilitated proliferation, mobility, migration, and invasion. | TWIST and SNAIL | [23] |
| ZHX1 | Increased | Oncogene | Glioblastoma | | Promoted proliferation and progression. | LncRNA MALAT1/ZHX1/miR-199a axis | [24] |
| ZHX1 | Increased | Oncogene | Glioblastoma | | Promoted proliferation, migration, and invasion. | LncRNA SNHG17/ZHX1/miR-23b-3p axis | [25] |
| ZHX1 | Increased | Oncogene | Glioblastoma | | Promoted glioma development. | LncRNA LINC01140/ZHX1/miR-199a-3p axis | [26] |
| ZHX1 | Increased | Oncogene | Cervical cancer | Cervical cancer patients with high DLG1-AS1 expression had worse prognosis. | Promoted proliferation of cervical cancer cells. | LncRNA DLG1-AS1/ZHX1/miR-107 axis | [27] |
| ZHX1 | Increased | Oncogene | Cholangiocarcinoma | | Promoted proliferation, migration, and invasion of cholangiocarcinoma cells. | Early growth response 1 (EGR1) | [28] |
| ZHX1 | Decreased | Tumor suppressor | Gastric cancer | | Promoted apoptosis, and inhibited cell proliferation and migration. | Cyclin D1 | [29] |
| ZHX1 | Decreased | Tumor suppressor | Gastric cancer | | Inhibited cell proliferation. | miRNA-199a-3p | [30] |
| ZHX1 | Decreased | Tumor suppressor | Gastric cancer | Increased ZHX1 expression was associated with better OS in patients with gastric cancer, without distant metastasis, received surgery, and poorly differentiated. | | | [31] |
| ZHX1 | Decreased | Tumor suppressor | Renal cancers | Reduction in ZHX1 mRNA expression was associated with the progression of clear cell renal cancers and poorer survival. | | | [18] |
| ZHX1 | Decreased | Tumor suppressor | Hepatocellular carcinoma | | Inhibited the proliferation of HCC SMMC-7721 cells. | | [32] |
| ZHX1 | Increased | Oncogene | Hepatocellular carcinoma | | Promoted cell growth and inhibited apoptosis. | BAX, caspase 3 and BCL2 | [33] |
| ZHX1 and ZHX2 | Decreased | Tumor suppressor | Chronic lymphocytic leukemia (CLL) | CLL patient with lower expression level of ZHX1 and ZHX2 had worse prognosis. | | | [34] |
| ZHX1 | Decreased | Tumor suppressor | T-cell acute lymphoblastic leukemia (T-ALL) | | | FOXN2 and FOXN3 | [35] |
| ZHX1 | Increased | Oncogene | Hodgkin lymphoma (HL) | | | | [36] |

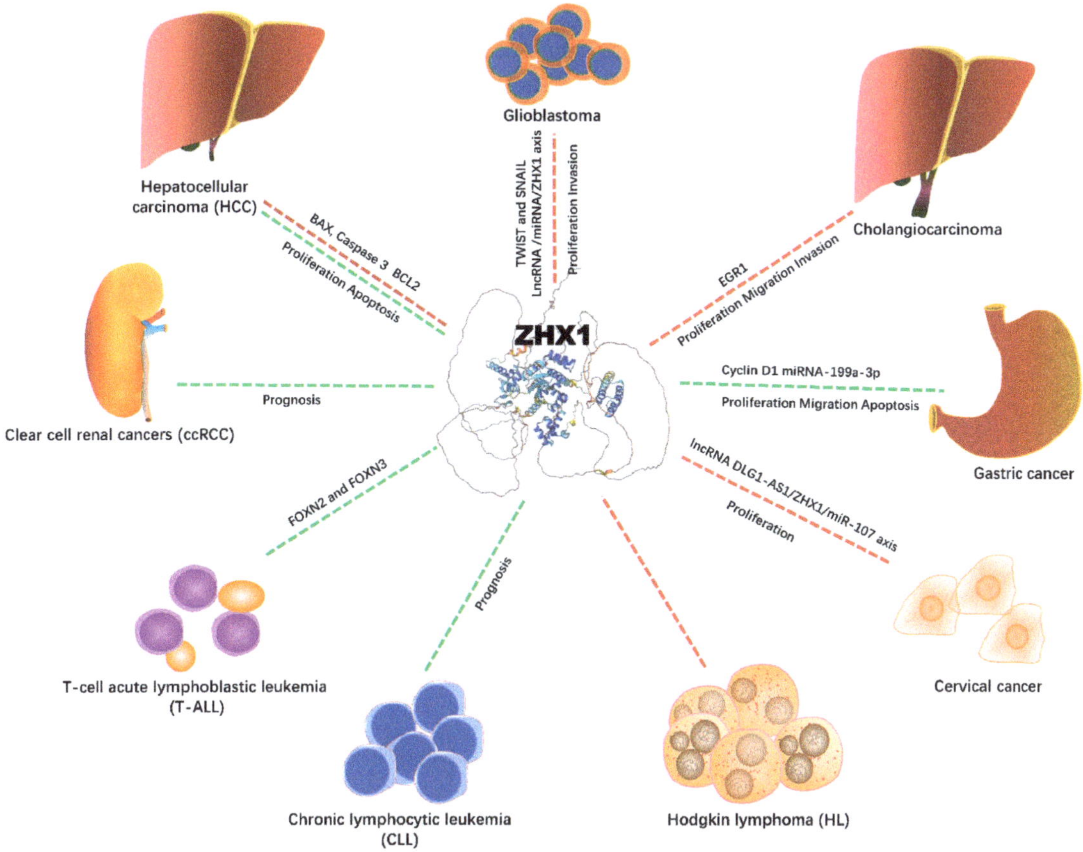


**Figure 1.** ZHX1 and cancers: biological functions and potential signaling networking.

In contrast, ZHX1 exerts tumor-suppressive functions in gastric cancer [29], in which ZHX1 expression was reduced. Interestingly, the reduced nuclear expression of ZHX1 was closely related to a larger tumor size, poorer differentiation, advanced TNM stages, and a deeper invasion. However, these features were not correlated with lymph node metastasis. In cultured gastric cancer cells, stable transfection with plasmids overexpressing ZHX1 significantly promoted apoptosis, and inhibited cell proliferation and migration. In nude mice, the ZHX1 overexpression inhibited the tumor growth that was associated with cell cycle arrest and a repressed expression of cyclin D1.

Another study on gastric cancers reported by Wang et al. also showed that ZHX1 was a tumor suppressor and a downstream target of tumor promoter miRNA-199a-3p [30]. MiR-199a-3p expression was increased in gastric cancer tissues and gastric cancer cell lines. The study also demonstrated that miR-199a-3p could negatively regulate ZHX1 gene expression in gastric cancer cells. Therefore, increasing the expression of miR-199a-3p inhibits ZHX1 expression, but reducing miR-199a-3p expression promotes ZHX1 expression. Furthermore, restoring ZHX1 expression inhibits cancer cell proliferation in vitro.

A more recent study indicated that an increased ZHX1 expression was associated with a better or worse prognosis for patients with gastric cancer in the clinical stages, including lymph node metastasis or distant metastasis, and surgical therapy [31]. For instance, a lower ZHX1 expression was related to a better prognosis of patients without lymph node metastasis, whereas a higher ZHX1 expression was related to a better prognosis of patients without distant metastasis. Furthermore, a higher ZHX1 expression was associated with a better prognosis of the patients who received surgery therapy and in the patients whose cancer showed poor differentiation.

A study on clear cell renal cancers (ccRCC) [18] based on an online database showed that both ZHX1 and ZHX3 expression levels were downregulated, while ZHX2 was upregulated. In particular, the ZHX1 and ZHX3 expression levels were remarkably lower, but ZHX2 expression was higher in the advanced stages compared with the early stages. In addition, ZHX1 and ZHX3 expression levels were associated with the T stage, while ZHX1 expression was associated with the M stage. These results suggest that a lower expression of ZHX1 and ZHX3 is related to the progression of clear cell renal cancers. Furthermore, Kaplan–Meier and multivariate regression analysis showed that a reduction in ZHX1 mRNA expression is associated with worse survival [18]. Altogether, these findings indicate that ZHX1 is a tumor suppressor in renal cancer.

In hepatocellular carcinoma (HCC), ZHX1 has also been reported as a tumor suppressor. One study [32] detected a decreased ZHX1 expression from a HCC tumor and cell sample. In vitro studies have indicated that an increased expression of ZHX1 suppresses HCC SMMC-7721 cell proliferation. Another study also showed the tumor-suppressive roles of ZHX1 in HCC, in which miR-199a-3p targeting the ZHX1/PUMA signal was reported to inhibit tumorigenesis. In this case, miR-199a-3p inhibited HCC HepG2 cell growth in vitro and induced apoptosis by upregulating the expression of ZHX1 and PUMA [33]. In contrast, the knockdown of ZHX1 or PUMA reversed miR-199a-3p-mediated tumor inhibition roles, as determined by a mechanistic investigation, suggesting that the miR-199a-3p/ZHX1/PUMA signaling could be linked to BCL2, BCL2-associated X, and cleaved caspase 3.

The prognostic impact of ZHX1 and ZHX2 in chronic lymphocytic leukemia (CLL) has recently been reported [34]. CLL patients with reduced ZHX1 and ZHX2 expression have a worse prognosis. In addition, the accumulation of chromosomal abnormalities was negatively associated with ZHX1 and ZHX2 expression in CLL patients.

More studies support the tumor-suppressive roles of ZHX1 in T-cell acute lymphoblastic leukemia (T-ALL), showing that ZHX1 was significantly reduced in T-ALL cells; moreover, other tumor suppressors, such as FOXN2 and FOXN3, were also concurrently downregulated in the T-ALL cell lines, in which both FOXN2 and FOXN3 formed a regulator network and directly activated the transcription of ZHX1 [35]. Furthermore, the physiological expression profile of normal hematopoietic cells revealed that ZHX1 was highly expressed in T-cells and was lower in B-cells, suggesting biased functions of ZHX1 in lymphopoiesis [36]. Lastly, the Hodgkin lymphoma cell lines that showed no expression of orthodenticle homeobox 1 and 2 (OTX1/2) (e.g., OTX 1 and 2 negative) overexpressed ZHX1, correlating with the genomic amplification of the 8q24 locus, supporting the oncogenic potential of ZHX1 in Hodgkin lymphoma [36].

### 2.2. ZHX2 and Cancer

Several studies have demonstrated the tumor-suppressor activities of ZHX2 in hepatocellular carcinoma (HCC) and other malignancies (Table 2), but accumulating evidence has also shown that ZHX2 plays oncogenic roles in carcinogenesis and progression in various cancers (Table 2).

**Table 2.** Studies on ZHX2 expression, function, clinical significance, and molecular mechanism.

| ZHX Family Member | Expression Levels | Functions | Cancer Types | Clinical Significance | Associated Effects | Molecular Mechanisms | References |
|---|---|---|---|---|---|---|---|
| ZHX2 | Decreased | Tumor suppressor | Hepatocellular carcinoma (HCC) | | Inhibited HCC cell proliferation and colony formation in vitro and xenograft growth in nude mice. | Cyclin A, Cyclin E and Ki-67 | [37] |
| ZHX2 | Decreased | Tumor suppressor | HCC | | | Hypermethylation of the ZHX2 promoter | [38] |
| ZHX2 | Decreased | Tumor suppressor | HCC | | ZHX2 expression in HCC was inversely correlated with serum AFP levels in HCC patients. | | [39] |
| ZHX2 | Decreased | Tumor suppressor | HCC | | | Repression of AFP | [40] |
| ZHX2 | Decreased | Tumor suppressor | HCC | | | Repression of PKM, HK2 | [16] |
| ZHX2 | Decreased | Tumor suppressor | HCC | | | Repression of GPC3 | [41] |
| ZHX2 | Decreased | Tumor suppressor | HBV-related HCC | | | HBx/miR-3188/ZHX2/Notch1 signaling pathway | [42] |
| ZHX2 | Decreased | Tumor suppressor | HBV-related HCC | | | miR-155 | [43] |
| ZHX2 | Decreased | Tumor suppressor | Lipogenesis-related HCC | | Inhibited de novo lipogenesis in HCC cells and HCC progression. | ZHX2/miR-24–3p/SREBP1c | [44] |
| ZHX2 | Decreased | Tumor suppressor | NAFLD-associated HCC | | Inhibited HCC cell proliferation, xenograft tumor growth, lipid deposition. | LPL | [45] |
| ZHX2 | Decreased | Tumor suppressor | Liver cancer stem cells (CSCs) | The lower ZHX2 expression and higher KDM2A were associated with poorer survival of patients. | Restricted tumor initiation, self-renewal and sorafenib-resistance of hepatic CSCs. | KDM2A, H3K36, NANOG, SOX4 and OCT4 | [46] |
| ZHX2 | Increased | Oncogene | Hepatocellular carcinoma (HCC) | Upregulation of ZHX2 was correlated with poor differentiation and cancer metastasis. | | | [47] |
| ZHX2 | Decreased | Tumor suppressor | Liver tumor | | Enhances the cytotoxicity of chemotherapeutic drugs. | Repressing multidrug resistance 1 (MDR1) via an interaction with NFYA | [48] |
| ZHX2 | Decreased | Tumor suppressor | lung cancer | | Suppressed cells proliferation, migration and invasion and promoted apoptosis. | p38MAPK signaling pathway | [49] |
| ZHX2 | Decreased | Tumor suppressor | Glioma | | Inhibited proliferation, migration, invasion and VM formation. | HNRNPD/linc00707/miR-651–3p/SP2 axis | [50] |
| ZHX2 | Decreased | Tumor suppressor | Thyroid cancer | Decreased ZHX2 expression was correlated with poor prognosis of thyroid cancer patients. | ZHX2 knockdown significantly promoted the migration of thyroid cancer cells. | S100A14 | [51] |

**Table 2.** *Cont.*

| ZHX Family Member | Expression Levels | Functions | Cancer Types | Clinical Significance | Associated Effects | Molecular Mechanisms | References |
|---|---|---|---|---|---|---|---|
| ZHX2 | Decreased | Tumor suppressor | Multiple myeloma (MM) | Increased ZHX2 expression with improved response to high-dose chemotherapy in multiple myeloma. | | | [52,53] |
| ZHX2 | Decreased | Tumor suppressor | Multiple myeloma (MM) | Low expression of ZHX2 is associated with poor outcome in multiple myeloma. | | | [54] |
| ZHX2 | Increased | Oncogene | Multiple myeloma (MM) | Multiple myeloma patients with higher ZHX2 expression showed poorer clinical outcomes of. | Knockdown of ZHX2 significantly enhanced the sensitivity of MM cells to Bortezomib (BTZ), inhibited nuclear translocation of NF-κB. | NF-κB. | [55] |
| ZHX2 | Decreased | Tumor suppressor | Hodgkin lymphoma | | Knockdown of ZHX2 led to inhibition of genes regulating differentiation and apoptosis. | STAT1 | [56] |
| ZHX2 | Decreased | Tumor suppressor | Hodgkin lymphoma | | | MSX1 and XBP1 | [57] |
| ZHX2 | Increased | Oncogene | Clear cell renal cell carcinoma (ccRCC) | | Depletion of ZHX2 inhibited VHL-deficient ccRCC cell growth and caused decreased expression of multiple genes linked with anti-apoptosis, cell proliferation, invasion/metastasis, and metabolism. | VHL, NF-κB | [21] |
| ZHX2 | Increased | Oncogene | Clear cell renal cell carcinoma (ccRCC) | | ZHX2 drove cell growth, migration and induced Sunitinib resistance by regulating self-protective autophagy. | VEGF, MEK/ERK1/2 signaling pathway | [58] |
| ZHX2 | Increased | Oncogene | Triple-negative breast cancer (TNBC) | | Depletion of ZHX2 inhibited TNBC cell growth and invasion in vitro, orthotopic tumor growth, and spontaneous lung metastasis in vivo. | HIF1α, H3K4me3 and H3K27ac; AP2B1, COX20, KDM3A, or PTGES3L; R491, R581, and R674 | [59] |
| ZHX2 and ZHX3 | Decreased | Tumor suppressor | Triple-negative breast cancer (TNBC) | High mRNA expression of ZHX2 and ZHX3 was significantly correlated with better OS of breast cancer patients. | | | [19] |
| ZHX2 | Increased | Oncogene | Gastric cancer (GC) | Upregulation of ZHX2 predicted poor prognosis in GC. | Promotes proliferation, invasion, migration, and inhibits cell apoptosis. | Immune infiltrating | [60] |
| ZHX2 and ZHX3 | Increased | Oncogene | Gastric cancer (GC) | Decreased mRNA expression of ZHX2 and ZHX3 was correlated with better rates of OS in patients with gastric cancer. | | | [31] |

ZHX2-involved tumor occurrence, development, progression in various tumors, and potential molecular mechanisms are summarized in Figure 2. Tumor-suppressive functions of ZHX2 were supported by the following studies, with most from hepatocellular carcinoma (HCC). Elevating ZHX2 expression can suppress HCC cell multiplication, colony formation, and mice tumor size. Conversely, reducing ZHX2 expression by siRNA significantly enhanced cell proliferation and colony formation [37]. Mechanistic studies demonstrated that tumor inhibition was through the suppression of Cyclins A and E via a direct binding between ZHX2 and the promoter domain of Cyclins A and E. The inverse correlation between ZHX2 and Cyclins A and E was further detected from liver cancer tissues. The overexpression of ZHX2, including nuclei and the cytoplasm, did not differ in tumor tissues and adjacent non-tumor tissues. Interestingly, a lower nuclear ZHX2 expression and a higher cytoplasmic ZHX2 expression were found in HCC tissues, but a higher nuclear ZHX2 expression and lower cytoplasmic ZHX2 expression were found in adjacent non-tumor tissue. This suggests that a reduced nuclear expression and increased cytoplastic expression of ZHX2 could play roles in hepatocarcinogenesis. This was further supported by cellular fragment analysis, showing that the low nuclear expression of ZHX2 was correlated with a high Cyclins A and E and Ki-67 expression level. Moreover, the patients with reduced ZHX2 nuclear expression have poor tumor differentiation, larger sizes, and a significantly shorter survival time [37]. This indicated the critical tumor-inhibitory roles of nuclear ZHX2 in HCC progression. Epigenetic analysis results have revealed that a reduced expression of ZHX2 was due to the hypermethylation of the ZHX2 promoter [38].

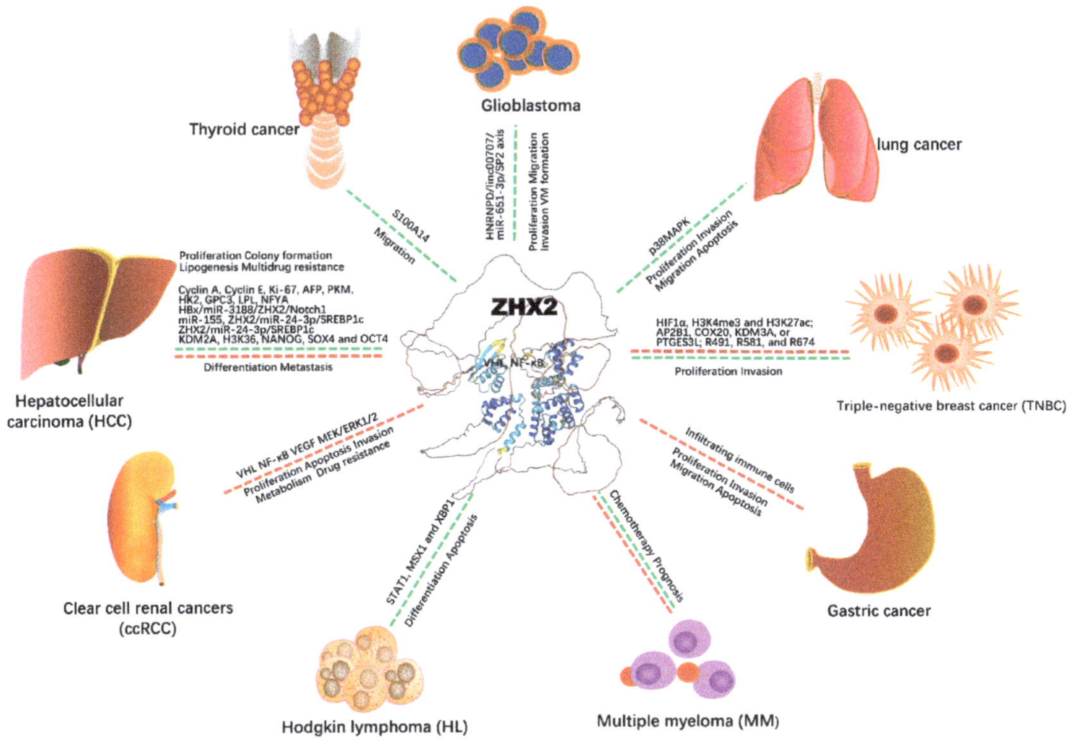


**Figure 2.** ZHX2 and cancers: biological functions and potential signaling networking.

Figure 2 A previous study demonstrated that the ZHX2 protein was involved in the postnatal repression of alpha fetoprotein (AFP) in mice [39]. Other studies have shown that ZHX2 expression in HCC was inversely correlated with serum AFP levels in HCC patients [14], suggesting a repressive role of ZHX2 on AFP. This was supported by in vitro experiments, showing that ZHX2 overexpression significantly decreased AFP secretion in HCC cells and that reducing the expression of ZHX2 restored APF expression in LO2 and SMMC7721 cells. Luciferase-based assays revealed that the repression of AFP by ZHX2 is regulated by the AFP promoter and requires intact HNF1 binding sites [40]. Besides the repression of AFP, ZHX2 also repressed the cancer biomarkers pyruvate kinase M1/2 (PKM) and hexokinase 2 (HK2) in HCC [16]. Moreover, ZHX2 suppressed oncogenic activation of glypican 3 (GPC3) in HCC [41]. Special attention has been given to GPC3 because of its diagnostic potential in HCC [61]. However, ZHX2 expression was negatively associated with GPC3 from HCC tissues and in cultured liver cell lines. In these studies, the increased expression of ZHX2 markedly reduced GPC3 expression, while the downregulation of ZHX2 by small molecules elevated GPC3 expression levels in vitro-cultured cells. Mechanistic studies have revealed that the suppressive action of ZHX2 is achieved through binding to the GPC3 promoter [41].

Epidemiology and clinical studies have demonstrated that virus-related hepatocellular carcinoma, particularly hepatitis B virus (HBV)-related HCC, is one of the main types of HCC [62]. During carcinogenesis of HCC, ZHX2 also plays a tumor-inhibitory role through dysregulating HBV X protein (HBx). In turn, HBx plays an important part in HBV-related HCC, activating miR-3188 and Notch signaling via CREB, and then inhibiting ZHX2 expression [42]. Conversely, the interaction between ZHX2 and NFYA can reduce Notch1. It has been reported that miR-3188 plays an oncogenic role and is highly expressed in HBV transgenic mice liver and HepG2.215 cells. Knockdown of miR-3188 inhibits cell proliferation, migration, invasion, and mouse tumor growth [42]. Thus, HBx-miR-3188-ZHX2-Notch1 is considered as the major signaling pathway during carcinogenesis and development in HBV-associated HCC. Song et al. also reported that ZHX2 is an HBV-related tumor suppressor gene in HCC [43]. ZHX2 expression is significantly decreased in tumor tissues of HBV-positive HCC and livers of HBV-transgenic mice. Further studies confirmed that ZHX2 expression levels and tumor-suppressive activities were inhibited by HBV-encoded proteins, particularly HBx [43]. Mechanistic studies showed that ZHX2 expression levels were affected by miR-155 via its seed sites in the ZHX2 3UTR. In fact, miR-155 levels were significantly increased in HBx-overexpressing HCC cell lines, HBV-positive HCC tissues, and HBV-transgenic mouse livers. Interestingly, ZHX2 levels were upregulated when miRNA-155 levels were blocked in vitro [43]. These findings suggest that HBV-mediated HCC-promoting properties could be attributed to the miR-155-induced silencing of ZHX2, and they suggest a novel therapy for HBV-related HCC by targeting the miR-155 pathway.

As an HCC-associated tumor suppressor, ZHX2 is found to be hepatoprotective by reducing liver lipid levels in a high-fat diet-fed rodent model [63]. In addition, ZHX2 could suppress lipoprotein lipase and indirectly prevent HCC formation [64]. It is known that lipogenesis plays an important role in carcinogenesis and the development of HCC. An experimental study has shown that the overexpression of ZHX2 in HCC cells significantly inhibited de novo lipogenesis [44]. ZHX2 expression was inversely correlated with the expression of SREBP1c. Further studies have revealed that ZHX2 inhibited liver tumorigenesis and development by inhibiting SREBP1c, which regulates de novo lipogenesis. The therapeutic strategy provided additional evidence showing that fatostatin, as a SREBP1c suppressor, attenuated liver tumorigenesis in Zhx2 conditional deletion in the liver of mouse models. Mechanistically, ZHX2 upregulated miR-24–3p expression at transcription levels, further leading to the degradation of SREBP1c targeted by miR-24–3p. Conclusively, these data provided a novel mechanism for ZHX2 inhibiting HCC progression, and targeting the ZHX2/SREBP1c axis could be a novel treatment method of liver cancer.

Numerous studies have proved that non-alcoholic fatty liver disease (NAFLD) is closely associated with hepatocellular carcinoma (HCC). ZHX2 was found to be decreased in NAFLD-HCC liver tissue [45]. This study showed that increasing ZHX2 expression disturbed lipid homeostasis and hepatocytes' lipid deposition and prevented exogenous lipids uptake via the suppression of lipid lipase (LPL), resulting in the inhibition of HCC cell proliferation. Vice versa, increasing LPL expression in HCC cells could reverse ZHX2-mediated cell proliferation inhibition, tumor growth in a xenograft mouse model, lipid metabolism, and tumor formation in mouse liver. Moreover, in an HCC cohort study, immunohistochemical staining demonstrated a negative correlation between ZHX2 and LPL expression. Taken together, ZHX2 protects hepatocytes from cell growth and NAFLD-HCC progression through regulating lipid deposition and the transcriptional repression of LPL [45].

ZHX2 expression levels have been significantly reduced in liver cancer stem cells (CSCs) from different origins [46]. Manipulating the deficiency of ZHX2 could enhance tumor progression and liver CSC stemness. Elevating ZHX2 expression attenuated liver CSCs transformation from the initiation of tumor formation, self-renewal, and sorafenib-resistance. Mechanical studies have demonstrated that ZHX2 suppresses liver CSCs via epigenetic regulation, that is, by suppressing the KDM2A-related demethylation of histone H3 lysine 36 (H3K36), because the H3K36 promoter has multiple binding sites of stemness-associated transcription factors, including NANOG, SOX4, and OCT4. Clinical studies have found that a lower expression of ZHX2 and a higher KDM2A expression were associated with a shorter survival time, which was linked to the transcriptionally repressing KDM2A by ZHX2 in HCC. These findings have improved our understanding of the molecular mechanisms of HCC relapse and drug resistance [46].

Therapeutic studies have also demonstrated that ZHX2 improved the cytotoxicity of chemotherapeutic drugs by inhibiting multidrug resistance 1 (MDR1) via an interaction with NFYA in liver tumor cells [48]. Firstly, an inverse correlation of ZHX2 and MDR1 expression levels was observed in HCC tissues by immunohistochemical staining. Secondly, luciferase reporter assays showed that ZHX2 repressed the promoter activity of the MDR1. However, the knockdown of NFYA or mutation of the NFY binding site eliminated the ZHX2-mediated repression of MDR1 at a transcriptional level. This suggests that ZHX2 was interacting with NFYA, thus, reducing NFY binding to the MDR1 promoter.

However, it has been reported that the ZHX2 protein was overexpressed in liver cancer tissues, and that its upregulation expression was related to poor differentiation and cancer metastasis [47], suggesting an oncogene role of ZHX2 in HCC.

In lung cancer, ZHX2 exhibits tumor suppressor functions. For example, ZHX2 expression was significantly reduced in the human lung cancer cell lines [49]. Compared to the control groups, cell growth, moving, and invasion were dramatically inhibited by ZHX2, in which cell apoptosis and apoptosis-related proteins were increased. In addition, ZHX2 inhibited tumor growth in terms of the reductions in growth rate, tumor size and weight, and PCNA-positive cell numbers. Mechanically, ZHX2 acts as a tumor suppressor in lung cancer by inhibiting the p38MAPK signaling pathway.

In glioma cells, it has been reported that ZHX2 interacts with HNRNPD and further regulates vasculogenic mimicry (VM) formation through the linc00707/miR-651–3p/SP2 pathway [50]. The expression levels of ZHX2 and miR-651–3p were remarkedly reduced, while HNRNPD, linc00707, and specific protein 2 (SP2) expression were remarkedly increased in glioma. Moreover, increasing the ZHX2 and miR-651–3p expression or decreasing the expression of HNRNPD, linc00707, and SP2 suppressed glioma cells' proliferation, migration, invasion, and VM formation. Interestingly, by reducing the HNRNPD expression, it caused an increase in ZHX2 mRNA stability, and ZHX2 negatively regulated the linc00707 expression by binding to its promoter region. Firstly, linc00707 binds with miR-651–3p, and the latter binds to the SP2 mRNA 3′ untranslated region (3′UTR) and negatively regulates its expression. Then, SP2 binds to the promoter regions of MMP2, MMP9, and VE-cadherin, which are VM formation-related proteins, and therefore, play a role in enhancing transcription for the regulation of glioma cell VM formation.

It was also reported that ZHX2 inhibited the progression of thyroid cancer [51]. The decreased ZHX2 expression in thyroid cancer tissues was correlated with poor outcomes. The knockdown of ZHX2 significantly promoted thyroid cancer cell migration. S100 calcium-binding protein (S100A14) was highly expressed in human thyroid cancers, showing a negative correlation with ZHX2. Mechanistically, ZHX2 binds to the promoter region of the S100 to decrease its transcription. The inhibition of S100A14 attenuated thyroid cancer cell metastasis induced by ZHX2 knockdown in cultured cells and in animal models.

Multiple myeloma (MM) is an incurable hematological malignancy. Compared to the low-risk or indolent disease, ZHX2 expression was remarkably reduced in the high-risk or active disease. The multiple myeloma patients with lower ZHX2 expression have worse outcomes [54]. Other studies have also reported that increasing the ZHX2 expression could improve the response to high-dose chemotherapy in multiple myeloma [52,53]. In contrast, another study by Jiang et al. reported that multiple myeloma patients with a higher ZHX2 expression showed poorer clinical outcomes, and the knockdown of ZHX2 in cancer cells caused MM cells to be more sensitive to Bortezomib (BTZ) by regulating the nuclear translocation of NF-κB, also affecting NF-κB and the corresponding target genes' expression at the mRNA levels [55].

ZHX2 has also been reported as a tumor suppressor in Hodgkin lymphoma, where the recurrent breakpoint at 8q24 targets ZHX2. This aberration broke the far upstream activation elements of ZHX2 and decreased ZHX2 expression [56]. The gene expression profiling results indicated the regulating function of ZHX2 on differentiation and apoptosis. STAT1 (signal transducer and activation of transcription 1) and several STAT1 target genes were included, suggesting that ZHX2 functions as a tumor suppressor in Hodgkin lymphoma [56]. Further studies demonstrated the transcriptional deregulation of ZHX2 in B cell malignancies. Two transcription factors, homeodomain protein msh homeobox 1 (MSX1) and bZIP protein X-box binding protein 1 (XBP1), were shown to directly regulate ZHX2 expression. Multiple mechanisms might be involved in the suppression of ZHX2 expression in Hodgkin lymphoma cell lines. These include the loss of activation elements of ZHX2 upstream and a decreased expression of activators MSX1 and XBP1 [57].

ZHX2 can drive tumorigenesis in clear cell renal cell carcinoma (ccRCC). The loss of VHL usually upregulates ZHX2 levels, in particular, the nuclear expression of ZHX2 in ccRCC tumors. Mechanistic studies showed that VHL-upregulating ZHX2 was achieved through prolyl hydroxylation and proteasomal degradation signaling, supported by an experiment showing that the inhibition of prolyl hydroxylation and the degradation of proteasome could increase ZHX2 expression. The engineered deletion of ZHX2 caused decreased anti-apoptosis-associated multiple gene expression and inhibited cancer cell growth, metastasis, and metabolism. Moreover, ZHX2 promoted NF-κB activation and drove renal carcinogenesis. These studies reveal the oncogenic functions of ZHX2 in renal cancer and uncover a potential therapeutic target for ccRCC [21]. The ccRCC study of Zhu et al. found an additional mechanism. ZHX2-driven cell proliferation and migration were achieved though the activation of the MEK/ERK1/2 signaling pathway and the increasing VEGF expression, and led to Sunitinib resistance through regulating self-protective autophagy, providing a new insight for advanced ccRCC treatment [58].

Triple-negative breast cancer (TNBC) belongs to a more aggressive breast cancer subtype with a higher mortality rate; therefore, it is an urgent requirement to find a novel treatment strategy. Studies have demonstrated that ZHX2 is amplified or increased in TNBC patient tissues. Functional studies showed that manipulating the deletion of ZHX2 inhibits TNBC tumor growth and metastasis. Chromosomal immunoprecipitation sequencing results have shown that ZHX2 binds and transcriptionally activates HIF1$\alpha$ (hypoxia-inducible factor alpha), finally promoting gene expression. AP2B1, COX20, KDM3A, and PTGES3L were regulated by both ZHX2 and HIF1$\alpha$—their overexpression could partially rescue ZHX2 depletion-caused cell growth in TNBC cells. These findings strongly suggest that these target genes synergistically help the oncogene function of ZHX2. Genomic studies further showed that three residues (R491, R581, and R674) on ZHX2 are important for ZHX2 to exert its transcriptional activity. It was suggested that ZHX2 acts as an oncogene via activating the HIF1$\alpha$ pathway in TNBC, and ZHX could be used as a novel therapeutic target for TNBC [59].

It has been reported that ZHX2 forms heteromeric complexes with ZHX3 [8,16]. Their effect on breast cancer prognosis is similar. Both ZHX2 and ZHX3 expressions were remarkably reduced in TNBC tissues, and a high mRNA expression of ZHX2 and ZHX3 were closely associated with a better prognosis of breast cancer patients, especially in luminal A subtype breast cancer. In addition, the mRNA expression of ZHX2 and ZHX3 were also positively associated with the estrogen receptor and progesterone receptor, but ZHX2 mRNA expression was inversely correlated with HER2 in breast cancers [19].

In gastric cancer (GC), ZHX2 also exerts an oncogenic function [60]. An increase in ZHX2 expression was closely correlated with clinical characteristics; for instance, a higher ZHX2 expression predicted worse outcomes and immune infiltrating. An in vitro experimental study showed that ZHX2 overexpression can facilitate gastric cancer cell proliferation and migration and can suppress cell apoptosis. These findings suggest ZHX2 as a prognostic biomarker, and immune infiltration could be associated with the effect of ZHX2 on gastric cancer.

By contrast, gastric cancer patients with a decreased ZHX2 and ZHX3 mRNA expression showed better outcomes [31]. Reduced ZHX2 and ZHX3 expression was remarkably correlated with a longer survival time in the subgroups of GC patients without lymph node metastases or distant metastasis, suggesting that the low expression of ZHX2 and ZHX3 predicts a better prognosis for early-stage gastric cancer patients. Subgroup analyses also found that the downregulation of ZHX2 was predictive of an improved survival time in HER2-positive gastric cancer patients, but not in HER2-negative gastric cancer patients. In addition, the reduced expression of ZHX2 was associated with a favorable overall survival rate in gastric cancer patients who only received surgery, but not in the patients who received additional therapies [31]. Taken together, these data suggest that ZHX2 and ZHX3 act as oncogenes in gastric cancer.

The expression levels of ZHX2 in nature killer (NK) cells have shown an important clinical significance [65]. Kaplan–Meier analysis of overall survival from a TCGA dataset showed that a lower ZHX2 expression level in NK cells was associated with better prognosis in HCC patients. The genetical deleting ZHX2 gene improves IL-15-mediated NK cell activity, maturation, and cell viability in vitro, resulting in better antitumor immunity. Therapeutically, the transfer of ZHX2-deficient NK cells inhibited hepatoma homograft tumor growth and metastasis. These studies reveal a novel regulatory function of ZHX2 in NK cell maturation and its therapeutic potential by enhancing NK cell-mediated cancer surveillance [65].

### 2.3. ZHX3 and Cancer

The biological functions of ZHX3 in cancer have not been well characterized, but limited studies have shown that ZHX3 is a tumor suppressor in HCC, non-small cell lung cancer, breast cancer, and renal cancer (Table 3). It has also been reported that an increased ZHX3 expression is associated with the progression of bladder carcinoma and gastric cancer, indicating that ZHX3 is involved in oncogenic programs (Table 3). The clinical significance of ZHX3 and the underlying potential mechanism are summarized in Figure 3.

ZHX3 was first identified as a suppressor of the AFP gene in HCC and was a good candidate for the tumor suppressor present at 16q22 [16]. As with ZHX2, ZHX3 is expressed at very low levels in HCC cells compared with rat hepatocytes. ZHX3 repressed the transcription of the luciferase reporter gene that was fused to the promoters of PKM and HK2 [16]. This suggests that the loss of expression of ZHX3 might be a critical factor during hepatocellular carcinogenesis.

**Table 3.** Studies on ZHX3 expression, function, clinical significance, and molecular mechanism.

| ZHX Family Member | Expression Levels | Functions | Cancer Types | Clinical Significance | Associated Effects | Molecular Mechanisms | References |
|---|---|---|---|---|---|---|---|
| ZHX3 | Decreased | Tumor suppressor | HCC | ZHX3 was first identified as a suppressor of the AFP gene. | | PKM and HK2 | [16] |
| ZHX3 | Decreased | Tumor suppressor | Non-small cell lung cancer (NSCLC) | Lower ZHX3 expression in the tumor had a significant greater risk of lymph node metastasis and was associated with poorer survival time. | | | [66] |
| ZHX1 and ZHX3 | Decreased | Tumor suppressor | Renal carcinoma | A lower ZHX3 expression in cancers was correlated with a high risk of lymph node metastasis and worse outcomes. | | NMI and ARPC5 | [18] |
| ZHX3 | Decreased | Tumor suppressor | Breast cancer | Patients with decreased ZHX3 protein levels had poorer outcomes. | | | [19] |
| ZHX3 | Increased | Oncogene | Gastric cancer | Over-expression of ZHX3 was associated with worse OS. | | | [31] |
| ZHX3 | Increased | Oncogene | Urothelial carcinoma of the bladder (UCB) | | Promote the migration and invasion capacities of UCB cells both in vitro and in vivo. | TRIM21 and RGS2 | [20] |

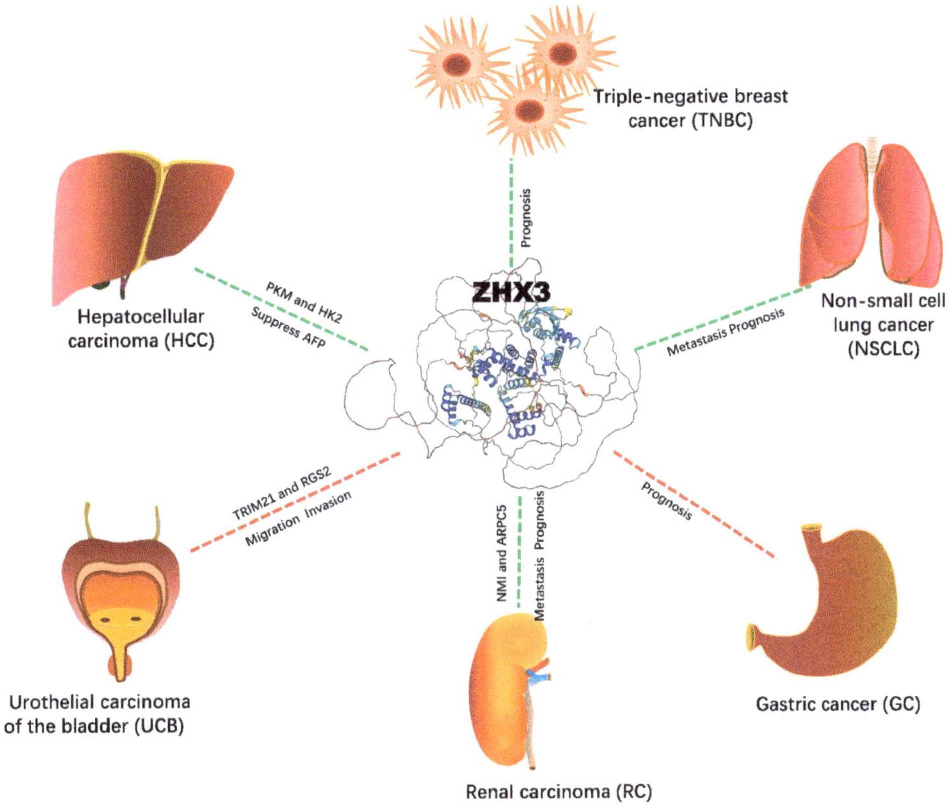


**Figure 3.** ZHX3 and cancers: biological functions and potential signaling networking.

ZHX3 expression was also remarkably decreased in non-small cell lung cancer (NSCLC) in comparison to the adjacent non-tumor tissues [66]. Interestingly, a lower ZHX3 expression in the tumor had a significantly greater risk of lymph node metastasis and was associated with a poorer survival time. Therefore, ZHX3 was an independent factor affecting metastasis and in predicting the 5-year overall survival rate among NSCLC patients.

A recent study on renal carcinoma showed a tumor-suppressive role for ZHX3 [18]. As seen with ZHX1, ZHX3 showed a lower expression in cancers, and was associated with a high-risk of lymph node metastasis and worse outcomes. Further analysis of mRNA co-occurrence using cBioPortal data (www.cbioportal.org) showed opposite relationships between the expressions of ZHX1 and ZHX3 and some well-known oncogenes [18]. For instance, ZHX3 expression was negatively associated with the expressions of N-myc, STAT Interactor (NMI), and actin-related protein 2/3 complex subunit 5 (ARPC5). Both NMI and ARPC5 play critical roles in initiating cancer formation and facilitating cancer cell proliferation, migration, and metastasis [67,68].

In breast cancer, a negative ZHX3 expression was correlated with lymph node metastasis, poor differentiation, advanced tumor stage, and positive estrogen receptor expression [19]. The immunohistochemical (IHC) staining analysis showed that the patients with decreased ZHX3 protein levels had a poor prognosis. The increased ZHX3 expression was associated with good outcomes in breast cancer patients, indicating that ZHX3 might act as a prognostic biomarker for breast cancer patients [19].

ZHX3 expression can also be used as a prognostic indicator for gastric cancer [31]. IHC staining confirmed that gastric cancer patients with a high ZHX3 expression have a worse prognosis [31].

In a recent study [20], ZHX3 was first screened in a urothelial carcinoma of the bladder (UCB) as a critical oncogenic factor associated with poor prognosis—this was based from The Cancer Genome Atlas dataset and a large cohort of UCB clinical samples. ZHX3 promoted the migration and invasion of UCB cell in vitro and in vivo. Mechanistically, ZHX3 repressed the expression of regulator of G protein signaling 2 (RGS2). Meanwhile, as a target, ZHX3 was regulated by tripartite motif 21 (TRIM21), which mediates its ubiquitination and subsequent degradation. These results indicate that ZHX3 is an oncogene and has a therapeutic potential for UCB [20].

### 2.4. ZHXs and Cancer Patient Survival

The association between ZHX family member expression levels and the survival time in multiple types of cancers were refined by us using the comprehensive datasets from the GEPIA (Gene Expression Profiling Interactive Analysis, http://gepia.cancer-pku.cn/). We selected overall survival (OS) or disease-free survival (DFS) as survival time indicators. Figure 4A shows that the cancer types with a low ZHX1, ZHX2, or ZHX3 expression have better OS or DFS, including ZHX1 in adrenocortical carcinoma and liver hepatocellular carcinoma, ZHX2 in bladder urothelial carcinoma, and ZHX3 in bladder urothelial carcinoma, esophageal carcinoma, uveal melanoma, mesothelioma, and ovarian serous cystadenocarcinoma. In contrast, the types of cancers with high ZHX1, ZHX2, or ZHX3 expressions have better OS or DFS, as shown in Figure 4B, including ZHX1 in cholangial carcinoma and kidney renal clear cell carcinoma, ZHX2 in brain lower grade glioma, mesothelioma and sarcoma, and ZHX3 in renal clear cell carcinoma and thyroid carcinoma. These results again have proven the roles of the ZHX genes as a double-edged sword in cancer patient survival.

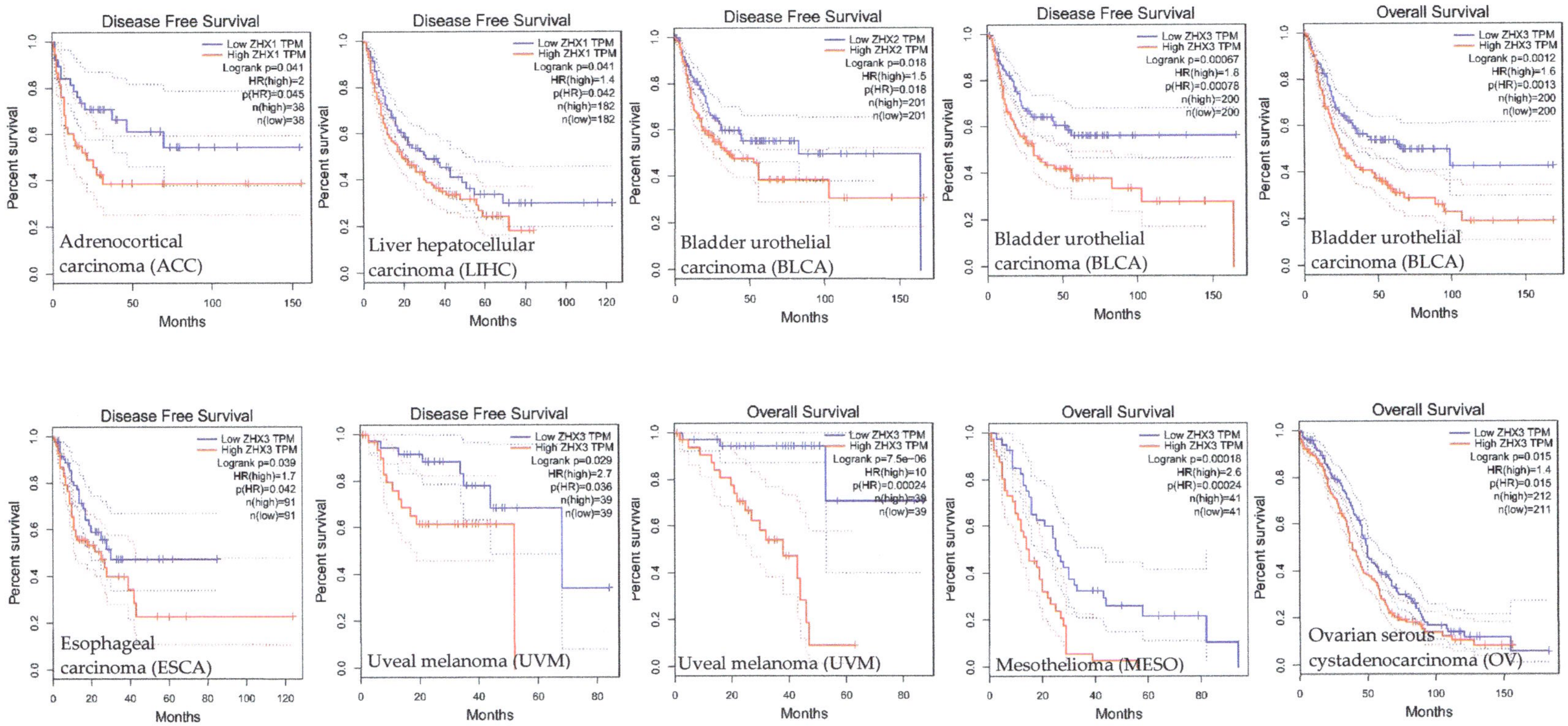


(A)

**Figure 4.** *Cont.*

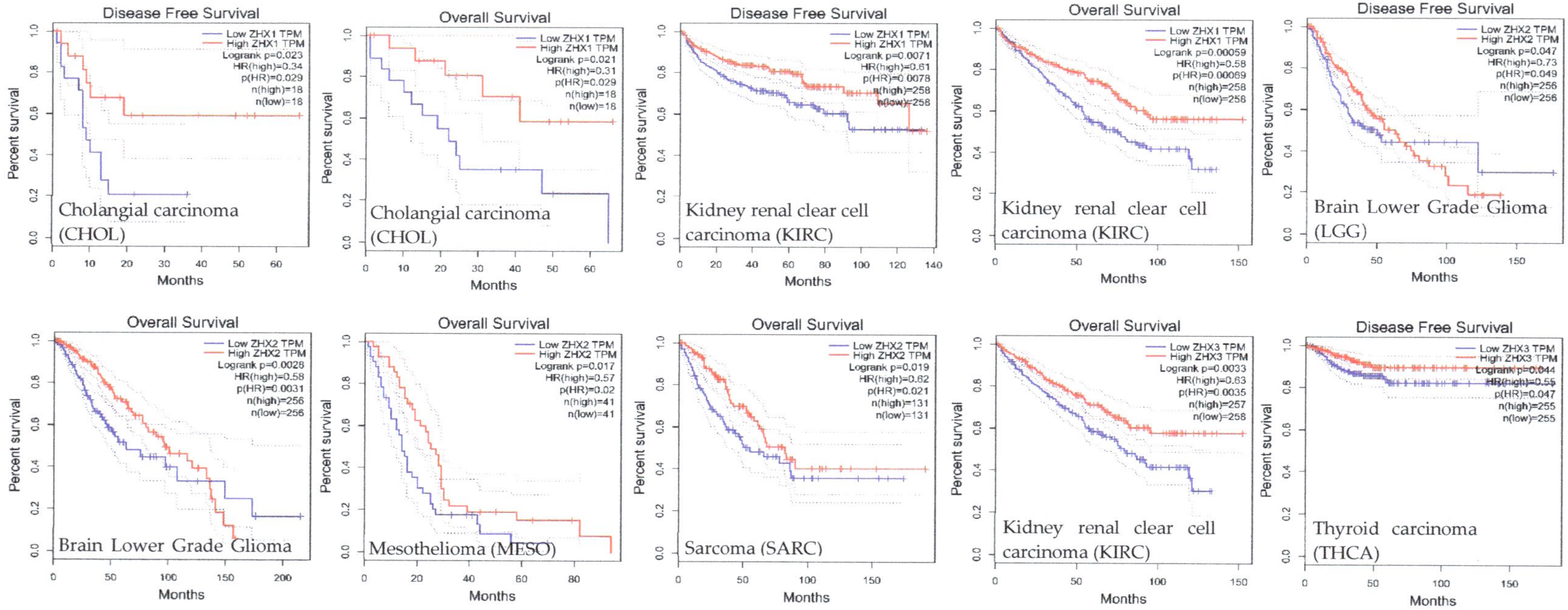


(B)

**Figure 4.** ZHX1, ZHX2, and ZHX3 gene expression level and survival analysis in 12 types of cancers. These panels were generated using the GEPIA database (http://gepia.cancer-pku.cn/) on 06/02/2022. (**A**): Patients with low ZHX1, ZHX2, or ZHX3 expressions have better OS or DFS. The words in red stood for ZHX1, ZHX2, and ZHX3 high expressions, the words in blue stood for ZHX1, ZHX2, and ZHX3 low expressions. (**B**): Patients with high ZHX1, ZHX2, or ZHX3 expressions have better OS or DFS. GEPIA performs overall survival (OS) or disease-free survival (DFS) analysis based on ZHX1, ZHX2, and ZHX3 gene expressions. The cox proportional hazard ratio and 95% confidence interval information can also be included in the survival plot. "Median" was used in group cutoff. Cutoff-high (%) indicates that samples with expression levels higher than this threshold are considered as the high-expression cohort and samples with expression levels lower than this threshold are considered the low-expression cohort. Cancer types of abbreviations: ACC, adrenocortical carcinoma; BLCA, bladder urothelial carcinoma; CHOL, cholangial carcinoma; ESCA, esophageal carcinoma; KIRC, kidney renal clear cell carcinoma; LGG, brain lower grade glioma; LIHC, liver hepatocellular carcinoma; MESO, mesothelioma; OV, ovarian serous cystadenocarcinoma; SARC, sarcoma; THCA, thyroid carcinnooma; UVM, uveal melanoma.

## 3. Conclusions and Perspectives

Currently, studies on the ZHX family members in cancer have demonstrated tumor-suppressive and oncogenic roles in carcinogenesis and progression by power BI tools (Figure 5). These opposing effects are also seen in the roles of ZHXs in the prediction of outcomes and responses to therapy. These differences are most likely specific to cancer type. However, the studies are still limited to a few cancer types. The biological functions of ZHXs in cancers of the respiratory system, gastrointestinal tract, head and neck, and reproductive system are largely unknown, and thus, require further investigation. Moreover, neither ZHX-related signaling pathways nor upstream and downstream regulators in carcinogenesis and therapy are clear, and thus, need further elucidation.

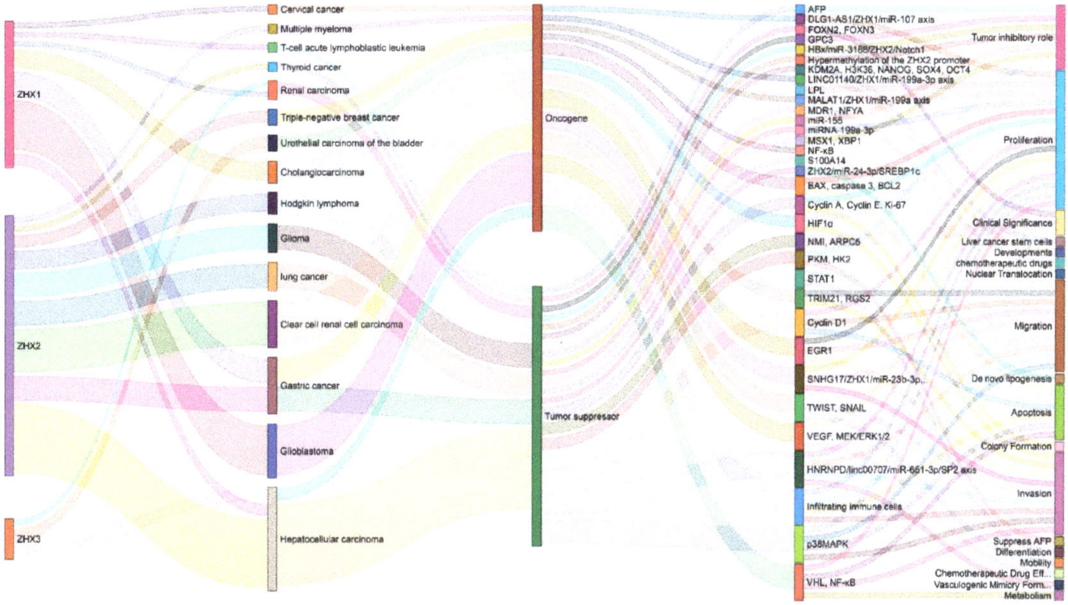


**Figure 5.** A comprehensive illustration of ZHXs as a double-edged sword in cancers: an oncogene or tumor suppressor, underlying mechanisms, and biological functions.

**Author Contributions:** Conceptualization, Y.G., Y.B. and W.Y.; Methodology, Y.G.; Data collection, Y.G.; Formal analysis, Y.G. and H.Z.; Data curation, Y.G. and Z.H.; Writing—original draft preparation, Y.B. and W.Y.; Funding acquisition, Y.G., Y.B. and W.Y. All authors have read and agreed to the published version of the manuscript.

**Funding:** This project was funded in part by the Doctoral Research Startup Fund from Mudanjiang Medical University (2021-MYBSKY-064 to Y.B. and 2021-MYBSKY-065 to Y.G.), and by the Cancer Center Support Grant from the National Cancer Institute (P30 CA016058) and the Clinical and Translational Science Shared Resource at Ohio State University Comprehensive Cancer Center.

**Institutional Review Board Statement:** Not applicable.

**Informed Consent Statement:** Not applicable.

**Data Availability Statement:** Not applicable.

**Conflicts of Interest:** The authors declare no conflict of interest.

## References


1. Kawata, H.; Yamada, K.; Shou, Z.; Mizutani, T.; Yazawa, T.; Yoshino, M.; Sekiguchi, T.; Kajitani, T.; Miyamoto, K. Zinc-fingers and homeoboxes (ZHX) 2, a novel member of the ZHX family, functions as a transcriptional repressor. *Biochem. J.* **2003**, *373*, 747–757. [CrossRef] [PubMed]
2. Shou, Z.; Yamada, K.; Inazu, T.; Kawata, H.; Hirano, S.; Mizutani, T.; Yazawa, T.; Sekiguchi, T.; Yoshino, M.; Kajitani, T.; et al. Genomic structure and analysis of transcriptional regulation of the mouse zinc-fingers and homeoboxes 1 (ZHX1) gene. *Gene* **2003**, *302*, 83–94. [CrossRef]
3. Wienk, H.; Lammers, I.; Hotze, A.; Wu, J.; Wechselberger, R.W.; Owens, R.; Stammers, D.K.; Stuart, D.; Kaptein, R.; Folkers, G.E. The tandem zinc-finger region of human ZHX adopts a novel C2H2 zinc finger structure with a C-terminal extension. *Biochemistry* **2009**, *48*, 4431–4439. [CrossRef]
4. Yamada, K.; Printz, R.L.; Osawa, H.; Granner, D.K. Human ZHX1: Cloning, chromosomal location, and interaction with transcription factor NF-Y. *Biochem. Biophys. Res. Commun.* **1999**, *261*, 614–621. [CrossRef]
5. Spear, B.T.; Jin, L.; Ramasamy, S.; Dobierzewska, A. Transcriptional control in the mammalian liver: Liver development, perinatal repression, and zonal gene regulation. *Cell. Mol. Life Sci.* **2006**, *63*, 2922–2938. [CrossRef] [PubMed]
6. Yamada, K.; Kawata, H.; Shou, Z.; Hirano, S.; Mizutani, T.; Yazawa, T.; Sekiguchi, T.; Yoshino, M.; Kajitani, T.; Miyamoto, K. Analysis of zinc-fingers and homeoboxes (ZHX)-1-interacting proteins: Molecular cloning and characterization of a member of the ZHX family, ZHX3. *Biochem. J.* **2003**, *373*, 167–178. [CrossRef] [PubMed]
7. Yamada, K.; Kawata, H.; Matsuura, K.; Shou, Z.; Hirano, S.; Mizutani, T.; Yazawa, T.; Yoshino, M.; Sekiguchi, T.; Kajitani, T.; et al. Functional analysis and the molecular dissection of zinc-fingers and homeoboxes 1 (ZHX1). *Biochem. Biophys. Res. Commun.* **2002**, *297*, 368–374. [CrossRef]
8. Kawata, H.; Yamada, K.; Shou, Z.; Mizutani, T.; Miyamoto, K. The mouse zinc-fingers and homeoboxes (ZHX) family; ZHX2 forms a heterodimer with ZHX3. *Gene* **2003**, *323*, 133–140. [CrossRef]
9. Barthelemy, I.; Carramolino, L.; Gutierrez, J.; Barbero, J.L.; Marquez, G.; Zaballos, A. zhx-1: A novel mouse homeodomain protein containing two zinc-fingers and five homeodomains. *Biochem. Biophys. Res. Commun.* **1996**, *224*, 870–876. [CrossRef]
10. Shou, Z.; Yamada, K.; Kawata, H.; Yokoyama, O.; Miyamoto, K. A mechanism of induction of the mouse zinc-fingers and homeoboxes 1 (ZHX1) gene expression by interleukin-2. *Biochem. Biophys. Res. Commun.* **2004**, *314*, 885–890. [CrossRef] [PubMed]
11. Kim, S.H.; Park, J.; Choi, M.C.; Kim, H.P.; Park, J.H.; Jung, Y.; Lee, J.H.; Oh, D.Y.; Im, S.A.; Bang, Y.J.; et al. Zinc-fingers and homeoboxes 1 (ZHX1) binds DNA methyltransferase (DNMT) 3B to enhance DNMT3B-mediated transcriptional repression. *Biochem. Biophys. Res. Commun.* **2007**, *355*, 318–323. [CrossRef] [PubMed]
12. Ogata-Kawata, H.; Yamada, K.; Uesaka-Yoshino, M.; Kagawa, N.; Miyamoto, K. BS69, a corepressor interacting with ZHX1, is a bifunctional transcription factor. *Front. Biosci.* **2007**, *12*, 1911–1926. [CrossRef] [PubMed]
13. Ovsyannikova, I.G.; Kennedy, R.B.; O'Byrne, M.; Jacobson, R.M.; Pankratz, V.S.; Poland, G.A. Genome-wide association study of antibody response to smallpox vaccine. *Vaccine* **2012**, *30*, 4182–4189. [CrossRef] [PubMed]
14. Peterson, M.L.; Ma, C.; Spear, B.T. Zhx2 and Zbtb20: Novel regulators of postnatal alpha-fetoprotein repression and their potential role in gene reactivation during liver cancer. *Semin. Cancer Biol.* **2011**, *21*, 21–27. [CrossRef] [PubMed]
15. Jiang, J.; Creasy, K.T.; Purnell, J.; Peterson, M.L.; Spear, B.T. Zhx2 (zinc fingers and homeoboxes 2) regulates major urinary protein gene expression in the mouse liver. *J. Biol. Chem.* **2017**, *292*, 6765–6774. [CrossRef] [PubMed]
16. Yamada, K.; Ogata-Kawata, H.; Matsuura, K.; Kagawa, N.; Takagi, K.; Asano, K.; Haneishi, A.; Miyamoto, K. ZHX2 and ZHX3 repress cancer markers in normal hepatocytes. *Front. Biosci. (Landmark Ed.)* **2009**, *14*, 3724–3732. [CrossRef] [PubMed]
17. Suehiro, F.; Nishimura, M.; Kawamoto, T.; Kanawa, M.; Yoshizawa, Y.; Murata, H.; Kato, Y. Impact of zinc fingers and homeoboxes 3 on the regulation of mesenchymal stem cell osteogenic differentiation. *Stem Cells Dev.* **2011**, *20*, 1539–1547. [CrossRef]
18. Kwon, R.J.; Kim, Y.H.; Jeong, D.C.; Han, M.E.; Kim, J.Y.; Liu, L.; Jung, J.S.; Oh, S.O. Expression and prognostic significance of zinc fingers and homeoboxes family members in renal cell carcinoma. *PLoS ONE* **2017**, *12*, e0171036. [CrossRef]
19. You, Y.; Ma, Y.; Wang, Q.; Ye, Z.; Deng, Y.; Bai, F. Attenuated ZHX3 expression serves as a potential biomarker that predicts poor clinical outcomes in breast cancer patients. *Cancer Manag. Res.* **2019**, *11*, 1199–1210. [CrossRef]
20. Deng, M.; Wei, W.; Duan, J.; Chen, R.; Wang, N.; He, L.; Peng, Y.; Ma, X.; Wu, Z.; Liu, J.; et al. ZHX3 promotes the progression of urothelial carcinoma of the bladder via repressing of RGS2 and is a novel substrate of TRIM21. *Cancer Sci.* **2021**, *112*, 1758–1771. [CrossRef]
21. Zhang, J.; Wu, T.; Simon, J.; Takada, M.; Saito, R.; Fan, C.; Liu, X.D.; Jonasch, E.; Xie, L.; Chen, X.; et al. VHL substrate transcription factor ZHX2 as an oncogenic driver in clear cell renal cell carcinoma. *Science* **2018**, *361*, 290–295. [CrossRef] [PubMed]
22. Liu, Y.; Ma, D.; Ji, C. Zinc fingers and homeoboxes family in human diseases. *Cancer Gene Ther.* **2015**, *22*, 223–226. [CrossRef] [PubMed]
23. Kwon, R.J.; Han, M.E.; Kim, Y.J.; Kim, Y.H.; Kim, J.Y.; Liu, L.; Heo, W.; Oh, S.O. Roles of zinc-fingers and homeoboxes 1 during the proliferation, migration, and invasion of glioblastoma cells. *Tumour Biol.* **2017**, *39*, 1010428317694575. [CrossRef]
24. Liao, K.; Lin, Y.; Gao, W.; Xiao, Z.; Medina, R.; Dmitriev, P.; Cui, J.; Zhuang, Z.; Zhao, X.; Qiu, Y.; et al. Blocking lncRNA MALAT1/miR-199a/ZHX1 Axis Inhibits Glioblastoma Proliferation and Progression. *Mol. Ther. Nucleic Acids* **2019**, *18*, 388–399. [CrossRef] [PubMed]
25. Ge, B.H.; Li, G.C. Long non-coding RNA SNHG17 promotes proliferation, migration and invasion of glioma cells by regulating the miR-23b-3p/ZHX1 axis. *J. Gene Med.* **2020**, *22*, e3247. [CrossRef] [PubMed]

26. Xin, Y.; Zhang, W.; Mao, C.; Li, J.; Liu, X.; Zhao, J.; Xue, J.; Li, J.; Ren, Y. LncRNA LINC01140 Inhibits Glioma Cell Migration and Invasion via Modulation of miR-199a-3p/ZHX1 Axis. *Onco Targets Ther.* **2020**, *13*, 1833–1844. [CrossRef]
27. Rui, X.; Xu, Y.; Huang, Y.; Ji, L.; Jiang, X. lncRNA DLG1-AS1 Promotes Cell Proliferation by Competitively Binding with miR-107 and Up-Regulating ZHX1 Expression in Cervical Cancer. *Cell Physiol. Biochem.* **2018**, *49*, 1792–1803. [CrossRef] [PubMed]
28. Kwon, R.J.; Han, M.E.; Kim, J.Y.; Liu, L.; Kim, Y.H.; Jung, J.S.; Oh, S.O. ZHX1 Promotes the Proliferation, Migration and Invasion of Cholangiocarcinoma Cells. *PLoS ONE* **2016**, *11*, e0165516. [CrossRef] [PubMed]
29. Ma, X.; Huang, M.; Wang, Z.; Liu, B.; Zhu, Z.; Li, C. ZHX1 Inhibits Gastric Cancer Cell Growth through Inducing Cell-Cycle Arrest and Apoptosis. *J. Cancer* **2016**, *7*, 60–68. [CrossRef] [PubMed]
30. Wang, Z.; Ma, X.; Cai, Q.; Wang, X.; Yu, B.; Cai, Q.; Liu, B.; Zhu, Z.; Li, C. MiR-199a-3p promotes gastric cancer progression by targeting ZHX1. *FEBS Lett.* **2014**, *588*, 4504–4512. [CrossRef] [PubMed]
31. You, Y.; Bai, F.; Li, H.; Ma, Y.; Yao, L.; Hu, J.; Tian, Y. Prognostic value and therapeutic implications of ZHX family member expression in human gastric cancer. *Am. J. Transl. Res.* **2020**, *12*, 3376–3388. [PubMed]
32. Wang, J.; Liu, D.; Liang, X.; Gao, L.; Yue, X.; Yang, Y.; Ma, C.; Liu, J. Construction of a recombinant eukaryotic human ZHX1 gene expression plasmid and the role of ZHX1 in hepatocellular carcinoma. *Mol. Med. Rep.* **2013**, *8*, 1531–1536. [CrossRef] [PubMed]
33. Guan, J.; Liu, Z.; Xiao, M.; Hao, F.; Wang, C.; Chen, Y.; Lu, Y.; Liang, J. MicroRNA-199a-3p inhibits tumorigenesis of hepatocellular carcinoma cells by targeting ZHX1/PUMA signal. *Am. J. Transl. Res.* **2017**, *9*, 2457–2465.
34. Maciel, N.I.G.; Filiu-Braga, L.D.C.; Neves, F.A.R.; Rego, E.M.; Lucena-Araujo, A.R.; Saldanha-Araujo, F. Low expression of ZHX1 and ZHX2 impacts on the prognosis of chronic lymphocytic leukemia. *Biomark. Res.* **2021**, *9*, 10. [CrossRef] [PubMed]
35. Nagel, S.; Pommerenke, C.; Meyer, C.; Kaufmann, M.; MacLeod, R.A.F.; Drexler, H.G. Identification of a tumor suppressor network in T-cell leukemia. *Leuk Lymphoma* **2017**, *58*, 2196–2207. [CrossRef] [PubMed]
36. Nagel, S.; Ehrentraut, S.; Meyer, C.; Kaufmann, M.; Drexler, H.G.; MacLeod, R.A. Aberrantly Expressed OTX Homeobox Genes Deregulate B-Cell Differentiation in Hodgkin Lymphoma. *PLoS ONE* **2015**, *10*, e0138416. [CrossRef]
37. Yue, X.; Zhang, Z.; Liang, X.; Gao, L.; Zhang, X.; Zhao, D.; Liu, X.; Ma, H.; Guo, M.; Spear, B.T.; et al. Zinc fingers and homeoboxes 2 inhibits hepatocellular carcinoma cell proliferation and represses expression of Cyclins A and E. *Gastroenterology* **2012**, *142*, 1559–1570.e1552. [CrossRef]
38. Lv, Z.; Zhang, M.; Bi, J.; Xu, F.; Hu, S.; Wen, J. Promoter hypermethylation of a novel gene, ZHX2, in hepatocellular carcinoma. *Am. J. Clin. Pathol.* **2006**, *125*, 740–746. [CrossRef] [PubMed]
39. Perincheri, S.; Dingle, R.W.; Peterson, M.L.; Spear, B.T. Hereditary persistence of alpha-fetoprotein and H19 expression in liver of BALB/cJ mice is due to a retrovirus insertion in the Zhx2 gene. *Proc. Natl. Acad. Sci. USA* **2005**, *102*, 396–401. [CrossRef] [PubMed]
40. Shen, H.; Luan, F.; Liu, H.; Gao, L.; Liang, X.; Zhang, L.; Sun, W.; Ma, C. ZHX2 is a repressor of alpha-fetoprotein expression in human hepatoma cell lines. *J. Cell. Mol. Med.* **2008**, *12*, 2772–2780. [CrossRef] [PubMed]
41. Luan, F.; Liu, P.; Ma, H.; Yue, X.; Liu, J.; Gao, L.; Liang, X.; Ma, C. Reduced nucleic ZHX2 involves in oncogenic activation of glypican 3 in human hepatocellular carcinoma. *Int. J. Biochem. Cell Biol.* **2014**, *55*, 129–135. [CrossRef] [PubMed]
42. Zhou, S.J.; Deng, Y.L.; Liang, H.F.; Jaoude, J.C.; Liu, F.Y. Hepatitis B virus X protein promotes CREB-mediated activation of miR-3188 and Notch signaling in hepatocellular carcinoma. *Cell Death Differ.* **2017**, *24*, 1577–1587. [CrossRef] [PubMed]
43. Song, X.; Tan, S.; Wu, Z.; Xu, L.; Wang, Z.; Xu, Y.; Wang, T.; Gao, C.; Gong, Y.; Liang, X.; et al. HBV suppresses ZHX2 expression to promote proliferation of HCC through miR-155 activation. *Int. J. Cancer* **2018**, *143*, 3120–3130. [CrossRef] [PubMed]
44. Yu, X.; Lin, Q.; Wu, Z.; Zhang, Y.; Wang, T.; Zhao, S.; Song, X.; Chen, C.; Wang, Z.; Xu, L.; et al. ZHX2 inhibits SREBP1c-mediated de novo lipogenesis in hepatocellular carcinoma via miR-24-3p. *J. Pathol.* **2020**, *252*, 358–370. [CrossRef] [PubMed]
45. Wu, Z.; Ma, H.; Wang, L.; Song, X.; Zhang, J.; Liu, W.; Ge, Y.; Sun, Y.; Yu, X.; Wang, Z.; et al. Tumor suppressor ZHX2 inhibits NAFLD-HCC progression via blocking LPL-mediated lipid uptake. *Cell Death Differ.* **2020**, *27*, 1693–1708. [CrossRef]
46. Lin, Q.; Wu, Z.; Yue, X.; Yu, X.; Wang, Z.; Song, X.; Xu, L.; He, Y.; Ge, Y.; Tan, S.; et al. ZHX2 restricts hepatocellular carcinoma by suppressing stem cell-like traits through KDM2A-mediated H3K36 demethylation. *EBioMedicine* **2020**, *53*, 102676. [CrossRef] [PubMed]
47. Hu, S.; Zhang, M.; Lv, Z.; Bi, J.; Dong, Y.; Wen, J. Expression of zinc-fingers and homeoboxes 2 in hepatocellular carcinogenesis: A tissue microarray and clinicopathological analysis. *Neoplasma* **2007**, *54*, 207–211. [PubMed]
48. Ma, H.; Yue, X.; Gao, L.; Liang, X.; Yan, W.; Zhang, Z.; Shan, H.; Zhang, H.; Spear, B.T.; Ma, C. ZHX2 enhances the cytotoxicity of chemotherapeutic drugs in liver tumor cells by repressing MDR1 via interfering with NF-YA. *Oncotarget* **2015**, *6*, 1049–1063. [CrossRef]
49. Tian, X.; Wang, Y.; Li, S.; Yue, W.; Tian, H. ZHX2 inhibits proliferation and promotes apoptosis of human lung cancer cells through targeting p38MAPK pathway. *Cancer Biomark.* **2020**, *27*, 75–84. [CrossRef] [PubMed]
50. Yu, S.; Ruan, X.; Liu, X.; Zhang, F.; Wang, D.; Liu, Y.; Yang, C.; Shao, L.; Liu, Q.; Zhu, L.; et al. HNRNPD interacts with ZHX2 regulating the vasculogenic mimicry formation of glioma cells via linc00707/miR-651-3p/SP2 axis. *Cell Death Dis.* **2021**, *12*, 153. [CrossRef]
51. Zhang, Y.; Sun, M.; Gao, L.; Liang, X.; Ma, C.; Lu, J.; Yue, X. ZHX2 inhibits thyroid cancer metastasis through transcriptional inhibition of S100A14. *Cancer Cell Int.* **2022**, *22*, 76. [CrossRef] [PubMed]
52. Harousseau, J.L.; Shaughnessy, J., Jr.; Richardson, P. Multiple myeloma. *Hematol. Am. Soc. Hematol. Educ. Prog.* **2004**, 237–256. [CrossRef]

53. Legartova, S.; Harnicarova-Horakova, A.; Bartova, E.; Hajek, R.; Pour, L.; Kozubek, S. Expression of RAN, ZHX-2, and CHC1L genes in multiple myeloma patients and in myeloma cell lines treated with HDAC and Dnmts inhibitors. *Neoplasma* **2010**, *57*, 482–487. [CrossRef] [PubMed]
54. Armellini, A.; Sarasquete, M.E.; Garcia-Sanz, R.; Chillon, M.C.; Balanzategui, A.; Alcoceba, M.; Fuertes, M.; Lopez, R.; Hernandez, J.M.; Fernandez-Calvo, J.; et al. Low expression of ZHX2, but not RCBTB2 or RAN, is associated with poor outcome in multiple myeloma. *Br. J. Haematol.* **2008**, *141*, 212–215. [CrossRef]
55. Jiang, J.; Sun, Y.; Xu, J.; Xu, T.; Xu, Z.; Liu, P. ZHX2 mediates proteasome inhibitor resistance via regulating nuclear translocation of NF-kappaB in multiple myeloma. *Cancer Med.* **2020**, *9*, 7244–7252. [CrossRef]
56. Nagel, S.; Schneider, B.; Rosenwald, A.; Meyer, C.; Kaufmann, M.; Drexler, H.G.; MacLeod, R.A. t(4;8)(q27;q24) in Hodgkin lymphoma cells targets phosphodiesterase PDE5A and homeobox gene ZHX2. *Genes Chromosomes Cancer* **2011**, *50*, 996–1009. [CrossRef]
57. Nagel, S.; Schneider, B.; Meyer, C.; Kaufmann, M.; Drexler, H.G.; Macleod, R.A. Transcriptional deregulation of homeobox gene ZHX2 in Hodgkin lymphoma. *Leuk. Res.* **2012**, *36*, 646–655. [CrossRef]
58. Zhu, L.; Ding, R.; Yan, H.; Zhang, J.; Lin, Z. ZHX2 drives cell growth and migration via activating MEK/ERK signal and induces Sunitinib resistance by regulating the autophagy in clear cell Renal Cell Carcinoma. *Cell Death Dis.* **2020**, *11*, 337. [CrossRef]
59. Fang, W.; Liao, C.; Shi, R.; Simon, J.M.; Ptacek, T.S.; Zurlo, G.; Ye, Y.; Han, L.; Fan, C.; Bao, L.; et al. ZHX2 promotes HIF1alpha oncogenic signaling in triple-negative breast cancer. *Elife* **2021**, *10*, e70412. [CrossRef]
60. Cheng, A.; Guo, X.; Dai, X.; Wang, Z. Upregulation of ZHX2 predicts poor prognosis and is correlated with immune infiltration in gastric cancer. *FEBS Open Bio* **2021**, *11*, 1785–1798. [CrossRef]
61. Morford, L.A.; Davis, C.; Jin, L.; Dobierzewska, A.; Peterson, M.L.; Spear, B.T. The oncofetal gene glypican 3 is regulated in the postnatal liver by zinc fingers and homeoboxes 2 and in the regenerating liver by alpha-fetoprotein regulator 2. *Hepatology* **2007**, *46*, 1541–1547. [CrossRef] [PubMed]
62. Castelli, G.; Pelosi, E.; Testa, U. Liver Cancer: Molecular Characterization, Clonal Evolution and Cancer Stem Cells. *Cancers* **2017**, *9*, 127. [CrossRef] [PubMed]
63. Clinkenbeard, E.L.; Turpin, C.; Jiang, J.; Peterson, M.L.; Spear, B.T. Liver size and lipid content differences between BALB/c and BALB/cJ mice on a high-fat diet are due, in part, to Zhx2. *Mamm. Genome* **2019**, *30*, 226–236. [CrossRef] [PubMed]
64. Gargalovic, P.S.; Erbilgin, A.; Kohannim, O.; Pagnon, J.; Wang, X.; Castellani, L.; LeBoeuf, R.; Peterson, M.L.; Spear, B.T.; Lusis, A.J. Quantitative trait locus mapping and identification of Zhx2 as a novel regulator of plasma lipid metabolism. *Circ. Cardiovasc. Genet* **2010**, *3*, 60–67. [CrossRef]
65. Tan, S.; Guo, X.; Li, M.; Wang, T.; Wang, Z.; Li, C.; Wu, Z.; Li, N.; Gao, L.; Liang, X.; et al. Transcription factor Zhx2 restricts NK cell maturation and suppresses their antitumor immunity. *J. Exp. Med.* **2021**, *218*, e20210009. [CrossRef] [PubMed]
66. Minamiya, Y.; Saito, H.; Ito, M.; Imai, K.; Konno, H.; Takahashi, N.; Motoyama, S.; Ogawa, J. Suppression of Zinc Finger Homeobox 3 expression in tumor cells decreases the survival rate among non-small cell lung cancer patients. *Cancer Biomark.* **2012**, *11*, 139–146. [CrossRef]
67. Pruitt, H.C.; Devine, D.J.; Samant, R.S. Roles of N-Myc and STAT interactor in cancer: From initiation to dissemination. *Int. J. Cancer* **2016**, *139*, 491–500. [CrossRef]
68. Kinoshita, T.; Nohata, N.; Watanabe-Takano, H.; Yoshino, H.; Hidaka, H.; Fujimura, L.; Fuse, M.; Yamasaki, T.; Enokida, H.; Nakagawa, M.; et al. Actin-related protein 2/3 complex subunit 5 (ARPC5) contributes to cell migration and invasion and is directly regulated by tumor-suppressive microRNA-133a in head and neck squamous cell carcinoma. *Int. J. Oncol.* **2012**, *40*, 1770–1778. [CrossRef]

International Journal of
*Molecular Sciences*

*Review*

# Role of RBMS3 Novel Potential Regulator of the EMT Phenomenon in Physiological and Pathological Processes

**Tomasz Górnicki [1,*], Jakub Lambrinow [1], Monika Mrozowska [2], Marzena Podhorska-Okołów [3], Piotr Dzięgiel [2] and Jędrzej Grzegrzółka [2]**

[1] Faculty of Medicine, Wroclaw Medical University, 50-368 Wroclaw, Poland
[2] Division of Histology and Embryology, Department of Human Morphology and Embryology, Wroclaw Medical University, 50-368 Wroclaw, Poland
[3] Division of Ultrastructure Research, Wroclaw Medical University, 50-368 Wroclaw, Poland
* Correspondence: tomasz.gornicki@student.umed.wroc.pl

**Abstract:** RNA-binding protein 3 (RBMS3) plays a significant role in embryonic development and the pathogenesis of many diseases, especially cancer initiation and progression. The multiple roles of RBMS3 are conditioned by its numerous alternative expression products. It has been proven that the main form of RBMS3 influences the regulation of microRNA expression or stabilization. The absence of RBMS3 activates the Wnt/β-catenin pathway. The expression of c-Myc, another target of the Wnt/β-catenin pathway, is correlated with the RBMS3 expression. Numerous studies have focused solely on the interaction of RBMS3 with the epithelial–mesenchymal transition (EMT) protein machinery. EMT plays a vital role in cancer progression, in which RBMS3 is a new potential regulator. It is also significant that RBMS3 may act as a prognostic factor of overall survival (OS) in different types of cancer. This review presents the current state of knowledge about the role of RBMS3 in physiological and pathological processes, with particular emphasis on carcinogenesis. The molecular mechanisms underlying the role of RBMS3 are not fully understood; hence, a broader explanation and understanding is still needed.

**Keywords:** target discovery; epithelial–mesenchymal transition (EMT); RNA-binding protein 3 (RBMS3); carcinogenesis; target therapy

**Citation:** Górnicki, T.; Lambrinow, J.; Mrozowska, M.; Podhorska-Okołów, M.; Dzięgiel, P.; Grzegrzółka, J. Role of RBMS3 Novel Potential Regulator of the EMT Phenomenon in Physiological and Pathological Processes. *Int. J. Mol. Sci.* **2022**, *23*, 10875. https://doi.org/10.3390/ijms231810875

Academic Editor: Laura Paleari

Received: 5 September 2022
Accepted: 14 September 2022
Published: 17 September 2022

**Publisher's Note:** MDPI stays neutral with regard to jurisdictional claims in published maps and institutional affiliations.



## 1. Introduction

RNA-binding motif single-stranded-interacting protein 3 (RBMS3) is a glycine-rich protein that was described for the first time by D. Penkov et al. in 2000 [1]. The gene encoding this protein is called *RBMS3*, and it is located in the short arm of chromosome 3, specifically in the 3p24.1 region. The discovery of RBMS3 was an effect of the screening of human fibroblast cDNA with an upstream element of the a2(I) collagen promoter Box 5A. It belongs to the family of *c-Myc* gene single-strand binding proteins (MSSPs) involved in DNA replication, transcription, apoptosis induction, and cell–cycle progression [2,3]. Published papers provide evidence of the wide range of processes in which RBMS3 takes part in, including regulation of embryogenesis, pathogenesis of liver fibrosis, and bisphosphonate-related osteonecrosis of the jaw (BRONJ) [4–6]. From 2008 onwards, RBMS3 has become a potential prognostic marker of different types of cancer and a factor regulating the process of carcinogenesis [7,8]. In addition, recent articles have provided evidence of RBMS3 taking part in the epithelial–mesenchymal transition (EMT), a key process responsible for the creation of distant metastases [9].

RBMS3's ability to suppress the growth and progression of different types of cancers makes it an interesting potential target for the development of novel anticancer therapies. There is still a need for a summary of the role of RBMS3 in physiology and pathology that would provide a synthetic evaluation of the information available about it. In this article, we are going to review and systematize the current state of knowledge about RBMS3 and its function in physiology and pathology, with a particular focus on its role in EMT.

## 2. Methods

The authors searched for topic-related materials in four big medical databases—PubMed, Embase, Ovid, and Scopus—on 12 December 2021. The searched keywords included: RBMS3, RNA-binding motif single-stranded-interacting protein 3, rbms3 cancer, rbms3 EMT, EMT, and epithelial–mesenchymal transition. The keywords were the same for all databases. The articles were screened for relevance and analyzed based on inclusion criteria. An article was presumed relevant if RBMS3 was directly mentioned by the authors of the research. References from all relevant articles were also reviewed to ensure the inclusion of all articles directly related to the topic of RBMS3 (Figure 1).

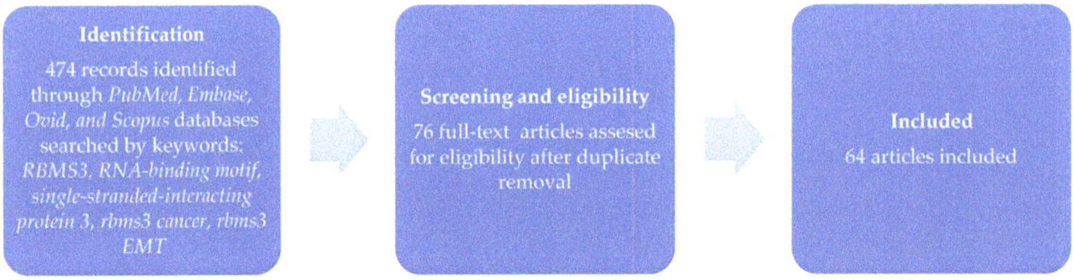


**Figure 1.** Workflow literature review.

## 3. Role in Development and Physiology

It is stated that the expression of RBMS3 may be a part of the regulatory mechanisms of pancreas embryonic development in mouse models. Authors have shown a restricted expression of protein in the embryonic pancreas, the neural tube, and the dorsal root ganglion, with peak expression in the pancreas taking place at E13.5. RBMS3 acts on a post-transcriptional level and is able to bind to the 3′-UTR of pancreas transcription factor 1, the alpha subunit (*Ptf1α*) mRNA, stabilizing it and increasing the level of Ptf1α protein in cells. Ptf1α is responsible for exocrine cell differentiation. Scientists have also provided evidence of RBMS3 expression affecting the expression of various digestive enzymes and its role in maintaining the function of mature exocrine cells in the pancreas [4].

Experiments conducted on the zebrafish model brought to light the potential impact of RBMS3 on craniofacial development and chondrogenesis [10]. RBMS3 was discovered to be expressed transiently in the cranial neural crest, and its knockdown results in severe craniofacial defects. The authors point to the TGF-β receptor pathway as the mechanism responsible for these abnormalities. RBMS3 binds to and stabilizes the transcripts of the Smad2 pathway. Further studies discovered that RBMS3 also has the ability to interact with *Smad1*, as well as cell cycle regulators, such as the TGF-β, receptor *cyclin D1* and *Rac1* transcripts, introducing RBMS3 as a global regulator of chondrogenesis [10].

RBMS3 was also discovered to take part in preventing the degeneration of the nucleus pulposus. The mechanism underlying this process consists of decreasing the activity of the Wnt/β-catenin signaling pathway and targets, such as metalloproteinase-13 (MMP13) [11]. The study also showed that RBMS3 increases nucleus pulposus cell proliferation and decreases apoptosis, inflammation, and extracellular matrix degradation levels [11].

## 4. RBMS3 in Pathological Noncancerous Processes

From the moment of the discovery of RBMS3, scientists pointed to the role of this protein and the gene that encodes it in a wide range of pathological processes, including liver fibrosis, osteonecrosis of the jaw (ONJ), and exfoliation syndrome [12–14].

Liver fibrosis is a wound-healing type of EMT process serving as a response to the injury. It can potentially develop into cirrhosis and lead to organ failure as a consequence [12]. The focal point of this process lies in the activation of hepatic stellate cells (HSCs), which are responsible for the storage of vitamin A during their quiescent state but produce an excessive amount of the extracellular matrix after activation. One of the factors involved in inducing the activation of HSCs is the pair-related homeobox transcription factor Prx1, also involved in the production of collagen type α1(I). In their work, Fritz and Stefanovic provided evidence of the role RBMS3 has in the regulation of Prx1 expression. By binding to the 3′-UTR of the *Prx1* mRNA, RBMS3 stabilizes the structure of the mRNA, increasing the effectiveness of translation and the level of the Prx1 protein, thus leading to the stimulation of collagen type α1(I) gene transcription in HSCs. These results together with the post-transcriptional regulation of collagen type α1(I) expressions show a probable mechanism of RBMS3's role in the onset of liver fibrosis [5].

Another domain of RBMS3 influence is its impact on bone density. There is evidence of a statistically significant interaction between the *RBMS3* and *ZNF516* genes that negatively impacts the hip bone mineral density (BMD). The study was conducted using the novel approach of genome-wide association studies (GWAS), a method that successfully unveiled a number of genetic loci that impact BMD [15,16]. Table 1 presents all the discovered single-nucleotide polymorphisms (SNPs) of the *RBMS3* gene discussed in this article. Although molecular mechanisms underlying this interaction are currently unknown, RBMS3's impact on collagen expression may influence the extracellular matrix of bone tissue.

**Table 1.** Single-nucleotide polymorphisms of the *RBMS3* gene and their correlation with pathological processes.

| RBMS3-Related Processes | Identified SNP | References |
|---|---|---|
| Bone-mineral-density-related disorders | rs6549904<br>rs7640046<br>rs17024608 | [15] |
| Osteonecrosis of the jaw (ONJ) | rs17024608 | [17] |
| Exfoliation glaucoma | rs12490863 | [18] |
| Exfoliation syndrome | rs12490863 | [19] |
| Primary Sjögren's syndrome | rs13079920<br>rs13072846 | [20] |
| Systemic sclerosis | rs1449292 | [21] |
| Periodontal disease | rs17718700 | [22] |
| Lymphocyte glucocorticoid sensitivity | rs6549965 | [23] |
| Short-acting bronchodilator response | rs1266115<br>rs150703870 | [24] |

Another pathological process that may, among others, involve alteration in collagen type α1(I) expression is osteonecrosis of the jaw (ONJ). It is a serious adverse effect mainly connected to the administration of bisphosphonates (BPs), which are antiosteoclastic drugs used, among others, in oncological therapy to control bone metastasis and hypercalcemia. The frequency of ONJ ranges from 0.6% in breast cancer to even 15% in multiple myeloma [13]. Research conducted with the help of GWAS discovered a relation between the variation in the *RBMS3* gene and 5.8 times higher probability of developing bisphosphonate-related osteonecrosis of the jaw (BRONJ) [17]. Even though other researchers were not able to confirm this relation [13], taking into consideration the impact of RBMS3 on bone density postulated in [15], there is a wide area for researchers to establish the exact role of RBMS3 in ONJ [18].

Exfoliation syndrome (XFS) is an age-related systemic disease that is the most common risk factor for open-angle glaucoma, which can cause irreversible blindness. Based on familial aggregation studies, XFS is suspected to be a genetic disease. Specific loci in the *RBMS3* gene are proven to be correlated with susceptibility to XFS and exfoliation glaucoma, although the exact mechanism of this impact is yet to be discovered [14,19,25].

RBMS3 was also found to be potentially involved in autoimmune diseases. Specifically, there is evidence that certain SNPs in the *RBMS3* gene are responsible for an increased susceptibility to systemic sclerosis (SSc) and primary Sjögren's syndrome (PSS) [20,21]. A weak correlation was also found between RBMS3 and periodontal disease [22].

The versatility of RBMS3 reaches even the field of psychiatric health care and neurodegenerative diseases, since various authors have linked it to resistance to antidepressant therapy and susceptibility to schizophrenia and amyotrophic lateral sclerosis (ALS) [26–28]. Gastrointestinal dysfunction is a common symptom in the autism spectrum disorder (ASD). The exact underlying mechanism of this process is unknown, but researchers revealed that in a specific group of patients with *FOXP1* haploinsufficiency, downstream targets of the Foxp1 protein are dysregulated in the mice model. One of these targets is RBMS3, thus providing additional data about its role in this disorder [29].

RBMS3's impact seems to not be restricted only to the pathogenesis of different diseases. It also determines the response to some forms of therapy, with two effects described in the literature: (1) the regulation of lymphocyte sensitivity to glucocorticoids by decreasing cellular proliferation of peripheral blood mononuclear cells and (2) the modulation of the response to inhaled short-acting bronchodilators (BD) [23,24].

A recent study using CRISPR interference (CRISPRi) tried to assess the molecular mechanisms connected to the genes associated with chronic obstructive pulmonary disease (COPD) and low lung function. After a GWAS analysis searching for genes related to COPD, the experiments were conducted on human-induced pluripotent stem cell (iPSC)–derived lung epithelium. The results of this study show that the knockdown of *RBMS3* enhances the proliferation of cells, which is the basis for later experiments clarifying the exact role of RBMS3 in COPD [30].

## 5. Role of RBMS3 in Carcinogenesis

In 2008, RBMS3 was mentioned in the context of neoplastic processes for the first time [31]. From that time onwards, RBMS3 has significantly grown in popularity and importance as a potential marker and regulator in many different types of cancer. The increasing amount of scientific data provided by researchers has started to unveil the specific mechanisms of RBMS3's impact on carcinogenesis and metastasis (Table 2).

**Table 2.** Role of RBMS3 in carcinogenesis.

| Tumor Type | Correlation with High or Low Expression of RBMS3 | Mechanism of Action | References |
|---|---|---|---|
| Bladder cancer | High expression correlates with poorer prognosis.<br>Low expression correlates with shorter OS. | Further research is needed. | [32,33] |
| Gallbladder carcinoma | High expression inhibits growth and promotes apoptosis in vitro. | Further research is needed. | [34] |
| Prostate cancer | Upregulation of *RBMS-AS3* correlates with faster tumor growth, angiogenesis, and migration. | *RBMS-AS3*/miR-4534/VASH1 axis. | [35–37] |
| Ovarian epithelial cancer | Loss of *RBMS3* gene is correlated with poorer prognosis.<br>Deletion of *RBMS3* promotes efflux and induces chemoresistance. | RBMS3 promotes efflux.<br>Lack of RBMS3 activates the Wnt/β-catenin pathway. | [38,39] |
| Nasopharyngeal cancer | Ectopic expression inhibits tumor growth and foci formation. | RBMS3 increases the level of p53, and thus p21 and MMP2 and MMP9.<br>c-Myc/Wnt/β-catenin axis. | [40,41] |
| Gastric cancer | Low expression correlates with poorer prognosis, poor histological grade, and angiogenesis. | Wnt/β-catenin pathway.<br>Low expression of RBMS3 induces overexpression of HIF1-A. | [8,42,43] |
| Esophageal squamous cell carcinoma | Low expression correlates with poorer prognosis.<br>Ectopic expression inhibits tumor growth. | RBMS3 induces downregulation of c-Myc and CDK4. | [44,45] |
| Lung cancer | Low expression correlates with worse OS. | Downregulation of RBMS3 and upregulation of c-Myc and β-catenin. | [46] |
| Papillary thyroid cancer | High expression of *RBMS3-AS1* correlates with shorter OS. | Further research is needed. | [47] |
| Breast cancer | High expression inhibits tumor growth, invasion, and migration.<br>Low expression correlates with poorer prognosis and shorter OS.<br>Levels of expression of ER and RBMS3 are correlated. | Wnt/β-catenin axis.<br>MEG3-miR-141-3p-RBMS3 axis. | [7,48–51] |

### 5.1. Bladder Cancer

The results showed that the downregulation of RBMS3 in bladder cancer was specifically related to a better overall survival (OS), with a higher expression of RBMS3 implicating a poorer prognosis. This was confirmed a few months later by Chen et al. The expression of RBMS3 was also significantly correlated with grade and stages T and M in the TNM scale [32,33].

### 5.2. Gallbladder Carcinoma (GBC)

The relationship between the expression of RBMS3 and bladder cancer is one of the most recently discussed in the literature. While studying the role of RBMS3 in gallbladder carcinoma, scientists found its downregulation at the mRNA, and protein levels in the tested specimens had an impact on their overall survival. A low expression correlated with a worse OS and acted as an independent negative prognostic factor. Moreover, the overexpression of RBMS3 successfully inhibits growth and promotes the apoptosis of GBC cell lines in in vitro studies. A low expression of RBMS3 also leads to increased angiogenesis, highlighting another process influenced by this protein [34].

### 5.3. Prostate Cancer

Studies on prostate cancer provided evidence of another biological mechanism of the role of RBMS3 in carcinogenesis. *RBMS3-AS3*, a long noncoding RNA (lncRNA), was found to play a significant role as an antitumor factor. LncRNAs are noncoding RNA fragments longer than 200 nucleotides with the ability to bind to different microRNAs (miRNAs) functioning as competing endogenous RNA (ceRNA) [35]. *RBMS3-AS3* binds competitively to miR-4534, increasing the level of its downstream target vasohibin 1 (VASH1), creating the molecular axis *RBMS3-AS3*/miR-4534/VASH1, which may play a pivotal role in prostate cancer development and treatment. *RBMS3-AS3* is downregulated in prostate cancer, which leads to an upregulation of miR-4534, which decreases the level of VASH1. Experimental upregulation of *RBMS-AS3* led to the inhibition of tumor growth, angiogenesis, and migration by the upregulation of VASH1. VASH1 as a downstream target is also important because it can work as an individual prognostic marker of prostate cancer, and recent studies have shown that its upregulation can inhibit lymphangiogenesis [36]. Another product of the *RBMS3* gene belongs to the group of circular RNAs (circRNAs) containing noncoding RNA with various functions. has_circ_0064644 was the most downregulated circRNA in prostate cancer. Its exact role in prostate cancer progression is yet to be revealed [37].

### 5.4. Epithelial Ovarian Cancer (EOC)

Managing patients with ovarian epithelial cancer is still an exceedingly challenging task for oncologists due to the high rate of relapses caused by chemoresistance. Platinum-based therapy, combined with surgical cytoreduction, is still one of the most effective methods of treatment in EOC. The studies conducted to elucidate the role of RBMS3 in EOC provided data to support the statement that the deletion of the region of chromosome 3 containing the gene for RBMS3 is correlated with a poorer prognosis and acts as an independent prognostic factor for relapse-free survival in this type of cancer. The deletion of *RBMS3* leads to the development of chemoresistance in the patient-derived xenograft (PDX) model and in EOC cell lines. The molecular mechanism underlying these results consists of several elements. First, the loss of *RBMS3* promotes efflux in EOC cells, preventing cytotoxic platinum from getting into the cells. The downregulation of RBMS3 significantly decreases platinum-induced DNA damage and apoptosis, indicating a potential role in restricting DNA damage repair. The lack of RBMS3 activates the Wnt/β-catenin pathway by allowing the strong negative regulator miR-126-5p to downregulate strong Wnt/β-catenin repressors. RBMS3 takes part in the competitive stabilization of many identified repressors, including DKK3, AXIN1, BACH1, and NFAT5 [38]. The *RBMS3* gene was also used in the creation of the tumor-mutation-burden-related signature model. This is a model that uses the total number of replacement and insertion/deletion (indel) mutations per basic group in the exon coding region of the assessed gene in the genome of a tumor cell to predict overall survival in a specific cancer, in this case, ovarian cancer [39].

### 5.5. Nasopharyngeal Cancer (NPC)

Studies conducted on nasopharyngeal cancer introduced RBMS3 as a potential regulator of the cell cycle. Researchers provided evidence of the significant downregulation of RBMS3 in NPC cell lines and postoperational tumor specimens. The ectopic expression of RBMS3 proved to have the ability to inhibit tumor growth and foci formation. As the reason for these abilities, scientists provided a number of molecular mechanisms related to the cell cycle, including apoptosis and microvessel formation. RBMS3 increased the level of p53, which plays a crucial role in promoting the cell cycle from the G1 phase to the S phase. The upregulation of p53 creates a cascade of effects that prevent cells from going further in the cell cycle. An increased expression of p53 increases the expression of p21, which has the ability to suppress the cell cycle by inhibiting the complex cyclin E/CDK2. This complex has an influence on retinoblastoma proteins (RBs), decreasing their phosphorylated inactive form in favor of the unphosphorylated one, which has the ability to stop cells from reaching the next stage of the cell cycle. The overexpression of p53 along with MMP2 and MMP9 may also have an impact on the inhibition of microvessel formation by RBMS3. Changes in the expression of MMP2, MMP9, MMP7, and c-Myc may be explained by the inhibited nuclear translocation of β-catenin. C-Myc is an important downstream target of the Wnt/β-catenin pathway in this case, since its expression correlates with a poorer prognosis, and there is evidence of RBMS3's abilities to bind to the promoter region of *c-Myc*. The role of RBMS3 in the increased apoptotic activity of NPC was explained with the activation of caspase 9 and PARP by RBMS3 [40,41].

### 5.6. Gastric Cancer (GC)

All studies concerning the connection between the expression of RBMS3 and gastric cancer provided information about the downregulation of RBMS3 in this type of cancer. RBMS3 was found to have an impact on the secreted frizzled-related protein 1 (SFRP1), playing a significant role in the downregulation of the Wnt/β-catenin pathway by the competitive inhibition of Wnt-frizzled membrane receptor (Fzs) complexes. The low expression of RBMS3 and SFRP1 was found to correlate with a poorer prognosis. The expression of both proteins is statistically related to a poor histological grade and prognosis. The combined expression of RBMS3 and SFP1 acts as an independent prognostic factor in GC. Another downstream target regulated by RBMS3 in GC is the basic helix-loop-helix-PAS transcription factor α (HIF1-A) subunit of the HIF-1 protein, responsible for the induction of VEGF expression in cancer cells. VEGF is a key factor responsible for angiogenesis in tumors. The expression of HIF1-A is increased in GC cells. This, combined with a decreased level of the RBMS3 expression, correlates with a poor histopathological differentiation and a stronger angiogenesis. The overexpression of RBMS3 in GC cells revealed an increased percentage of cells in the G0/1 phase and a lower number of cells in the S phase of the cell cycle, but it had no statistically significant influence on cells in the stage G2/M. Additionally, lower expressions of CDK1, CDK6, E2F1, and MYC were observed, providing evidence of RBMS3's impact on the cell cycle in GC. RBMS3 also has an impact on circular RNA (circRNA) single-stranded enclosed RNAs, which are common regulators of carcinogenesis. *CircRBMS3* is postulated to be tied with an advanced TNM stage, poor differentiation, larger tumor size, and lymph node metastasis positivity by the regulation of miR-153 and SNAIL1. The overexpression of *circRBMS3* was also shown to be connected to a lower OS. The artificial knockdown of *circRBMS3* results in the inhibition of tumor growth and invasiveness [8,42,43].

### 5.7. Esophageal Squamous Cell Carcinoma (ESCC)

The loss of the 3p fragment of chromosome 3 is one of the most common chromosomal alterations in esophageal squamous cell carcinoma. One of the frequently lost genes is

*RBMS3* [44]. The downregulation of RBMS3 significantly correlates with poorer outcomes in patients with ESCC. The ectopic expression of RBMS3 results in tumor growth impairment confirmed by foci formation and tumor xenograft formation tests. Experimental data point to the downregulation of c-Myc and CDK4 as the mechanism mediating RBMS3's tumor suppressive gene (TSG) abilities. Interestingly, other cell-cycle-related proteins, such as CDK2 or cyclin E or D1, dysregulated in other types of cancer, do not seem to be involved in RBMS3's role in ESCC. Further studies showed that Rb, the downstream target of CDK2, was also found to be altered by the expression of RBMS3. A decreased level of CDK2 increases the level of inactivated phosphorylated Rb at Ser807/811 and Ser780 [45].

### 5.8. Lung Cancer

Depending on the type of lung cancer, different approaches to the role of RBMS3 were taken, highlighting different aspects of RBMS3's effect on lung cancer progression. Lung squamous cell carcinoma (LSCC) was characterized by the downregulation of RBMS3 and the upregulation of c-Myc and β-catenin. Oddly enough, there was only a statistically significant correlation of RBMS3's expression with c-Myc. The combined positive expression of RBMS3 and negative expression of c-Myc act as an independent prognostic factor of shorter OS [46]. As for small-cell lung cancer (SCLC), Xiuwei Li et al. provided evidence of the downregulation of RBMS3 and its upstream miRNA hsa-miR-7-5p by using bioinformatic methods. Hsa-miR-7-5p was previously reported to display tumor-suppressive properties in glioma and glioblastoma by the regulation of the EGFR, PI3K/ATK, Raf/MEK/ERK, and IGF-1R pathways [52,53]. Another type of lung cancer discussed in the context of RBMS3 expression was non-small-cell lung cancer (NSCLC). By using computational methods, scientists identified RBMS3 as a core transcription factor regulating lung-adenocarcinoma-associated genes [54]. Other bioinformatic analyses provided evidence of RBMS3 belonging to the group of genes most negatively correlated with tumorigenesis and being dysregulated in precancer cells. Furthermore, this dysregulation advances through cancer progression [55].

### 5.9. Papillary Thyroid Cancer

The analysis of lncRNA in papillary thyroid cancer revealed that another product of *RBMS3*'s expression, *RBMS3-AS1*, is closely associated with a patient's shorter OS, broadening the variety of tumors in which RBMS3 has the potential to be a diagnostic marker [47].

### 5.10. Hepatocellular Carcinoma (HCC)

Studies conducted on hepatocellular carcinoma present RBMS3 in a position of effector instead of regulator. In this case, an upregulated miR-1269 is responsible for altering the expression of *RBMS3* and eight other genes: *AGAP1*, *AGK*, *BMPER*, *BPTF*, *C16orf74*, *DACT1*, *LIX1L*, and *ZNF706* [56].

### 5.11. Neuroblastoma

The potential role of RBMS3 in the carcinogenesis of the neuroblastoma was discovered through high-resolution array copy number analyses that showed the presence of homozygous deletion on 3p. However, there are no further studies on this issue [31].

### 5.12. Breast Cancer (BC)

The role of RBMS3 has been most extensively explored in breast cancer among all types of cancer. The expression profile of *RBMS3* at the protein and RNA levels is downregulated. The overexpression of *RBMS3* inhibits the growth, invasion, and migration of BC cells. In vivo experiments conducted in mice also showed an attenuation of tumor growth. As for the clinicopathological characteristics, the downregulation of *RBMS3* correlates with a poor prognosis and a shorter OS. A negative ER status corresponding with the expression of RBMS3 and the combined expression of both these parameters act as independent prognostic factors. The molecular mechanisms underlying these effects include the impact on the Wnt/β-catenin pathway and the cell cycle, confirmed by the inhibited expression of β-catenin, c-Myc, and cyclin D1 in RBMS3 expressing cancer cells [7,48]. Another point of regulation lies in the lncRNA (long noncoding RNA) maternally expressed gene 3 (MEG3)-miR-141-3p-RBMS3 axis. LncRNA encoded by *MEG3* was found to have tumor-suppressive abilities in different types of tumors, including glioma, gastric cancer, and melanoma. MiR-141-3p is a microRNA (miRNA) belonging to the miR-200 family dysregulated in many tumors. An overexpression of miR-141-3p was found in bladder cancer and esophageal squamous cell carcinoma. A low expression of MEG3 upregulates miR-141-3p, which anterogradely downregulates RBMS3 in BC. MEG3 is a tumor-suppressive gene regulating AKT and NF-κB signaling pathways, inducing apoptosis through its impact on Bcl-2 and C casp-3 and p53 signaling. MiR-141-3p is a miRNA whose role depends on the type of tumor, with capabilities ranging from tumor-suppressive abilities to overexpression correlated with poor prognosis and chemoresistance [49–51]. Moreover, a recent study showed that the *RBMS3* gene expression in the tumor-associated stromal cells of breast tumor was gradually downregulated among grade I, II, and III of breast cancer. The downregulation of this gene was also correlated with worse clinical outcome and poorer survival prognosis [57].

## 6. Epithelial–Mesenchymal Transition and Role of RBMS3 in This Process

Epithelial–mesenchymal transition (EMT) is a biological process that allows epithelial cells to switch their phenotype to quasi-mesenchymal [58–60]. EMT causes epithelial cells to lose characteristic features, such as tight cell–cell junctions [61] and cell polarity [59], and acquire mesenchymal properties instead [62]. This is first observed during embryogenesis, in gastrulation or tissue morphogenesis [63]. Furthermore, the process plays a crucial role in wound healing, fibrosis, and tumor progression [60–66]. The reverse process is called MET, from mesenchymal–epithelial transition, and it occurs when the mesenchymal cells acquire epithelial characteristics [67].

Typically, epithelial cells appear as cells attached to basal lamina, with tight cell–cell junctions and apical–basal polarity [68]. When it comes to EMT, epithelial cells lose these properties and the ability of the expression of E-cadherin—a molecule that is essential to maintaining the epithelial phenotype [58]. The loss of E-cadherin is considered to be a hallmark of EMT along with the acquisition of the expression of vimentin of N-cadherin [68]. During EMT, the epithelial cells, which have a typical cobblestone morphology, transform into quasi-mesenchymal cells, which have a rather fibroblastic-like phenotype [69]. This transition allows cells to acquire a migratory phenotype and become more invasive [70]. These changes in phenotype require rearrangements of the cytoskeleton and the cell metabolism [71]. Due to the acquisition of these properties, EMT plays a significant role in tumor progression, metastasis, and malignancy [58,71].

Three types of EMT processes can be distinguished. Type 1 describes an EMT that occurs in the development of tissues. The EMT subtype that occurs in fibrosis and wound healing is type 2, with type 3 being observed in cancer cells [72]. Although, historically, EMT was discovered by developmental biologists [63], modern studies focus on the link between EMT and cancer [73]. Recent observations suggest that EMT is also involved in the therapeutic resistance of various tumors [67,74,75].

EMT is a process that is strictly determined by genetic mechanisms. Several transcription factors involved in this phenotype change have been discovered [76]. Some well-described EMT-TFs (epithelial–mesenchymal transition transcription factors) are SNAIL1, SNAIL2, TWIST1, and ZEB1. However, the list of EMT-TFs is way longer, and there are many more transcription factors involved in EMT, for instance, FOX- or SOX-TF [76]. The crucial signaling pathways of EMT are Wnt and TGF-β, but other pathways, such as Notch or Hedgehog, are also involved [68]. Some of the descriptions of the molecular mechanisms seem to be quite preposterous; thus, there is still a lot of speculation and uncertainty surrounding the topic.

As it has already been mentioned, EMT plays a major role in cancer progression. EMT allows cancer cells to become more mesenchymal-like. EMT is probably responsible for the creation of circulating tumor cells (CTCs), which are strictly connected to the ability to metastasize [77]. CTCs are an element of the invasion-metastatic cascade, and EMT is believed to be involved it this type of tumor progression [58]. It is worth noticing that the reverse process, mesenchymal–epithelial transition (MET), is also important for the ability of cancer cells to metastasize [67,78,79]. EMT is considered to be a relevant process in the development of cancer steam cells (CSCs). Therefore, it could be responsible for therapeutic resistance [58,78].

With a better understanding of EMT's complexity and its importance and vital role in cancer progression, invasion, and the development of metastases and CSCs comes the necessity to find and describe the key regulators of this process. RBMS3 is a novel potential regulator of EMT, with an increasing amount of data trying to unveil its molecular role in this process. Figure 2 and Table 3 present the currently proposed mechanisms of the impact of RBMS3 on the EMT process.

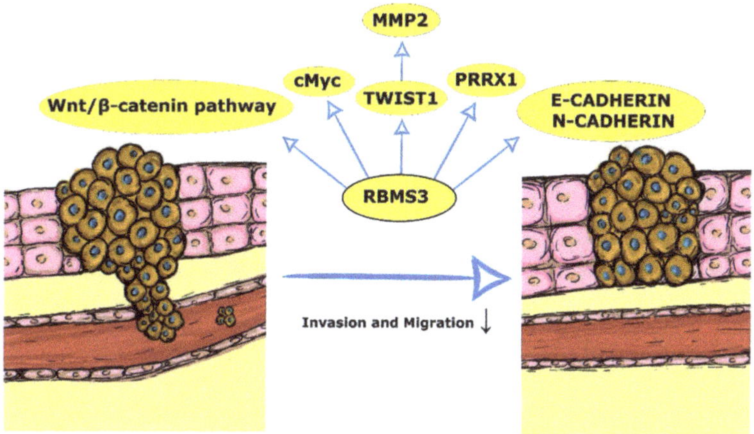


**Figure 2.** Mechanisms of the impact of RBMS3 on the EMT process.

The Wnt/β-catenin signaling pathway is a critical molecular mechanism regulating the EMT process. Downstream targets of Wnt include, among others, Twist, Snail, and MMP7 genes facilitating EMT [9]. In several of the previously discussed types of cancer, the Wnt/β-catenin pathway was inhibited by RBMS3's expression. The expression of c-Myc, another downstream target of the Wnt/β-catenin pathway, was also investigated and found statistically correlated with RBMS3's expression.

Several studies focus solely on RBMS3's interaction with the EMT machinery. While studying breast cancer in 2019, Zhu L et al. identified a regulatory axis consisting of RBMS3, TWIST1, and matrix metalloproteinase 2 (MMP2), responsible for the migration and invasion of the tumor. The expression of RBMS3 downregulated the expression of TWIST1, one of the key factors of EMT, and consecutively, its downstream target MMP2, leading to EMT impairment and invasion and migration inhibition [9]. Another study conducted on breast cancer provided interesting data stating that the expression of RBMS3 is a required factor for EMT induction in immortalized mammary epithelial cell lines. In the triple negative breast cancer (TNBC) model, RBMS3 was essential for maintaining the mesenchymal phenotype, invasiveness, and migration ability. In vivo experiments showed the loss of RBMS3 to impair the growth of the tumor and its ability to create metastasis. As the potential molecular basis of this process, the authors indicated RBMS3's ability to influence expression and stabilize *PRRX1* mRNA, a transcription factor regulating EMT [80]. Research conducted on gastric cancer by Zhao seems to be coherent with Zhu's results, showing an increased expression of E-cadherin and a decreased expression of

N-cadherin and β-catenin in RBMS3-overexpressing gastric cancer cells. Moreover, an increased expression of RBMS3 significantly decreased the invasive abilities of cells [81].

Taking into consideration all the information contained in this chapter, there is convincing evidence of RBMS3 being one of the regulators involved in the EMT process, even though the exact mechanism of this regulation requires further investigation and may differ depending on the molecular subtype of cancer.

**Table 3.** Proposed molecular mechanisms of RBMS3's impact on EMT.

| Type of Cancer | Currently Proposed Mechanisms of RBMS3' Impact on EMT |
| --- | --- |
| Breast cancer | 1. The expression of RBMS3 downregulates the expression of TWIST1 and, consecutively, its downstream target MMP2, leading to EMT impairment [9]<br>2. Loss of RBMS3 impairs the growth of the tumor and its ability to create metastasis by influencing expression and stabilizing *PRRX1* mRNA, a transcription factor regulating EMT [80] |
| Gastric cancer | 1. Increased expression of E-cadherin and decreased expression of N-cadherin and β-catenin in RBMS3-overexpressing cancer cells [81] |

## 7. Conclusions

All the information provided in this review depicts *RBMS3* as a functionally versatile gene that uses its main and multiple alternative products of expression to play a significant role in embryonic development and the pathogenesis of many different diseases, especially the induction and progression of cancers. The main ways in which RBMS3 impacts cells are the regulation of the expression or stabilization of miRNA, inhibiting Wnt/β-catenin signaling pathway and other EMT-related transcription factors. These molecular characteristics make RBMS3 a promising biomarker of OS and a prognostic factor in neoplastic processes, where statistical data support this statement for many different types of cancer. Another potential use of RBMS3 is as a target for anticancer drugs, thanks to its function as TSG and its proven ability to suppress cancer migration and invasive abilities. Artificially increased expression of RBMS3 utilizing genome editing techniques may potentially improve the outcome of standard therapies in many types of cancers. Increased expression of RBMS3 may prevent the creation of micrometastases that are too small to be picked up in diagnostic imaging and may lead to relapse of tumor.

There are some limitations to targeting RBMS3 mainly concerning the lack of a deep understanding of molecular mechanisms that are responsible for RBMS3 tumor suppressive abilities and the regulation of this properties. Additionally, currently, there are not enough data concerning the role of RBMS3 expression in different types of healthy human tissues and the consequences of RBMS3 level alteration. However, the year-on-year increasing amount of data and the incoherencies of some of the results indicate that the molecular role of RBMS3, especially in the regulation of cancer development, is a good subject for further research that may lead to the development of novel diagnostic and therapeutic strategies that will improve the outcome of patients with neoplastic diseases.

**Author Contributions:** Conceptualization, T.G. (Tomasz Górnicki), J.L. and M.M.; methodology, T.G.; software, T.G.; validation, J.G., M.P.-O. and P.D.; formal analysis, J.G. and M.M.; investigation, T.G. and J.L.; writing—original draft preparation, T.G., J.L. and M.M.; writing—review and editing, J.G., M.P.-O. and P.D.; visualization, T.G.; supervision, J.G., P.D. and M.P.-O.; project administration, J.G., P.D. and M.P.-O. All authors have read and agreed to the published version of the manuscript.

**Funding:** This research received no external funding.

**Institutional Review Board Statement:** Not applicable.

**Informed Consent Statement:** Not applicable.

**Data Availability Statement:** Not applicable.

**Conflicts of Interest:** The authors declare no conflict of interest.

## References


1. Penkov, D.; Ni, R.; Else, C.; Piñol-Roma, S.; Ramirez, F.; Tanaka, S. Cloning of a human gene closely related to the genes coding for the c-myc single-strand binding proteins. *Gene* **2000**, *243*, 27–36. [CrossRef]
2. Niki, T.; Izumi, S.; Saëgusa, Y.; Taira, T.; Takai, T.; Iguchi-Ariga, S.M.; Ariga, H. MSSP promotes ras/myc cooperative cell transforming activity by binding to c-Myc. *Genes Cells* **2000**, *5*, 127–141. [CrossRef] [PubMed]
3. Fujimoto, M.; Matsumoto, K.; Iguchi-Ariga, S.M.; Ariga, H. Disruption of MSSP, c-myc single-strand binding protein, leads to embryonic lethality in some homozygous mice. *Genes Cells* **2001**, *6*, 1067–1075. [CrossRef] [PubMed]
4. Lu, C.K.; Lai, Y.C.; Chen, H.R.; Chiang, M.K. Rbms3, an RNA-binding protein, mediates the expression of Ptf1a by binding to its 3'UTR during mouse pancreas development. *DNA Cell Biol.* **2012**, *31*, 1245–1251. [CrossRef] [PubMed]
5. Fritz, D.; Stefanovic, B. RNA-binding protein RBMS3 is expressed in activated hepatic stellate cells and liver fibrosis and increases expression of transcription factor Prx1. *J. Mol. Biol.* **2007**, *371*, 585–595. [CrossRef] [PubMed]
6. RBMS3: A novel gene implicated in the risk of BRONJ. *Bonekey Rep.* **2012**, *1*, 118, PMCID:3727801. [CrossRef] [PubMed]
7. Yang, Y.; Quan, L.; Ling, Y. RBMS3 Inhibits the Proliferation and Metastasis of Breast Cancer Cells. *Oncol. Res.* **2018**, *26*, 9–15. [CrossRef]
8. Zhang, T.; Wu, Y.; Fang, Z.; Yan, Q.; Zhang, S.; Sun, R.; Khaliq, J.; Li, Y. Low expression of RBMS3 and SFRP1 are associated with poor prognosis in patients with gastric cancer. *Am. J. Cancer Res.* **2016**, *6*, 2679–2689.
9. Zhu, L.; Xi, P.W.; Li, X.X.; Sun, X.; Zhou, W.B.; Xia, T.S.; Shi, L.; Hu, Y.; Ding, Q.; Wei, J.F. The RNA binding protein RBMS3 inhibits the metastasis of breast cancer by regulating Twist1 expression. *J. Exp. Clin. Cancer Res.* **2019**, *38*, 105, Erratum in: *J. Exp. Clin. Cancer Res.* **2020**, *39*, 21. [CrossRef]
10. Jayasena, C.S.; Bronner, M.E. Rbms3 functions in craniofacial development by posttranscriptionally modulating TGF-β signaling. *J. Cell. Biol.* **2012**, *199*, 453–466. [CrossRef]
11. Wang, J.J.; Liu, X.Y.; Du, W.; Liu, J.Q.; Sun, B.; Zheng, Y.P. RBMS3 delays disc degeneration by inhibiting Wnt/β-catenin signaling pathway. *Eur. Rev. Med. Pharmacol. Sci.* **2020**, *24*, 499–507. [CrossRef]
12. Aydın, M.M.; Akçalı, K.C. Liver fibrosis. *Turk. J. Gastroenterol.* **2018**, *29*, 14–21. [CrossRef]
13. Hoff, A.O.; Toth, B.; Hu, M.; Hortobagyi, G.N.; Gagel, R.F. Epidemiology and risk factors for osteonecrosis of the jaw in cancer patients. *Ann. N. Y. Acad. Sci.* **2011**, *1218*, 47–54. [CrossRef]
14. Zukerman, R.; Harris, A.; Vercellin, A.V.; Siesky, B.; Pasquale, L.R.; Ciulla, T.A. Molecular Genetics of Glaucoma: Subtype and Ethnicity Considerations. *Genes* **2020**, *12*, 55. [CrossRef]
15. Yang, T.L.; Guo, Y.; Li, J.; Zhang, L.; Shen, H.; Li, S.M.; Li, S.K.; Tian, Q.; Liu, Y.J.; Papasian, C.J.; et al. Gene-gene interaction between RBMS3 and ZNF516 influences bone mineral density. *J. Bone Miner. Res.* **2013**, *28*, 828–837. [CrossRef]
16. Rivadeneira, F.; Styrkársdottir, U.; Estrada, K.; Halldórsson, B.V.; Hsu, Y.H.; Richards, J.B.; Zillikens, M.C.; Kavvoura, F.K.; Amin, N.; Aulchenko, Y.S.; et al. Twenty bone-mineral-density loci identified by large-scale meta-analysis of genome-wide association studies. *Nat. Genet.* **2009**, *41*, 1199–1206. [CrossRef]
17. Nicoletti, P.; Cartsos, V.M.; Palaska, P.K.; Shen, Y.; Floratos, A.; Zavras, A.I. Genomewide pharmacogenetics of bisphosphonate-induced osteonecrosis of the jaw: The role of RBMS3. *Oncologist* **2012**, *17*, 279–287. [CrossRef]
18. Yang, G.; Singh, S.; Chen, Y.; Hamadeh, I.S.; Langaee, T.; McDonough, C.W.; Holliday, L.S.; Lamba, J.K.; Moreb, J.S.; Katz, J.; et al. Pharmacogenomics of osteonecrosis of the jaw. *Bone* **2019**, *124*, 75–82. [CrossRef]
19. Aung, T.; Ozaki, M.; Lee, M.C.; Schlötzer-Schrehardt, U.; Thorleifsson, G.; Mizoguchi, T.; Igo RPJr Haripriya, A.; Williams, S.E.; Astakhov, Y.S.; Orr, A.C.; et al. Genetic association study of exfoliation syndrome identifies a protective rare variant at LOXL1 and five new susceptibility loci. *Nat. Genet.* **2017**, *49*, 993–1004. [CrossRef]
20. Song, I.W.; Chen, H.C.; Lin, Y.F.; Yang, J.H.; Chang, C.C.; Chou, C.T.; Lee, M.M.; Chou, Y.C.; Chen, C.H.; Chen, Y.T.; et al. Identification of susceptibility gene associated with female primary Sjögren's syndrome in Han Chinese by genome-wide association study. *Hum. Genet.* **2016**, *135*, 1287–1294. [CrossRef]
21. Tyler, A.; Mahoney, J.M.; Carter, G.W. Genetic Interactions Affect Lung Function in Patients with Systemic Sclerosis. *G3 Genes Genomes Genet.* **2020**, *10*, 151–163. [CrossRef]
22. Offenbacher, S.; Divaris, K.; Barros, S.P.; Moss, K.L.; Marchesan, J.T.; Morelli, T.; Zhang, S.; Kim, S.; Sun, L.; Beck, J.D.; et al. Genome-wide association study of biologically informed periodontal complex traits offers novel insights into the genetic basis of periodontal disease. *Hum. Mol. Genet.* **2016**, *25*, 2113–2129. [CrossRef]
23. Maranville, J.C.; Baxter, S.S.; Witonsky, D.B.; Chase, M.A.; Di Rienzo, A. Genetic mapping with multiple levels of phenotypic information reveals determinants of lymphocyte glucocorticoid sensitivity. *Am. J. Hum. Genet.* **2013**, *93*, 735–743. [CrossRef]
24. Genome-wide association study of short-acting bronchodilator response identifies novel pharmacogenetic loci in spiromics. *Am. J. Respir. Crit. Care Med.* **2020**, *201*, 1.
25. Aung, T.; Chan, A.S.; Khor, C.C. Genetics of Exfoliation Syndrome. *J. Glaucoma* **2018**, *27*, S12–S14. [CrossRef]
26. Marchetti, L.; Lauria, M.; Caberlotto, L.; Musazzi, L.; Popoli, M.; Mathé, A.A.; Domenici, E.; Carboni, L. Gene expression signature of antidepressant treatment response/non-response in Flinders Sensitive Line rats subjected to maternal separation. *Eur. Neuropsychopharmacol.* **2020**, *31*, 69–85. [CrossRef]
27. Martins-Silva, T.; Salatino-Oliveira, A.; Genro, J.P.; Meyer, F.D.T.; Li, Y.; Rohde, L.A.; Hutz, M.H.; Tovo-Rodrigues, L. Host genetics influences the relationship between the gut microbiome and psychiatric disorders. *Prog. Neuropsychopharmacol. Biol. Psychiatry* **2021**, *106*, 110153. [CrossRef]

28. Bakkar, N.; Kovalik, T.; Lorenzini, I.; Spangler, S.; Lacoste, A.; Sponaugle, K.; Ferrante, P.; Argentinis, E.; Sattler, R.; Bowser, R. Artificial intelligence in neurodegenerative disease research: Use of IBM Watson to identify additional RNA-binding proteins altered in amyotrophic lateral sclerosis. *Acta Neuropathol.* **2018**, *135*, 227–247. [CrossRef]
29. Fröhlich, H.; Kollmeyer, M.L.; Linz, V.C.; Stuhlinger, M.; Groneberg, D.; Reigl, A.; Zizer, E.; Friebe, A.; Niesler, B.; Rappold, G. Gastrointestinal dysfunction in autism displayed by altered motility and achalasia in Foxp1+/- mice. *Proc. Natl. Acad. Sci. USA* **2019**, *116*, 22237–22245. [CrossRef]
30. Werder, R.; Cho, M.H.; Zhou, A.X.; Kotton, D.N.; Wilson, A.A. A CRISPRi Approach to Investigate GWAS Genes in iPS-Derived Alveolar Epithelial Cells. *Am. J. Respir. Crit. Care Med.* **2021**, *203*, A1042. [CrossRef]
31. Carén, H.; Erichsen, J.; Olsson, L.; Enerbäck, C.; Sjöberg, R.M.; Abrahamsson, J.; Kogner, P.; Martinsson, T. High-resolution array copy number analyses for detection of deletion, gain, amplification and copy-neutral LOH in primary neuroblastoma tumors: Four cases of homozygous deletions of the CDKN2A gene. *BMC Genom.* **2008**, *9*, 353. [CrossRef] [PubMed]
32. Wu, Y.; Liu, Z.; Wei, X.; Feng, H.; Hu, B.; Liu, B.; Luan, Y.; Ruan, Y.; Liu, X.; Liu, Z.; et al. Identification of the Functions and Prognostic Values of RNA Binding Proteins in Bladder Cancer. *Front. Genet.* **2021**, *12*, 574196. [CrossRef] [PubMed]
33. Chen, F.; Wang, Q.; Zhou, Y. The construction and validation of an RNA binding protein-related prognostic model for bladder cancer. *BMC Cancer* **2021**, *21*, 244. [CrossRef]
34. Wu, Y.; Meng, D.; You, Y.; Sun, R.; Yan, Q.; Bao, J.; Sun, Y.; Yun, D.; Li, Y.; Sun, D. Increased expression of RBMS3 predicts a favorable prognosis in human gallbladder carcinoma. *Oncol. Rep.* **2020**, *44*, 55–68. [CrossRef]
35. Zhang, Y.; Xu, Y.; Feng, L.; Li, F.; Sun, Z.; Wu, T.; Shi, X.; Li, J.; Li, X. Comprehensive characterization of lncRNA-mRNA related ceRNA network across 12 major cancers. *Oncotarget* **2016**, *7*, 64148–64167. [CrossRef]
36. Jiang, Z.; Zhang, Y.; Chen, X.; Wu, P.; Chen, D. Long noncoding RNA RBMS3-AS3 acts as a microRNA-4534 sponge to inhibit the progression of prostate cancer by upregulating VASH1. *Gene Ther.* **2020**, *27*, 143–156. [CrossRef]
37. Greene, J.; Baird, A.M.; Lim, M.; Flynn, J.; McNevin, C.; Brady, L.; Sheils, O.; Gray, S.G.; McDermott, R.; Finn, S.P. Differential CircRNA Expression Signatures May Serve as Potential Novel Biomarkers in Prostate Cancer. *Front. Cell Dev. Biol.* **2021**, *9*, 605686. [CrossRef]
38. Wu, G.; Cao, L.; Zhu, J.; Tan, Z.; Tang, M.; Li, Z.; Hu, Y.; Yu, R.; Zhang, S.; Song, L.; et al. Loss of RBMS3 Confers Platinum Resistance in Epithelial Ovarian Cancer via Activation of miR-126-5p/β-catenin/CBP signaling. *Clin. Cancer Res.* **2019**, *25*, 1022–1035. [CrossRef]
39. Bi, F.; Chen, Y.; Yang, Q. Significance of tumor mutation burden combined with immune infiltrates in the progression and prognosis of ovarian cancer. *Cancer Cell Int.* **2020**, *20*, 373. [CrossRef]
40. Chen, J.; Kwong, D.L.; Zhu, C.L.; Chen, L.L.; Dong, S.S.; Zhang, L.Y.; Tian, J.; Qi, C.B.; Cao, T.T.; Wong, A.M.; et al. RBMS3 at 3p24 inhibits nasopharyngeal carcinoma development via inhibiting cell proliferation, angiogenesis, and inducing apoptosis. *PLoS ONE* **2012**, *7*, e44636. [CrossRef]
41. Chen, J.; Fu, L.; Zhang, L.Y.; Kwong, D.L.; Yan, L.; Guan, X.Y. Tumor suppressor genes on frequently deleted chromosome 3p in nasopharyngeal carcinoma. *Chin. J. Cancer* **2012**, *31*, 215–222. [CrossRef]
42. Wu, Y.; Yun, D.; Zhao, Y.; Wang, Y.; Sun, R.; Yan, Q.; Zhang, S.; Lu, M.; Zhang, Z.; Lu, D.; et al. Down regulation of RNA binding motif, single-stranded interacting protein 3, along with up regulation of nuclear HIF1A correlates with poor prognosis in patients with gastric cancer. *Oncotarget* **2017**, *8*, 1262–1277. [CrossRef]
43. Ghafouri-Fard, S.; Honarmand Tamizkar, K.; Jamali, E.; Taheri, M.; Ayatollahi, S.A. Contribution of circRNAs in gastric cancer. *Pathol. Res. Pract.* **2021**, *227*, 153640. [CrossRef]
44. Qin, Y.R.; Fu, L.; Sham, P.C.; Kwong, D.L.; Zhu, C.L.; Chu, K.K.; Li, Y.; Guan, X.Y. Single-nucleotide polymorphism-mass array reveals commonly deleted regions at 3p22 and 3p14.2 associate with poor clinical outcome in esophageal squamous cell carcinoma. *Int. J. Cancer* **2008**, *123*, 826–830. [CrossRef]
45. Li, Y.; Chen, L.; Nie, C.J.; Zeng, T.T.; Liu, H.; Mao, X.; Qin, Y.; Zhu, Y.H.; Fu, L.; Guan, X.Y. Downregulation of RBMS3 is associated with poor prognosis in esophageal squamous cell carcinoma. *Cancer Res.* **2011**, *71*, 6106–6115. [CrossRef]
46. Liang, Y.N.; Liu, Y.; Meng, Q.; Li, X.; Wang, F.; Yao, G.; Wang, L.; Fu, S.; Tong, D. RBMS3 is a tumor suppressor gene that acts as a favorable prognostic marker in lung squamous cell carcinoma. *Med. Oncol.* **2015**, *32*, 459. [CrossRef]
47. Zhao, Y.; Wang, H.; Wu, C.; Yan, M.; Wu, H.; Wang, J.; Yang, X.; Shao, Q. Construction and investigation of lncRNA-associated ceRNA regulatory network in papillary thyroid cancer. *Oncol. Rep.* **2018**, *39*, 1197–1206. [CrossRef]
48. Wang, C.; Wu, Y.; Liu, Y.; Pan, F.; Zeng, H.; Li, X.; Yu, L. Tumor Suppressor Effect of RBMS3 in Breast Cancer. *Technol. Cancer Res. Treat.* **2021**, *20*, 15330338211004921. [CrossRef]
49. Dong, S.; Ma, M.; Li, M.; Guo, Y.; Zuo, X.; Gu, X.; Zhang, M.; Shi, Y. LncRNA MEG3 regulates breast cancer proliferation and apoptosis through miR-141-3p/RBMS3 axis. *Genomics* **2021**, *113*, 1689–1704. [CrossRef]
50. Jin, Y.Y.; Chen, Q.J.; Xu, K.; Ren, H.T.; Bao, X.; Ma, Y.N.; Wei, Y.; Ma, H.B. Involvement of microRNA-141-3p in 5-fluorouracil and oxaliplatin chemo-resistance in esophageal cancer cells via regulation of PTEN. *Mol. Cell. Biochem.* **2016**, *422*, 161–170. [CrossRef]
51. Wang, M.; Hu, M.; Li, Z.; Qian, D.; Wang, B.; Liu, D.X. miR-141-3p functions as a tumor suppressor modulating activating transcription factor 5 in glioma. *Biochem. Biophys. Res. Commun.* **2017**, *490*, 1260–1267. [CrossRef] [PubMed]
52. Li, G.; Huang, M.; Cai, Y.; Yang, Y.; Sun, X.; Ke, Y. Circ-U2AF1 promotes human glioma via derepressing neuro-oncological ventral antigen 2 by sponging hsa-miR-7-5p. *J. Cell. Physiol.* **2019**, *234*, 9144–9155. [CrossRef] [PubMed]

53. Li, X.; Ma, C.; Luo, H.; Zhang, J.; Wang, J.; Guo, H. Identification of the differential expression of genes and upstream microRNAs in small cell lung cancer compared with normal lung based on bioinformatics analysis. *Medicine* **2020**, *99*, e19086. [CrossRef] [PubMed]
54. Liu, C.; Zhang, Y.H.; Huang, T.; Cai, Y. Identification of transcription factors that may reprogram lung adenocarcinoma. *Artif. Intell. Med.* **2017**, *83*, 52–57. [CrossRef]
55. Huang, J.; Li, Y.; Lu, Z.; Che, Y.; Sun, S.; Mao, S.; Lei, Y.; Zang, R.; Li, N.; Zheng, S.; et al. Analysis of functional hub genes identifies CDC45 as an oncogene in non-small cell lung cancer—A short report. *Cell. Oncol.* **2019**, *42*, 571–578. [CrossRef]
56. Gan, T.Q.; Tang, R.X.; He, R.Q.; Dang, Y.W.; Xie, Y.; Chen, G. Upregulated MiR-1269 in hepatocellular carcinoma and its clinical significance. *Int. J. Clin. Exp. Med.* **2015**, *8*, 714–721.
57. Uddin, M.N.; Wang, X. Identification of key tumor stroma-associated transcriptional signatures correlated with survival prognosis and tumor progression in breast cancer. *Breast Cancer* **2022**, *29*, 541–561. [CrossRef]
58. Dongre, A.; Weinberg, R.A. New insights into the mechanisms of epithelial-mesenchymal transition and implications for cancer. *Nat. Rev. Mol. Cell. Biol.* **2019**, *20*, 69–84. [CrossRef]
59. Nieto, M.A. Epithelial-Mesenchymal Transitions in development and disease: Old views and new perspectives. *Int. J. Dev. Biol.* **2009**, *53*, 1541–1547. [CrossRef]
60. Thiery, J.P.; Acloque, H.; Huang, R.Y.; Nieto, M.A. Epithelial-mesenchymal transitions in development and disease. *Cell* **2009**, *139*, 871–890. [CrossRef]
61. Lee, J.M.; Dedhar, S.; Kalluri, R.; Thompson, E.W. The epithelial-mesenchymal transition: New insights in signaling, development, and disease. *J. Cell. Biol.* **2006**, *172*, 973–981. [CrossRef]
62. Thiery, J.P. Epithelial-mesenchymal transitions in tumour progression. *Nat. Rev. Cancer* **2002**, *2*, 442–454. [CrossRef]
63. Hay, E.D. An overview of epithelio-mesenchymal transformation. *Acta Anat.* **1995**, *154*, 8–20. [CrossRef]
64. Kalluri, R.; Neilson, E.G. Epithelial-mesenchymal transition and its implications for fibrosis. *J. Clin. Investig.* **2003**, *112*, 1776–1784. [CrossRef]
65. Kalluri, R.; Weinberg, R.A. The basics of epithelial-mesenchymal transition. *J. Clin. Investig.* **2009**, *119*, 1420–1428. [CrossRef]
66. Stone, R.C.; Pastar, I.; Ojeh, N.; Chen, V.; Liu, S.; Garzon, K.I.; Tomic-Canic, M. Epithelial-mesenchymal transition in tissue repair and fibrosis. *Cell Tissue Res.* **2016**, *365*, 495–506. [CrossRef]
67. Pei, D.; Shu, X.; Gassama-Diagne, A.; Thiery, J.P. Mesenchymal-epithelial transition in development and reprogramming. *Nat. Cell Biol.* **2019**, *21*, 44–53. [CrossRef]
68. Cho, E.S.; Kang, H.E.; Kim, N.H.; Yook, J.I. Therapeutic implications of cancer epithelial-mesenchymal transition (EMT). *Arch. Pharm. Res.* **2019**, *42*, 14–24. [CrossRef]
69. Kong, D.; Li, Y.; Wang, Z.; Sarkar, F.H. Cancer Stem Cells and Epithelial-to-Mesenchymal Transition (EMT)-Phenotypic Cells: Are They Cousins or Twins? *Cancers* **2011**, *3*, 716–729. [CrossRef]
70. Gonzalez, D.M.; Medici, D. Signaling mechanisms of the epithelial-mesenchymal transition. *Sci. Signal.* **2014**, *7*, re8. [CrossRef]
71. Ye, X.; Tam, W.L.; Shibue, T.; Kaygusuz, Y.; Reinhardt, F.; Ng Eaton, E.; Weinberg, R.A. Distinct EMT programs control normal mammary stem cells and tumour-initiating cells. *Nature.* **2015**, *525*, 256–260. [CrossRef] [PubMed]
72. Kalluri, R. EMT: When epithelial cells decide to become mesenchymal-like cells. *J. Clin. Investig.* **2009**, *119*, 1417–1419. [CrossRef] [PubMed]
73. Babaei, G.; Aziz, S.G.; Jaghi, N.Z.Z. EMT, cancer stem cells and autophagy; The three main axes of metastasis. *Biomed. Pharmacother.* **2021**, *133*, 110909. [CrossRef] [PubMed]
74. Montanari, M.; Rossetti, S.; Cavaliere, C.; D'Aniello, C.; Malzone, M.G.; Vanacore, D.; Di Franco, R.; La Mantia, E.; Iovane, G.; Piscitelli, R.; et al. Epithelial-mesenchymal transition in prostate cancer: An overview. *Oncotarget* **2017**, *8*, 35376–35389. [CrossRef]
75. Fiori, M.E.; Di Franco, S.; Villanova, L.; Bianca, P.; Stassi, G.; De Maria, R. Cancer-associated fibroblasts as abettors of tumor progression at the crossroads of EMT and therapy resistance. *Mol. Cancer* **2019**, *18*, 70. [CrossRef]
76. Lamouille, S.; Xu, J.; Derynck, R. Molecular mechanisms of epithelial-mesenchymal transition. *Nat. Rev. Mol. Cell. Biol.* **2014**, *15*, 178–196. [CrossRef]
77. Thiery, J.P.; Lim, C.T. Tumor dissemination: An EMT affair. *Cancer Cell.* **2013**, *23*, 272–273. [CrossRef]
78. Nieto, M.A.; Huang, R.Y.; Jackson, R.A.; Thiery, J.P. EMT: 2016. *Cell* **2016**, *166*, 21–45. [CrossRef]
79. Lei, Y.; Chen, L.; Zhang, G.; Shan, A.; Ye, C.; Liang, B.; Sun, J.; Liao, X.; Zhu, C.; Chen, Y.; et al. MicroRNAs target the Wnt/β-catenin signaling pathway to regulate epithelial-mesenchymal transition in cancer (Review). *Oncol. Rep.* **2020**, *44*, 1299–1313. [CrossRef]
80. Block, C.J.; Mitchell, A.V.; Wu, L.; Glassbrook, J.; Craig, D.; Chen, W.; Dyson, G.; DeGracia, D.; Polin, L.; Ratnam, M.; et al. RNA binding protein RBMS3 is a common EMT effector that modulates triple-negative breast cancer progression via stabilizing PRRX1 mRNA. *Oncogene* **2021**, *40*, 6430–6442. [CrossRef]
81. Zhao, P.; Liu, D.; Zhang, H.; Song, Y. RBMS3 inhibits invasion and epithelial-mesenchymal transition of gastric cancer cells via regulating Wnt/β-catenin signal pathway. *Tumor* **2017**, *37*, 1032–1040. [CrossRef]

International Journal of *Molecular Sciences*

*Article*

# Mass Spectrometric Metabolic Fingerprinting of 2-Deoxy-D-Glucose (2-DG)-Induced Inhibition of Glycolysis and Comparative Analysis of Methionine Restriction versus Glucose Restriction under Perfusion Culture in the Murine L929 Model System

**Julian Manuel Volland [1,†], Johannes Kaupp [1,†], Werner Schmitz [2], Anna Chiara Wünsch [1], Julia Balint [1], Marc Möllmann [3], Mohamed El-Mesery [4], Kyra Frackmann [1], Leslie Peter [1], Stefan Hartmann [1], Alexander Christian Kübler [1] and Axel Seher [1,*]**

1 Department of Oral and Maxillofacial Plastic Surgery, University Hospital Wuerzburg, D-97070 Wuerzburg, Germany
2 Department of Biochemistry and Molecular Biology, Biocenter, D-97074 Wuerzburg, Germany
3 Fraunhofer ISC, Translational Center RT, D-97070 Wuerzburg, Germany
4 Department of Biochemistry, Faculty of Pharmacy, Mansoura University, Mansoura 35516, Egypt
* Correspondence: seher_a@ukw.de; Tel.: +49-931-201-74841
† These authors contributed equally to this work.

**Citation:** Volland, J.M.; Kaupp, J.; Schmitz, W.; Wünsch, A.C.; Balint, J.; Möllmann, M.; El-Mesery, M.; Frackmann, K.; Peter, L.; Hartmann, S.; et al. Mass Spectrometric Metabolic Fingerprinting of 2-Deoxy-D-Glucose (2-DG)-Induced Inhibition of Glycolysis and Comparative Analysis of Methionine Restriction versus Glucose Restriction under Perfusion Culture in the Murine L929 Model System. *Int. J. Mol. Sci.* **2022**, 23, 9220. https://doi.org/10.3390/ijms23169220

Academic Editor: Laura Paleari

Received: 15 July 2022
Accepted: 14 August 2022
Published: 16 August 2022



**Abstract:** All forms of restriction, from caloric to amino acid to glucose restriction, have been established in recent years as therapeutic options for various diseases, including cancer. However, usually there is no direct comparison between the different restriction forms. Additionally, many cell culture experiments take place under static conditions. In this work, we used a closed perfusion culture in murine L929 cells over a period of 7 days to compare methionine restriction (MetR) and glucose restriction (LowCarb) in the same system and analysed the metabolome by liquid chromatography mass spectrometry (LC-MS). In addition, we analysed the inhibition of glycolysis by 2-deoxy-D-glucose (2-DG) over a period of 72 h. 2-DG induced very fast a low-energy situation by a reduced glycolysis metabolite flow rate resulting in pyruvate, lactate, and ATP depletion. Under perfusion culture, both MetR and LowCarb were established on the metabolic level. Interestingly, over the period of 7 days, the metabolome of MetR and LowCarb showed more similarities than differences. This leads to the conclusion that the conditioned medium, in addition to the different restriction forms, substantially reprogramm the cells on the metabolic level.

**Keywords:** amino acid restriction; glucose restriction; mass spectrometry; low carb; 2-deoxy-D-glucose; 2-DG; methionine; perfusion culture; energy restriction; caloric restriction

## 1. Introduction

A cell needs two basic elements—mass and energy. The mass consists mainly of amino acids. They enable basic structures and functions in the form of proteins and enzymes and constitute the largest component of a cell's mass [1]. For energy, the most important representatives are ATP and NAD(P)H, which are essentially obtained from carbohydrates and lipids. Glycolysis, and with it glucose, is established as one of the oldest reaction pathways for energy production, with the help of which ATP and NADH can be produced very fast [2]. For this reason, and due to its high water solubility, glucose is the central energy resource for the cells of an organism.

Mass and energy are essential for the maintenance of the cell and, to an even greater extent, for proliferation. In the case of neoplastic cells, defined by space-occupying proliferation that is almost unlimited, they, therefore, play an even greater role. For this reason, different forms of restriction are potential strategies for tumour therapy. What they all have in common is that a resource is limited to such an extent that it is no longer sufficiently available to supply the tumour cells. In the case of glucose, the approach is based on the so-called Warburg effect, which was described by Warburg and colleagues in 1924 [3]. They were able to show that tumour cells take up more glucose under aerobic conditions and metabolise it to lactate. This metabolic pathway is normally used only under an oxygen deficiency, i.e., under anaerobic conditions. At first glance, this seems inefficient, since glycolysis provides only two ATP as a net gain, while complete degradation to $CO_2$ provides another 32 ATP. Although the causes of the Warburg effect are not yet understood in detail, there are a number of possibilities. For example, aerobic glycolysis enables the very rapid synthesis of ATP to maintain a high ATP level, increased synthesis of NADPH, which is increasingly needed for the biosynthesis of lipids, among other molecules, and the acidification of the tumour microenvironment, which can suppress the immune response to the tumour [4]. It is now known that the Warburg effect is generally characteristic of proliferating cells, but it plays a quantitatively and qualitatively greater role in neoplastic cells due to the nonlimited proliferation. For this reason, glucose restriction, which prevents the tumour from using glucose in any way, offers a fundamental approach to tumour therapy [5].

However, the possibilities for attacking tumours with the help of restriction have expanded considerably in the past 10 years. Although glucose is one of the central molecules in metabolism, especially in tumours, the restriction of energy and/or mass also offers a very good approach. Energy can be reduced with the help of calorie restriction, and mass can be reduced by amino acid or protein restriction [6–9]. Central to all forms of restriction is the induction of a low-energy metabolism (LEM), which is significantly mediated by the protein complex mTOR. The intracellular activity of the mTOR determines proliferation and growth. A multitude of sensors pass information about the energy and mass content to the mTOR switch point. For example, the energy equivalents ATP and NADH can be measured via the protein AMPK (5′ AMP-activated protein kinase) and with the help of sirtuins [10]. A lack of energy leads to the inhibition of the mTOR via the two proteins mentioned and results in a halt to proliferation and the induction of autophagy for the purpose of energy recycling. The content of selected amino acids is measured with the help of protein complexes; in the case of methionine, SAMTOR plays a decisive role. Here, methionine is not measured directly, but an intermediate product, S-adenosylmethionine (SAM), whose content is an indicator of the intracellular methionine concentration. However, a low methionine concentration also leads to the inhibition of the mTOR [11].

In the long term, all forms of restriction lead to astonishing results in almost all organism types, from yeast to nematodes to Drosophila to humans. In addition to the extension of the absolute lifespan, the prevention of heart and circulatory diseases, type II diabetes and cancer are prominent [7,12,13]. The restriction of glucose in the form of a reduced increase (low carb) or as absolute an avoidance as possible (ketogenic) results in similar mechanisms to calorie restriction [14].

All forms of restriction have much in common, as the reactions either use basic mechanisms of the cell or rely on evolutionarily conserved mechanisms. However, there will be equally striking differences due to the complexity. Especially with regard to the evaluation of the advantages and disadvantages of individual forms of restriction, the level of scientific knowledge is very low. This is mainly because scientific working groups usually focus only on one particular form of restriction, and thus, it is very difficult to compare the individual forms of restriction. In the case of glucose, this is further complicated by the fact that many studies do not work with a constant glucose concentration. Physiologically, in mammals, the glucose concentration in the blood is kept at a certain threshold value, which it does not fall below—in humans, approximately 3 mM, and in mice, approximately 6 mM [15]. Many experiments with glucose carried out in cell culture start with a concentration that is in the low-carb range. However, this value is not static, but continues to drop over the course of the experiment. This can lead to inconsistent results that are difficult to interpret.

Experiments cannot always be carried out over a longer period of time in this way, as cells undergo apoptosis or simply die if the glucose deficiency remains below the physiological threshold [16].

The aim of our research group is to enable the comparison of individual restriction forms at the molecular level. First, we use the murine model system L929, which enables the comparison of different restrictions in one and the same system [17–19]. Second, we use closed, circular perfusion cultures that allow metabolite concentrations to be kept constant over longer periods of time. In this work, we used liquid chromatography mass spectrometry (LC-MS) to analyse the effect of the glycolysis inhibitor 2-deoxy-glucose (2-DG) at time points 2, 5, 8, 24, 48, and 72 h under proliferative conditions in our L929 model system. In addition, we compared methionine restriction (MetR) with glucose restriction under LowCarb conditions (3 mM) in a perfusion culture over a period of 7 days.

## 2. Results

In previous work, we already used the murine cell line L929 as a model system for the analysis of different forms of amino acid restriction [17–19]. The use of murine cells has several advantages. First, many studies in the field of restriction have been conducted in rodent models, and the mouse has been established as a model system for the study of energy metabolism [20]. Second, murine metabolism is up to 100× faster than human metabolism [21–23]. Differences in metabolism can be better analysed. Third, many of the metabolic pathways are strongly conserved in evolution. Especially in the area of restriction, many similarities can be seen from yeast to nematodes, Drosophila, rodents, primates, and humans [6,9]. mTOR and sirtuins are just two examples of these strongly conserved mechanisms [24,25].

### 2.1. Cell Division Frequency and Methionine Sensitivity of L929 Cells

Figure 1 shows the proliferation behaviour of L929 cells over a period of 120 h (Figure 1a). As described in the Materials and Methods, one can calculate the cell division frequency (f), i.e., how long a cell population takes to divide once. Figure 1b shows the values from Figure 1a, which reveal exponential growth and were used for the determination of f. For the period from 24 h to 72 h, the cell division frequency was 0.99, i.e., the cell number doubled approximately once per day. L929 thus showed very rapid growth in cell culture. The interval 72/80 was not included in the evaluation because the 80 h value no longer showed exponential growth, which can also be clearly seen in the cell division frequency of f = 0.39 for this interval (Figure 1b). The response rate to MetR was also very good (Figure 1c). Proliferation was already significantly reduced after 24 h and did not increase further during the rest of the study period.

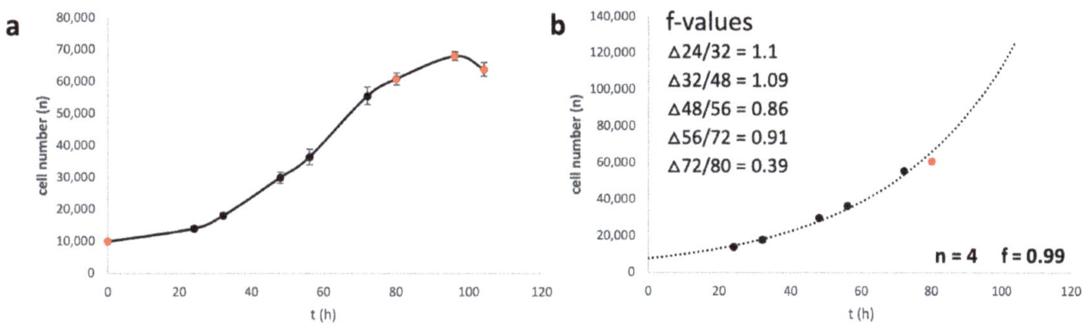


**Figure 1.** *Cont.*

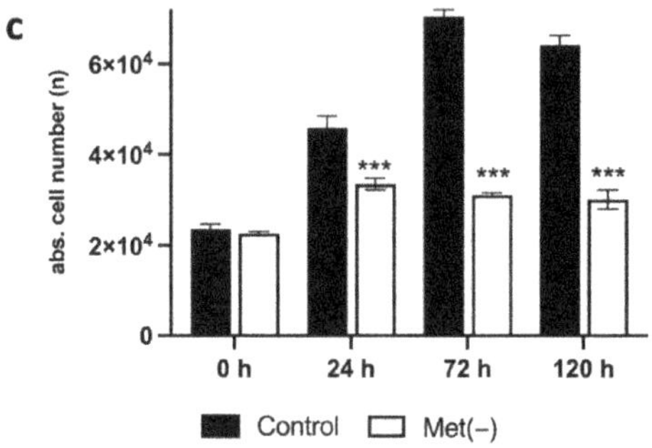


**Figure 1.** (**a**–**c**) Proliferation analysis of L929 cells. A total of 10,000 cells were seeded per well and incubated overnight. (**a**) Absolute cell numbers were analysed via ImageXpress digital microscopy as described in the Materials and Methods. The figures show a summary of the results obtained from four independent experiments (six values for every group per experiment); (**b**) trend curve of the values from figure (**a**) showing exponential growth. Values that do not represent exponential growth are shown as red dots; (**c**) L929 cells were stimulated for 24, 72, and 120 h with or without methionine. Cell proliferation was analysed via ImageXpress digital microscopy analysis as described in the Materials and Methods. The figure shows one representative experiment (five values for every group), as published previously [17]. Comparisons between the control and Met(−) groups were performed by applying one-way ANOVA followed by the Tukey–Kramer multiple comparison test (*** $p < 0.001$).

### 2.2. Glucose Restriction and 2-DG Are Strong Inhibitors of Proliferation in L929 Cells

The main aim of this work was to establish and analyse glucose restriction in the model system L929. In the first experiment, the glucose sensitivity of the cell line and the response to the glycolysis inhibitor 2-deoxy-D-glucose (2-DG) were tested. Over a period of 96 h, log2 dilution series were generated, and absolute cell numbers were obtained every 24 h using automated digital microscopy (Pico). The results are summarised in Figure 2.

At low glucose concentrations, the proliferation of L929 cells is strongly restricted (Figure 2a). In the range of 3 mM, the proliferation rate increases strongly. The concentration of 11 mM corresponds to the standard concentration of the RPMI medium used and was, therefore, chosen as the starting concentration.

The glycolysis inhibitor 2-DG is taken up via glucose transporters and can be phosphorylated by hexokinase to 2-deoxy-glucose-6-phosphate, which cannot be further processed by glucose-6-phosphate isomerase and competitively inhibits it. In addition, hexokinase is not competitively inhibited by 2-deoxy-glucose-6-phosphate, which means that less glucose-6-phosphate is already being formed [26]. As a result, the cellular concentration of ATP decreases. In addition, 2-DG increases oxidative stress, inhibits N-linked glycosylation, and induces autophagy. It can efficiently slow cell growth and potently facilitate apoptosis in specific cancer cells. Although 2-DG itself has limited therapeutic effects in many types of cancers, it may be combined with other therapeutic agents or radiotherapy to exert a synergistic anticancer effect [27]. In the L929 cell line, 2-DG showed a clear antiproliferative effect at low concentrations (Figure 2b). Up to a concentration of 1.3 mM, the proliferation rate decreased drastically. At further increasing concentrations up to 20 mM, the cell number decreased only slightly.

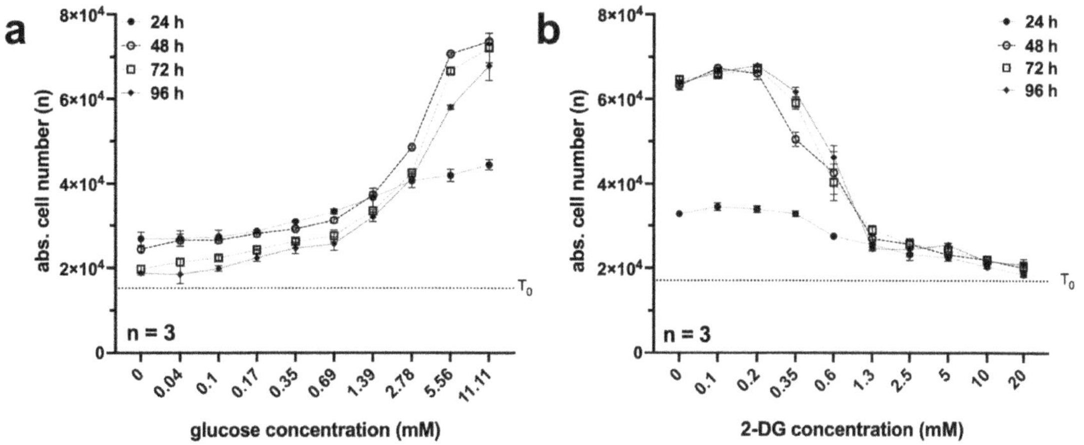


**Figure 2.** (**a**,**b**) Analysis of proliferation under glucose restriction or with the 2-DG glycolysis inhibitor in L929 cells. A total of $1 \times 10^4$ cells per well were seeded on 96-well plates. On the following day, the cells were incubated in triplicate with (**a**) 11 mM glucose or (**b**) 20 mM 2-DG with a log(2)-dilution series. After 24, 48, 72, and 96 h, absolute cell numbers (n) were analysed via automated digital microscopy (Pico). The results and standard deviation of three independent experiments are shown. The cell number ($T_0$) at the beginning of the experiment was also determined and is shown as a dashed line in the diagram.

### *2.3. 2-DG Works in L929 Cells at the Metabolic Level as an Energy Restriction Mimetic*

In a previous work, we showed that amino acid restriction by methionine in L929 cells can induce low energy metabolism (LEM) [18]. The effect of the inhibition of glycolysis by 2-DG on the metabolome should now be analysed by liquid chromatography mass spectrometry (LC-MS). It is known that the inhibition of glycolytic processes by 2-DG is followed by a limitation of the cellular energy supply; ATP levels are reduced, and the AMP/ATP ratio is subsequently increased, leading to AMPK activation, which increases the NAD+/NADH ratio and activates SIRT1. Indeed, in rodents, adding 2-DG to the diet led to phenotypes including decreased body weight, blood glucose, insulin, body temperature, and heart rate. Moreover, 2DG was shown to induce protection against oxidative stress (overview in [28]). Last but not least, 2-DG also seems to have the potential of a caloric restriction mimetic (CRM), a substance that can induce the same effects in organisms as caloric or amino acid/protein restriction [29].

The L929 cells were analysed over a period of 72 h. Since 2-DG inhibits glycolysis very rapidly, we selected three early time points (2, 5, and 8 h) and three mid-term time points (24, 48, and 72 h). Cells were stimulated with 625 μM 2-DG. We decided on this concentration based on the results of the 2-DG experiments (Figure 2b). At this concentration, a strong antiproliferative effect can already be seen, but the inhibition is not so strong that proliferation is completely prevented, and the cells may run the risk of dying within the experimental period due to the lack of ATP. Figure 3a–c shows selected liquid chromatography mass spectrometry results summarised in groups. The complete LC-MS results, including the individual measurements, raw data, and standard deviation, are provided as the Supplementary File S1.

As previously described, the effectiveness of the inhibitor 2-DG should be evident from the metabolites of glycolysis as well as at the level of the energy currencies. 2-DG was already effective after 2 h, and glycolysis was inhibited in the first steps (Figure 3a). Both the phosphorylated hexoses and the aldohexoses accumulated over the entire period of the experiment. In addition, the throughput rate of the subsequent metabolites up to pyruvate decreased drastically, which can be seen particularly well in the metabolites diphosphoglycerate, 3-phosphoglycerate, and phosphoenolpyruvate after 48 h and 72 h. The lactate value also decreased significantly to 51%.

**a**

| | | Control 2 h | 5 h | 8 h | 24 h | 48 h | 72 h | 2-DG 2 h | 5 h | 8 h | 24 h | 48 h | 72 h |
|---|---|---|---|---|---|---|---|---|---|---|---|---|---|
| Glycolysis | Aldohexose | 41 | 45 | 34 | 52 | 64 | 82 | 83 | 100 | 70 | 63 | 95 | 87 |
| | Hexose-P | 8 | 8 | 7 | 10 | 10 | 8 | 50 | 58 | 71 | 93 | 100 | 78 |
| | Fructose-1,6-P2 | 45 | 79 | 83 | 75 | 65 | 42 | 100 | 80 | 71 | 99 | 50 | 45 |
| | Glyceral/-on-P | 76 | 100 | 61 | 83 | 65 | 33 | 33 | 47 | 39 | 36 | 23 | 26 |
| | Diphosphoglycerate | 29 | 92 | 94 | 100 | 58 | 46 | 54 | 39 | 36 | 46 | 2 | 4 |
| | 3-Phosphoglycerate | 59 | 73 | 77 | 63 | 55 | 71 | 100 | 39 | 34 | 80 | 15 | 15 |
| | Phosphoenolpyruvate | 85 | 100 | 85 | 90 | 80 | 79 | 63 | 51 | 39 | 45 | 15 | 18 |
| | Pyruvate | 88 | 84 | 100 | 98 | 71 | 57 | 76 | 62 | 66 | 78 | 53 | 37 |
| | Lactate | 60 | 70 | 67 | 88 | 100 | 79 | 68 | 68 | 61 | 62 | 54 | 51 |
| Energy | AMP | 72 | 71 | 61 | 70 | 82 | 72 | 87 | 99 | 78 | 82 | 100 | 96 |
| | ADP | 80 | 81 | 75 | 71 | 97 | 47 | 92 | 100 | 68 | 63 | 33 | 42 |
| | ATP | 79 | 96 | 100 | 46 | 98 | 21 | 69 | 75 | 40 | 14 | 0 | 1 |
| | NAD+ | 61 | 66 | 57 | 35 | 40 | 48 | 73 | 84 | 88 | 97 | 100 | 86 |
| | NADP+ | 75 | 86 | 90 | 79 | 69 | 68 | 87 | 74 | 74 | 93 | 100 | 95 |
| | Creatine | 70 | 61 | 88 | 63 | 39 | 54 | 68 | 64 | 87 | 100 | 65 | 86 |
| | Phosphocreatine | 94 | 85 | 87 | 49 | 27 | 54 | 81 | 87 | 98 | 40 | 52 | 100 |
| | Creatinine | 79 | 74 | 73 | 70 | 72 | 86 | 100 | 94 | 80 | 81 | 77 | 90 |

**b**

| | | Control 2 h | 5 h | 8 h | 24 h | 48 h | 72 h | 2-DG 2 h | 5 h | 8 h | 24 h | 48 h | 72 h |
|---|---|---|---|---|---|---|---|---|---|---|---|---|---|
| AAs | Cysteine | 8 | 13 | 21 | 15 | 32 | 49 | 11 | 18 | 21 | 39 | 82 | 100 |
| | Glutamine | 43 | 43 | 51 | 85 | 85 | 100 | 69 | 71 | 67 | 37 | 41 | 59 |
| Glu-AAs | GluAla | 56 | 50 | 47 | 44 | 51 | 68 | 51 | 42 | 48 | 42 | 100 | 100 |
| | GluArg | 18 | 20 | 22 | 30 | 40 | 51 | 24 | 26 | 26 | 32 | 78 | 100 |
| | GluAsn | 37 | 39 | 43 | 62 | 67 | 63 | 43 | 40 | 53 | 45 | 88 | 100 |
| | GluAsp | 32 | 32 | 41 | 34 | 28 | 46 | 39 | 35 | 39 | 55 | 100 | 95 |
| | GluCys | 60 | 65 | 82 | 46 | 87 | 100 | 61 | 70 | 64 | 51 | 66 | 91 |
| | GluGln | 40 | 42 | 39 | 64 | 81 | 100 | 47 | 50 | 52 | 36 | 57 | 97 |
| | GluGlu | 34 | 43 | 45 | 48 | 60 | 67 | 35 | 42 | 49 | 51 | 91 | 100 |
| | GluGly | 33 | 35 | 34 | 41 | 58 | 69 | 27 | 28 | 33 | 39 | 100 | 98 |
| | GluHis | 17 | 22 | 27 | 41 | 64 | 75 | 24 | 21 | 27 | 31 | 73 | 100 |
| | GluIle | 59 | 55 | 50 | 58 | 59 | 51 | 48 | 46 | 53 | 47 | 83 | 100 |
| | GluLeu | 26 | 26 | 32 | 38 | 44 | 42 | 34 | 32 | 39 | 38 | 77 | 100 |
| | GluLys | 25 | 30 | 33 | 46 | 48 | 51 | 38 | 36 | 41 | 41 | 71 | 100 |
| | GluMet | 11 | 10 | 12 | 11 | 31 | 53 | 16 | 16 | 15 | 16 | 59 | 100 |
| | GluPhe | 20 | 24 | 29 | 33 | 37 | 37 | 21 | 15 | 22 | 21 | 66 | 100 |
| | GluPro | 100 | 92 | 95 | 89 | 84 | 86 | 100 | 77 | 71 | 71 | 75 | 70 |
| | GluSer | 42 | 49 | 47 | 65 | 56 | 63 | 43 | 45 | 46 | 47 | 77 | 100 |
| | GluThr | 56 | 50 | 46 | 62 | 68 | 67 | 51 | 44 | 45 | 33 | 78 | 100 |
| | GluTyr | 16 | 17 | 27 | 38 | 52 | 63 | 20 | 14 | 21 | 28 | 78 | 100 |
| | GluVal | 38 | 36 | 44 | 46 | 49 | 51 | 58 | 40 | 41 | 39 | 70 | 100 |

**c**

| | | Control 2 h | 5 h | 8 h | 24 h | 48 h | 72 h | 2-DG 2 h | 5 h | 8 h | 24 h | 48 h | 72 h |
|---|---|---|---|---|---|---|---|---|---|---|---|---|---|
| EAs & Cholines | Phosphoethanolamine | 37 | 41 | 40 | 14 | 5 | 5 | 53 | 50 | 61 | 63 | 63 | 100 |
| | CDP-Ethanolamine | 100 | 95 | 93 | 52 | 39 | 33 | 100 | 86 | 93 | 73 | 60 | 77 |
| | Choline | 87 | 86 | 69 | 55 | 72 | 87 | 87 | 89 | 88 | 92 | 100 | 98 |
| | Acetylcholine | 18 | 18 | 24 | 25 | 32 | 61 | 26 | 26 | 31 | 39 | 59 | 100 |
| | Phosphocholine | 64 | 67 | 85 | 63 | 35 | 8 | 82 | 85 | 99 | 100 | 64 | 68 |
| | Glycerophosphocholine | 42 | 46 | 59 | 73 | 49 | 38 | 56 | 48 | 57 | 100 | 73 | 81 |
| Carnitines | Carnitine | 59 | 56 | 75 | 65 | 66 | 100 | 75 | 66 | 81 | 80 | 75 | 88 |
| | AC-(02:0) | 28 | 29 | 16 | 7 | 36 | 67 | 15 | 21 | 21 | 30 | 68 | 100 |
| | AC-(03:0) | 48 | 43 | 34 | 37 | 63 | 100 | 72 | 78 | 52 | 36 | 59 | 70 |
| | AC-(04:0) | 36 | 38 | 29 | 38 | 49 | 100 | 48 | 50 | 37 | 27 | 29 | 50 |
| | AC-(05:0) | 13 | 12 | 13 | 64 | 53 | 100 | 13 | 11 | 14 | 21 | 24 | 39 |
| | AC-(05:1) | 35 | 39 | 25 | 66 | 100 | 87 | 29 | 35 | 26 | 27 | 33 | 26 |
| | AC-(06:0) | 91 | 94 | 66 | 44 | 88 | 49 | 100 | 99 | 74 | 51 | 32 | 63 |
| | AC-(08:0) | 63 | 42 | 45 | 29 | 100 | 19 | 63 | 68 | 49 | 13 | 2 | 5 |

| 0 | 10 | 20 | 30 | 40 | 50 | 60 | 70 | 80 | 90 | 100 |
|---|---|---|---|---|---|---|---|---|---|---|

**Figure 3.** (**a**–**c**) Overview of metabolic classes and metabolic groups between control and 2-DG-stimulated cells. The metabolism of the murine cell line L929 was analysed via liquid chromatography mass spectrometry in complete medium (control) and with 625 µM 2-DG. For each time point of the

experiment, the preparation was performed in triplicate. After 2 h, 5 h, 8 h, 24 h, 48 h, and 72 h, the cell lysates (intracellular) were analysed by LC-MS. The results were reproduced in three independent experiments and finally summarised. This figure shows the results of selected classes of substances and metabolic pathways for (**a**) glycolysis and energy, (**b**) amino acids (AAs), glutamine-linked amino acids (Glu-AAs), and (**c**) ethanolamines (EAs), cholines, and carnitines. For the cell pellets, the highest measured value in each test series was defined as 100%. The colour scaling is shown below the results as a legend.

The efficacy of 2-DG can also be demonstrated using different energy currencies. The AMP level was continuously high over the entire period. The ATP level collapsed after 8 h at the latest and dropped continuously to a value of approximately 1%. After 72 h, the AMP/ATP ratio, a critical parameter for the energetic state of a cell, was 1/96. In addition, the NAD(+) and NADP(+) levels remained continuously high. The levels of creatine and phosphocreatine were significantly higher than those in the control over the entire period. Phosphocreatine can be used as an additional phosphate buffer in the case of ATP deficiency, which is a classic physiological situation in the case of heavy stress within muscles.

It is noticeable that most amino acids do not show any significant difference. Exceptions are the two amino acids cysteine and glutamine (Figure 3b). Glutamine reached the maximum value of 100% after 72 h in the control and cysteine after 72 h under 2-DG.

A much clearer picture emerges with the glutamate-coupled gamma-glutamyl amino acids (Glu-Aas) (Figure 3b). These are dipeptides formed extracellularly from glutamate and another amino acid (AA), which can then be imported into the cell via different transporters. The mechanism of Glu-AA formation has become controversial. According to the classical thesis, this happens by means of the glutathione cycle. Glutathione is exported and hydrolysed, and the resulting glutamate is transferred to a free amino acid by a transferase. This produces Glu-AA and Cys-Gly. These dipeptides can then be taken up by the cell and hydrolysed to free amino acids. New glutathione is then formed from glutamate, cysteine, and glycine, and the next cycle can start [30]. However, the role of glutathione in this cycle is now critically discussed [31]. Nevertheless, Glu-AAs remain extracellularly formed dipeptides, which essentially serve to import the amino acid supply. The metabolic profile clearly shows that in the control, the two valuable resources glutamine and cysteine are imported in the form of GluGln and GluCys, whereas in the control, apart from proline (70%), all other amino acids are imported up to the maximum value of 100%.

A significant difference can also be seen in the fatty acids (Figure 3c). In the group of ethanolamines and cholines (EAs and cholines), the precursors for phospholipids and sphingomyelin, the total amount under 2-DG increases steadily after 24 h and reaches 100% for phosphocholines and glycerophosphocholines after 24 h, cholines after 48 h, and acetylcholines and phospoethanolamines after 72 h.

Basically, the analysis of the metabolome under 2-DG shows the effective and efficient blocking of glycolysis in L929 cells and the normal, more or less expected reaction of this system with regard to the inhibition of glycolysis.

### 2.4. Comparative Analysis of MetR and LowCarb Treatments of L929 in Perfusion Culture

In the next step, MetR and LowCarb were compared in the first experimental setup in the model system L929 in otherwise identical conditions. For this purpose, cells were confluently seeded in a 35 mm Petri dish, further cultured for 48 h, and then incubated in a closed perfusion system with complete medium, Met(−) (0 mg/L) or LowCarb (3 mM glucose) for 7 days in a bioreactor at 37 degrees Celsius and 5% $CO_2$. Subsequently, for LC-MS analysis, 1,000,000 cells were harvested in each case, and 1 mL was taken from the respective depot media bottle of the perfusion culture. The experiment was repeated four times, and the results were summarised. Only selected results are presented below. A presentation of the entire results, including the raw data, individual measurements and the standard deviations, is attached as a Supplementary File S2.

Perfusion culture offers great advantages. Cultivation is possible over a much longer period of time, as the nutrients are permanently supplied fresh (open perfusion) or kept constantly high (closed perfusion). Many of the results within the restriction field come from animal experiments conducted over weeks, months, and sometimes even years [7,8]. Although it is possible to analyse the early processes very well at the molecular level in simple cell culture, many mechanisms of restriction are based on longer-lasting or even permanent incubation. In addition to the time factor, however, the maintenance of the concentration of metabolites plays a decisive role. This is very well illustrated by the example of glucose. A certain glucose level is essential both in the body (blood glucose level) and within the cell culture. If the level is maintained below this threshold over a long period of time, cells undergo apoptosis [32]. In a static cell culture, the initial glucose concentration is in the low-carb range and then decreases with an increasing duration of the experiment and eventually becomes critical. In a perfusion culture, the necessary metabolite concentration can be maintained over a long period of time. In this experiment, we decided to use a closed perfusion culture. The open system offers the advantage that the cells are permanently supplied with fresh medium and the concentrations of more or less all metabolites remain constant. The closed system, on the other hand, offers the advantage that secreted metabolites accumulate and finally reach concentrations above the detection limit. In addition, communication among cells also plays a major role. Messenger substances such as growth factors and interleukins, but also simple metabolites such as lactate condition the medium and can play a decisive role. In the case of MetR, these substances include polyamines, specifically spermine or spermidine, which can be advantageous for the conversion of an amino acid restriction (AAR) [33].

One of the most striking results of comparing MetR and LowCarb in perfusion culture was that the similarities of the metabolic profile were much greater than the significant differences. Of over 170 metabolites analysed both intracellularly and extracellularly, many showed similar tendencies. This is exemplified by the essential and nonessential amino acids (Figure 4a). Most of the amino acids were at a similar level in both restriction conditions and differed only slightly even from the control, usually increasing by a few percent or between 20% and 40%. Methionine, as expected, was drastically reduced in MetR (5%), but also dropped to 73% in LowCarb. The most significant difference compared to the control was seen with cysteine, which dropped to 33% and 27%, respectively. Based on the amino acids in the supernatant, it can be seen that a large proportion of the amino acids were absorbed from the medium under each type of cultivation. The exception here was cysteine, which was secreted in large amounts. Under MetR, serine (244%), glutamine (177%), glycine (153%), alanine (151%), and threonine (150%) were also secreted to a greater extent. This was also reflected in the very high number of Glu-AAs "activated" for intracellular transport in the medium (Figure 4b). Intracellularly, the level in both restriction conditions was again quite similar, and a large proportion of the amino acids were imported more strongly than in the control. Exceptions were the GluAA of the amino acids alanine, arginine, asparagine, and, at approximately 10% less, glutamine. The lack of GluMet (0%) provided very good evidence of the efficiency of MetR. Another conspicuous feature was the acetylated cysteine (Ac-Cys) under MetR, which occurred more frequently (229%) than under the other two culture conditions.

| | Class | Name | Cells | | Supernatant | | |
|---|---|---|---|---|---|---|---|
| | | | MetR | LowCarb | Ctrl. | MetR | LowCarb |
| **a** | non-ess. AA | Ala | 108 | 133 | 101 | 151 | 95 |
| | | Arg | 115 | 111 | 20 | 18 | 15 |
| | | Asn | 110 | 121 | 25 | 57 | 26 |
| | | Asp | 110 | 114 | 53 | 102 | 51 |
| | | Cys | 33 | 27 | 1.841 | 3.890 | 2.036 |
| | | Glu | 106 | 114 | 99 | 112 | 106 |
| | | Gln | 113 | 110 | 136 | 177 | 117 |
| | | Gly | 103 | 137 | 101 | 153 | 107 |
| | | Pro | 105 | 88 | 48 | 67 | 46 |
| | | Ser | 106 | 113 | 94 | 244 | 79 |
| | | Tyr | 105 | 85 | 43 | 44 | 41 |
| | ess. AA | His | 124 | 143 | 30 | 44 | 22 |
| | | Ile | 103 | 102 | 33 | 42 | 33 |
| | | Leu | 111 | 108 | 36 | 45 | 34 |
| | | Lys | 130 | 128 | 34 | 34 | 23 |
| | | Met | 5 | 73 | 31 | 10 | 33 |
| | | Phe | 107 | 109 | 44 | 56 | 42 |
| | | Thr | 105 | 119 | 94 | 150 | 93 |
| | | Trp | 109 | 117 | 44 | 59 | 44 |
| | | Val | 103 | 100 | 49 | 53 | 46 |
| **b** | Ac-AA | Ac-Cys | 229 | 86 | 900 | 1.203 | 1.106 |
| | Glu-AA | GluAla | 94 | 96 | 1.491 | 2.227 | 1.233 |
| | | GluArg | 64 | 76 | 311 | 228 | 271 |
| | | GluAsn | 94 | 56 | 1.084 | 1.821 | 1.057 |
| | | GluAsp | 315 | 336 | 4.780 | 10.620 | 5.467 |
| | | GluGln | 89 | 91 | 856 | 1.078 | 655 |
| | | GluGlu | 112 | 152 | 7.085 | 10.482 | 9.231 |
| | | GluGly | 110 | 175 | 1.881 | 2.165 | 1.954 |
| | | GluHis | 163 | 255 | 1.713 | 1.494 | 1.415 |
| | | GluIle | 105 | 106 | 194 | 216 | 168 |
| | | GluLeu | 114 | 112 | 318 | 416 | 338 |
| | | GluLys | 148 | 131 | 759 | 766 | 562 |
| | | GluMet | 0 | 116 | 1.034 | 167 | 1.026 |
| | | GluPro | 104 | 98 | 293 | 256 | 155 |
| | | GluSer | 139 | 107 | 852 | 1.814 | 784 |
| | | GluThr | 125 | 115 | 742 | 754 | 654 |
| | | GluVal | 110 | 114 | 78 | 105 | 68 |
| **c** | "S"-Metabs | Cystine | 119 | 114 | 4 | 3 | 6 |
| | | Homocysteine | 107 | 92 | 5 | 4 | 8 |
| | | Homocystine | 113 | 26 | 0 | 0 | 0 |
| | | Cysteine-S-sulfate | 102 | 102 | 59 | 48 | 82 |
| | | Methionine sulfoxide | 19 | 97 | 97 | 33 | 147 |
| | | Cystathionine | 106 | 118 | 3 | 1 | 5 |
| | | GSH | 109 | 50 | 8.606 | 45.223 | 10.118 |
| | | GSSG | 154 | 58 | 47 | 130 | 116 |
| | | Hypotaurine | 174 | 154 | 146.347 | 246.103 | 133.440 |
| | | Taurine | 112 | 63 | 1.986 | 2.155 | 1.925 |

| | | | |
|---|---|---|---|
| Cells | 10 | 100 | 200 |
| Supernatant | 10 | 100 | 500 |

**Figure 4.** (**a–c**) Overview of metabolic classes and metabolic groups of liquid chromatography mass spectrometry of L929 cells in perfusion culture under MetR and LowCarb conditions. The metabolism of the murine cell line L929 was analysed via LC-MS in complete medium (control), MetR (0 mg/L), and Low Carb (3 mM glucose) for 7 days in a closed perfusion cell culture. After 7 days, the cell

lysates (intracellular) and the medium (supernatant) were analysed by LC-MS. The results were reproduced in three independent experiments and finally summarised. This figure shows the results of selected classes of substances and metabolic pathways for (**a**) nonessential (non-ess. AA) and essential (ess. AA) amino acids; (**b**) acyl-linked amino acids (Ac-AA) and glutamate-linked amino acids (Glu-AA); (**c**) molecules of sulphur metabolism ("S"-Metabos). For the cell pellets, the value of the control in each test series was defined as 100%. For the values in the medium, the measurement of the control medium used was defined as 100%. The colour scaling is shown below the results as a legend.

Sulphur-based metabolism was definitely prominent in MetR. Due to the lack of a sulphur source, the entire "S" metabolism was under pressure. Within the perfusion culture, however, it was seen after 7 days that with the exception of methionine sulfoxide (19%), the cells could maintain a sulphur-based metabolism very well (Figure 4c). The situation was different under LowCarb. Here, homocysteine, a precursor of methionine or an intermediate product in the breakdown of cysteine, decreased. Taurine also decreased and hypotaurine increased; both are products of cysteine degradation. Two central metabolites are the different forms of glutathione. Both the reduced form GSH and the oxidised form GSSG decreased by approximately half compared to the control. Under MetR, GSH remained relatively constant at 109%, and GSSG increased to 154%.

Glycolysis was also regulated differently under the two forms of restriction (Figure 5a). Under MetR, the intermediates hexose-P, fructose-1,6-bisphosphate, and phosphoenolpyruvate, and the glycolysis metabolites were increased overall compared to the control; under LowCarb, hexose-P, fructose-1,6-bisphosphate, and pyruvate were significantly decreased. The two intermediates, 3-phosphoglycerate and phosphoenolpyruvate, were increased to 142% and 239%, respectively, indicating stagnation within glycolysis under LowCarb.

The differences in carbohydrates, which are involved in glycolysis processes, among other things, were also clear (Figure 5b). The metabolites UDP-glucose (UDP-Glc), UDP-glucuronate, and UDP-N-acetylglucosamine (UDP-GlcNAc) were increased between approx. 60% and 90% under MetR and reduced between approx. 30% and 45% under LowCarb. UDP-Glc is derived from the glucose-6-phosphate of glycolysis and is then modified to UDP-glucoronate. Both molecules are, therefore, fundamentally dependent on the activity of glycolysis. UDP-GlcNAc is thought to be part of the glucose sensing mechanism. There is also evidence that it plays a part in insulin sensitivity in other cells [34].

Interestingly, the TCA cycle was only marginally affected under all conditions (Figure 5c). Citrate/isocitrate was significantly higher (195%) under LowCarb than in the control and MetR. However, all other metabolites were at the same level. Under neither MetR nor LowCarb did metabolites of the citrate cycle seem to be massively used for other metabolic pathways.

Significant differences between MetR and LowCarb were also seen in the phospholipids (Figure 5d). While the phospholipid phosphoethanolamine decreased under both conditions compared to the control, the precursors of phosphatidylcholine, CDP-choline and CDP-ethanolamine, were increased under MetR by 297% and 154%, respectively, while they were strongly reduced under LowCarb by 32% and 39%, respectively. In the area of lipid metabolism, carboxylic acid acetoacetate is still conspicuous and is strongly reduced under LowCarb by 39%. Lipid metabolism seems to be fundamentally lower under LowCarb.

| | Class | Name | Cells MetR | Cells LowCarb |
|---|---|---|---|---|
| a | Glycolysis | Hexose-P | 150 | 43 |
| | | Fructose-1,6-P2 | 135 | 69 |
| | | 3-Phosphoglycerate | 102 | 142 |
| | | Phosphoenolpyruvate | 146 | 239 |
| | | Pyruvate | 102 | 76 |
| | | Lactate | 107 | 105 |
| b | Carbohydrates | UDP-Glc | 191 | 55 |
| | | UDP-GlcNAc | 162 | 71 |
| | | UDP-Glucuronate | 181 | 54 |
| c | TCA Cycle | Citrate/Isocitr. | 136 | 195 |
| | | Itaconic acid | 87 | 102 |
| | | Aconitate | 112 | 120 |
| | | a-Ketoglutarate | 101 | 90 |
| | | Succinate | 97 | 94 |
| | | Fumarate | 91 | 92 |
| | | Malate | 92 | 91 |
| | | Oxaloacetate | 117 | 122 |
| d | Phospholipid Prec. | Phosphoethanolamine | 31 | 64 |
| | | CDP-Choline | 297 | 41 |
| | | CDP-Ethanolamine | 154 | 32 |
| | Carboxylic Acids | Acetoacetate | 112 | 39 |
| e | Polyamines | N-Acetylputrescine | 72 | 49 |
| | | Spermidine | 76 | 38 |
| | | N-Acetylspermidine | 20 | 142 |
| | | Spermine | 86 | 70 |
| f | Pyrimidines | UMP | 152 | 67 |
| | | UDP | 98 | 21 |
| | | CMP | 141 | 59 |
| | Purines | IMP | 42 | 19 |
| | | AMP | 110 | 61 |
| | | GMP | 64 | 38 |
| | Nicotinamides | Nicotinamide | 97 | 114 |
| | | mNAM | 177 | 71 |
| | | NAD(+) | 189 | 69 |

| Cells | 10 | 100 | 200 |
|---|---|---|---|

**Figure 5.** (**a–c**) Overview of metabolic classes and metabolic groups of liquid chromatography mass spectrometry of L929 cells in perfusion culture under MetR and LowCarb conditions. The metabolism of the murine cell line L929 was analysed via LC-MS in complete medium (control), MetR (0 mg/L),

and Low Carb (3 mM glucose) for 7 days in a closed perfusion cell culture. After 7 days, the cell lysates (intracellular) and the medium (supernatant) were analysed by LC-MS. The results were reproduced in three independent experiments and finally summarised. This figure shows the results of selected classes of substances and metabolic pathways for (**a**) glycolysis; (**b**) carbohydrates; (**c**) molecules of the tricarboxylic acid (TCA) cycle; (**d**) phospholipid precursors; (**e**) polyamines; (**f**) pyrimidines, purines, and nicotinamides. The value of the control in each test series was defined as 100%. The colour scaling is shown below the results as a legend.

The polyamines in the form of spermidines can perform different functions within the cell. As already mentioned, they can be advantageous for the implementation of restriction since secreted spermidine can trigger autophagy [35]. In addition, polyamines are an indicator of the energy state of the cells. At high energy levels, large amounts of acetyl-Co are produced via glycolysis and other metabolic pathways. The degree of acetylation of, e.g., proteins, is, therefore, proportional to the energy level of the cell [36]. This is then also shown by different polyamines that are increasingly acetylated [37]. In addition, high intracellular spermine or spermidine concentrations promote protein synthesis and proliferation [38]. Basically, the different polyamines are reduced under both MetR and LowCarb and indicate a lower energy state than in the control (Figure 5e). N-acetylspermidine, at 142% under LowCarb, is an exception.

The last group presented here includes the purines, pyrimidines, and nicotinamides (Figure 5f). Here, too, many metabolites were at similar levels in all groups. In the case of purines, UMP, UDP, and CMP were significantly reduced under LowCarb conditions, while under MetR, UMP and CMP were increased by between 50% and 40%. Among purines, some metabolites were reduced under both restrictions, such as IMP and GMP. AMP was slightly increased under MetR with 110%. The nicotinamides showed the opposite trend. While mNAM and NAD(+) were increased by approximately 80–90% under MetR, both metabolites were reduced by approximately 30% under LowCarb.

### 2.5. Unlimited Cell Proliferation of L929 Cells in Closed Perfusion Culture

At the end of each experiment, the total cell numbers were determined (Figure 6). It was noticeable that in the control and under LowCarb, the cell line L929 was able to grow further and form multiple layers in the cell culture. We originally chose the experimental conditions so that the cell lawn was confluent at the beginning of the experiment and the cells were prevented from further proliferation by contact inhibition. Therefore, the analysis under this nonproliferative condition essentially examined only the effects of the different metabolite concentrations of methionine and glucose. In a previous analysis, we examined MetR under proliferative and nonproliferative conditions in normal cell culture [17]. Cell proliferation was not observed under static conditions. Instead, as shown in Figure 2c, the absolute cell number in the control even decreased slightly after 120 h. When the medium in the perfusion culture was kept at a constant high level, the L929 cells could continue to grow beyond contact inhibition under both control and LowCarb conditions. The restriction of methionine continued to severely limit growth even under perfusion conditions.

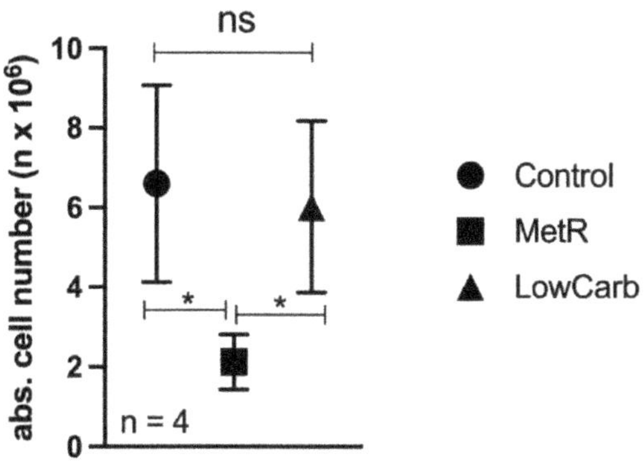


**Figure 6.** Final cell number in perfusion culture under different conditions. Towards the end of the experiment, the absolute cell number of each Petri dish was determined using an EVE$^{TM}$ automated cell counter. The mean values of the four experiments and the respective range of cell counts are plotted here. Comparisons between the control and MetR groups were performed by applying one-way ANOVA followed by the Tukey–Kramer multiple comparison test (* $p < 0.05$, ns = not significant).

## 3. Discussion

In this work, we used liquid chromatography mass spectrometry to analyse the influence of glucose restriction using the glycolysis inhibitor 2-DG in our L929 model system to demonstrate that the system is suitable for analyses in the LowCarb condition. In a further analysis, we then compared the metabolome analyses of the two restriction conditions MetR and LowCarb in a perfusion system.

Analysis of the glycolysis inhibitor 2-DG showed a clear antiproliferative effect on L929 cells (Figure 2). In addition, as expected, the inhibitor was shown to slow glycolysis at the metabolic level downstream of the actual target, glucose-6-phosphate isomerase (Figure 3) [26]. The clearest marker for the effectiveness of 2-DG is the extreme drop in the intracellular ATP concentration to 1% (Figure 3) [27]. The conclusion is that the cell line L929 is very well suited for analysing glycolysis metabolism.

Using perfusion culture, the two restriction conditions MetR and LowCarb were compared over a period of 7 days in a closed perfusion culture. Analysis of the metabolome showed that both restriction conditions cause observable changes at the molecular level. In the case of MetR, in addition to the low methionine and GluMet contents, it is particularly noticeable that some amino acids are increasingly secreted. This is a phenomenon that we have already defined in an earlier paper as characteristic of L929 under MetR [18]. Glucose restriction was also shown to be specific at the molecular level and clearly differed from the effects induced by 2-DG, as here the concentration of the first two metabolites of glycolysis—hexose phosphate and fructose 1,6-bisphosphate—strongly decreased and then accumulated to a greater degree than 3-phosphoglycerate. The decrease in UDP-glucose, a product that, as already mentioned, is strongly dependent on the turnover rate of glycolysis, also clearly showed the changes induced by LowCarb.

Nevertheless, it is noticeable that the vast majority of the metabolites remained at similar levels, even compared to the control. However, this is also evidence of the quality of the analysis. For example, if one analyses gene expression across the transcriptome of 10,000 genes, the absolute majority of genes will not be differentially regulated. If, for example, 9000 genes were differentially regulated, this would be a strong indication of a faulty analysis. Thus, on the one hand, the many metabolites at the same level show the quality of the analysis and the stability of the perfusion culture, but on the other hand, the question arises as to what extent the different forms of restriction have been implemented consistently? In the following discussion, we want to assess to what extent the results reflect the implementation of the two restriction forms MetR and LowCarb in our system, which factors have a decisive influence on the profile, and which further analyses are useful.

The incubation period has a significant influence on the metabolic profile. It is possible that the system must be incubated for a longer period than 7 days so that the specific restriction profiles can be more extreme. On the other hand, the different forms of restriction also have many common molecular mechanisms, as already described [14]. Thus, the results can also be interpreted as a slow but sure alignment of the metabolites over time, while in the short term, the differences are much larger. Looking at the overall profile of the 2-DG analysis (Supplement File S1), for example, it is noticeable that at the early time points within the first 8 h, most of the metabolites are significantly different compared to those in the control.

The metabolites selected for analysis also play a role. In LC-MS analysis, we chose a setup to analyse characteristic metabolites of individual metabolic pathways, such as glycolysis or the TCA cycle. However, these approximately 170 metabolites represent only a small part of the real existing metabolites of a cell. Thus, it is possible that other metabolites not used in this analysis show a much more differentiated picture than the products we selected. Again, the argument applies that the most important metabolic pathways align over time.

The growth conditions can also influence the results. In this work, we decided to start the analyses under confluent conditions to obtain sufficient cell material for the LC-MS analyses. Furthermore, we know from previous work that the low energy status under MetR is essentially independent of proliferation or cell contact inhibition [17]. Figure 6 clearly shows that the cells continued to grow steadily within the perfusion culture. We could not observe this effect in the static cell culture. However, we assume that this effect has no negative influence on the LC-MS analyses. Again, the already mentioned argument applies that the high number of metabolites at similar levels is a quality criterion. If proliferation affected the results, the differences when comparing MetR to the control and LowCarb should be much more striking. However, the question arises as to why the cells manage to continue growing even under the LowCarb condition. Figure 2a shows the analyses performed to determine the glucose concentration used in perfusion culture. Under static conditions, the proliferation rate decreased dramatically after 96 h with an initial concentration of 3 mM glucose, but under these experimental conditions, the glucose level also continued to decrease during the experimental period. Basically, this experiment served to confirm that we can work with a level of 3 mM glucose in the perfusion culture without running the risk of cell death. After 96 h at 3 mM, the cells were observed to be alive under a microscope and proliferated significantly better than under even lower glucose concentrations. We used a medium tank in the perfusion culture that provided 60 times the volume of the contents of the Petri dish. We assumed that these conditions were suitable for L929 cells and that the cells were well supplied over this period. However, we did not want to lower the glucose concentration below 3 mM because this concentration does not correspond to the physiological conditions of the blood glucose level. In mice, 3 mM is already a very low value. Therefore, in the perfusion culture, the constant supply of fresh medium seems to be sufficient for the cells to grow steadily both in the control and under LowCarb.

The closed perfusion culture is another factor that has a strong influence on the metabolome. Basically, the closed version offers the advantage that the medium is conditioned over the experimental period. Thus, on the one hand, the medium can be included in the LC-MS analysis; on the other hand, secreted messengers may well be essential to implement the individual restriction forms. However, the exact opposite can also occur. The active substances in the medium influence the cellular events more than MetR or LowCarb. Analysis of the supernatants showed that a large number of metabolites accumulated over the experimental period. Thus, it is possible that the concentration of the secreted metabolites influences the metabolism and overlaps both MetR and LowCarb in the metabolic profile, leading to an alignment of the LC-MS profiles.

However, perfusion culture offers extreme advantages. Basically, the main advantage lies in the simulation of the blood flow of an organism, in that, on the one hand, fresh nutrients can be continuously supplied and, on the other hand, waste and possibly toxins can be removed. The differences between open and closed perfusion systems will be shown in further studies.

In principle, L929 cells are also suitable for analysing glucose metabolism. They react to 2-DG inhibition accordingly. The cells are also suitable for perfusion culture, and we were able to show that after 7 days in a closed system, the cells implemented a MetR- or LowCarb-specific metabolism. However, a large number of metabolites were at similar levels even compared to the control. For this reason, the question arises as to whether the restriction forms result in a largely similar metabolism or whether the profile is influenced more significantly by certain factors than by the restriction itself. Based on the above arguments, in future work, the cells should be analysed over a longer period of time (up to 21 days) to determine whether the metabolism continues to differentiate or remains similar. Furthermore, the experiments should be conducted in an open perfusion system to limit the effect on the restrictive factors methionine and glucose. We are sure that future analyses using the perfusion culture technique will reveal fundamental insights and mechanisms under the different forms of restriction.

## 4. Materials and Methods

### 4.1. Cell Culture

The murine fibroblast cell line L929 was purchased from the Leibniz Institute, DSMZ-German Collection of Microorganisms and Cell Cultures GmbH (Braunschweig, Germany). Cells were cultured in RPMI 1640 medium (Gibco, Life Technologies; Darmstadt, Germany) with 10% FCS (Sigma-Aldrich, Darmstadt, Germany) and 1% penicillin/streptomycin (P/S; 100 U/mL penicillin and 100 µg/mL streptomycin (Thermo Fisher Scientific, Darmstadt, Germany)) at 37 °C in a humidified atmosphere containing 5% $CO_2$.

### 4.2. ImageXpress Pico Automated Cell Imaging System—Digital Microscopy (Pico Assay)

Cells were seeded at 10,000 cells in 100 µL of culture medium per well of a 96-well plate and incubated overnight. The following day, the cells were incubated in complete, methionine-free, cysteine-free, or methionine- and cysteine-free medium. The incubation time is stated in the corresponding figure legend. For staining, 10 µL of Hoechst staining solution (1:200 dilution of Hoechst 33342 (Thermo Fisher, Darmstadt, Germany) (10 mg/mL in $H_2O$) in medium) was added to each well. After a 20–30-min incubation period, wells were analysed with an ImageXpress Pico automated cell imaging system (Molecular Devices, San Jose, CA, USA) via automated digital microscopy. The cells were analysed with transmitted light in the DAPI channel at 4x magnification. The complete area of every well was screened. The focus and exposure time were set via an auto setup and controlled by analysing 3–4 test wells. Finally, every result was confirmed visually, and 95% of the cells were counted and analysed.

### 4.3. Analysis of the Cell Progression Rate Using the Pico Assay

Cells were seeded at 10,000 cells in 100 µL of culture medium per well in a 96-well plate. After 24, 32, 48, 56, 72, 80, 96, 104, and 120 h, cell numbers were measured with six values for every time point, as described under Pico Assay (Section 4.2). The growth of a cell population can be described with the following formula:

$$N_t = N_0 \times 2^{(t \times f)}$$

($N_t$ = cell number at time t; $N_0$ = cell number at time 0; t = time in days (d); f = cell division frequency (1/d)).

To determine f, the formula is rearranged as follows:

$$f = \frac{\left(\log\left(\frac{N_t}{N_0}\right) / \log(2)\right)}{t}$$

To obtain an overview, the measured values were first plotted as a simple diagram. From this, it was possible to see at what point the growth entered the plateau phase. Then, the values were plotted as an exponential curve, and only values in the exponential growth phase were used to determine f. In the case of L929, these were the time points 24, 32, 48, 56, and 72 h. From these values, the individual f values were calculated for the intermediate periods (Δ24/32, Δ32/48, Δ48/56, and Δ56/72). The total value f was then calculated as the mean of the four Δf values.

### *4.4. Experiments under 2-DG for LC-MS*

L929 cells were seeded in 20 mL of medium in 15 cm Petri dishes and incubated overnight. A total of $2 \times 10^6$ cells/Petri dish were seeded under proliferative conditions for 2, 5, and 8 h, and $1 \times 10^6$ cells were seeded at 24, 48, and 72 h to prevent confluence during the test period. Every value was measured in triplicate. All media contained 10% FCS and 1% P/S [100 U/mL penicillin and 100 µg/mL streptomycin (Thermo Fisher Scientific, Darmstadt, Germany)]. After seeding, the cells were incubated with 20 mL of complete RPMI medium or 20 mL of RPMI medium with 625 µM 2-deoxy-D-glucose (2-DG) (Sigma-Aldrich, Darmstadt, Germany). Before harvesting, 1 mL of the supernatant was stored for analysis. The remaining medium was then removed, and the cells were washed with 10 mL of PBS and detached with 3 mL of trypsin/EDTA (Thermo Fisher Scientific, Darmstadt, Germany). After the addition of 7 mL of the appropriate medium, the absolute cell number in the suspensions was analysed with the automated cell counter EVE™ [NanoEntek (VWR, Darmstadt, Germany)]. Each sample was measured three times, and the mean value was calculated to obtain an accurate result. Pellets with $1 \times 10^6$ cells were produced by centrifugation (5 min at 1200 rpm at RT). Until the LC-MS analysis, all samples were stored at −20 °C.

### *4.5. Closed Perfusion Culture Experiments for LC-MS*

For perfusion culture (Figure 7), we used RPMI 1640 (Genaxxon Bioscience, Ulm, Germany), which lacks methionine and glucose, as the basal medium. Every medium was prepared from the basal medium. All media contained 10% FCS (Sigma-Aldrich, Darmstadt, Germany) and 1% P/S [100 U/mL penicillin and 100 µg/mL streptomycin (Thermo Fisher Scientific, Darmstadt, Germany)]. For the control (complete medium), the amino acid methionine (Sigma-Aldrich, Darmstadt, Germany) was added at 15 mg/L, and glucose (Sigma-Aldrich, Darmstadt, Germany) was added at 11.1 mM. For MetR medium, glucose was added at the same concentration, and for LowCarb medium, methionine was added at 15 mg/L and glucose was added at a final concentration of 3 mM.

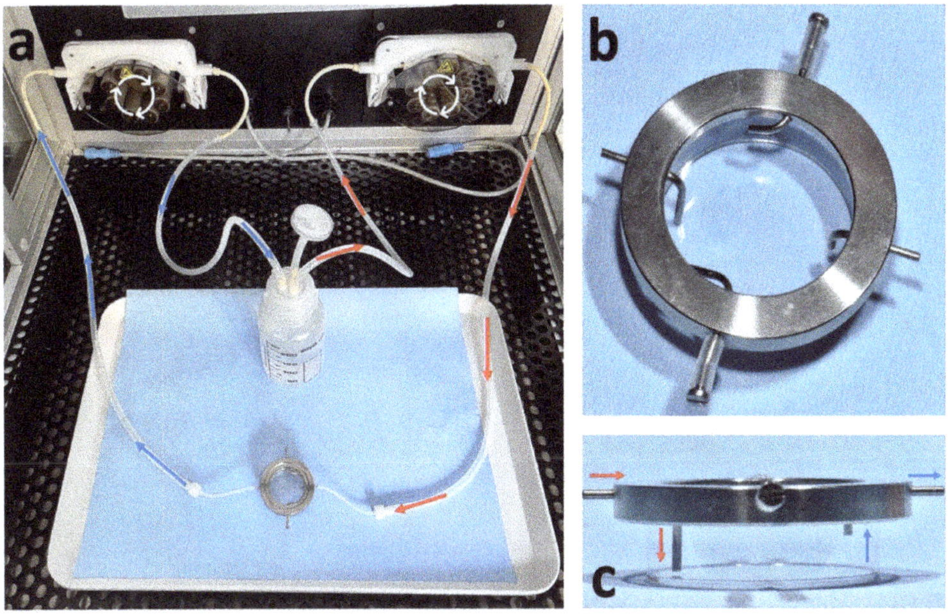


**Figure 7.** (**a**–**c**) Closed perfusion culture and dual perfusion set for 35 mm petri dishes. (**a**) Examples for a closed perfusion for one Petri dish. The inside of the bioreactor is shown. The red line shows the inlet, and the blue line shows the outlet. The tubes run through separate pump systems. This allows the outlet to be faster than the inlet and thus prevents the petri dish from overflowing; (**b**) top view; (**c**) side view of the dual perfusion set for 35 mm petri dishes. The system has a total of 4 inlets and outlets. In our case, only two were used, and the other two were closed. The red arrows show the influx, and the blue arrows show the outflux, including the height of the medium in the petri dish.

A total of $1 \times 10^6$ cells were seeded in 35 mm dishes (Greiner Bio-One, Frickenhausen, Germany). After 48 h, cells reached confluence, the medium was removed, and the cells were stimulated with complete medium (control), MetR, or LowCarb medium. Petri dishes were incubated in a bioreactor at 37 °C in a humidified atmosphere containing 5% $CO_2$. Every medium tank contained 180 mL (60-fold excess compared to the contents of the Petri dish (3 mL)). The flow rate was 0.8 mL/min, with the inflow being slower than the outflow to prevent the Petri dish from overflowing. During incubation, the Petri dish was supplied with medium using a dual perfusion set for 35 mm petri dishes (PeCon GmbH, Erbach, Germany). This consisted of a perfusion ring, which had an inlet and outlet for the medium, and a metal lid with a glass interior to prevent contamination Figure 1a–c. After 7 days, 1 mL of the supernatant was removed before harvesting and stored for analysis. The remaining medium was then removed, and the cells were washed with 10 mL of PBS and detached with 3 mL of trypsin/EDTA (Thermo Fisher Scientific, Darmstadt, Germany). After the addition of 7 mL of the appropriate medium, the absolute cell number in the suspensions was analysed with the automated cell counter EVE$^{TM}$ [NanoEntek (VWR, Darmstadt, Germany)]. Each sample was measured four times, and the mean value was calculated to obtain an accurate result. Pellets with $1 \times 10^6$ cells were produced by centrifugation (5 min at 1200 rpm at RT). Until the LC-MS analysis, all samples were stored at −20 °C.

*4.6. LC-MS*

Analysis of water-soluble metabolites in cell extracts and culture media.

4.6.1. Cells

After the addition of 0.5 mL of MeOH/$CH_3CN$/$H_2O$ (50/30/20, *v*/*v*/*v*) containing 10 µM lamivudine, cell pellets were homogenised by ultrasound treatment (10 × 1 s, 250 W output energy). Medium: One hundred microlitres of culture medium was combined with 0.4 mL of MeOH/$CH_3CN$ (50/30, *v*/*v*) containing 10 µM lamivudine. The external standard lamivudine was not used for absolute metabolite quantification, but was used as a quality control to compensate for eventually occurring technical issues. As quality control and for the determination of the corresponding retention times, most of the annotated metabolites (which are commercially available) were run as mixtures of pure compounds under identical experimental conditions. General procedure: The resulting suspension was centrifuged (20 kRCF for 2 min in an Eppendorf centrifuge 5424), and the supernatant was applied to a C18-SPE column that was activated with 0.5 mL of $CH_3CN$ and equilibrated with 0.5 mL of MeOH/$CH_3CN$/$H_2O$ (50/30/20, *v*/*v*/*v*). The SPE eluate was evaporated in a vacuum concentrator. The resulting pellet was dissolved in 50 µL (cell extracts) or 500 µL (media extracts) of 5 mM $NH_4OAc$ in $CH_3CN$/(25%/75%, *v*/*v*).

4.6.2. LC Parameters

Mobile phase A consisted of 5 mM $NH_4OAc$ in $CH_3CN$/$H_2O$ (5/95, *v*/*v*), and mobile phase B consisted of 5 mM $NH_4OAc$ in $CH_3CN$/$H_2O$ (95/5, *v*/*v*).

After application of 3 µL of the sample to a ZIC-HILIC column (at 30 °C), the LC gradient programme was as follows: 100% solvent B for 2 min, a linear decrease to 40% solvent B over 16 min, maintenance at 40% solvent B for 9 min, and an increase to 100% solvent B over 1 min. The column was maintained at 100% solvent B for 5 min for column equilibration before each injection. The flow rate was maintained at 200 µL/min. The eluent was directed to the ESI source of the QE-MS from 1.85 min to 20.0 min after sample injection.

The MS parameters were as follows: scan type, full MS in positive-and-negative mode (alternating); scan range, 69–1000 *m*/*z*; resolution, 70,000; AGC-target, 3E6; maximum injection time, 200 ms; sheath gas, 30; auxiliary gas, 10; sweep gas, 3; spray voltage, 3.6 kV (positive mode) or 2.5 kV (negative mode); capillary temperature, 320 °C; S-lens RF level, 55.0; and auxiliary gas heater temperature, 120 °C. Annotation and data evaluation: peaks corresponding to the calculated monoisotopic masses (MIM +/− $H^+$ ± 2 mMU) were integrated using TraceFinder software (Thermo Scientific, Bremen, Germany). Materials: Ultrapure water was obtained from a Millipore water purification system (Milli-Q Merck Millipore, Darmstadt, Germany). HPLC–MS solvents, LC–MS $NH_4OAc$, and lamivudine were purchased from Merck (Darmstadt, Germany). The RP18-SPE columns were 50 mg of Strata C18-E (55 µm) in 1-mL tubes (Phenomenex, Aschaffenburg, Germany). The sonifier was a Branson Ultrasonics 250 equipped with a 13-mm sonotrode (Thermo Scientific, Bremen, Germany).

4.6.3. LC-MS System

A Thermo Scientific Dionex UltiMate 3000 UHPLC system linked to a Q Exactive mass spectrometer (QE-MS) equipped with a HESI probe (Thermo Scientific, Bremen, Germany) was used. The samples were analysed with a high-resolution mass spectrometer, allowing the generation of XIC data that were analysed by applying a very narrow *m*/*z* margin (+/−3 mMU). The particle filter was a Javelin filter with an ID of 2.1 mm (Thermo Scientific, Bremen, Germany). The UPLC-precolumn was a SeQuant ZIC-HILIC column (5-µm particles, 20 × 2 mm) (Merck, Darmstadt, Germany). The UPLC column was a SeQuant ZIC-HILIC column (3.5-µm particles, 100 × 2.1 mm) (Merck, Darmstadt, Germany).

Raw data analysis and value generation (in short):

LC-MS analyses were carried out in four independent experiments at 24 h, 48 h, 72 h, 96 h, and 120 h, with each value obtained from triplicate measurements. Metabolites were quantified in cell pellets and corresponding supernatants (media) under methionine-supplemented and methionine-free conditions (12 samples per time point in total). The resulting peak areas were normalised against that of lamivudine as an external standard.

From this, the mean value and standard deviation were calculated for each triplicate. For better comparisons, the values were converted to percentages. For the values of the media, the control measurement of the medium used was defined as 100%. For the cell pellets, the highest measured value in each test series within an experiment was defined as 100%. From these values, the average mean values from the four experiments were then summarised in the individual tables. For a better overview, the results were rounded to natural numbers and shown as a heatmap. The corresponding colour range is indicated individually under each table. The raw data and results for the two profiles are provided in the Supplementary Files S1 and S2.

### *4.7. Statistical Analysis*

Statistical analysis was performed using GraphPad Prism 8.0 (GraphPad Software, San Diego, CA, USA). One-way ANOVA was used to compare and analyse the data of different groups, followed by the Tukey–Kramer multiple comparison test (ns; nonsignificant; * $p < 0.05$, *** $p < 0.001$).

**Supplementary Materials:** The supporting information can be downloaded at https://www.mdpi.com/article/10.3390/ijms23169220/s1.

**Author Contributions:** Conceptualisation, A.S. and W.S.; methodology, A.S., W.S., J.M.V. and J.K.; software, W.S.; validation, M.E.-M., A.C.W., J.B., K.F. and L.P.; formal analysis, A.S. and W.S.; statistical analysis, M.E.-M.; investigation, A.S., W.S., M.M. and J.M.V.; data curation, A.S., M.E.-M., S.H. and W.S.; writing—original draft preparation, M.E.-M. and A.S.; writing—review and editing, W.S., S.H. and A.C.K.; visualisation, A.S. and W.S.; supervision, A.S. and W.S.; project administration, A.S. All authors have read and agreed to the published version of the manuscript.

**Funding:** This research received no external funding.

**Acknowledgments:** This publication was supported by the Open Access Publication Fund of the University of Wuerzburg.

**Conflicts of Interest:** The authors declare no conflict of interest.

## References

1. Hosios, A.M.; Hecht, V.C.; Danai, L.V.; Johnson, M.O.; Rathmell, J.C.; Steinhauser, M.L.; Manalis, S.R.; Vander Heiden, M.G. Amino acids rather than glucose account for the majority of cell mass in proliferating mammalian cells. *Dev. Cell* **2016**, *36*, 540–549. [CrossRef] [PubMed]
2. Fothergill-Gilmore, L.A.; Michels, P.A. Evolution of glycolysis. *Prog. Biophys. Mol. Biol.* **1993**, *59*, 105–235. [CrossRef]
3. Warburg, O. Über den stoffwechsel der carcinomzelle. *Naturwissenschaften* **1924**, *12*, 1131–1137. [CrossRef]
4. Liberti, M.V.; Locasale, J.W. The warburg effect: How does it benefit cancer cells? *Trends Biochem. Sci.* **2016**, *41*, 211–218. [CrossRef]
5. Vergati, M.; Krasniqi, E.; Monte, G.D.; Riondino, S.; Vallone, D.; Guadagni, F.; Ferroni, P.; Roselli, M. Ketogenic diet and other dietary intervention strategies in the treatment of cancer. *Curr. Med. Chem.* **2017**, *24*, 1170–1185. [CrossRef]
6. Ball, Z.B.; Barnes, R.H.; Visscher, M.B. The effects of dietary caloric restriction on maturity and senescence, with particular reference to fertility and longevity. *Am. J. Physiol.* **1947**, *150*, 511–519. [CrossRef]
7. Fontana, L.; Partridge, L. Promoting health and longevity through diet: From model organisms to humans. *Cell* **2015**, *161*, 106–118. [CrossRef]
8. Fontana, L.; Partridge, L.; Longo, V.D. Extending healthy life span–from yeast to humans. *Science* **2010**, *328*, 321–326. [CrossRef]
9. Mirzaei, H.; Suarez, J.A.; Longo, V.D. Protein and amino acid restriction, aging and disease: From yeast to humans. *Trends Endocrinol. Metab.* **2014**, *25*, 558–566. [CrossRef]
10. Cantó, C.; Gerhart-Hines, Z.; Feige, J.N.; Lagouge, M.; Noriega, L.; Milne, J.C.; Elliott, P.J.; Puigserver, P.; Auwerx, J. AMPK regulates energy expenditure by modulating $NAD^+$ metabolism and SIRT1 activity. *Nature* **2009**, *458*, 1056–1060. [CrossRef]
11. Gu, X.; Orozco, J.M.; Saxton, R.A.; Condon, K.J.; Liu, G.Y.; Krawczyk, P.A.; Scaria, S.M.; Harper, J.W.; Gygi, S.P.; Sabatini, D.M. SAMTOR is an S-adenosylmethionine sensor for the mTORC1 pathway. *Science* **2017**, *358*, 813–818. [CrossRef] [PubMed]
12. Zou, K.; Rouskin, S.; Dervishi, K.; McCormick, M.A.; Sasikumar, A.; Deng, C.; Chen, Z.; Kaeberlein, M.; Brem, R.B.; Polymenis, M.; et al. Life span extension by glucose restriction is abrogated by methionine supplementation: Cross-talk between glucose and methionine and implication of methionine as a key regulator of life span. *Sci. Adv.* **2020**, *6*, eaba1306. [CrossRef]
13. Mirzaei, H.; Raynes, R.; Longo, V.D. The conserved role of protein restriction in aging and disease. *Curr. Opin. Clin. Nutr. Metab. Care* **2016**, *19*, 74–79. [CrossRef] [PubMed]

14. Veech, R.L.; Bradshaw, P.C.; Clarke, K.; Curtis, W.; Pawlosky, R.; King, M.T. Ketone bodies mimic the life span extending properties of caloric restriction. *IUBMB Life* **2017**, *69*, 305–314. [CrossRef] [PubMed]
15. Palliyaguru, D.L.; Shiroma, E.J.; Nam, J.K.; Duregon, E.; Teixeira, C.V.L.; Price, N.L.; Bernier, M.; Camandola, S.; Vaughan, K.L.; Colman, R.J.; et al. Fasting blood glucose as a predictor of mortality: Lost in translation. *Cell Metab.* **2021**, *33*, 2189–2200.e3. [CrossRef]
16. Moley, K.H.; Mueckler, M.M. Glucose transport and apoptosis. *Apoptosis* **2000**, *5*, 99–105. [CrossRef]
17. Koderer, C.; Schmitz, W.; Wünsch, A.C.; Balint, J.; El-Mesery, M.; Volland, J.M.; Hartmann, S.; Linz, C.; Kübler, A.C.; Seher, A. Low energy status under methionine restriction is essentially independent of proliferation or cell contact inhibition. *Cells* **2022**, *11*, 551. [CrossRef]
18. Schmitz, W.; Koderer, C.; El-Mesery, M.; Gubik, S.; Sampers, R.; Straub, A.; Kübler, A.C.; Seher, A. Metabolic fingerprinting of murine L929 fibroblasts as a cell-based tumour suppressor model system for methionine restriction. *Int. J. Mol. Sci.* **2021**, *22*, 3039. [CrossRef]
19. Schmitz, W.; Ries, E.; Koderer, C.; Völter, M.F.; Wünsch, A.C.; El-Mesery, M.; Frackmann, K.; Kübler, A.C.; Linz, C.; Seher, A. Cysteine restriction in murine L929 fibroblasts as an alternative strategy to methionine restriction in cancer therapy. *Int. J. Mol. Sci.* **2021**, *22*, 11630. [CrossRef]
20. Azzu, V.; Valencak, T.G. Energy metabolism and ageing in the mouse: A mini-review. *Gerontology* **2017**, *63*, 327–336. [CrossRef]
21. Melzer, K.; Heydenreich, J.; Schutz, Y.; Renaud, A.; Kayser, B.; Mäder, U. Metabolic equivalent in adolescents, active adults and pregnant women. *Nutrients* **2016**, *8*, 438. [CrossRef] [PubMed]
22. National Research Council Subcommittee on Laboratory Animal Nutrition. *Nutrient Requirements of Laboratory Animals: Fourth Revised Edition, 1995*; National Academies Press: Washington, DC, USA, 1995. [CrossRef]
23. Speakman, J.R. Body size, energy metabolism and lifespan. *J. Exp. Biol.* **2005**, *208*, 1717–1730. [CrossRef] [PubMed]
24. Kim, J.; Guan, K.L. mTOR as a central hub of nutrient signalling and cell growth. *Nat. Cell Biol.* **2019**, *21*, 63–71. [CrossRef]
25. Longo, V.D.; Kennedy, B.K. Sirtuins in aging and age-related disease. *Cell* **2006**, *126*, 257–268. [CrossRef] [PubMed]
26. Xi, H.; Kurtoglu, M.; Lampidis, T.J. The wonders of 2-deoxy-D-glucose. *IUBMB Life* **2014**, *66*, 110–121. [CrossRef]
27. Zhang, D.; Li, J.; Wang, F.; Hu, J.; Wang, S.; Sun, Y. 2-deoxy-D-glucose targeting of glucose metabolism in cancer cells as a potential therapy. *Cancer Lett.* **2014**, *355*, 176–183. [CrossRef]
28. Nikolai, S.; Pallauf, K.; Huebbe, P.; Rimbach, G. Energy restriction and potential energy restriction mimetics. *Nutr. Res. Rev.* **2015**, *28*, 100–120. [CrossRef]
29. Saraswat, K.; Kumar, R.; Rizvi, S.I. Glycolytic inhibitor 2-deoxy-D-glucose at chronic low dose mimics calorie restriction in rats through mitohormetic induction of reactive oxygen species. *Rejuvenation Res.* **2019**, *22*, 377–384. [CrossRef]
30. Griffith, O.W.; Bridges, R.J.; Meister, A. Transport of gamma-glutamyl amino acids: Role of glutathione and gamma-glutamyl transpeptidase. *Proc. Natl. Acad. Sci. USA* **1979**, *76*, 6319–6322. [CrossRef]
31. Bachhawat, A.K.; Yadav, S. The glutathione cycle: Glutathione metabolism beyond the $\gamma$-glutamyl cycle. *IUBMB Life* **2018**, *70*, 585–592. [CrossRef]
32. Menendez, J.A.; Oliveras-Ferraros, C.; Cufí, S.; Corominas-Faja, B.; Joven, J.; Martin-Castillo, B.; Vazquez-Martin, A. Metformin is synthetically lethal with glucose withdrawal in cancer cells. *Cell Cycle* **2012**, *11*, 2782–2792. [CrossRef] [PubMed]
33. Madeo, F.; Eisenberg, T.; Pietrocola, F.; Kroemer, G. Spermidine in health and disease. *Science* **2018**, *359*, eaan2788. [CrossRef] [PubMed]
34. Hanover, J.A. Glycan-dependent signaling: O-linked N-acetylglucosamine. *FASEB J.* **2001**, *15*, 1865–1876. [CrossRef] [PubMed]
35. Chen, Y.; Zhuang, H.; Chen, X.; Shi, Z.; Wang, X. Spermidine-induced growth inhibition and apoptosis via autophagic activation in cervical cancer. *Oncol. Rep.* **2018**, *39*, 2845–2854. [CrossRef] [PubMed]
36. Madeo, F.; Carmona-Gutierrez, D.; Hofer, S.J.; Kroemer, G. Caloric restriction mimetics against age-associated disease: Targets, mechanisms, and therapeutic potential. *Cell Metab.* **2019**, *29*, 592–610. [CrossRef] [PubMed]
37. Pegg, A.E. Spermidine/spermine-N(1)-acetyltransferase: A key metabolic regulator. *Am. J. Physiol. Endocrinol. Metab.* **2008**, *294*, E995–E1010. [CrossRef]
38. Mandal, S.; Mandal, A.; Johansson, H.E.; Orjalo, A.V.; Park, M.H. Depletion of cellular polyamines, spermidine and spermine, causes a total arrest in translation and growth in mammalian cells. *Proc. Natl. Acad. Sci. USA* **2013**, *110*, 2169–2174. [CrossRef]

International Journal of *Molecular Sciences*

*Review*

# Nutrition as Personalized Medicine against SARS-CoV-2 Infections: Clinical and Oncological Options with a Specific Female Groups Overview

**Miriam Dellino [1,2,*,†], Eliano Cascardi [3,4,*,†], Marina Vinciguerra [1], Bruno Lamanna [1,5], Antonio Malvasi [1], Salvatore Scacco [6], Silvia Acquaviva [6], Vincenzo Pinto [1], Giovanni Di Vagno [2], Gennaro Cormio [7], Raffaele De Luca [8], Miria Lafranceschina [8], Gerardo Cazzato [9], Giuseppe Ingravallo [9], Eugenio Maiorano [9], Leonardo Resta [9], Antonella Daniele [8,‡] and Daniele La Forgia [8,‡]**

1 Department of Biomedical Sciences and Human Oncology, University of Bari, 70100 Bari, Italy
2 Clinic of Obstetrics and Gynecology, "San Paolo" Hospital, 70123 Bari, Italy
3 Department of Medical Sciences, University of Turin, 10124 Turin, Italy
4 Pathology Unit, FPO-IRCCS Candiolo Cancer Institute, Str. Provinciale 142, Km 3.95, 10060 Candiolo, Italy
5 Fetal Medicine Research Institute, King's College Hospital, London SE5 9RS, UK
6 Department of Basic Medical Sciences and Neurosciences, University of Bari "Aldo Moro", 70121 Bari, Italy
7 Gynecologic Oncology Unit, IRCCS Istituto Tumori Giovanni Paolo II, Department of Interdisciplinary Medicine (DIM), University of Bari "Aldo Moro", 70121 Bari, Italy
8 IRCCS Istituto Tumori Giovanni Paolo II, 70124 Bari, Italy
9 Department of Emergency and Organ Transplantation, University of Bari "Aldo Moro", 70121 Bari, Italy
* Correspondence: miriamdellino@hotmail.it (M.D.); eliano20@hotmail.it (E.C.)
† These authors contributed equally to this work.
‡ These authors contributed equally to this work.

**Citation:** Dellino, M.; Cascardi, E.; Vinciguerra, M.; Lamanna, B.; Malvasi, A.; Scacco, S.; Acquaviva, S.; Pinto, V.; Di Vagno, G.; Cormio, G.; et al. Nutrition as Personalized Medicine against SARS-CoV-2 Infections: Clinical and Oncological Options with a Specific Female Groups Overview. *Int. J. Mol. Sci.* **2022**, *23*, 9136. https://doi.org/10.3390/ijms23169136

Academic Editor: Laura Paleari

Received: 18 July 2022
Accepted: 12 August 2022
Published: 15 August 2022



**Abstract:** Coronavirus disease 2019 (COVID-19) is a respiratory disease caused by severe acute respiratory syndrome coronavirus-2 (SARS-CoV-2). It is acknowledged that vulnerable people can suffer from mortal complications of COVID-19. Therefore, strengthening the immune system particularly in the most fragile people could help to protect them from infection. First, general nutritional status and food consumption patterns of everyone affect the effectiveness of each immune system. The effects of nutrition could impact the level of intestinal and genital microbiota, the adaptive immune system, and the innate immune system. Indeed, immune system cells and mediators, which are crucial to inflammatory reaction, are in the structures of fats, carbohydrates, and proteins and are activated through vitamins (vit) and minerals. Therefore, the association of malnutrition and infection could damage the immune response, reducing the immune cells and amplifying inflammatory mediators. Both amount and type of dietary fat impact on cytokine biology, that consequently assumes a crucial role in inflammatory disease. This review explores the power of nutrition in the immune response against COVID-19 infection, since a specific diet could modify the cytokine storm during the infection phase. This can be of vital importance in the most vulnerable subjects such as pregnant women or cancer patients to whom we have deemed it necessary to dedicate personalized indications.

**Keywords:** COVID-19; SARS-CoV-2; ketogenic diet; nutrition in pregnancy; microbiota; cancer; carcinoma; precision medicine; health promotion; metabolomics

## 1. Pathobiology and Metabolic Dysfunction of SARS-CoV-2

Coronavirus 2019 (COVID-19) has become a global threat with a mortality rate around 6% [1]. In adults, the clinical manifestations of COVID-19 may or may not appear for the entire duration of the incubation period (2–14 days) and beyond, or they may be severe. Children are less likely to develop symptomatic infections and are less prone to serious illness [2], although there are reports of children who have a disease called Multisystem

Inflammatory Syndrome [3,4]. COVID-19 belongs to the beta family of Coronaviridae and is known for the spike protein used to hook and infect the host cell [5–7], conferring COVID-19 easy transmissivity and high pathogenicity, [8] allowing the virus to: (i) fuse cell–cell and RNA release, (ii) start the replication cycle, and (iii) spread further to infect more cells [9,10], ensuring greater transmissibility (R0 > 2). From the ultrastructural point of view, the virus, having a higher specificity than other viruses of the same family [11], binds to the host via the binding domains of the angiotensin 2 converting enzyme receptor (ACE2), expressed more on the luminal surface of type II alveolar epithelial cells [12], resulting in an increase in angiotensin-2 with relative increase in pulmonary vascular permeability and subsequent severe acute respiratory syndrome or multiorgan dysfunction [13]. This clinical picture is due to the presence of acute inflammation, mainly CD4 and CD8 positive T lymphomonocytes, responsible for the recall of cytokines (interleukins such as IL-1$\beta$, IL-2R, IL-6, as well as interferon and tumor necrosis factor) and chemokines such as CCL-2, CCL-3, and CCL-10, which determine the hyperinflammatory state in COVID-19-positive patients [14]. Following this process, there is therefore the desquamation of pneumocytes, the formation of fibrinoid exudates and pulmonary edema, and finally formation of hyaline membrane with consequent irreversible alveolar damage. Conversely, during the recovery phase, there is an increase in pro-coagulation factors and subsequent activation of the coagulation cascade with the formation of small and large peripheral vascular thrombi also known as disseminated intravascular coagulation [13,15]. In more critically ill patients, laboratory tests also show hyperferritinaemia, elevated lactate dehydrogenase, and erythrocyte sedimentation rate [16]. This pathological structure can also be associated with further complications at the olfactory level [17] or at the visual level (eye pain, redness, and conjunctivitis), splenic and hepatic level where drug toxicity and immune-mediated damage play a role [18], renal, neurological, and skin level such as erythematous rashes and urticaria, lympho-haematological [19–26], and even at the level of the cardiovascular system [27–29] where patients with SARS-CoV-2 have an increased risk of developing acute myocardial infarction following coronary spasm, hypoxic damage, microthrombi, direct vascular endothelial damage, hypercoagulability, and instability of the atherosclerotic plaque [30]. It is also known that the downregulation of ACE2 and the increased stimulation of the angiotensin II receptor are associated with systemic hypotension, hypokalaemia, and lung damage [31]. As is well known, iron represents one of the main elements in the fight against pathogens, given its ability to support the immune system during the viral replication phase. In fact, since high availability of iron is required for COVID-19 to allow the hydrolysis of the ATP required in this phase, the innate immune system will modify the bioavailability of iron downwards, limiting its ability to replicate and helping the body to fight the virus. Ferritin is the most important component of iron metabolism, and its main role is to store iron during the ferric state during infections. During the inflammatory phase, the serum iron concentration decreases and ferritin increases, causing hyperferritinemia and activation of macrophages. Regulating this mechanism is hepcidin, capable of regulating the concentration of intra- and extracellular iron thus depriving the pathogen of iron for the replication of COVID-19 [32,33]. The metabolic processes of iron are also generated by the generation of reactive oxygen species (ROS) [34], which can cause mitochondrial dysfunction favoring multiple-organ damage as occurs during COVID-19 infection [34,35]. Alterations at this level lead to an upregulation of mitochondrial genes and genes that respond to oxidative stress with further accumulations of intracellular iron. These generate ROS and reactive nitrogen and sulphur species, further contributing to the increase in the inflammatory response in COVID-19 disease. In fact, the same mitochondrial ROS also directly activates the production of proinflammatory cytokines that can alter mitochondrial homeostasis and mitochondrial respiration thus causing various systemic alterations [36]. The inflammatory phase also leads to a depletion of the body's natural antioxidant agents such as vit C, which in favorable physiological conditions protects cells, decreases oxidative stress and ROS, and improves the body's circulatory function and its immune system. In critical conditions, its concentration drops precipitously so as to require intravenous administration to reach a quantity sufficient for the body. Although still in need of robust confirmation, on this basis clinical studies have been conducted on the treatment of COVID-19-positive patients, which have shown how this supplement can improve the critical condition of some patients [37,38]. Similarly to vit C, vit D also has the same effect on oxidative stress and provides a protective effect from COVID-19 but, also in this case, further studies should be encouraged to shed light on the effects of vit D deficiency in COVID-19-positive patients [39–41].

## 2. Microbiota and SARS-CoV-2 Infection

Lately, different authors have reported in literature the possibility, more than suggestive, of a connection between COVID-19 infection and intestinal microbiota dysbiosis, considering that, in COVID-19 patients with severe intestinal dysfunction, the presence of virions in fecal, oral, and gastrointestinal samples was found in a range between 2% and 36% [42–51]. The most commonly reported intestinal symptoms related to COVID-19 could be mild as nausea and stomach discomfort or more intense such as vomiting and diarrhea [27,52–55]. Moreover, variations of the intestinal microbiota, with an increase in opportunistic pathogenic germs and reduction of protective commensal bacteria, relate to fecal levels of SARS-CoV-2 and severity of symptoms from COVID-19. Furthermore, this microbiota alteration could continue even after the eradication of SARS-CoV-2 and after the remission of disease symptoms [56]. Recent scientific evidence provided that the gravity of COVID-19 symptomatology is linked to comorbidity and advanced age since both aspects are connected to inflammation and alteration of the intestinal microbiota, in consideration of both the nature of the intestinal flora and its bacterial composition [57]. Consequently, it has been assumed that an intervention directed at supporting the intestinal barrier and reducing the inflammatory stimulus by recommending a diet rich in fiber and fermenting foods could be suitable to reduce the infection and gastrointestinal symptoms linked to COVID-19 [58]. Therefore, it is supposed that COVID-19 infection might be capable of modifying and undermining the integrity of the gut microbiota with a consequent higher gravity of the disease and complications [58]. Consequently, the intestinal biological components and its related dysbiosis are considered a potentially fundamental element to influence the adaptive response versus respiratory pathogens [59]. Indeed, recent studies have reported that patients with COVID-19 infection and underlying comorbidity with consequently reduced gut microbiota diversity and greater intestinal permeability showed a worse prognosis [52]. Confirmation of this evidence was the evaluation of how food microbes, such as probiotics or prebiotics, develop an antiviral effect against coronaviruses and could strengthen host immune functions. Moreover, a sensible decrease of Lactobacillus and Bifidobacterium species has been found in patients affected by COVID-19, even if the clinical meaning of this result is not yet defined [60]. Therefore, intestinal microbiota [61] could lead healthy subjects to an hyperinflammatory condition [62], which enhances the harmful effects of COVID-19. Indeed, composition of the intestinal microbiota is involved in the production of inflammatory cytokines and has an important role in the induction, development, and correct function of the host immune system and its activity [63]. In addition, ACE2 protein is widely distributed on the luminal surface of intestinal epithelial cell [64,65]. Indeed, recent evidence has reported that the "cytokine storm" may be the key that aggravates the morbidity and mortality of the COVID-19 infection [66] (Figure 1), and there is even evidence that COVID-19 affects women less and causes a stronger T-cell response than males, leading to more effective viral clearance [67–69].

This condition of hyperinflammation is also documented in histological specimens of the colon and mesentery [70]. Therefore, anti-cytokine therapy for reducing patients' hyperinflammatory status could be recommended for the treatment of patients with severe COVID-19 and particularly in immunosuppressed subjects [71]. Furthermore, the prescription of a balanced diet, able at the same time to provide the organism with an adequate nutritional supply, can favor the correct functioning of the immune system [72]. In this regard, specific micronutrients such as vit (C and D) or elements such as zinc, taken directly from food through a correct diet or prescribed in the form of supplements, represent for this purpose a valid support having the dual function of guaranteeing a contribution to the organism for its physiological functions and allowing it to have a continuous reserve in pathological conditions, ensuring a constant and strong immune function even in critical situations [73]. Moreover, the recent literature is unable to provide data so concrete and reliable as to consider the clinical use of probiotics in patients with symptoms related to COVID-19, and therefore their administration cannot be officially recommended until there are experimental results that will be clearer on their relationship with the intestinal microbiota in COVID-19 patients, as well as on their functionality and usefulness in the fight against COVID-19 [74]. However, it is supposed that an approach that modulates the intestinal microbiota could represent one of the therapeutic approaches of COVID-19 and its complications [75]. This is especially important for our vulnerable populations, such as the elderly, and oncological and immunosuppressed patients, and it requires further, in-depth research. On the other hand, the role of the microbiota linked to COVID-19 has not yet been studied in newborns, but it is shown that through breastfeeding, vaccinated women [76–82] confer immunity to their children; an increase in IgG and IgA antibodies against SARS-CoV-2 [83–86] and consequent decrease in the risk of hospitalization was demonstrated [87,88]. Preliminary studies on these positive effects had concerned the analysis of these antibodies also at the level of the umbilical cord and placenta [89–91]. This research represents a driving force for future studies in this patient class.

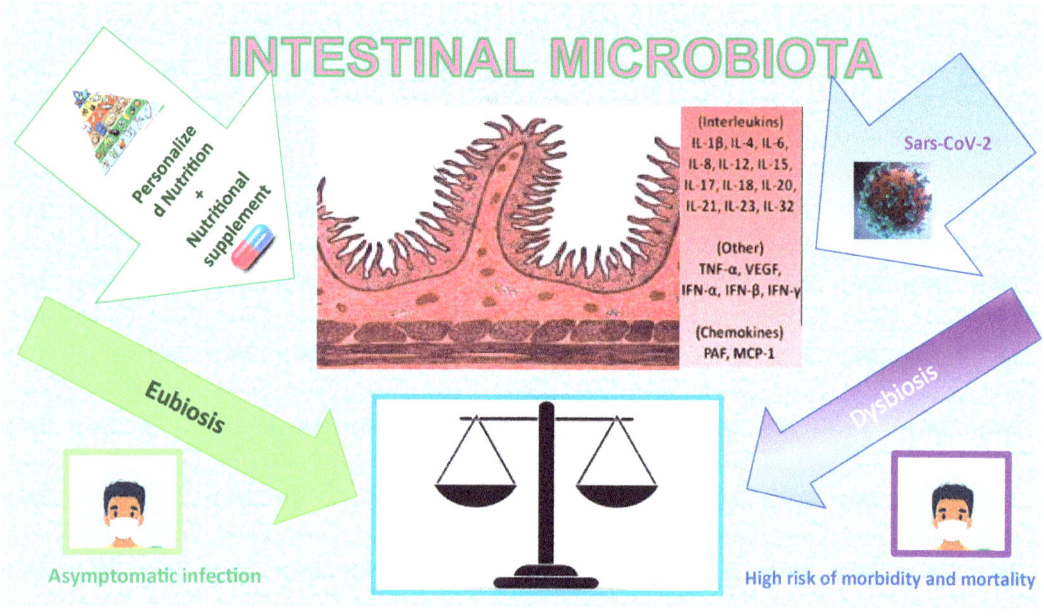


**Figure 1.** Impact of COVID-19, nutrition, and nutritional supplement on Intestinal Microbiota. The figure is a graphic illustration of how the correct use of macro- and micronutrients can be the ideal balance of our organism in the defence or rapid recovery to COVID-19, having a positive effect on the Intestinal Microbiota (green side), mainly for their antioxidant and anti-inflammatory properties.

## 3. Nutrition in Pregnancy during COVID-19 Infection

Pregnancy represents a special biological phase of the life of a woman and determines a lot of indicatory physiological changes involving all organ systems in the body with the main aim to sustain the growing fetus [92]. Nutrition represents a pivotal point for the physiological changes of pregnancy and, in this regard, childbearing causes an increased water need in order to expand maternal blood water and increase cardiac output from

4 to 6 L/min and, thus, blood flow to the kidneys and utero-placental unit, decreasing blood pressure overall [93]. The increase in tidal volume requires an average increase in oxygen demand in pregnancy of about 20% of minute ventilation. Breathing and nutrition are closely related processes. Through the digestive system, oxygen is used to oxidize the nutrients obtained from ingested foods and to obtain the necessary energy. Physiological changes in respiratory function in pregnancy allow the umbilical vein blood amount to be rich enough in oxygen for the fetal cellular respiration process [94]. Nutrition means fetal growth; especially in the third trimester of pregnancy, hormones of placental origin determine a state of insulin resistance in the pregnant woman; a condition of hyperglycaemia is established, which maintains fetal growth and lipolysis to satisfy maternal needs [95]. The mother's requirement for increased insulin could be a biological solution to promote fetal growth, while insulin itself can remain the main growth factor alongside conception product development [96]. It is known that nutrition plays an essential role in the development and maintenance of the immune system. Its deficiency can negatively affect the risk of infections and the maternal organism's acceptance of the product of conception, characterized by different genetic and haplotypic heritages [97]. If innate immunity (NK cells and monocytes) is preserved, acquired immunity (T cells and B cells) is downregulated, harming any specific immune response, hence causing an immunosuppressed state [98]. In addition, the inflammatory response differs alongside pregnancy according to the three main stages: the pro-inflammatory stage (implantation and placentation in the first trimester), the anti-inflammatory stage (fetal growth in the second trimester), and the second pro-inflammatory stage (initiation of parturition) [99]. Intuitively, nutritional deficiencies can compromise the immune response by greatly interfering with the body's response to external inputs, leading to an amplification of the pathophysiological pathways and, thus, conditioning the corresponding outcomes [92]. To date, the SARS-CoV-2 infection represents one of the most interesting examples of how peculiar the interactions of the systemic range of the action pathogen are, i.e., a peculiar biological environment [97]. From this point of view, nutrition might act as a synergic co-factor, which may be able to influence organism response, while taking into consideration its significance is still under scientific evaluation. It is established that macronutrient intake variably influences the immune system response and inflammatory pathways. Thus, it may be used as a predictive factor for response to SARS-CoV-2 in pregnant women. To date, only a few papers are available. Moreover, these existing papers (Table 1) might be used to discriminate between food and macronutrients useful or harmful in COVID-19 infection management [100].

**Table 1.** All the studies existing to date investigating the effects of specific diet regimen, foods, micro-, or macronutrients on the response to COVID-19 viral infection in pregnant women.

| Paper ID | Year of Publication | Type of Study | Cohort ($n°$) | Diet Regimen | COVID-19 Outcomes |
|---|---|---|---|---|---|
| Eskenzi et al. [95] | 2021 | Prospective, multinational | 2071 | Low-calorie diet in glucose altered metabolism or overweight | Improving |
| Chen et al. [101] | 2022 | Prospective, nationwide (China), multicenter | 3678 | Diet based on vegetables and fruit | Improving |
| Erol et al. [102] | 2021 | Prospective case-control (Turkey) | 60 | Lower vit E<br>Higher Afamin serum levels | Worsening<br>Worsening |

**Table 1.** *Cont.*

| Paper ID | Year of Publication | Type of Study | Cohort ($n^{\circ}$) | Diet Regimen | COVID-19 Outcomes |
|---|---|---|---|---|---|
| Anuk et al. [103] | 2021 | Prospective case-control (Turkey) | 200 | Lower Zinc<br>Higher Copper<br>Lower Zn/Cu ratio<br>Higher Magnesium serum levels | Worsening<br>Worsening<br>Worsening<br>Worsening |
| Uta et al. [104] | 2022 | Prospective case-control (Romania) | 446 | Iron deficiency<br>Iron and folate supplement (in anemic women) | Worsening<br>Improving |
| Nawsherwan et al. [92] | 2020 | Review | - | Micronutrient deficiency and/or low intake (vit A, C, D, E, Fe, Se, Zn) | Worsening |
| Mate et al. [100] | 2021 | Review | - | "Functional food" diet (vegetables, omega3 fatty acid, micronutrient, vit A, B, D, E, choline) | Improving |

The need of FA is not the same for the whole population and reaches its maximum levels during pregnancy and breastfeeding. Since the 1990s, in order to prevent NTDs (neural tubal defects), a daily intake of FA (folic acid) between 400 and 600 mcgr has been highly recommended during pregnancy by most health systems around the world; this is to compensate the insufficient average dietary intake of FA. A nutrition that includes fresh fruit and vegetables is therefore essential to meet the daily requirements of folate. FA also contributes to the physiological anemia of pregnancy, promoting hematopoiesis and therefore the increase in Hb levels [105]. This works only if there is enough iron deposited to be used. In a case-control study during the COVID-19 pandemic, the authors [104] demonstrated that it is mandatory to screen for and treat anemia with both FA and iron. They found that among pregnant women with iron deficiency anemia, the COVID-19 group had a higher risk of puerperal infection, emergency c-section and SGA (small for gestational age). Low birth weight, prematurity, and lower APGAR scores were also more frequent in the COVID-19 group. Specifically, a daily combination of FA and iron could help to normalize weight of the newborn in this setting [104]. According to the latest international guidelines, it is mandatory to increase the daily FA dose up to 4–5 mg in case of: previous NTD-affected pregnancy, neurological disease and malformations, contemporary therapy with anti-epileptic drugs, or, with malabsorption disease and high maternal BMI and/or impaired stages of glucose metabolism. This latter group might be more vulnerable to contracting the COVID-19 viral infection, as reported in (Eskenazy et. al., the INTERCOVID Study, 2021). According to this study, GDM (gestational diabetes mellitus), pre-existing DM, and being overweight or obese are all independent risk factors for SARS-CoV-2. An appropriate low-calorie dietary regimen besides eventual pharmacotherapy is needed for weight control during pregnancy and to improve pregnant women's immune-receptivity. Indeed, in the DM and GDM groups, the obese and overweight women were at a higher risk of having COVID-19 than the normal weight women, and also the insulin-dependent diabetes group was at higher risk of COVID-19 than the diet-therapy group both for DM and GDM [95]. Mate et. al. emphasized the role of a well-balanced dietary regimen for pregnant woman in the SARS-CoV-2 setting, while focusing on the main nutrient useful to boost the immune system against systemic viral infection [100,106]. It should be mandatory to adopt an adequate nutrition and eventual feeding supplement not only during pregnancy, but also from preconception to the early post-natal period. The model of "food safety" which has been proposed is based on the experience of other viral and/or respiratory infections during pregnancy, which are well-known in the relevant literature. The core concept is to base the dietary regimen on functional foods, i.e., fruit and vegetables, including sufficient amounts of recommended micronutrients such as vit A, B, D, E, Omega-3 Fatty Acids, Choline, Zn, Fe, Se, and, as mentioned above, FA. This helps the immune system and has a positive impact on pregnancy outcomes in case of viral infection, including SARS-CoV-2. Omega-3 poly-unsaturated fatty acids down-regulate the excessive inflammatory response triggered by some systemic viral infections. Choline, in the case of respiratory virus infections, improves adverse fetal effects. The main adverse effects linked to not recommended food during pregnancy in case of systemic infection are reported in the existing literature: IUGR (intrauterine growth restriction), increased infant mortality, congenital diseases, miscarriage, preeclampsia, and nervous system anomalies or dysfunction [100]. It is important to mention the national cross-sectional study conducted in China by Chen et al. in 2022 on the sample of 3678 pregnant women during the COVID-19 pandemic. The Authors reported that a median daily intake of so-called functional food, like vegetables and fruit, significantly decreased to low, moderate, and high severity of pandemic, and, in this sense making immune system more vulnerable. In addition to the low quality of the foods consumed, the perinatal outcomes got worse [101]. Interestingly, researchers in Northern New England have proposed a model to identify and address questionable food choices among pregnant women during the pandemic period, highlighting the need for specific food programs [107]. Erol et al. discussed that vit E and Afamin are significant predictors of adverse perinatal outcomes in COVID-19 infected women. Specifically, vit E levels [102] were significantly lower than in the healthy group of women. In addition, Afamin levels were significantly higher, while positively correlated with CRP (C reactive protein) levels. The higher adverse perinatal outcomes in COVID-19 groups are due to higher oxidative stress [102]. A subsequent study conducted by Erol et al., 2021 enlightened readers on the possible correlation between maternal selenium status and clinical outcomes of pregnant women with the SARS-CoV-2 infection. They found that selenium levels negatively correlate with gestational week, D-dimer, and interleukin-6 (IL-6). As the infection got worse, researchers could observe lower selenium levels and higher inflammatory factors. This vicious circle leads to a higher maternal need for selenium, and it makes a selenium supplement mandatory [102]. According to the study conducted by Anuk et al. on pregnant women affected by SARS-CoV-2, serum zinc levels also negatively correlate with IL-6, and with other acute phase markers, i.e., erythrocyte sedimentation rate, PCT (procalcitonin), and CRP [103]. On the other hand, the serum copper level showed a positive correlation. Therefore, serum Zn/Cu levels were a reliable predictor of viral infection severity in pregnancy. Similarly, in the case when the increased serum magnesium levels were found, there were predictors of pancytopenia and higher CPR [103].

The appropriate assessment of the nutritional status of pregnant patients during and in post-COVID-19, is one of the pillars in the management of these patients. On the other hand, personalized dietary recommendations for these patients represent the best strategy to ensure their recovery. Unfortunately, data on SARS-CoV-2 and nutrition during pregnancy are still in the process of collection, and models are in the developing phase. However, a recommended feeding protocol should be developed in the future.

## 4. Nutrition in Oncologic Patients during COVID-19 Infection

The different female carcinomas [108–118] can influence nutritional status, with weight loss seen in 30–80% of early cases (observed in the 6 months prior to diagnosis) and correlated with the stage of the disease. Weight loss and malnutrition are the result of a reduction in food intake, in the presence of symptoms such as mycosis of the oral cavity, anorexia, nausea, vomiting, dysphagia, and dysgeusia related both to the site of the tumor and to specific treatments; undernutrition and loss of weight and muscle mass can lead to a greater risk of chemotherapy toxicity, while in other cases anticancer therapies and chemotherapy can lead to weight gain resulting in excess malnutrition, and then overweight/obesity, which represent risk factors for metabolic syndrome. In these patients with a compromised nutritional status, COVID-19 infection can further worsen the prognosis as the virus acts on the nutritional status by producing hypercatabolism and rapid muscle wasting [119]. The leading cause of death among infected patients is basically an acute respiratory distress syndrome caused by generalized inflammation. Several studies have analyzed the nutritional status and mortality among cancer patients diagnosed with COVID-19, demonstrating that some immune inflammatory parameters such as the number of lymphocytes could be associated with lower survival. Among other factors, being overweight/obese contributes to the overproduction of proinflammatory IL-6 and reduces cytotoxicity of Natural Killer cells [120]. The assessment of nutritional status in these patients should also take into consideration the possibility of sarcopenia development. The decrease in concentration of some proteins present in the serum such as albumin is often associated with malnutrition, inflammation, and lymphopenia that are negative prognostic factors in these patients [121]. Sarcopenia, defined as the depletion of skeletal muscle mass and muscle strength, was associated with the proliferation of peripheral mononuclear cells, and damaged homeostasis of natural killer lymphocytes [122]; the main mechanism underlying impaired immunity refers the abnormal myokines such as interleukin IL-6, IL-15, and IL-17 [123]. Regarding metabolic stress during severe infection by COVID-19, skeletal muscle is catabolized to provide the immune system, liver, and gut with amino acids [124]. In light of the above, the nutritional management of these patients is essential. Evaluation is a priority in order to be able to assign a correct nutritional therapy considering that these oncological patients affected by COVID-19 have an increased protein-caloric requirement due to the infection and fever. In a recent study of 2021, Sukkar SG et al. showed that the administration of a ketogenic diet characterized by a low carbohydrate (20–120 g) content, high lipid content, and normoprotein in patients with COVID-19 results in a reduction in mortality and artificial ventilation in these subjects. Ketosis protects healthy tissues against oxidative stress as it decreased production of reactive oxygen species and increased endogenous antioxidant capacity. The ketogenic diet can also inhibit inflammation, reducing circulating inflammatory markers. A KD, via hydroxybutyrate, is capable of activating hydroxycarboxylic acid receptor 2, which inhibits nuclear factor kb in macrophages, dendritic cells, and microglia reducing neuroinflammation [125]. As part of a ketogenic diet, it is possible to combine an intermittent fasting regime that many authors have proposed to alleviate the symptoms in the course of COVID-19 infection. Intermittent fasting is a practice that involves the restriction of eating and there are several approaches. For example, with 16/8, food is consumed only in 8 h or the 5:2 approach, which involves eating regularly five days a week and limiting daily calorie intake on the other two days, to 500–600 calories; it has gained popularity in recent years and shows promise as a possible new paradigm in the approach to weight loss and the reduction of inflammation and has many potential long term health benefits. There is good evidence that IF can benefit cardiometabolic health by decreasing blood pressure, insulin resistance, and oxidative stress. According to a study published in BMJ Nutrition, Prevention & Health, intermittent fasting could reduce the risk of hospitalization or death in patients who have contracted COVID-19 [126]. A metabolic change in IF is the increase in linoleic acid-enriched triacylglycerol species in the liver and serum during fasting. It has been hypothesized that linoleic acid (a polyunsaturated fatty acid of the omega 6) locks the spike protein of SARS-CoV-2 in a confirmation that is not conducive to the effective binding to ACE2. Besides elevated concentration of linoleic acid during fasting, it may lessen the number of infected cells and the number of SARS-CoV-2 virions in cells and so decrease the severity of the symptoms. Another benefit of IF is the increased levels of galectin-3, and this protein has been shown to bind directly to pathogens and have various effects on the functions of immune cells. Particularly, a randomized human study of low-frequency (once-per week) 24-h water-only intermittent fasting showed that fasting increased galectin-3 level over a moderate term and this protein involved in host defence to infectious diseases [127,128] in that it stimulates anti-inflammatory effects by modulating nuclear factor kappa-light-chain-enhancer of activated B cells (NF-κB) and the NLR family pyrin domain containing 3 (NLRP3) inflammasome [129], which should inhibit the hyper-inflammation associated with COVID-19. On the others hand, a number of micronutrients, including vitamin D, zinc, and omega-3 fatty acids have been shown to play key roles in supporting immune function [130–132] and in reducing risk of respiratory infection [131]. These nutrients can be obtained from the diet and are available as dietary supplements either alone or as part of multivitamin or multi-nutrient mixtures. A biologically plausible role exists for certain vitamins and minerals in immune pathways. Vitamin D is a steroid hormone involved in the modulation of the innate and acquired immune system as well as in the production of the antimicrobial peptides such as human β-defensin-2 and cathelicidin, in addition to the expression of genes responsible for the destruction of the intracellular pathogens. Many studies have consistently suggested that vitamin D deficiency is associated with increased risk of respiratory tract infections, especially in influenza and now COVID-19 [133–136], and some studies have suggested that it reduce SARS-CoV-2 transmission by enhancing antiviral immunity and to reduce mortality by mitigating the cytokine storm linked with severe COVID-19 [137]. Zinc is a trace element with potent immunoregulatory and antiviral properties and is utilized in the treatment of COVID-19. In a study in 2020, Jothimani D. et al. demonstrated that zinc deficient patients developed more complications, and the deficiency was associated with a prolonged hospital stay and increased mortality [138]. Other protective elements against infection have been hypothesized to be the very-long chain omega-3 fatty acids (EPA and DHA) that have anti-inflammatory properties that may help reduce morbidity and mortality from COVID-19 infection. These omega-3 s are associated with lower levels of circulating inflammatory cytokines, and intervention with fish oils reduces levels. EPA and DHA are precursors to a suite of inflammation-resolving mediators (IRMs [139]) that actively regulate the resolution of acute inflammation. IRMs down-regulate cytokine production and promote a return to homeostasis via monocyte/macrophage-mediated uptake of debris, apoptosis of neutrophils, and clearing of microbes. Accordingly, higher intakes of EPA and DHA (which result in higher RBC EPA+DHA levels, hereafter called the Omega-3 Index, O3I, have been proposed to lower the risk for adverse outcomes from COVID-19 [140–146], and case reports suggesting benefit have been published. An excessive inflammatory response, called a 'cytokine storm', is a frequent occurrence of the severe form of COVID-19. Omega-3 fatty acids have potent anti-inflammatory activities, and these fatty acids can mitigate the cytokine storm of COVID-19 such as suggested in a study pilot by Arash Asher et al. [147]. Maintaining a good nutritional status is essential for the functioning of the immune system and for defence against viral diseases. According to the European Society for Clinical nutrition and Metabolism (ESPEN) guidelines of good nutrition in patients with SARS-CoV-2, it depends on an adequate level of intake of nutrients such as vit D, A, C, B, and zinc, copper, iron, and selenium, which play a key role in the prevention of malnutrition [148].

## 5. Discussion and Potential Future Approaches

For about two years, the main symptom investigated by the vast majority of world researchers in the fight against COVID-19 was pulmonary [149–154]. Effort was due first of all to the impossibility of knowing how to manage a pandemic with the health complications that have arisen and, secondly, to the absence of targeted COVID-19 drugs, which in the first pandemic phase saw the experimentation of therapies already existing for others purposes and, in principle, were at least adaptable to the pulmonary and systemic symptoms presented by patients COVID-19 affected [155–160]. The advent of vaccines [85,161–163] has in fact revolutionized this perspective as they have guaranteed not only a preventive and protective effect in the contagion phase, but also a therapeutic effect as severe symptoms have increasingly given way to milder symptoms or even a lack of symptoms in the vaccinated population. This aspect must also be taken into consideration for extra-pulmonary symptoms, even in the long-term, assuming future viral variants and future vaccines. Proof of this is the study by Ran Barzilay et al., which, in a cohort of 3000 patients, showed that extrapulmonary symptoms were reduced by more than 50% in vaccinated patients, highlighting how vaccinated subjects have lower risks than unvaccinated ones [164]. Preliminary data from other studies go in this direction, reinforcing the concept that vaccination tends to reduce the most common symptoms related to COVID-19 [165]. Comparisons between these two classes also showed that on average there is a higher likelihood of having no symptoms among vaccinated than unvaccinated and reversed when considering the total number of symptoms or hospitalization [166,167]. For the future, supplementary comparative studies between these two classes would be desirable in relation also to the use of micronutrients, in addition to those specific to the gender inherent in the adverse effects of vaccination. In this sense, the data are in fact discordant considering that some authors report comparable data between males and females, while others do not, also in relation to the side effects after the first dose of the vaccine [168–171].

Probably today one of the least investigated aspects of the pathogenesis and development of COVID-19 manifestations is related to nutrition and how this can positively influence both in the short and long term. This topic is therefore both current and of great social impact considering that both physical and psychological manifestations related to COVID-19 have been identified, which can persist long-term even beyond recovery [172]. In fact, several authors around the world have also reported symptoms of psychological stress such as: sleep loss, nervousness, fatigue, depression, anxiety, [173–177] especially in the female population, very often linked to other cofactors such as the social and economic status or the patient's health. Fortunately, these are often of a transitory type, especially if considered short-term [178], but, in the literature, there are several manuscripts that [172,179–183] report an up to three times increased risk in women of experiencing long-term symptoms. Unfortunately, on these analyses, there does not seem to be uniformity of results as other authors report discordant data [184–186]. These discrepancies suggest that additional factors, perhaps little studied to date, may contribute to these mechanisms. In fact, ethnic or hormonal causes have been seldom investigated in women with COVID-19. The latter, in fact, seem to influence the state of hyperinflammation even after healing [187,188] and consequently on the production of IgG antibodies [189]. These studies highlight some critical issues and consequently some perspectives still to be explored on COVID-19. The first of these is linked to the impossibility of having long or very long-term studies, capable of evaluating the effects of the COVID-19 infection in a prolonged manner; this lack may be the reason for conflicting data, which could in part be filled by a wider sharing of data by different research groups through the development of global rather than national studies or single research groups. This effort could actually confirm or dispel contributing causes such as ethnicity or the role of hormones in long-term effects. New confirmations on the role of gender could imply targeted therapeutic solutions or personalized prevention models; targeted research looking in this direction would therefore be desirable [190]. Another important aspect that deserves more study is the evaluation of long-term effects in patients with already established chronic diseases such as diabetes, arterial hypertension, or pulmonary dysfunction. In these pathologies, patients very often present with immune system alterations and an increase in pro-inflammatory factors, which together with COVID-19 could exacerbate the immune response or delay its effectiveness, resulting in slower or more difficult healing. Food supplementation could represent a valid support in these patients, especially if in old age or with slower recovery capacity, as already evaluated by some authors but only in a short time [179,182]. This aspect is even more important in those patients who, being bedridden for very long periods, risk going into sarcopenia; dietary implementation with the use of food supplements, or probiotics such as creatine or phosphocreatine could be an important aid in slowing down muscle resorption and facilitating its anabolism in this class of patients. Moreover, it is possible that this dietary approach may be useful in association with anticancer drugs such as Bevacizumab [117], used in the fight against COVID-19 as antiangiogenics for its treatment [37,191]. It is also no coincidence that, albeit slowly, the first studies on natural antioxidants such as vit C or D begin to appear in the literature, which demonstrate an improvement in critically ill patients by supporting their immune system, even if, in this regard, data should be more robust to be widely used in routine clinical practice [38,40]. Another possible aspect to be investigated with particular interest could be inherent to hepcidin agonists that can be used to increase endogenous production, allowing their use to decrease ferroportin and iron levels [192]. Furthermore, natural hepcidin inducers are also present in fruits and vegetables [193]. Finally, interesting approaches could also be described in the future regarding the relationship between COVID-19, nutrition, and changes in the intestinal and vaginal microbiota [118].

## 6. Conclusions

In recent years, COVID-19 and the resulting global pandemic have raised many both direct and indirect issues. To date, global efforts have been directed not only at improving vaccines against the virus but also at developing ever more targeted and specific medical therapies. It is therefore evident that the study of the mechanisms of action of the virus but above all of the excipients or supplements that can interact with the viral inflammatory process is a fundamental pillar in the fight against COVID-19. From this point of view, diet, through the intake of vit and micronutrients, represents the first bulwark in the prevention and fight against the virus and its associated symptoms, necessary for hospitalization or not. Furthermore, nutrition takes on the character of personalized medicine based on the patient's condition. This is more significant in women where one must also consider other possible factors than the normal physiological or oncological state, such as pregnancy. Based on these assessments, the characteristics and needs of the immune system change together with the ability to react to the infection and regulate its spread at a multi-organ level. This highlights how diet can change and be reshaped as needed through assessments and personalized indications based on the state of health of COVID-19 positive patients.

**Author Contributions:** Conceptualization, E.C. and M.D.; literature review, E.C., M.V., B.L., A.D. and M.D.; writing—original draft preparation, E.C., M.V., B.L., A.D. and M.D.; writing—review and editing, E.C., G.I., G.C. (Gerardo Cazzato), G.C. (Gennaro Cormio), V.P., A.M., S.S., E.M., S.A., D.L.F., R.D.L., M.L., L.R. and G.D.V.; funding acquisition, A.D. and D.L.F. All authors have read and agreed to the published version of the manuscript.

**Funding:** This work has been supported by funding from the Italian Ministry of Health "Ricerca corrente 2022".

**Institutional Review Board Statement:** Not applicable.

**Informed Consent Statement:** Not applicable.

**Data Availability Statement:** Not applicable.

**Acknowledgments:** Thanks to Antonietta Cimmino, a memorable example of faith, dedication to work and passion for science and research.

**Conflicts of Interest:** The authors declare no conflict of interest.

## References

1. Dong, L.; Hu, S.; Gao, J. Discovering drugs to treat coronavirus disease 2019 (COVID-19). *Drug Discov. Ther.* **2020**, *14*, 58–60. [CrossRef] [PubMed]
2. Ludvigsson, J.F. Systematic review of COVID-19 in children shows milder cases and a better prognosis than adults. *Acta Paediatr.* **2020**, *109*, 1088–1095. [CrossRef] [PubMed]
3. Mahase, E. COVID-19: Concerns grow over inflammatory syndrome emerging in children. *BMJ* **2020**, *369*, m1710. [CrossRef] [PubMed]
4. Verdoni, L.; Mazza, A.; Gervasoni, A.; Martelli, L.; Ruggeri, M.; Ciuffreda, M.; Bonanomi, E.; D'Antiga, L. An outbreak of severe Kawasaki-like disease at the Italian epicentre of the SARS-CoV-2 epidemic: An observational cohort study. *Lancet* **2020**, *395*, 1771–1778. [CrossRef]
5. Li, F. Structure, Function, and Evolution of Coronavirus Spike Proteins. *Annu. Rev. Virol.* **2016**, *3*, 237–261. [CrossRef] [PubMed]
6. van Hemert, M.J.; van den Worm, S.H.; Knoops, K.; Mommaas, A.M.; Gorbalenya, A.E.; Snijder, E.J. SARS-coronavirus replication/transcription complexes are membrane-protected and need a host factor for activity in vitro. *PLoS Pathog.* **2008**, *4*, e1000054. [CrossRef] [PubMed]
7. Ochani, R.; Asad, A.; Yasmin, F.; Shaikh, S.; Khalid, H.; Batra, S.; Sohail, M.R.; Mahmood, S.F.; Ochani, R.; Arshad, M. COVID-19 pandemic: From origins to outcomes. A comprehensive review of viral pathogenesis, clinical manifestations, diagnostic evaluation, and management. *Infez. Med.* **2021**, *29*, 20–36.

8. Manning-Sayi, J.; Leonard, B.E. Studies on the nootropic potential of some angiotensin converting enzyme inhibitors in the mouse. *Prog. Neuropsychopharmacol. Biol. Psychiatry* **1989**, *13*, 953–962. [CrossRef]
9. Hoffmann, M.; Kleine-Weber, H.; Schroeder, S.; Kruger, N.; Herrler, T.; Erichsen, S.; Schiergens, T.S.; Herrler, G.; Wu, N.H.; Nitsche, A.; et al. SARS-CoV-2 Cell Entry Depends on ACE2 and TMPRSS2 and Is Blocked by a Clinically Proven Protease Inhibitor. *Cell* **2020**, *181*, 271–280.e278. [CrossRef]
10. Krijnse-Locker, J.; Ericsson, M.; Rottier, P.J.; Griffiths, G. Characterization of the budding compartment of mouse hepatitis virus: Evidence that transport from the RER to the Golgi complex requires only one vesicular transport step. *J. Cell. Biol.* **1994**, *124*, 55–70. [CrossRef] [PubMed]
11. Bourgonje, A.R.; Abdulle, A.E.; Timens, W.; Hillebrands, J.L.; Navis, G.J.; Gordijn, S.J.; Bolling, M.C.; Dijkstra, G.; Voors, A.A.; Osterhaus, A.D.; et al. Angiotensin-converting enzyme 2 (ACE2), SARS-CoV-2 and the pathophysiology of coronavirus disease 2019 (COVID-19). *J. Pathol.* **2020**, *251*, 228–248. [CrossRef] [PubMed]
12. Wan, Y.; Shang, J.; Graham, R.; Baric, R.S.; Li, F. Receptor Recognition by the Novel Coronavirus from Wuhan: An Analysis Based on Decade-Long Structural Studies of SARS Coronavirus. *J. Virol.* **2020**, *17*, e00127-20. [CrossRef] [PubMed]
13. Zhang, H.; Penninger, J.M.; Li, Y.; Zhong, N.; Slutsky, A.S. Angiotensin-converting enzyme 2 (ACE2) as a SARS-CoV-2 receptor: Molecular mechanisms and potential therapeutic target. *Intensive Care Med.* **2020**, *46*, 586–590. [CrossRef]
14. Ye, Q.; Wang, B.; Mao, J. The pathogenesis and treatment of the 'Cytokine Storm' in COVID-19. *J. Infect.* **2020**, *80*, 607–613. [CrossRef]
15. Wichmann, D.; Sperhake, J.P.; Lutgehetmann, M.; Steurer, S.; Edler, C.; Heinemann, A.; Heinrich, F.; Mushumba, H.; Kniep, I.; Schroder, A.S.; et al. Autopsy Findings and Venous Thromboembolism in Patients With COVID-19: A Prospective Cohort Study. *Ann. Intern. Med.* **2020**, *173*, 268–277. [CrossRef]
16. World Health Organization. *COVID-19 Weekly Epidemiological Update*, 99th ed.; World Health Organization: Geneva, Switzerland, 2022.
17. Jin, Y.; Yang, H.; Ji, W.; Wu, W.; Chen, S.; Zhang, W.; Duan, G. Virology, Epidemiology, Pathogenesis, and Control of COVID-19. *Viruses* **2020**, *12*, 372. [CrossRef]
18. Zhang, C.; Shi, L.; Wang, F.S. Liver injury in COVID-19: Management and challenges. *Lancet Gastroenterol. Hepatol.* **2020**, *5*, 428–430. [CrossRef]
19. Cazzato, G.; Romita, P.; Foti, C.; Cimmino, A.; Colagrande, A.; Arezzo, F.; Sablone, S.; Barile, A.; Lettini, T.; Resta, L. Purpuric skin rash in a patient undergoing Pfizer-BioNTech COVID-19 vaccination: Histological evaluation and perspectives. *Vaccines* **2021**, *9*, 760. [CrossRef]
20. Cazzato, G.; Foti, C.; Colagrande, A.; Cimmino, A.; Scarcella, S.; Cicco, G.; Sablone, S.; Arezzo, F.; Romita, P.; Lettini, T. Skin manifestation of SARS-CoV-2: The italian experience. *J. Clin. Med.* **2021**, *10*, 1566. [CrossRef]
21. Grandolfo, M.; Romita, P.; Bonamonte, D.; Cazzato, G.; Hansel, K.; Stingeni, L.; Conforti, C.; Giuffrida, R.; Foti, C. Drug reaction with eosinophilia and systemic symptoms syndrome to hydroxychloroquine, an old drug in the spotlight in the COVID-19 era. *Dermatol. Ther.* **2020**, *33*, e13499. [CrossRef]
22. Luers, J.C.; Rokohl, A.C.; Loreck, N.; Wawer Matos, P.A.; Augustin, M.; Dewald, F.; Klein, F.; Lehmann, C.; Heindl, L.M. Olfactory and Gustatory Dysfunction in Coronavirus Disease 2019 (COVID-19). *Clin. Infect. Dis.* **2020**, *71*, 2262–2264. [CrossRef] [PubMed]
23. Wiersinga, W.J.; Rhodes, A.; Cheng, A.C.; Peacock, S.J.; Prescott, H.C. Pathophysiology, Transmission, Diagnosis, and Treatment of Coronavirus Disease 2019 (COVID-19): A Review. *JAMA* **2020**, *324*, 782–793. [CrossRef] [PubMed]
24. Recalcati, S. Cutaneous manifestations in COVID-19: A first perspective. *J. Eur. Acad. Dermatol. Venereol.* **2020**, *34*, e212–e213. [CrossRef] [PubMed]
25. Aggarwal, K.; Agarwal, A.; Jaiswal, N.; Dahiya, N.; Ahuja, A.; Mahajan, S.; Tong, L.; Duggal, M.; Singh, M.; Agrawal, R.; et al. Ocular surface manifestations of coronavirus disease 2019 (COVID-19): A systematic review and meta-analysis. *PLoS ONE* **2020**, *15*, e0241661. [CrossRef]
26. Yao, X.H.; Li, T.Y.; He, Z.C.; Ping, Y.F.; Liu, H.W.; Yu, S.C.; Mou, H.M.; Wang, L.H.; Zhang, H.R.; Fu, W.J.; et al. A pathological report of three COVID-19 cases by minimal invasive autopsies. *Zhonghua Bing Li Xue Za Zhi* **2020**, *49*, 411–417. [CrossRef]
27. Wang, D.; Hu, B.; Hu, C.; Zhu, F.; Liu, X.; Zhang, J.; Wang, B.; Xiang, H.; Cheng, Z.; Xiong, Y.; et al. Clinical Characteristics of 138 Hospitalized Patients With 2019 Novel Coronavirus-Infected Pneumonia in Wuhan, China. *JAMA* **2020**, *323*, 1061–1069. [CrossRef]
28. Zheng, Y.Y.; Ma, Y.T.; Zhang, J.Y.; Xie, X. COVID-19 and the cardiovascular system. *Nat. Rev. Cardiol.* **2020**, *17*, 259–260. [CrossRef]
29. Long, B.; Brady, W.J.; Koyfman, A.; Gottlieb, M. Cardiovascular complications in COVID-19. *Am. J. Emerg. Med.* **2020**, *38*, 1504–1507. [CrossRef]
30. Ackermann, M.; Verleden, S.E.; Kuehnel, M.; Haverich, A.; Welte, T.; Laenger, F.; Vanstapel, A.; Werlein, C.; Stark, H.; Tzankov, A.; et al. Pulmonary Vascular Endothelialitis, Thrombosis, and Angiogenesis in COVID-19. *N. Engl. J. Med.* **2020**, *383*, 120–128. [CrossRef]
31. Silhol, F.; Sarlon, G.; Deharo, J.C.; Vaisse, B. Downregulation of ACE2 induces overstimulation of the renin-angiotensin system in COVID-19: Should we block the renin-angiotensin system? *Hypertens. Res.* **2020**, *43*, 854–856. [CrossRef]
32. Taneri, P.E.; Gomez-Ochoa, S.A.; Llanaj, E.; Raguindin, P.F.; Rojas, L.Z.; Roa-Diaz, Z.M.; Salvador, D., Jr.; Groothof, D.; Minder, B.; Kopp-Heim, D.; et al. Anemia and iron metabolism in COVID-19: A systematic review and meta-analysis. *Eur. J. Epidemiol.* **2020**, *35*, 763–773. [CrossRef] [PubMed]

33. Banchini, F.; Vallisa, D.; Maniscalco, P.; Capelli, P. Iron overload and Hepcidin overexpression could play a key role in COVID infection, and may explain vulnerability in elderly, diabetics, and obese patients. *Acta Biomed.* **2020**, *91*, e2020013. [CrossRef] [PubMed]
34. Schofield, J.H.; Schafer, Z.T. Mitochondrial Reactive Oxygen Species and Mitophagy: A Complex and Nuanced Relationship. *Antioxid. Redox Signal* **2021**, *34*, 517–530. [CrossRef] [PubMed]
35. Shenoy, S. Coronavirus (COVID-19) sepsis: Revisiting mitochondrial dysfunction in pathogenesis, aging, inflammation, and mortality. *Inflamm. Res.* **2020**, *69*, 1077–1085. [CrossRef]
36. Saleh, J.; Peyssonnaux, C.; Singh, K.K.; Edeas, M. Mitochondria and microbiota dysfunction in COVID-19 pathogenesis. *Mitochondrion* **2020**, *54*, 1–7. [CrossRef]
37. Rosa, S.G.V.; Santos, W.C. Clinical trials on drug repositioning for COVID-19 treatment. *Rev. Panam. Salud Publica* **2020**, *44*, e40. [CrossRef]
38. Carr, A.C. A new clinical trial to test high-dose vitamin C in patients with COVID-19. *Crit. Care* **2020**, *24*, 133. [CrossRef]
39. Ali, N. Role of vitamin D in preventing of COVID-19 infection, progression and severity. *J. Infect. Public Health* **2020**, *13*, 1373–1380. [CrossRef]
40. Panarese, A.; Shahini, E. Letter: COVID-19, and vitamin D. *Aliment. Pharmacol. Ther.* **2020**, *51*, 993–995. [CrossRef]
41. Grant, W.B.; Lahore, H.; McDonnell, S.L.; Baggerly, C.A.; French, C.B.; Aliano, J.L.; Bhattoa, H.P. Evidence that Vitamin D Supplementation Could Reduce Risk of Influenza and COVID-19 Infections and Deaths. *Nutrients* **2020**, *12*, 988. [CrossRef]
42. Pagnini, C.; Urgesi, R.; Di Paolo, M.C.; Graziani, M.G. Fighting the Battle against SARS-CoV-2 as Gastroenterologists in Italy. *Gastroenterology* **2020**, *159*, 1619. [CrossRef] [PubMed]
43. Bonato, G.; Dioscoridi, L.; Mutignani, M. Fecal-Oral Transmission of SARS-CoV-2: Practical Implications. *Gastroenterology* **2020**, *159*, 1621–1622. [CrossRef] [PubMed]
44. Fu, B.; Qian, K.; Fu, X. SARS-CoV-2-Induced Vomiting as Onset Symptom in a Patient with COVID-19. *Dig. Dis. Sci.* **2020**, *65*, 1568–1570. [CrossRef]
45. Garg, M.; Christensen, B.; Lubel, J.S. Gastrointestinal ACE2, COVID-19 and IBD: Opportunity in the Face of Tragedy? *Gastroenterology* **2020**, *159*, 1623–1624.e3. [CrossRef] [PubMed]
46. El Hajra Martinez, I.; Relea Perez, L.; Calvo Moya, M. Presence of SARS-Coronavirus-2 in the Ileal Mucosa: Another Evidence for Infection of GI Tract by This Virus. *Gastroenterology* **2020**, *159*, 1624–1625. [CrossRef]
47. Song, J.; Patel, J.; Khatri, R.; Nadpara, N.; Malik, Z.; Parkman, H.P. Gastrointestinal symptoms in patients hospitalized with COVID-19: Prevalence and outcomes. *Medicine* **2022**, *101*, e29374. [CrossRef]
48. Milano, A.; Efthymakis, K.; D'Ardes, D.; Tana, M.; Mazzotta, E.; De Febis, G.; Laterza, F.; Tarquini, P.; Marini, E.; Porreca, E.; et al. Gastrointestinal manifestations of SARS-CoV-2 infection in an Italian population of hospitalized patients. *Ther. Adv. Gastroenterol.* **2022**, *15*, 17562848221104610. [CrossRef]
49. Chen, T.H.; Hsu, M.T.; Lee, M.Y.; Chou, C.K. Gastrointestinal Involvement in SARS-CoV-2 Infection. *Viruses* **2022**, *14*, 1188. [CrossRef]
50. Tabesh, E.; Soheilipour, M.; Sami, R.; Mansourian, M.; Tabesh, F.; Soltaninejad, F.; Dehghan, M.; Nikgoftar, N.; Gharavinia, A.; Ghasemi, K.; et al. Gastrointestinal manifestations in patients with coronavirus disease-2019 (COVID-19): Impact on clinical outcomes. *J. Res. Med. Sci.* **2022**, *27*, 32. [CrossRef]
51. Wang, B.; Zhang, L.; Wang, Y.; Dai, T.; Qin, Z.; Zhou, F.; Zhang, L. Alterations in microbiota of patients with COVID-19: Potential mechanisms and therapeutic interventions. *Signal Transduct. Target Ther.* **2022**, *7*, 143. [CrossRef]
52. Huang, C.; Wang, Y.; Li, X.; Ren, L.; Zhao, J.; Hu, Y.; Zhang, L.; Fan, G.; Xu, J.; Gu, X.; et al. Clinical features of patients infected with 2019 novel coronavirus in Wuhan, China. *Lancet* **2020**, *395*, 497–506. [CrossRef]
53. Zhou, F.; Yu, T.; Du, R.; Fan, G.; Liu, Y.; Liu, Z.; Xiang, J.; Wang, Y.; Song, B.; Gu, X.; et al. Clinical course and risk factors for mortality of adult inpatients with COVID-19 in Wuhan, China: A retrospective cohort study. *Lancet* **2020**, *395*, 1054–1062. [CrossRef]
54. Guan, W.J.; Ni, Z.Y.; Hu, Y.; Liang, W.H.; Ou, C.Q.; He, J.X.; Liu, L.; Shan, H.; Lei, C.L.; Hui, D.S.C.; et al. Clinical Characteristics of Coronavirus Disease 2019 in China. *N. Engl. J. Med.* **2020**, *382*, 1708–1720. [CrossRef] [PubMed]
55. Robba, C.; Battaglini, D.; Pelosi, P.; Rocco, P.R.M. Multiple organ dysfunction in SARS-CoV-2: MODS-CoV-2. *Expert. Rev. Respir. Med.* **2020**, *14*, 865–868. [CrossRef] [PubMed]
56. Xiao, F.; Tang, M.; Zheng, X.; Liu, Y.; Li, X.; Shan, H. Evidence for Gastrointestinal Infection of SARS-CoV-2. *Gastroenterology* **2020**, *158*, 1831–1833.e1833. [CrossRef] [PubMed]
57. Zuo, T.; Zhang, F.; Lui, G.C.Y.; Yeoh, Y.K.; Li, A.Y.L.; Zhan, H.; Wan, Y.; Chung, A.C.K.; Cheung, C.P.; Chen, N.; et al. Alterations in Gut Microbiota of Patients With COVID-19 During Time of Hospitalization. *Gastroenterology* **2020**, *159*, 944–955.e948. [CrossRef]
58. Kalantar-Zadeh, K.; Ward, S.A.; Kalantar-Zadeh, K.; El-Omar, E.M. Considering the Effects of Microbiome and Diet on SARS-CoV-2 Infection: Nanotechnology Roles. *ACS Nano* **2020**, *14*, 5179–5182. [CrossRef] [PubMed]
59. Di Renzo, L.; Gualtieri, P.; Romano, L.; Marrone, G.; Noce, A.; Pujia, A.; Perrone, M.A.; Aiello, V.; Colica, C.; De Lorenzo, A. Role of Personalized Nutrition in Chronic-Degenerative Diseases. *Nutrients* **2019**, *11*, 1707. [CrossRef]
60. Mehta, P.; McAuley, D.F.; Brown, M.; Sanchez, E.; Tattersall, R.S.; Manson, J.J.; Hlh Across Speciality Collaboration, U.K. COVID-19: Consider cytokine storm syndromes and immunosuppression. *Lancet* **2020**, *395*, 1033–1034. [CrossRef]

61. Monteleone, G.; Sarzi-Puttini, P.C.; Ardizzone, S. Preventing COVID-19-induced pneumonia with anticytokine therapy. *Lancet Rheumatol.* **2020**, *2*, e255–e256. [CrossRef]
62. Chen, L.; Deng, H.; Cui, H.; Fang, J.; Zuo, Z.; Deng, J.; Li, Y.; Wang, X.; Zhao, L. Inflammatory responses and inflammation-associated diseases in organs. *Oncotarget* **2018**, *9*, 7204–7218. [CrossRef] [PubMed]
63. Galmes, S.; Serra, F.; Palou, A. Current State of Evidence: Influence of Nutritional and Nutrigenetic Factors on Immunity in the COVID-19 Pandemic Framework. *Nutrients* **2020**, *12*, 2738. [CrossRef]
64. Hamming, I.; Timens, W.; Bulthuis, M.L.; Lely, A.T.; Navis, G.; van Goor, H. Tissue distribution of ACE2 protein, the functional receptor for SARS coronavirus. A first step in understanding SARS pathogenesis. *J. Pathol.* **2004**, *203*, 631–637. [CrossRef] [PubMed]
65. Hashimoto, T.; Perlot, T.; Rehman, A.; Trichereau, J.; Ishiguro, H.; Paolino, M.; Sigl, V.; Hanada, T.; Hanada, R.; Lipinski, S.; et al. ACE2 links amino acid malnutrition to microbial ecology and intestinal inflammation. *Nature* **2012**, *487*, 477–481. [CrossRef] [PubMed]
66. Ben-Eltriki, M.; Hopefl, R.; Wright, J.M.; Deb, S. Association between Vitamin D Status and Risk of Developing Severe COVID-19 Infection: A Meta-Analysis of Observational Studies. *J. Am. Coll. Nutr.* 2021; *online ahead of print*. [CrossRef] [PubMed]
67. Takahashi, T.; Ellingson, M.K.; Wong, P.; Israelow, B.; Lucas, C.; Klein, J.; Silva, J.; Mao, T.; Oh, J.E.; Tokuyama, M.; et al. Sex differences in immune responses that underlie COVID-19 disease outcomes. *Nature* **2020**, *588*, 315–320. [CrossRef]
68. Scully, E.P.; Haverfield, J.; Ursin, R.L.; Tannenbaum, C.; Klein, S.L. Considering how biological sex impacts immune responses and COVID-19 outcomes. *Nat. Rev. Immunol.* **2020**, *20*, 442–447. [CrossRef]
69. Russo, C.; Morello, G.; Malaguarnera, R.; Piro, S.; Furno, D.L.; Malaguarnera, L. Candidate genes of SARS-CoV-2 gender susceptibility. *Sci. Rep.* **2021**, *11*, 21968. [CrossRef]
70. Varga, Z.; Flammer, A.J.; Steiger, P.; Haberecker, M.; Andermatt, R.; Zinkernagel, A.S.; Mehra, M.R.; Schuepbach, R.A.; Ruschitzka, F.; Moch, H. Endothelial cell infection and endotheliitis in COVID-19. *Lancet* **2020**, *395*, 1417–1418. [CrossRef]
71. Chousterman, B.G.; Swirski, F.K.; Weber, G.F. Cytokine storm and sepsis disease pathogenesis. *Semin. Immunopathol.* **2017**, *39*, 517–528. [CrossRef]
72. Mizgerd, J.P. Acute lower respiratory tract infection. *N. Engl. J. Med.* **2008**, *358*, 716–727. [CrossRef] [PubMed]
73. Tsoupras, A.; Lordan, R.; Zabetakis, I. Thrombosis and COVID-19: The Potential Role of Nutrition. *Front. Nutr.* **2020**, *7*, 583080. [CrossRef] [PubMed]
74. Name, J.J.; Souza, A.C.R.; Vasconcelos, A.R.; Prado, P.S.; Pereira, C.P.M. Zinc, Vitamin D and Vitamin C: Perspectives for COVID-19 With a Focus on Physical Tissue Barrier Integrity. *Front. Nutr.* **2020**, *7*, 606398. [CrossRef] [PubMed]
75. Harrison, C. Sepsis: Calming the cytokine storm. *Nat. Rev. Drug Discov.* **2010**, *9*, 360–361. [CrossRef] [PubMed]
76. Golan, Y.; Prahl, M.; Cassidy, A.G.; Gay, C.; Wu, A.H.B.; Jigmeddagva, U.; Lin, C.Y.; Gonzalez, V.J.; Basilio, E.; Chidboy, M.A.; et al. COVID-19 mRNA Vaccination in Lactation: Assessment of Adverse Events and Vaccine Related Antibodies in Mother-Infant Dyads. *Front. Immunol.* **2021**, *12*, 777103. [CrossRef] [PubMed]
77. Juncker, H.G.; Mulleners, S.J.; Coenen, E.R.M.; van Goudoever, J.B.; van Gils, M.J.; van Keulen, B.J. Comparing Human Milk Antibody Response After 4 Different Vaccines for COVID-19. *JAMA Pediatr.* **2022**, *176*, 611–612. [CrossRef] [PubMed]
78. Yeo, K.T.; Chia, W.N.; Tan, C.W.; Ong, C.; Yeo, J.G.; Zhang, J.; Poh, S.L.; Lim, A.J.M.; Sim, K.H.Z.; Sutamam, N.; et al. Neutralizing Activity and SARS-CoV-2 Vaccine mRNA Persistence in Serum and Breastmilk After BNT162b2 Vaccination in Lactating Women. *Front. Immunol.* **2021**, *12*, 783975. [CrossRef] [PubMed]
79. Kelly, J.C.; Carter, E.B.; Raghuraman, N.; Nolan, L.S.; Gong, Q.; Lewis, A.N.; Good, M. Anti-severe acute respiratory syndrome coronavirus 2 antibodies induced in breast milk after Pfizer-BioNTech/BNT162b2 vaccination. *Am. J. Obstet. Gynecol.* **2021**, *225*, 101–103. [CrossRef] [PubMed]
80. Perl, S.H.; Uzan-Yulzari, A.; Klainer, H.; Asiskovich, L.; Youngster, M.; Rinott, E.; Youngster, I. SARS-CoV-2-Specific Antibodies in Breast Milk After COVID-19 Vaccination of Breastfeeding Women. *JAMA* **2021**, *325*, 2013–2014. [CrossRef] [PubMed]
81. Rosenberg-Friedman, M.; Kigel, A.; Bahar, Y.; Werbner, M.; Alter, J.; Yogev, Y.; Dror, Y.; Lubetzky, R.; Dessau, M.; Gal-Tanamy, M.; et al. BNT162b2 mRNA vaccine elicited antibody response in blood and milk of breastfeeding women. *Nat. Commun.* **2021**, *12*, 6222. [CrossRef] [PubMed]
82. Young, B.E.; Seppo, A.E.; Diaz, N.; Rosen-Carole, C.; Nowak-Wegrzyn, A.; Cruz Vasquez, J.M.; Ferri-Huerta, R.; Nguyen-Contant, P.; Fitzgerald, T.; Sangster, M.Y.; et al. Association of Human Milk Antibody Induction, Persistence, and Neutralizing Capacity With SARS-CoV-2 Infection vs. mRNA Vaccination. *JAMA Pediatr.* **2022**, *176*, 159–168. [CrossRef]
83. Hunagund, S.; Golan, Y.; Asiodu, I.V.; Prahl, M.; Gaw, S.L. Effects of Vaccination Against Influenza, Pertussis, and COVID-19 on Human Milk Antibodies: Current Evidence and Implications for Health Equity. *Front. Immunol.* **2022**, *13*, 910383. [CrossRef]
84. Vimercati, A.; Dellino, M.; Crupano, F.M.; Gargano, G.; Cicinelli, E. Ultrasonic assessment of cesarean section scar to vesicovaginal fold distance: An instrument to estimate pre-labor uterine rupture risk. *J. Matern. Fetal. Neonatal Med.* **2021**, 1–5. [CrossRef]
85. Wang, Z.; Schmidt, F.; Weisblum, Y.; Muecksch, F.; Barnes, C.O.; Finkin, S.; Schaefer-Babajew, D.; Cipolla, M.; Gaebler, C.; Lieberman, J.A.; et al. mRNA vaccine-elicited antibodies to SARS-CoV-2 and circulating variants. *Nature* **2021**, *592*, 616–622. [CrossRef] [PubMed]
86. Selma-Royo, M.; Bauerl, C.; Mena-Tudela, D.; Aguilar-Camprubi, L.; Perez-Cano, F.J.; Parra-Llorca, A.; Lerin, C.; Martinez-Costa, C.; Collado, M.C. Anti-SARS-CoV-2 IgA and IgG in human milk after vaccination is dependent on vaccine type and previous SARS-CoV-2 exposure: A longitudinal study. *Genome Med.* **2022**, *14*, 42. [CrossRef] [PubMed]

87. Halasa, N.B.; Olson, S.M.; Staat, M.A.; Newhams, M.M.; Price, A.M.; Boom, J.A.; Sahni, L.C.; Cameron, M.A.; Pannaraj, P.S.; Bline, K.E.; et al. Effectiveness of Maternal Vaccination with mRNA COVID-19 Vaccine During Pregnancy Against COVID-19-Associated Hospitalization in Infants Aged <6 Months-17 States, July 2021–January 2022. *MMWR Morb. Mortal. Wkly. Rep.* **2022**, *71*, 264–270. [CrossRef] [PubMed]
88. Carlsen, E.O.; Magnus, M.C.; Oakley, L.; Fell, D.B.; Greve-Isdahl, M.; Kinge, J.M.; Haberg, S.E. Association of COVID-19 Vaccination During Pregnancy With Incidence of SARS-CoV-2 Infection in Infants. *JAMA Intern. Med.* **2022**, *182*, 825–831. [CrossRef] [PubMed]
89. Collier, A.Y.; McMahan, K.; Yu, J.; Tostanoski, L.H.; Aguayo, R.; Ansel, J.; Chandrashekar, A.; Patel, S.; Apraku Bondzie, E.; Sellers, D.; et al. Immunogenicity of COVID-19 mRNA Vaccines in Pregnant and Lactating Women. *JAMA* **2021**, *325*, 2370–2380. [CrossRef] [PubMed]
90. Gray, K.J.; Bordt, E.A.; Atyeo, C.; Deriso, E.; Akinwunmi, B.; Young, N.; Baez, A.M.; Shook, L.L.; Cvrk, D.; James, K.; et al. Coronavirus disease 2019 vaccine response in pregnant and lactating women: A cohort study. *Am. J. Obstet. Gynecol.* **2021**, *225*, 303.e1–303.e17. [CrossRef] [PubMed]
91. Prahl, M.; Golan, Y.; Cassidy, A.G.; Matsui, Y.; Li, L.; Alvarenga, B.; Chen, H.; Jigmeddagva, U.; Lin, C.Y.; Gonzalez, V.J.; et al. Evaluation of transplacental transfer of mRNA vaccine products and functional antibodies during pregnancy and early infancy. *medRxiv* **2021**, *30*, 4422. [CrossRef]
92. Nawsherwan; Khan, S.; Zeb, F.; Shoaib, M.; Nabi, G.; Ul Haq, I.; Xu, K.; Li, H. Selected Micronutrients: An Option to Boost Immunity against COVID-19 and Prevent Adverse Pregnancy Outcomes in Pregnant Women: A Narrative Review. *Iran. J. Public Health* **2020**, *49*, 2032–2043. [CrossRef]
93. Kazma, J.M.; van den Anker, J.; Allegaert, K.; Dallmann, A.; Ahmadzia, H.K. Anatomical and physiological alterations of pregnancy. *J. Pharmacokinet. Pharmacodyn.* **2020**, *47*, 271–285. [CrossRef] [PubMed]
94. Soma-Pillay, P.; Nelson-Piercy, C.; Tolppanen, H.; Mebazaa, A. Physiological changes in pregnancy. *Cardiovasc. J. Afr* **2016**, *27*, 89–94. [CrossRef] [PubMed]
95. Eskenazi, B.; Rauch, S.; Iurlaro, E.; Gunier, R.B.; Rego, A.; Gravett, M.G.; Cavoretto, P.I.; Deruelle, P.; Garcia-May, P.K.; Mhatre, M.; et al. Diabetes mellitus, maternal adiposity, and insulin-dependent gestational diabetes are associated with COVID-19 in pregnancy: The INTERCOVID study. *Am. J. Obstet. Gynecol.* **2022**, *227*, 74.e1–74.e16. [CrossRef] [PubMed]
96. Kampmann, U.; Knorr, S.; Fuglsang, J.; Ovesen, P. Determinants of Maternal Insulin Resistance during Pregnancy: An Updated Overview. *J. Diabetes Res.* **2019**, *2019*, 5320156. [CrossRef] [PubMed]
97. Chen, M.; Zeng, J.; Liu, X.; Sun, G.; Gao, Y.; Liao, J.; Yu, J.; Luo, X.; Qi, H. Changes in physiology and immune system during pregnancy and coronavirus infection: A review. *Eur. J. Obstet. Gynecol. Reprod. Biol.* **2020**, *255*, 124–128. [CrossRef] [PubMed]
98. Aghaeepour, N.; Ganio, E.A.; McIlwain, D.; Tsai, A.S.; Tingle, M.; Van Gassen, S.; Gaudilliere, D.K.; Baca, Q.; McNeil, L.; Okada, R.; et al. An immune clock of human pregnancy. *Sci. Immunol.* **2017**, *1*, eaan2946. [CrossRef]
99. Mor, G.; Aldo, P.; Alvero, A.B. The unique immunological and microbial aspects of pregnancy. *Nat. Rev. Immunol.* **2017**, *17*, 469–482. [CrossRef]
100. Mate, A.; Reyes-Goya, C.; Santana-Garrido, A.; Sobrevia, L.; Vazquez, C.M. Impact of maternal nutrition in viral infections during pregnancy. *Biochim. Biophys. Acta Mol. Basis Dis.* **2021**, *1867*, 166231. [CrossRef] [PubMed]
101. Chen, H.; Li, H.; Cao, Y.; Qi, H.; Ma, Y.; Bai, X.; Zhao, Y.; Wu, L.; Liu, C.; Wei, J.; et al. Food Intake and Diet Quality of Pregnant Women in China During the COVID-19 Pandemic: A National Cross-Sectional Study. *Front. Nutr.* **2022**, *9*, 853565. [CrossRef] [PubMed]
102. Erol, S.A.; Tanacan, A.; Anuk, A.T.; Tokalioglu, E.O.; Biriken, D.; Keskin, H.L.; Moraloglu, O.T.; Yazihan, N.; Sahin, D. Evaluation of maternal serum afamin and vitamin E levels in pregnant women with COVID-19 and its association with composite adverse perinatal outcomes. *J. Med. Virol.* **2021**, *93*, 2350–2358. [CrossRef] [PubMed]
103. Anuk, A.T.; Polat, N.; Akdas, S.; Erol, S.A.; Tanacan, A.; Biriken, D.; Keskin, H.L.; Moraloglu Tekin, O.; Yazihan, N.; Sahin, D. The Relation Between Trace Element Status (Zinc, Copper, Magnesium) and Clinical Outcomes in COVID-19 Infection During Pregnancy. *Biol. Trace Elem. Res.* **2021**, *199*, 3608–3617. [CrossRef]
104. Uta, M.; Neamtu, R.; Bernad, E.; Mocanu, A.G.; Gluhovschi, A.; Popescu, A.; Dahma, G.; Dumitru, C.; Stelea, L.; Citu, C.; et al. The Influence of Nutritional Supplementation for Iron Deficiency Anemia on Pregnancies Associated with SARS-CoV-2 Infection. *Nutrients* **2022**, *14*, 836. [CrossRef] [PubMed]
105. Vinciguerra, M.; Lamanna, B.; Pititto, F.; Cicinelli, R.; Dellino, M.; Picardi, N.; Ricci, I.; Cicinelli, E.; Vimercati, A.; Malvasi, A. Peri-Conceptional Intake of Folic Acid Supplement to Date: A Medical-Legal Issue. *Biomed. J. Sci. Tech. Res.* **2022**, *43*, 34324–34329. [CrossRef]
106. Mate, A.; Reyes-Goya, C.; Santana-Garrido, Á.; Vázquez, C.M. Lifestyle, maternal nutrition and healthy pregnancy. *Curr. Vasc. Pharmacol.* **2021**, *19*, 132–140. [CrossRef] [PubMed]
107. Canavan, C.R.; D'Cruze, T.; Kennedy, M.A.; Hatchell, K.E.; Boardman, M.; Suresh, A.; Goodman, D.; Dev, A. Missed opportunities to improve food security for pregnant people: A qualitative study of prenatal care settings in Northern New England during the COVID-19 pandemic. *BMC Nutr.* **2022**, *8*, 8. [CrossRef] [PubMed]
108. Annaratone, L.; Cascardi, E.; Vissio, E.; Sarotto, I.; Chmielik, E.; Sapino, A.; Berrino, E.; Marchio, C. The Multifaceted Nature of Tumor Microenvironment in Breast Carcinomas. *Pathobiology* **2020**, *87*, 125–142. [CrossRef] [PubMed]

109. Grassini, D.; Cascardi, E.; Sarotto, I.; Annaratone, L.; Sapino, A.; Berrino, E.; Marchiò, C. Unusual patterns of HER2 expression in breast cancer: Insights and perspectives. *Pathobiology* **2022**, *2*, 1–19. [CrossRef] [PubMed]
110. Tamma, R.; Limongelli, L.; Maiorano, E.; Pastore, D.; Cascardi, E.; Tempesta, A.; Carluccio, P.; Mastropasqua, M.G.; Capodiferro, S.; Covelli, C.; et al. Vascular density and inflammatory infiltrate in primary oral squamous cell carcinoma and after allogeneic hematopoietic stem cell transplantation. *Ann. Hematol.* **2019**, *98*, 979–986. [CrossRef] [PubMed]
111. Limongelli, L.; Cascardi, E.; Capodiferro, S.; Favia, G.; Corsalini, M.; Tempesta, A.; Maiorano, E. Multifocal Amelanotic Melanoma of the Hard Palate: A Challenging Case. *Diagnostics* **2020**, *10*, 424. [CrossRef]
112. Dellino, M.; Carriero, C.; Silvestris, E.; Capursi, T.; Paradiso, A.; Cormio, G. Primary Vaginal Carcinoma Arising on Cystocele Mimicking Vulvar Cancer. *J. Obstet. Gynaecol. Can.* **2020**, *42*, 1543–1545. [CrossRef] [PubMed]
113. Dellino, M.; Minoia, C.; Paradiso, A.V.; De Palo, R.; Silvestris, E. Fertility Preservation in Cancer Patients During the Coronavirus (COVID-19) Pandemic. *Front. Oncol.* **2020**, *10*, 1009. [CrossRef] [PubMed]
114. Dellino, M.; Silvestris, E.; Loizzi, V.; Paradiso, A.; Loiacono, R.; Minoia, C.; Daniele, A.; Cormio, G. Germinal ovarian tumors in reproductive age women: Fertility-sparing and outcome. *Medicine* **2020**, *99*, e22146. [CrossRef]
115. Silvestris, E.; Dellino, M.; Cafforio, P.; Paradiso, A.V.; Cormio, G.; D'Oronzo, S. Breast cancer: An update on treatment-related infertility. *J. Cancer Res. Clin. Oncol.* **2020**, *146*, 647–657. [CrossRef]
116. Tamma, R.; Annese, T.; Ruggieri, S.; Brunetti, O.; Longo, V.; Cascardi, E.; Mastropasqua, M.G.; Maiorano, E.; Silvestris, N.; Ribatti, D. Inflammatory cells infiltrate and angiogenesis in locally advanced and metastatic cholangiocarcinoma. *Eur. J. Clin. Investig.* **2019**, *49*, e13087. [CrossRef]
117. Tamma, R.; Rutigliano, M.; Lucarelli, G.; Annese, T.; Ruggieri, S.; Cascardi, E.; Napoli, A.; Battaglia, M.; Ribatti, D. Microvascular density, macrophages, and mast cells in human clear cell renal carcinoma with and without bevacizumab treatment. *Urol. Oncol.* **2019**, *37*, 355.e11–355.e19. [CrossRef]
118. Cascardi, E.; Cazzato, G.; Daniele, A.; Silvestris, E.; Cormio, G.; Di Vagno, G.; Malvasi, A.; Loizzi, V.; Scacco, S.; Pinto, V. Association between Cervical Microbiota and HPV: Could This Be the Key to Complete Cervical Cancer Eradication? *Biology* **2022**, *11*, 1114. [CrossRef]
119. Wang, P.Y.; Li, Y.; Wang, Q. Sarcopenia: An underlying treatment target during the COVID-19 pandemic. *Nutrition* **2021**, *84*, 111104. [CrossRef]
120. Hawrylkowicz, V.; Lietz-Kijak, D.; Kazmierczak-Siedlecka, K.; Solek-Pastuszka, J.; Stachowska, L.; Folwarski, M.; Parczewski, M.; Stachowska, E. Patient Nutrition and Probiotic Therapy in COVID-19: What Do We Know in 2021? *Nutrients* **2021**, *13*, 3385. [CrossRef]
121. Bafunno, D.; Romito, F.; Lagattolla, F.; Delvino, V.A.; Minoia, C.; Loseto, G.; Dellino, M.; Guarini, A.; Catino, A.; Montrone, M.; et al. Psychological well-being in cancer outpatients during COVID-19. *J. BUON* **2021**, *26*, 1127–1134.
122. Chen, L.K.; Woo, J.; Assantachai, P.; Auyeung, T.W.; Chou, M.Y.; Iijima, K.; Jang, H.C.; Kang, L.; Kim, M.; Kim, S.; et al. Asian Working Group for Sarcopenia: 2019 Consensus Update on Sarcopenia Diagnosis and Treatment. *J. Am. Med. Dir. Assoc.* **2020**, *21*, 300–307.e302. [CrossRef] [PubMed]
123. Nelke, C.; Dziewas, R.; Minnerup, J.; Meuth, S.G.; Ruck, T. Skeletal muscle as potential central link between sarcopenia and immune senescence. *EBioMedicine* **2019**, *49*, 381–388. [CrossRef] [PubMed]
124. Weimann, A.; Braga, M.; Carli, F.; Higashiguchi, T.; Hubner, M.; Klek, S.; Laviano, A.; Ljungqvist, O.; Lobo, D.N.; Martindale, R.; et al. ESPEN guideline: Clinical nutrition in surgery. *Clin. Nutr.* **2017**, *36*, 623–650. [CrossRef]
125. Sukkar, S.G.; Muscaritoli, M. A Clinical Perspective of Low Carbohydrate Ketogenic Diets: A Narrative Review. *Front. Nutr.* **2021**, *8*, 642628. [CrossRef]
126. Horne, B.D.; Muhlestein, J.B.; May, H.T.; Le, V.T.; Bair, T.L.; Knowlton, K.U.; Anderson, J.L.; Investigators, I.R. Association of Periodic Fasting with Lower Severity of COVID-19 Outcomes in the SARS-CoV-2 Pre-Vaccine Era: An Observational Cohort from the INSPIRE Registry. *medRxiv* **2022**. [CrossRef]
127. Wang, W.-H.; Lin, C.-Y.; Chang, M.R.; Urbina, A.N.; Assavalapsakul, W.; Thitithanyanont, A.; Chen, Y.-H.; Liu, F.-T.; Wang, S.-F. The role of galectins in virus infection-A systemic literature review. *J. Microbiol. Immunol. Infect.* **2020**, *53*, 925–935. [CrossRef]
128. Li, S.-W.; Yang, T.-C.; Lai, C.-C.; Huang, S.-H.; Liao, J.-M.; Wan, L.; Lin, Y.-J.; Lin, C.-W. Antiviral activity of aloe-emodin against influenza A virus via galectin-3 up-regulation. *Eur. J. Pharmacol.* **2014**, *738*, 125–132. [CrossRef]
129. Pugliese, G.; Iacobini, C.; Pesce, C.M.; Menini, S. Galectin-3: An emerging all-out player in metabolic disorders and their complications. *Glycobiology* **2015**, *25*, 136–150. [CrossRef]
130. Gombart, A.F.; Pierre, A.; Maggini, S. A review of micronutrients and the immune system–working in harmony to reduce the risk of infection. *Nutrients* **2020**, *12*, 236. [CrossRef]
131. Martineau, A.R.; Jolliffe, D.A.; Hooper, R.L.; Greenberg, L.; Aloia, J.F.; Bergman, P.; Dubnov-Raz, G.; Esposito, S.; Ganmaa, D.; Ginde, A.A. Vitamin D supplementation to prevent acute respiratory tract infections: Systematic review and meta-analysis of individual participant data. *BMJ* **2017**, *356*, i6583. [CrossRef]
132. Calder, P.C. Nutrition, immunity and COVID-19. *BMJ Nutr. Prev. Health* **2020**, *3*, 74–92. [CrossRef]
133. Macaya, F.; Espejo Paeres, C.; Valls, A.; Fernández-Ortiz, A.; González Del Castillo, J.; Martín-Sánchez, F.J.; Runkle, I.; Rubio Herrera, M.Á. Interaction between age and vitamin D deficiency in severe COVID-19 infection. *Nutr. Hosp.* **2020**, *37*, 1039–1042. [CrossRef] [PubMed]

134. Abdollahi, A.; Kamali Sarvestani, H.; Rafat, Z.; Ghaderkhani, S.; Mahmoudi-Aliabadi, M.; Jafarzadeh, B.; Mehrtash, V. The association between the level of serum 25 (OH) vitamin D, obesity, and underlying diseases with the risk of developing COVID-19 infection: A case-control study of hospitalized patients in Tehran, Iran. *J. Med. Virol.* **2021**, *93*, 2359–2364. [CrossRef] [PubMed]
135. Biesalski, H.K. Obesity, vitamin D deficiency and old age a serious combination with respect to coronavirus disease-2019 severity and outcome. *Curr. Opin. Clin. Nutr. Metab. Care* **2021**, *24*, 18–24. [CrossRef]
136. Cooper, I.D.; Crofts, C.A.; DiNicolantonio, J.J.; Malhotra, A.; Elliott, B.; Kyriakidou, Y.; Brookler, K.H. Relationships between hyperinsulinaemia, magnesium, vitamin D, thrombosis and COVID-19: Rationale for clinical management. *Open. Heart* **2020**, *7*, e001356. [CrossRef]
137. Alvarez-Rodriguez, L.; Lopez-Hoyos, M.; Garcia-Unzueta, M.; Amado, J.A.; Cacho, P.M.; Martinez-Taboada, V.M. Age and low levels of circulating vitamin D are associated with impaired innate immune function. *J. Leukoc. Biol.* **2012**, *91*, 829–838. [CrossRef]
138. Jothimani, D.; Kailasam, E.; Danielraj, S.; Nallathambi, B.; Ramachandran, H.; Sekar, P.; Manoharan, S.; Ramani, V.; Narasimhan, G.; Kaliamoorthy, I.; et al. COVID-19: Poor outcomes in patients with zinc deficiency. *Int. J. Infect. Dis.* **2020**, *100*, 343–349. [CrossRef]
139. So, J.; Wu, D.; Lichtenstein, A.H.; Tai, A.K.; Matthan, N.R.; Maddipati, K.R.; Lamon-Fava, S. EPA and DHA differentially modulate monocyte inflammatory response in subjects with chronic inflammation in part via plasma specialized pro-resolving lipid mediators: A randomized, double-blind, crossover study. *Atherosclerosis* **2021**, *316*, 90–98. [CrossRef]
140. Panigrahy, D.; Gilligan, M.M.; Huang, S.; Gartung, A.; Cortes-Puch, I.; Sime, P.J.; Phipps, R.P.; Serhan, C.N.; Hammock, B.D. Inflammation resolution: A dual-pronged approach to averting cytokine storms in COVID-19? *Cancer Metastasis Rev.* **2020**, *39*, 337–340. [CrossRef]
141. Silvestris, E.; Dellino, M.; Depalo, R. Fertility preservation in cancer patients at the time of COVID-19 pandemic. *J. Gynecol. Obstet. Hum. Reprod.* **2020**, 101910. [CrossRef]
142. Hathaway, D.; Pandav, K.; Patel, M.; Riva-Moscoso, A.; Singh, B.M.; Patel, A.; Min, Z.C.; Singh-Makkar, S.; Sana, M.K.; Sanchez-Dopazo, R.; et al. Omega 3 Fatty Acids and COVID-19: A Comprehensive Review. *Infect. Chemother.* **2020**, *52*, 478–495. [CrossRef]
143. Weill, P.; Plissonneau, C.; Legrand, P.; Rioux, V.; Thibault, R. May omega-3 fatty acid dietary supplementation help reduce severe complications in COVID-19 patients? *Biochimie* **2020**, *179*, 275–280. [CrossRef]
144. Chang, J.P.; Pariante, C.M.; Su, K.P. Omega-3 fatty acids in the psychological and physiological resilience against COVID-19. *Prostaglandins Leukot Essent Fat. Acids* **2020**, *161*, 102177. [CrossRef] [PubMed]
145. Regidor, P.A.; Santos, F.G.; Rizo, J.M.; Egea, F.M. Pro resolving inflammatory effects of the lipid mediators of omega 3 fatty acids and its implication in SARS COVID-19. *Med. Hypotheses* **2020**, *145*, 110340. [CrossRef]
146. Rogero, M.M.; Leao, M.C.; Santana, T.M.; Pimentel, M.; Carlini, G.C.G.; da Silveira, T.F.F.; Goncalves, R.C.; Castro, I.A. Potential benefits and risks of omega-3 fatty acids supplementation to patients with COVID-19. *Free. Radic. Biol. Med.* **2020**, *156*, 190–199. [CrossRef]
147. Asher, A.; Tintle, N.L.; Myers, M.; Lockshon, L.; Bacareza, H.; Harris, W.S. Blood omega-3 fatty acids and death from COVID-19: A pilot study. *Prostaglandins Leukot Essent Fat. Acids* **2021**, *166*, 102250. [CrossRef]
148. Barazzoni, R.; Bischoff, S.C.; Breda, J.; Wickramasinghe, K.; Krznaric, Z.; Nitzan, D.; Pirlich, M.; Singer, P.; endorsed by the ESPEN Council. ESPEN expert statements and practical guidance for nutritional management of individuals with SARS-CoV-2 infection. *Clin. Nutr.* **2020**, *39*, 1631–1638. [CrossRef]
149. Peng, Q.-Y.; Wang, X.-T.; Zhang, L.-N. Findings of lung ultrasonography of novel corona virus pneumonia during the 2019–2020 epidemic. *Intensive Care Med.* **2020**, *46*, 849–850. [CrossRef]
150. Xie, Y.; Wang, X.; Yang, P.; Zhang, S. COVID-19 complicated by acute pulmonary embolism. *Radiol. Cardiothorac. Imaging* **2020**, *2*, e200067. [CrossRef]
151. Lai, C.-C.; Shih, T.-P.; Ko, W.-C.; Tang, H.-J.; Hsueh, P.-R. Severe acute respiratory syndrome coronavirus 2 (SARS-CoV-2) and coronavirus disease-2019 (COVID-19): The epidemic and the challenges. *Int. J. Antimicrob. Agents* **2020**, *55*, 105924. [CrossRef]
152. Martinez, M.A. Compounds with therapeutic potential against novel respiratory 2019 coronavirus. *Antimicrob. Agents Chemother.* **2020**, *64*, e00399-20. [CrossRef] [PubMed]
153. Lu, H.; Stratton, C.W.; Tang, Y.W. Outbreak of pneumonia of unknown etiology in Wuhan, China: The mystery and the miracle. *J. Med. Virol.* **2020**, *92*, 401. [CrossRef]
154. Gorbalenya, A.E.; Baker, S.C.; Baric, R.S.; de Groot, R.J.; Drosten, C.; Gulyaeva, A.A.; Haagmans, B.L.; Lauber, C.; Leontovich, A.M.; Neuman, B.W. Severe acute respiratory syndrome-related coronavirus: The species and its viruses—A statement of the Coronavirus Study Group. *BioRxiv* **2020**, *5*, 536–544. [CrossRef]
155. Chen, N.; Zhou, M.; Dong, X.; Qu, J.; Gong, F.; Han, Y.; Qiu, Y.; Wang, J.; Liu, Y.; Wei, Y. Epidemiological and clinical characteristics of 99 cases of 2019 novel coronavirus pneumonia in Wuhan, China: A descriptive study. *Lancet* **2020**, *395*, 507–513. [CrossRef]
156. Daniele, A.; Divella, R.; Pilato, B.; Tommasi, S.; Pasanisi, P.; Patruno, M.; Digennaro, M.; Minoia, C.; Dellino, M.; Pisconti, S.; et al. Can harmful lifestyle, obesity and weight changes increase the risk of breast cancer in BRCA 1 and BRCA 2 mutation carriers? A Mini review. *Hered. Cancer Clin. Pract.* **2021**, *19*, 45. [CrossRef] [PubMed]
157. Daniele, A.; Guarini, A.; Summa, S.; Dellino, M.; Lerario, G.; Ciavarella, S.; Ditonno, P.; Paradiso, A.V.; Divella, R.; Casamassima, P.; et al. Body Composition Change, Unhealthy Lifestyles and Steroid Treatment as Predictor of Metabolic Risk in Non-Hodgkin's Lymphoma Survivors. *J. Pers. Med.* **2021**, *11*, 215. [CrossRef] [PubMed]

158. Felsenstein, S.; Herbert, J.A.; McNamara, P.S.; Hedrich, C.M. COVID-19: Immunology and treatment options. *Clin. Immunol.* **2020**, *215*, 108448. [CrossRef] [PubMed]
159. Pascarella, G.; Strumia, A.; Piliego, C.; Bruno, F.; Del Buono, R.; Costa, F.; Scarlata, S.; Agrò, F.E. COVID-19 diagnosis and management: A comprehensive review. *J. Intern. Med.* **2020**, *288*, 192–206. [CrossRef] [PubMed]
160. Sanders, J.M.; Monogue, M.L.; Jodlowski, T.Z.; Cutrell, J.B. Pharmacologic treatments for coronavirus disease 2019 (COVID-19): A review. *JAMA* **2020**, *323*, 1824–1836. [CrossRef]
161. Le, T.T.; Andreadakis, Z.; Kumar, A.; Román, R.G.; Tollefsen, S.; Saville, M.; Mayhew, S. The COVID-19 vaccine development landscape. *Nat. Rev. Drug Discov.* **2020**, *19*, 305–306. [CrossRef] [PubMed]
162. Mulligan, M.J.; Lyke, K.E.; Kitchin, N.; Absalon, J.; Gurtman, A.; Lockhart, S.; Neuzil, K.; Raabe, V.; Bailey, R.; Swanson, K.A. Phase I/II study of COVID-19 RNA vaccine BNT162b1 in adults. *Nature* **2020**, *586*, 589–593. [CrossRef] [PubMed]
163. Vimercati, A.; Dellino, M.; Suma, C.; Damiani, G.R.; Malvasi, A.; Cazzato, G.; Cascardi, E.; Resta, L.; Cicinelli, E. Spontaneous Uterine Rupture and Adenomyosis, a Rare but Possible Correlation: Case Report and Literature Review. *Diagnostics* **2022**, *12*, 1574. [CrossRef] [PubMed]
164. Barzilay, R.; Moore, T.M.; Greenberg, D.M.; DiDomenico, G.E.; Brown, L.A.; White, L.K.; Gur, R.C.; Gur, R.E. Resilience, COVID-19-related stress, anxiety and depression during the pandemic in a large population enriched for healthcare providers. *Transl. Psychiatry* **2020**, *10*, 291. [CrossRef]
165. Kreier, F. Long-COVID symptoms less likely in vaccinated people, Israeli data say. *Nature* **2022**. [CrossRef]
166. Antonelli, M.; Penfold, R.S.; Merino, J.; Sudre, C.H.; Molteni, E.; Berry, S.; Canas, L.S.; Graham, M.S.; Klaser, K.; Modat, M.; et al. Risk factors and disease profile of post-vaccination SARS-CoV-2 infection in UK users of the COVID Symptom Study app: A prospective, community-based, nested, case-control study. *Lancet Infect. Dis.* **2022**, *22*, 43–55. [CrossRef]
167. Haas, E.J.; Angulo, F.J.; McLaughlin, J.M.; Anis, E.; Singer, S.R.; Khan, F.; Brooks, N.; Smaja, M.; Mircus, G.; Pan, K.; et al. Impact and effectiveness of mRNA BNT162b2 vaccine against SARS-CoV-2 infections and COVID-19 cases, hospitalisations, and deaths following a nationwide vaccination campaign in Israel: An observational study using national surveillance data. *Lancet* **2021**, *397*, 1819–1829. [CrossRef]
168. El Sahly, H.M.; Baden, L.R.; Essink, B.; Doblecki-Lewis, S.; Martin, J.M.; Anderson, E.J.; Campbell, T.B.; Clark, J.; Jackson, L.A.; Fichtenbaum, C.J.; et al. Efficacy of the mRNA-1273 SARS-CoV-2 Vaccine at Completion of Blinded Phase. *N. Engl. J. Med.* **2021**, *385*, 1774–1785. [CrossRef]
169. Bignucolo, A.; Scarabel, L.; Mezzalira, S.; Polesel, J.; Cecchin, E.; Toffoli, G. Sex Disparities in Efficacy in COVID-19 Vaccines: A Systematic Review and Meta-Analysis. *Vaccines* **2021**, *9*, 825. [CrossRef] [PubMed]
170. Omeish, H.; Najadat, A.; Al-Azzam, S.; Tarabin, N.; Abu Hameed, A.; Al-Gallab, N.; Abbas, H.; Rababah, L.; Rabadi, M.; Karasneh, R.; et al. Reported COVID-19 vaccines side effects among Jordanian population: A cross sectional study. *Hum. Vaccin Immunother.* **2022**, *18*, 1981086. [CrossRef] [PubMed]
171. Rabaan, A.A.; Bakhrebah, M.A.; Mutair, A.A.; Alhumaid, S.; Al-Jishi, J.M.; AlSihati, J.; Albayat, H.; Alsheheri, A.; Aljeldah, M.; Garout, M.; et al. Systematic Review on Pathophysiological Complications in Severe COVID-19 among the Non-Vaccinated and Vaccinated Population. *Vaccines* **2022**, *10*, 985. [CrossRef] [PubMed]
172. Bai, F.; Tomasoni, D.; Falcinella, C.; Barbanotti, D.; Castoldi, R.; Mule, G.; Augello, M.; Mondatore, D.; Allegrini, M.; Cona, A.; et al. Female gender is associated with long COVID syndrome: A prospective cohort study. *Clin. Microbiol. Infect.* **2022**, *28*, 611.e9–611.e16. [CrossRef] [PubMed]
173. Abdul Latif, N.I.; Mohamed Ismail, N.A.; Loh, S.Y.E.; Nur Azurah, A.G.; Midin, M.; Shah, S.A.; Kalok, A. Psychological Distress and COVID-19 Related Anxiety among Malaysian Women during the COVID-19 Pandemic. *Int. J. Environ. Res. Public Health* **2022**, *19*, 4590. [CrossRef] [PubMed]
174. Xia, C.L.; Wei, A.P.; Huang, Y.T. The COVID-19 Lockdown and Mental Wellbeing of Females in China. *Int. J. Environ. Res. Public Health* **2022**, *19*, 4960. [CrossRef]
175. Kavvadas, D.; Kavvada, A.; Karachrysafi, S.; Papaliagkas, V.; Cheristanidis, S.; Chatzidimitriou, M.; Papamitsou, T. Stress, Anxiety and Depression Prevalence among Greek University Students during COVID-19 Pandemic: A Two-Year Survey. *J. Clin. Med.* **2022**, *11*, 4263. [CrossRef] [PubMed]
176. Casagrande, M.; Favieri, F.; Tambelli, R.; Forte, G. The enemy who sealed the world: Effects quarantine due to the COVID-19 on sleep quality, anxiety, and psychological distress in the Italian population. *Sleep. Med.* **2020**, *75*, 12–20. [CrossRef] [PubMed]
177. Feng, H.; Gan, C.C.R.; Leiva, D.; Zhang, B.L.; Davies, S.E. COVID-19, sex, and gender in China: A scoping review. *Glob. Health* **2022**, *18*, 9. [CrossRef] [PubMed]
178. Qattan, A.M.N. Symptoms of psychological distress amongst women during the COVID-19 pandemic in Saudi Arabia. *PLoS ONE* **2022**, *17*, e0268642. [CrossRef] [PubMed]
179. Huang, C.; Huang, L.; Wang, Y.; Li, X.; Ren, L.; Gu, X.; Kang, L.; Guo, L.; Liu, M.; Zhou, X.; et al. 6-month consequences of COVID-19 in patients discharged from hospital: A cohort study. *Lancet* **2021**, *397*, 220–232. [CrossRef]
180. Venturelli, S.; Benatti, S.V.; Casati, M.; Binda, F.; Zuglian, G.; Imeri, G.; Conti, C.; Biffi, A.M.; Spada, M.S.; Bondi, E.; et al. Surviving COVID-19 in Bergamo province: A post-acute outpatient re-evaluation. *Epidemiol. Infect.* **2021**, *149*, e32. [CrossRef]
181. Mazza, M.G.; De Lorenzo, R.; Conte, C.; Poletti, S.; Vai, B.; Bollettini, I.; Melloni, E.M.T.; Furlan, R.; Ciceri, F.; Rovere-Querini, P.; et al. Anxiety and depression in COVID-19 survivors: Role of inflammatory and clinical predictors. *Brain Behav. Immun.* **2020**, *89*, 594–600. [CrossRef] [PubMed]

182. Xiong, Q.; Xu, M.; Li, J.; Liu, Y.; Zhang, J.; Xu, Y.; Dong, W. Clinical sequelae of COVID-19 survivors in Wuhan, China: A single-centre longitudinal study. *Clin. Microbiol. Infect.* **2021**, *27*, 89–95. [CrossRef] [PubMed]
183. Townsend, L.; Dyer, A.H.; Jones, K.; Dunne, J.; Mooney, A.; Gaffney, F.; O'Connor, L.; Leavy, D.; O'Brien, K.; Dowds, J.; et al. Persistent fatigue following SARS-CoV-2 infection is common and independent of severity of initial infection. *PLoS ONE* **2020**, *15*, e0240784. [CrossRef] [PubMed]
184. Carfi, A.; Bernabei, R.; Landi, F.; Gemelli Against COVID-19 Post-Acute Care Study Group. Persistent Symptoms in Patients After Acute COVID-19. *JAMA* **2020**, *324*, 603–605. [CrossRef]
185. Moreno-Perez, O.; Merino, E.; Leon-Ramirez, J.M.; Andres, M.; Ramos, J.M.; Arenas-Jimenez, J.; Asensio, S.; Sanchez, R.; Ruiz-Torregrosa, P.; Galan, I.; et al. Post-acute COVID-19 syndrome. Incidence and risk factors: A Mediterranean cohort study. *J. Infect.* **2021**, *82*, 378–383. [CrossRef] [PubMed]
186. Daher, A.; Balfanz, P.; Cornelissen, C.; Muller, A.; Bergs, I.; Marx, N.; Muller-Wieland, D.; Hartmann, B.; Dreher, M.; Muller, T. Follow up of patients with severe coronavirus disease 2019 (COVID-19): Pulmonary and extrapulmonary disease sequelae. *Respir. Med.* **2020**, *174*, 106197. [CrossRef]
187. Mohamed, M.S.; Moulin, T.C.; Schioth, H.B. Sex differences in COVID-19: The role of androgens in disease severity and progression. *Endocrine* **2021**, *71*, 3–8. [CrossRef]
188. Bienvenu, L.A.; Noonan, J.; Wang, X.; Peter, K. Higher mortality of COVID-19 in males: Sex differences in immune response and cardiovascular comorbidities. *Cardiovasc. Res.* **2020**, *116*, 2197–2206. [CrossRef]
189. Zeng, F.; Dai, C.; Cai, P.; Wang, J.; Xu, L.; Li, J.; Hu, G.; Wang, Z.; Zheng, F.; Wang, L. A comparison study of SARS-CoV-2 IgG antibody between male and female COVID-19 patients: A possible reason underlying different outcome between sex. *J. Med. Virol.* **2020**, *92*, 2050–2054. [CrossRef]
190. Shapiro, J.R.; Klein, S.L.; Morgan, R. Stop 'controlling' for sex and gender in global health research. *BMJ Glob. Health* **2021**, *6*, e005714. [CrossRef]
191. Lythgoe, M.P.; Middleton, P. Ongoing Clinical Trials for the Management of the COVID-19 Pandemic. *Trends Pharmacol. Sci.* **2020**, *41*, 363–382. [CrossRef]
192. Casu, C.; Nemeth, E.; Rivella, S. Hepcidin agonists as therapeutic tools. *Blood* **2018**, *131*, 1790–1794. [CrossRef] [PubMed]
193. Hawula, Z.J.; Wallace, D.F.; Subramaniam, V.N.; Rishi, G. Therapeutic Advances in Regulating the Hepcidin/Ferroportin Axis. *Pharmaceuticals* **2019**, *12*, 170. [CrossRef] [PubMed]

International Journal of *Molecular Sciences*

MDPI

*Article*

# Discovery of Long Non-Coding RNA MALAT1 Amplification in Precancerous Colorectal Lesions

**Anna Siskova [1,2,*], Jan Kral [3], Jana Drabova [4], Klara Cervena [1,2], Kristyna Tomasova [1,5], Jiri Jungwirth [6,7,8], Tomas Hucl [3], Pavel Kohout [9], Sandra Summerova [9], Ludmila Vodickova [1,2,5], Pavel Vodicka [1,2,5] and Veronika Vymetalkova [1,2,5,*]**

1 Department of Molecular Biology of Cancer, Institute of Experimental Medicine of the Czech Academy of Sciences, Videnska 1083, 142 00 Prague, Czech Republic; klara.cervena@iem.cas.cz (K.C.); kristyna.tomasova@iem.cas.cz (K.T.); ludmila.vodickova@iem.cas.cz (L.V.); pavel.vodicka@iem.cas.cz (P.V.)
2 Institute of Biology and Medical Genetics, 1st Faculty of Medicine, Charles University and General University Hospital in Prague, Albertov 4, 128 00 Prague, Czech Republic
3 Clinic of Hepatogastroenterology, Institute for Clinical and Experimental Medicine, Videnska 1958/9, 140 21 Prague, Czech Republic; jan.kral@ikem.cz (J.K.); tomas.hucl@ikem.cz (T.H.)
4 Department of Biology and Medical Genetics, 2nd Faculty of Medicine, Charles University and University Hospital Motol, V Uvalu 84, 150 06 Prague, Czech Republic; jana.drabova@fnmotol.cz
5 Biomedical Centre, Faculty of Medicine in Pilsen, Charles University, Alej Svobody 76, 323 00 Pilsen, Czech Republic
6 Institute of Physiology, 1st Faculty of Medicine Charles University, Katerinska 1660/32, 121 08 Praha, Czech Republic; jiri.jungwirth@kliniken-nordoberpfalz.ag
7 Department of Gastroenterology, Libera Scientia, Lucni 7a, 130 00 Prague, Czech Republic
8 Department of General, Visceral and Thoracic Surgery, Klinikum Weiden, Söllnerstraße 16, 92637 Weiden, Germany
9 Department of Internal Medicine, 3rd Faculty of Medicine University and Faculty Thomayer Hospital Prague, Ruska 87, 100 00 Prague, Czech Republic; pavel.kohout@ftn.cz (P.K.); sandra.summerova@ftn.cz (S.S.)
* Correspondence: anna.siskova@iem.cas.cz (A.S.); veronika.vymetalkova@iem.cas.cz (V.V.); Tel.: +420-241062251 (A.S.); +420-241062699 (V.V.)

**Citation:** Siskova, A.; Kral, J.; Drabova, J.; Cervena, K.; Tomasova, K.; Jungwirth, J.; Hucl, T.; Kohout, P.; Summerova, S.; Vodickova, L.; et al. Discovery of Long Non-Coding RNA MALAT1 Amplification in Precancerous Colorectal Lesions. *Int. J. Mol. Sci.* **2022**, *23*, 7656. https://doi.org/10.3390/ijms23147656

Academic Editor: Laura Paleari

Received: 16 June 2022
Accepted: 8 July 2022
Published: 11 July 2022



**Abstract:** A colorectal adenoma, an aberrantly growing tissue, arises from the intestinal epithelium and is considered as precursor of colorectal cancer (CRC). In this study, we investigated structural and numerical chromosomal aberrations in adenomas, hypothesizing that chromosomal instability (CIN) occurs early in adenomas. We applied array comparative genomic hybridization (aCGH) to fresh frozen colorectal adenomas and their adjacent mucosa from 16 patients who underwent colonoscopy examination. In our study, histologically similar colorectal adenomas showed wide variability in chromosomal instability. Based on the obtained results, we further stratified patients into four distinct groups. The first group showed the gain of *MALAT1* and *TALAM1*, long non-coding RNAs (lncRNAs). The second group involved patients with numerous microdeletions. The third group consisted of patients with a disrupted karyotype. The fourth group of patients did not show any CIN in adenomas. Overall, we identified frequent losses in genes, such as *TSC2*, *COL1A1*, *NOTCH1*, *MIR4673*, and *GNAS*, and gene gain containing *MALAT1* and *TALAM1*. Since long non-coding RNA MALAT1 is associated with cancer cell metastasis and migration, its gene amplification represents an important event for adenoma development.

**Keywords:** colorectal cancer; adenomas; array comparative genomic hybridization; long non-coding RNA; MALAT1

## 1. Introduction

Colorectal adenomas are abnormally growing intestinal epithelium that originates from colon crypts, where the stem cell profiling and differentiation have been disrupted due to changes in DNA. Some adenomas display altered genetic background (chromosomal rearrangement, mutation in tumor suppressor genes and oncogenes or epigenetic modifications) that predispose them to develop into colorectal cancer (CRC) [1]. CRC is the third most common cancer and second leading cause of death due to cancer worldwide [2]. Early detection of adenomas due to their potential to evolve into CRC substantially improves the patient's prognosis. For instance, CRC with stage I exhibits five-year overall survival (OS) of 90%, while the diagnosis in the late stages, III and IV, predicts OS of only 13% [3]. The risk of cancer development from adenomas increases with age, from 25% at age 55 years to 40% at age 80 years [4]. The process of adenoma transformation into cancer can last 5 to 10 years, depending on the accumulation of genetic and epigenetic alterations [5]. Malignant progression of adenoma to cancer is a multifactorial process that includes chromosomal instability (CIN), microsatellite instability (MSI), the influence of epigenetic factors, such as methylation of CpG islands (CIMP), and mutations in driver genes [6]. Mutations in driver genes responsible for adenoma–carcinoma progression have been identified by human genome profiling, and these patterns are referred to as driver mutations. Among the well-known genes whose driver mutations contribute to CRC development are, for example, *APC, KRAS, BRAF, TP53, PTEN, SMAD4, GNAS, NOTCH1, POLD1, POLE* and *MUTYH* [7–11].

CIN is caused by aberrant segregation of chromosomes during mitosis and is found in 65–70% of sporadic CRC cases. CIN is defined by losses or gains of loci on short or long arms of chromosomes or even losses or gains of whole chromosomes [12]. However, CIN is not a frequent subject of research in colorectal adenomas. We assume that CIN will be present already in adenomas. Clonal expansion of aberrant cells forms the basis for the intertumoral heterogeneity within the adenoma and consequently within the tumor. Therefore, each adenoma of an individual may display a unique genetic background [13]. The genetically diverse cell population results in somatic mosaicism, a common phenomenon in tumors, while in adenomas it is an unexplored area. Somatic mosaicism in tumors or adenomas is understood as the occurrence two or more genetically distinct cell populations within one tissue [14]. We assume, based on the 70% incidence of CIN in tumors, that CIN may appear already in the precancerous lesions.

Epigenetics has been revealed to be a major player in current cancer research. Regulation of gene expression can be influenced by non-coding RNAs (long or micro RNAs). Transcripts of long non-coding RNAs have more than 200 nucleotides and may function as tumor suppressors as well as oncogenes during cancer development [15]. Metastasis-associated lung adenocarcinoma transcript 1 (MALAT1) was firstly described in relation to lung cancer aggressiveness [16]. Long non-coding RNA MALAT1 affects cell proliferation by upregulation of the Wnt/β-catenin signaling pathway, regulates transcription and post-transcriptional modification of many genes, and acts as a microRNA sponge [17]. Our hypothesis is that amplification of MALAT1 already in adenoma tissue could be a precursor of cancer development.

This study aimed to describe chromosomal aberrations in colorectal adenoma tissues with similar histological background and clinical characteristics. This study stands out by demonstrating the array-based comparative genomic hybridization (aCGH) method over a wider number of precancerous colorectal stages using DNA from fresh frozen samples, in contrast to other studies dealing mainly with colorectal tumors using DNA predominantly from formalin-fixed paraffin-embedded (FFPE) samples. This study provides a novel insight into CIN in the precancerous stages, a poorly explored area that is nonetheless crucial to understanding tumorigenesis.

## 2. Results

The aCGH method was successfully performed on all 16 pairs of samples. Enzymatic labeling of DNA was fully achieved in the same 1:1 concentration ratio within each pair of samples. Derivative log ratio standard deviation (DLRSD) ranged from 0.1 to 0.2, which reflected low probe-to-probe log-ratio noise.

All 16 pairs of adenoma tissue samples showed varying degrees of chromosome-level instability. After a more detailed examination of these profiles, we observed that certain changes were repeated many times. Based on these obtained data, we were able to identify several groups gathering similar characteristics, and these served as a basis for creation of our four groups. Briefly, a general feature of the first group was the gain in adenoma tissue of chromosome 11, 11q13.1, encoding long non-coding RNA (lncRNA) *MALAT1* and its antisense transcript *TALAM1*, in patients P1, P2, P3, P5, and P16. The second group consisted of patients with numerous chromosomal microdeletions in adenoma tissues compared to the adjacent tissue with no aberrations in the region encoding for *MALAT1* or *TALAM1* (P6, P7, and P10). The third group included those patients with a disrupted karyotype with many losses and gains. The only common feature of this group was the relatively young age of the patients (P4, P5, and P16). In the last group, no differences were found between adenoma and adjacent tissue (P8, P9, P11, P12, P13, P14, and P15). Patients P5 and P16 overlapped between the first and third groups. The detailed results are described in the following subsections.

In addition, mosaicism was also detected in all nine samples with aberrations (P1, P2, P3, P4, P5, P6, P7, P10, P16) in adenoma tissues compared to the adjacent tissues. The percentage of mosaicism for each aberration in individual patients is given in the tables (Tables 1–3).

Beyond the four groups defined above, patients P1, P4, P6, P7, and P16 were also associated with the loss of the *TSC2* gene, which plays the role of a tumor suppressor. The loss of *TSC2* in 5 of 16 patients deserved our attention and is marked bold in the tables (Tables 1–3). Some microdeletions occurred more than once in the entire study population of patients, e.g., *COL1A1* (found in P6 and P7), *NOTCH1* and *MIR4673* (found in P4 and P16) and *GNAS* (found in P4 and P7).

### 2.1. The First Group with MALAT1 and TALAM1 Gain

This group (P1, P2, P3, P5, and P16) was characterized by the gain on chromosome 11 of loci encoding region corresponding to *MALAT1* and *TALAM1* lncRNAs and, in one case (P16), also *MALAT1*-associated small cytoplasmic RNA (MASCRNA). The comparison of gained region encoding *MALAT1*, *TALAM1*, and *MASCRNA* between five patients is displayed in Figure 1. This region was not completely amplified in any patient in this group, so its expression function was questionable. In addition, several aberrations, e.g., giant losses on 6q ($\hat{\alpha}$ = 65%) in P2, 5q ($\hat{\alpha}$ = 59%) in P3, and microdeletion on 16p ($\hat{\alpha}$ = 38%) along with gain on 1p ($\hat{\alpha}$ = 39%) in P1, were also detected in adenoma tissues compared to the adjacent tissues, as shown in Table 1 for P1, P2, and P3. Patients P5 and P16 were described in detail in the separate section depicting the third group of patients with whom they shared common characteristics. This group included both male (P2 and P16) and female (P1, P3 and P5) patients ranging in age from 43 years to 61 years. The adenomas in this group were both tubular (P1, P3 and P5) and tubulo-villous (P2 and P16) in nature with low-grade dysplasia (P1, P3, P5 and P16) and one case of high-grade dysplasia (P2). They were ranked grade 3 according to the Vienna classification (P1, P3, P5 and P16), with one case of grade 4.1 (P2). Adenomas were from both the colon (P1 and P2) and the rectum (P3, P5 and P16).

**Table 1.** Additional aberrations found in patients P1, P2 and P3.

| Patient | Type | Chromosome | Location | Cytoband | Size | $\hat{\alpha}$ [a] | Gene Name |
|---|---|---|---|---|---|---|---|
| P1 | loss | 16 | 2,110,696–2,136,380 | p13.3 | 25.685 kb | 38% | *TSC2* |
| | gain | 1 | 2,260,756–3,080,894 | p36.33–p36.32 | 820.139 kb | 39% | *PEX10, PLCH2, PANK4, HES5, TNFRSF14, MMEL1, ACTRT2, PRDM16, MORN1, LOC100129534, RER1, TNFRSF14-AS1, LOC100996583, PRXL2B, TTC34, PRDM16-DT, MIR4251* |
| | | 11 | 65,267,014–65,273,869 | q13.1 | 6.856 kb | 81% | *MALAT1, TALAM1* |
| P2 | loss | 6 | 107,338,10–127,407,686 | q21–q22.33 | 20,069.584 kb | 65% | *MTRES1, BEND3, PDSS2, SOBP, SEC63, OSTM1, NR2E1, SNX3, AFG1L, FOXO3, ARMC2, SESN1, CD164, SMPD2, MICAL1, ZBTB24, AK9, FIG4, GPR6, WASF1, CDC40, DDO, SLC22A16, CDK19, AMD1, GTF3C6, RPF2, SLC16A10, MFSD4B, REV3L, TRAF3IP2, FYN, CCN6, TUBE1, LAMA4, MARCKS, HDAC2, HS3ST5, FRK, COL10A1, DSE, TSPYL1, CALHM6, TRAPPC3L, RSPH4A, KPNA5, GPRC6A, RFX6, VGLL2, ROS1, GOPC, NUS1, PLN, MCM9, ASF1A, MAN1A1, TBC1D32, GJA1, HSF2, SERINC1, PKIB, FABP7, SMPDL3A, CLVS2, TRDN, NKAIN2, RNF217-AS1, RNF217, TPD52L1, HEY2, NCOA7, HINT3, CENPW, SCML4, OSTM1-AS1, LINC00222, ARMC2-AS1, CEP57L1, CCDC162P, PPIL6, METTL24, SNORA40C, GSTM2P1, SNORD166, TRAF3IP2-AS1, LINC02527, FAM229B, LOC101927640, RFPL4B, LINC02518, LINC02541, MROCKI, FLJ34503, HDAC2-AS2, LINC02534, TPI1P3, NT5DC1, TSPYL4, LOC100287467, CALHM5, CALHM4, RWDD1, ZUP1, FAM162B, DCBLD1, LOC101927919, SLC35F1, LOC105377967, CEP85L, BRD7P3, SELENOKP3, FAM184A, MIR548B, LOC285762, LOC105377975, MIR3144, TRDN-AS1, LOC100126584, HDDC2, LINC02523, NCOA7-AS1, TRMT11, MIR588* |
| | gain | 11 | 65,267,014–65,272,199 | q13.1 | 5.186 kb | 70% | *MALAT1, TALAM1* |
| P3 | loss | 5 | 102,026,08–127,375,136 | q21.1–q23.3 | 25,349.057 kb | 59% | *PAM, PPIP5K2, C5orf30, NUDT12, EFNA5, FBXL17, FER, MAN2A1, SLC25A46, TSLP, WDR36, CAMK4, STARD4, NREP, EPB41L4A, APC, SRP19, REEP5, DCP2, MCC, TSSK1B, YTHDC2, KCNN2, TRIM36, PGGT1B, FEM1C, TICAM2, CDO1, ATG12, AP3S1, LVRN, COMMD10, SEMA6A, DMXL1, TNFAIP8, HSD17B4, FAM170A, PRR16, FTMT, SRFBP1, LOX, SNCAIP, SNX2, PPIC, PRDM6, CEP120, CSNK1G3, ALDH7A1, PHAX, LMNB1, MARCHF3, MEGF10, GIN1, LINC02115, RAB9BP1, LINC01950, LINC01023, LOC285638, PJA2, LINC01848, TMEM232, MIR548F3, STARD4-AS1, NREP-AS1, EPB41L4A-AS1, SNORA13, LOC101927023, EPB41L4A-DT, LINC02200, LOC102467216, LOC101927078, LINC01957, CCDC112, TMED7-TICAM2, LOC101927100, TMED7, LOC102467217, LINCADL, ARL14EPL, MIR12130, LOC101927190, SEMA6A-AS1, SEMA6A-AS2, LINC02214, LINC00992, LINC02147, LINC02148, LINC02208, LINC02215, DTWD2, MIR1244-1, MIR1244-2, MIR1244-3, MIR1244-4, LOC105379143, MIR5706, LOC102467226, ZNF474, LOC100505841, MGC32805, LOC101927357, LINC02201, SNX24, LOC105379152, LINC01170, ZNF608, LOC101927421, LINC02240, LINC02039, LOC101927488, GRAMD2B, TEX43, LMNB1-DT, C5orf63, PRRC1, CTXN3, CCDC192, LINC01184* |
| | gain | 11 | 65,267,014–65,272,199 | q13.1 | 5.186 kb | 70% | *MALAT1, TALAM1* |

[a] percentage of mosaicism.

**Table 2.** List of losses in adenoma tissues of patients P6, P7, and P10 (second group).

| Patient | Type | Chromosome | Location | Cytoband | Size | ά [a] | Gene Name |
|---|---|---|---|---|---|---|---|
| P6 | loss | 16 | 2,103,321–2,138,073 | p13.3 | 34.753 kb | 51% | *TSC2* |
| | | 17 | 48,263,792–48,273,321 | q21.33 | 9.530 kb | 34% | *COL1A1* |
| P7 | loss | 7 | 73,442,449–73,481,111 | q11.23 | 38.663 kb | 35% | *ELN* |
| | | 16 | 2,105,434–2,138,073 | p13.3 | 32.640 kb | 58% | *TSC2* |
| | | 17 | 48,263,792–48,273,777 | q21.33 | 9.986 kb | 40% | *COL1A1* |
| | | 20 | 57,407,840–57,495,925 | q13.32 | 88.086 kb | 38% | *GNAS-AS1, GNAS, LOC1019* |
| P10 | loss | 9 | 21,549,338–23,792,459 | p21.3 | 2243.122 kb | 46% | *MIR31HG, MTAP, CDKN2A, CDKN2B-AS1, CDKN2B, DMRTA1, ELAVL2, CDKN2A-DT, LINC01239, LOC101929563* |

[a] percentage of mosaicism.

**Table 3.** List of aberrations found in P4, P5, and P16 (third group).

| Patient | Type | Chromosome | Location | Cytoband | Size | ά [a] | Gene Name |
|---|---|---|---|---|---|---|---|
| P4 | loss | 1 | 55,108,604–61,921,519 | p32.3–p31.3 | 6812.92 kb | 33% | *TTC4, PARS2, LEXM, DHCR24, BSND, PCSK9, USP24, PLPP3, PRKAA2, FYB2, C8A, C8B, DAB1, OMA1, TACSTD2, MYSM1, JUN, FGGY, HOOK1, CYP2J2, NFIA, MROH7, MROH7-TTC4, TTC22, TMEM61, LOC100507634, MIR4422HG, MIR4422, LINC01753, LINC01755, LINC01767, LOC101929935, DAB1-AS1, LINC01135, LINC01358, HSD52, MIR4711, LOC101926944, C1orf87, LINC01748, LOC101926964, NFIA-AS2, NFIA-AS1* |
| | | 4 | 1,802,707–1,809,469 | p16.3 | 6.763 kb | 65% | *FGFR3* |
| | | 9 | 139,394,991–139,418,283 | q34.3 | 23.293 kb | 50% | *NOTCH1, MIR4673* |
| | | 11 | 65,265,673–65,273,325 | q13.1 | 7.653 kb | 100% | *MALAT1, TALAM1* |
| | | 16 | 2,105,434–2,138,073 | p13.3 | 32.640 kb | 50% | *TSC2* |
| | gain | 6 | 107,068,675–109,166,111 | q21 | 2097.437 kb | 47% | *RTN4IP1, QRSL1, MTRES1, BEND3, PDSS2, SOBP, SEC63, OSTM1, NR2E1, SNX3, AFG1L, FOXO3, LINC02526, LINC02532, MIR587, SCML4, OSTM1-AS1, LINC00222* |
| | | 7 | 83,325–2,737,748 | p22.3 | 2654.42 kb | 40% | *FAM20C, PDGFA, PRKAR1B, DNAAF5, SUN1, GET4, ADAP1, COX19, CYP2W1, MIR339, GPER1, ZFAND2A, INTS1, MAFK, PSMG3, ELFN1, MAD1L1, MRM2, NUDT1, SNX8, EIF3B, CHST12, LFNG, BRAT1, IQCE, TTYH3, AMZ1, LOC102723672, LOC100507642, LOC105375115, LOC442497, HRAT92, LOC101927000, LOC101926963, C7orf50, GPR146, LOC101927021, UNCX, MICALL2, LOC100128653, TMEM184A, PSMG3-AS1, TFAMP1, ELFN1-AS1, MIR4655, SNORA114, MIR6836, GRIFIN, MIR4648* |
| P4 | monosomy | 8q | | - | | 25% | - |
| | | 13 | | | | 14% | |
| | | 17q | | | | 20% | |
| | | 20q | | | | 31% | |
| | | X | | | | 12% | |
| | trisomy | 8p | | - | | 23% | - |
| | | 10p | | | | 19% | |
| | | 14 | | | | 23% | |
| | | 17p | | | | 20% | |
| | | 18 | | | | 23% | |
| | | 20p | | | | 31% | |

**Table 3.** *Cont.*

| Patient | Type | Chromosome | Location | Cytoband | Size | $\hat{\alpha}$ [a] | Gene Name |
|---|---|---|---|---|---|---|---|
| P5 | gain | 11 | 65,267,014–65,273,869 | q13.1 | 6.856 kb | 100% | *MALAT1, TALAM1* |
| | trisomy | 7 | | - | | 23% | - |
| | | 13 | | | | 22% | |
| | | X | | | | 11% | |
| P16 | loss | 1 | 185,274268–199,118,773 | q25.3–q32.1 | 13,828 kb | 24% | *HMCN1, PRG4, TPR, ODR4, OCLM, PDC, PTGS2, PACERR, PLA2G4A, BRINP3, RGS18, RGS21, RGS1, RGS13, RGS2, UCHL5, RO60, GLRX2, CDC73, B3GALT2, KCNT2, CFH, CFHR3, CFHR1, CFHR4, CFHR2, CFHR5, F13B, ASPM, CRB1, DENND1B, LHX9, NEK7, ATP6V1G3, PTPRC, MIR181B1, MIR181A1, LOC102724919, LINC01036, LINC01037, LINC01351, LINC01720, LINC01680, MIR4426, LINC01032, SCARNA18B, MIR1278, LINC01031, LINC01724, MIR4735, ZBTB41, C1orf53, MIR181A1HG, LINC01222, LINC01221, LINC02789* |
| | | 1 | 890,945–3,729,090 | p36.33–p36.32 | 2838.146 kb | 40% | *NOC2L, PERM1, HES4, ISG15, AGRN, MIR200B, MIR200A, MIR429, TNFRSF18, TNFRSF4, SDF4, B3GALT6, C1QTNF12, SCNN1D, INTS11, CPTP, TAS1R3, DVL1, MXRA8, AURKAIP1, CCNL2, MRPL20, VWA1, ATAD3C, ATAD3B, ATAD3A, TMEM240, SSU72, MIB2, MMP23B, MMP23A, CDK11B, CDK11A, NADK, GNB1, CALML6, GABRD, PRKCZ, FAAP20, SKI, PEX10, PLCH2, PANK4, HES5, TNFRSF14, MMEL1, ACTRT2, PRDM16, MEGF6, MIR551A, TPRG1L, WRAP73, TP73, SMIM1, CEP104, KLHL17, PLEKHN1, LOC100288175, RNF223, C1orf159, LINC01342, TTLL10, UBE2J2, ACAP3, MIR6726, SNORD167, PUSL1, MIR6727, MIR6808, MRPL20-AS1, ANKRD65, TMEM88B, LINC01770, FNDC10, LOC105378586, SLC35E2B, SLC35E2A, TMEM52, CFAP74, LOC105378591, PRKCZ-AS1, MORN1, LOC100129534, RER1, TNFRSF14-AS1, LOC100996583, PRXL2B, TTC34, PRDM16-DT, MIR4251, ARHGEF16, TP73-AS1, CCDC27, LRRC47* |
| | loss | 9 | 139,389,744–139,440,753 | q34.3 | 51.010 kb | 70% | *NOTCH1, MIR4673, MIR4674* |
| | | 16 | 2,044,093–2,263,638 | p13.3 | 219.546 kb | 52% | *SYNGR3, ZNF598, NPW, SLC9A3R2, NTHL1, TSC2, PKD1, MIR1225, RAB26, TRAF7, CASKIN1, MLST8, PGP, LOC105371049, MIR6511B1, MIR6511B2, MIR4516, MIR3180-5, SNHG19, SNORD60, BRICD5* |
| | | 16 | 88,365,786–89,383,369 | q24.2–q24.3 | 1017.584 kb | 43% | *ZNF469, ZFPM1, IL17C, CYBA, MVD, SNAI3, RNF166, CTU2, PIEZO1, CDT1, APRT, GALNS, TRAPPC2L, CBFA2T3, ACSF3, CDH15, ANKRD11, MIR5189, LOC100128882, ZC3H18-AS1, ZC3H18, SNAI3-AS1, MIR4722, LOC100289580, LOC339059, PABPN1L, LOC101927793, LOC100129697, LINC00304, LINC02138, SLC22A31, ZNF778, LOC105371414* |
| | loss | 17 | 36,861,875–36,896,355 | q12 | 34.481 kb | 54% | *MLLT6, CISD3, PCGF2, MIR4726* |
| | | 22 | 19,702,774–19,851,138 | q11.21 | 148.365 kb | 47% | *SEPTIN5, SEPT5-GP1BB, GP1BB, TBX1, GNB1L, RTL10* |
| | gain | 11 | 65,268,530–65,276,115 | q13.1 | 7.586 kb | 100% | *MALAT1, TALAM1, MASCRNA* |
| | monosomy | 18 | | - | | 31% | - |
| P16 | trisomy | 3 | | - | | 23% | - |
| | | 5 | | | | 20% | |
| | | 6 | | | | 20% | |
| | | 7 | | | | 20% | |
| | | 8 | | | | 20% | |
| | | 12 | | | | 20% | |
| | | 13 | | | | 25% | |
| | | 15 | | | | 26% | |
| | | 19 | | | | 30% | |
| | | 20 | | | | 32% | |
| | | 21 | | | | 20% | |
| | | X | | | | 11% | |
| | | Y | | | | 10% | |

[a] percentage of mosaicism.

**Figure 1.** The representative example of gain in region encoding *MALAT1*, *TALAM1*, and *MASCRNA* is pictured as a blue rectangle in the region at chr11, q13.1, in adenoma of P1 at the top of the image. Thin horizontal red and blue lines represent values corresponding to the non-mosaic state of deletions (−1) or duplications (0.58). A comparison of this gained region between patients from the first group (P1, P2, P3, P5, and P16) is shown at the bottom of the image. Mosaicism of each patient is expressed by the symbol $\hat{\alpha}$.

*2.2. The Second Group with Microdeletions*

Common features of patients P6, P7 and P10 in this second group were many microdeletions at 7q, 9p, 16p, 17q, and 20q in the genome of adenoma tissue compared to that of adjacent tissue, and likewise without the presence of *MALAT1* or *TALAM1* gain on chromosome 11 or any other gain. The list of losses in the adenoma genome of patients is shown in Table 2. Similar aberrations were found in patients P6 and P7, such as loss at 16p of *TSC2* and loss at 17q of *COL1A1*. In other losses, patients differed from each other. Adenomas from these patients were all tubular in nature with low grade dysplasia, ranked by grade 3 according to the Vienna classification and originated from both the colon (P6, P7) and the rectum (P10). The age of the patients ranged from 63 to 68 years, and all were male.

*2.3. The Third Group with Affected Karyotype*

Three patients from the analyzed set showed significant disruption of the adenoma tissue karyotype compared to that of adjacent tissue (Figure 2). There were two females (P4 and P5) and one male (P16) in this group. Rather young age was the common feature shared by the patients: 29 years (P4), 43 years (P5), 43 years (P16). Histologically, they had tubular (P4, P5) to tubulo-villous (P16) adenomas with low grade dysplasia, according to the Vienna classification with grade 3, originating from both the colon (P4) and the rectum (P5 and P16). Since these are serious changes in karyotype that deserve increased attention, we have described the patients in this group in more detail. A section below is dedicated to each patient, including the family and personal history and the reason for the colonoscopy examination. These patients continue to be monitored after adenoma resection.

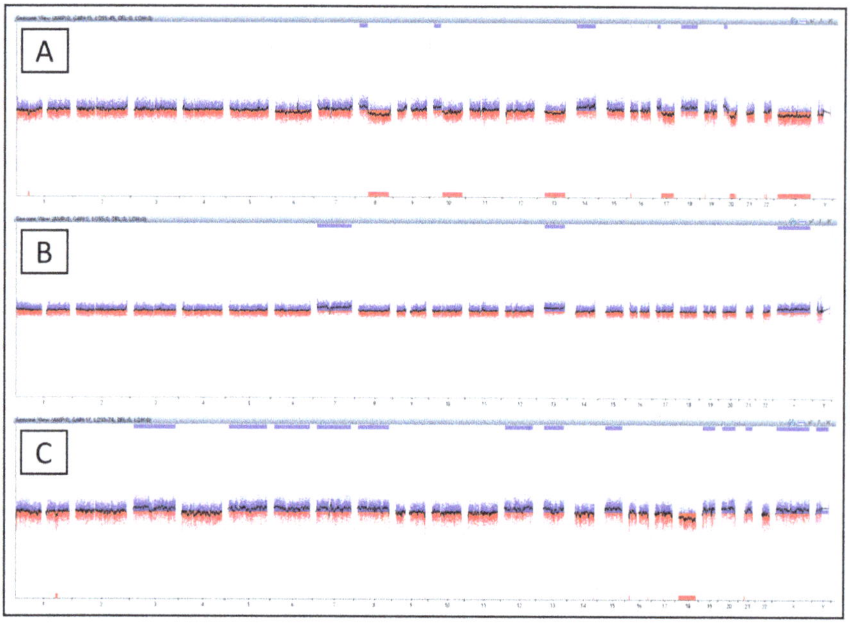


**Figure 2.** The entire genome of patient P4 (**A**), P5 (**B**), P16 (**C**). (**A**) Trisomy of chromosomes 14 and 18, monosomy of chromosomes 13 and X, aberrations on chromosomes 1, 6, 8, 10, 11, 17, 20 at P4. (**B**) Trisomy of chromosomes 7, 13, X and aberration on chromosome 11 at P5. (**C**) Trisomy of chromosomes 3, 5, 6, 7, 8, 12, 13, 15, 19, 20, 21, X, Y, monosomy of chromosome 18, and aberration on chromosome 11 at P16.

2.3.1. Patient No. 4

Patient P4 exhibited short arm gains along with long arm losses, as found on chromosomes 8 ($\hat{\alpha}$ = 23%), 10 ($\hat{\alpha}$ = 19%), 17 ($\hat{\alpha}$ = 20%), and 20 ($\hat{\alpha}$ = 32%) and including breaks in the centromeres of chromosomes 10 and 17. Chromosomes 8 and 20 had breaks outside their centromeres. On chromosome 8, the break was on the short arms. The short arm of chromosome 20 was partially duplicated together with a partial deletion of the long arm of this chromosome. Characteristics found in the genome of the pathological clone (Figure 2A) corresponded to the occurrence of isochromosomes or unbalanced translocation (chr: 8, 10, 17 and 20). In addition to these findings, this patient was found to have monosomy of chromosomes 13 ($\hat{\alpha}$ = 14%) and X ($\hat{\alpha}$ = 12%) and trisomy of chromosomes 14 ($\hat{\alpha}$ = 14%) and 18 ($\hat{\alpha}$ = 23%). Other aberrations such as microdeletion on chromosomes 1p, 4p, 9q, 16p, and gain on chromosomes 6q and 7p are listed in Table 3. Interestingly, deletion of loci bearing *MALAT1* and *TALAM1* ($\hat{\alpha}$ = 100%) also occurred in this patient. The patient had no family history either of CRC or other cancer and underwent colonoscopy due to intestinal discomforts such as diarrhea and flatulence. Furthermore, the patient suffered from cholelithiasis, a disease of the bile ducts. The patient was recommended to have a follow-up colonoscopy three years later.

2.3.2. Patient No. 5

In the adenoma genome of patient P5 (Figure 2B), trisomy of three chromosomes (chr: 7 ($\hat{\alpha}$ = 23%), 13 ($\hat{\alpha}$ = 22%) and X ($\hat{\alpha}$ = 11%)) was found, and at the same time gain of the region encoding for *MALAT1* and *TALAM1* ($\hat{\alpha}$ = 100%) (Table 3). The patient underwent a colonoscopy examination based on previous occurrences of adenomas in the colon and family history, and the patient's sibling suffered from Crohn's disease. The patient underwent urea dilatation and cervical conization in the past. The patient was recommended to have a follow-up colonoscopy three years later.

#### 2.3.3. Patient No. 16

Based on these data for P16 patient, we believe the pathological clone originally had three sets of chromosomes and one of them was gradually lost (Figure 2C). This patient was found to have trisomy of chromosomes 3 ($\hat{\alpha}$ = 23%), 5 ($\hat{\alpha}$ = 20%), 6 ($\hat{\alpha}$ = 20%), 7 ($\hat{\alpha}$ = 20%), 8 ($\hat{\alpha}$ = 20%), 12 ($\hat{\alpha}$ = 20%), 13 ($\hat{\alpha}$ = 25%), 15 ($\hat{\alpha}$ = 26%), 19 ($\hat{\alpha}$ = 30%), 20 ($\hat{\alpha}$ = 32%), 21 ($\hat{\alpha}$ = 20%), X ($\hat{\alpha}$ = 11%), and Y ($\hat{\alpha}$ = 10%), and one monosomy of chromosome 18 ($\hat{\alpha}$ = 31%). The gain of *MALAT1* and *TALAM1* ($\hat{\alpha}$ = 100%) and many microdeletions on chromosomes 1q, 1p, 9q, 16p, 16q, 17q, and 22q were also found in this sample (Table 3). The patient underwent a planned colonoscopy and was only treated for hemorrhoids. This patient had a family history of CRC and was recommended to have a follow-up colonoscopy three years later.

### *2.4. The Fourth Group with a Negative Finding*

The adenoma tissues of patients P8, P9, P11, P12, P13, P14, and P15 showed no changes at the chromosomal level compared to adjacent tissue. No relevant link between patients was found. The age of the group ranged from 44 to 67 years. This group included two women and five men. From a histological point of view, adenomas showed a tubular to tubulo-villous character with low-grade dysplasia (except for P9, who showed high-grade dysplasia), and tissues were taken from both the colon (P8, P12, P13, P14, and P15) and the rectum (P9, P11).

## 3. Discussion

Examination of CIN in precancerous stages is crucial to understand the development of colorectal adenoma. While most studies focus their investigation of CIN in carcinomas, this study took a step back and searched for possible causes of cancer already in adenomas. Carcinoma evolution from colorectal adenoma has been reported in relation with 8p, 17p, 15q, 18p and 18q losses and 5q, 7p, 7q, 8q, 13q, 20p and 20q gains [18,19]. In our study, we observed losses across the whole genome (1p, 1q, 4p, 5q, 6q, 7q, 9p, 9q, 11q, 16p, 16q, 17q, 20q, 22q), whereas gains were pronounced less frequently (1p, 6q, 7p, 11q). In this study, the most significant aberrations were the gain of *MALAT1* and *TALAM1* lncRNAs (found in P1, P2, P3, P5, and P16) and losses of genes such as *TSC2* (in P1, P4, P6, P7 and P16 individuals), *COL1A1* (P6 and P7), *NOTCH1* and *MIR4673* (in P4 and P16) and *GNAS* (in P4 and P7). The genes described above either play a role in signaling pathways (*NOTCH1*), affect transcription (*GNAS*, *MIR4673*, *MALAT1* and *TALAM1*), or are involved in cell structure (*COL1A1*) and proliferation (*TSC2*). Their importance in the cancer process has been previously evidenced and, therefore, their gain or loss in adenoma tissue deserves further attention [20–34].

When analyzing the data, we encountered a substantial variety of chromosomal aberrations across all samples. For better orientation in the data, we divided the patients into four groups based on similar features observed in the results. The first group of patients was characterized by one major loss in the genome along with the gain in the region encoding *MALAT1* and *TALAM1* lncRNAs. The physiological functions of MALAT1, referred to also as NEAT2 (nuclear-enriched abundant transcript 2), are in alternative splicing, transcriptional and post-transcriptional regulation, synapse formation, and myogenesis [33]. MALAT1, as an epigenetic player, has been often described in connection with cancer progression [34] or as an inflammatory regulator in diabetic retinopathy [35]. The increased presence of this transcript in cancer tissues compared to non-malignant tissues has been reported not only in lung cancer [36], but also in breast [37], bladder [38], cervical [39], and liver cancer [40], and CRC [40–43], and was associated with poor OS [44]. MALAT1 has been found to promote cell proliferation and migration of cancer cells by regulating the expression of genes promoting metastases, e.g., *RASSF6*, *HNF4G*, *CA2*, *ROBO1*, *MIA2* [36].

LncRNA MALAT1 acts primarily as an epigenetic modulator through small endogenous non-coding RNAs (miRNAs) that are no more than 22 nucleotides in size and control translation and post-transcriptional modifications. MiRNAs have been widely described in the process of carcinogenesis as negatively regulating gene expression in the target gene [45]. The role of the lncRNA MALAT1 in carcinogenesis is through interaction with miRNAs via a "sponge" event, a process whereby competing endogenous RNAs (ceRNAs) share recognition elements (MREs) with miRNAs and thus influence each other's function [46]. For example, MALAT1 overexpression reduces the expression of miR-145, which under normal conditions inhibits *SOX9*, the gene responsible for differentiation and skeleton development. The inhibition pathway MALAT1/miR-145/*SOX9* thus promotes colorectal cancer cell growth, migration, and invasion [47]. Among other miRNAs, that are target of MALAT1 in relation to colorectal cancer progression, are also miR-508–5p, miR-324-3p, miR-363–3p, and miR-129-5 [48]. The role of MALAT1 in inflammation and cancer progression was demonstrated by Huang et al. in hepatocellular carcinoma, where MALAT1 promotes cancer cell growth by binding Brahma-related gene 1 (BRG1) and recruiting it into the promoter region of *IL-6* and *CXCL8*, thus enabling transcription factors to start the expression of these pro-inflammatory mediators [48]. In addition, Qing et al. discovered that CRC patients with lower expression of *MALAT1* in primary tumors had a better prognosis [44,49]. Another study confirmed increased levels of *MALAT1* in colorectal adenomas and a significant difference between the type and number of polyps compared to unaffected colon epithelium [50]. Therefore, we concluded that amplification of the region encoding *MALAT1* and *TALAM1* in 5 out of the 16 adenoma samples revealed cancer potential in these samples. However, as illustrated in Figure 1, this region is not completely amplified in all patients. The question, therefore, remains whether the resulting product is fully functional. In the first group, in addition to the gain of *MALAT1* and *TALAM1*, losses on chromosomes 5, 6, and 16 were detected. The regions where the losses occurred contained genes associated with CRC: oncogenes (*ROS1*, *FER*) [51–53] and tumor suppressors (*APC*, *MCC*, *LOX*) [54–56]. The losses may also have contributed to the onset or development of the adenoma.

Microdeletions at 7q, 9p, 16p, 17q, and 20q together with no gains were observed in the intestinal adenoma genome of the second group of patients (P6, P7 and P10). Similar changes in the genome have been observed in other publications studying CIN in colorectal adenomas. Hirsch et al. applied aCGH to 13 FFPE-derived colorectal adenomas rising in high-grade adenomas and revealed losses at 1p, 1, 5q, 8p, 10q, 11q, 16p, 17p, 18q, 18, and 20p, and gains at 4q, 6, 7, 8q, 12p, 12q, 14q, 12, 13, 19, 20q, 20, and X. The study was compromised by high DNA fragmentation due to paraffin fixation [57]. Degraded DNA (length of fragment < 1000 bp) results in biased labeling and inaccurate results. An earlier study from 2002 by Hermsen et al. reported the most frequent 8p, 15q, 17p, and 18q losses and 8q, 13q, and 20q gains in FFPE-derived samples of 66 non-progressed adenomas, 46 progressed adenomas, and 36 colorectal adenomas using chromosomal CGH. However, the DNA extracted from the paraffin was partially degraded, which could have affected the results. The authors of this study observed predominantly losses of chromosomal regions in small non-progressed adenomas, while progressed adenomas were characterized by an increased incidence of gains in the genome [58]. Although the specific losses in our data did not exactly correlate with the already published results, we lean towards the theory that the dominance of microdeletions over gains in genome occurs mainly in non-progressed adenomas, as observed in our samples consisting of tubular, low-grade adenomas without signs of malignancy. The advantage of our study was that we used fresh frozen samples instead of FFPE samples where DNA is often highly degraded. At the same time, we applied the most modern approach to the current CGH using high-resolution array CGH. Thanks to these modifications, we obtained more accurate and reliable results.

The third group included patients with a severely disrupted karyotype. In patient P4 abnormal formations of chromosomes 8, 10, 17 and 20 were found. We assume these could be either isochromosomes or unbalanced translocations. The gain of short arms and loss of long arms, which we could see at chromosomes 10 and 17, suggested that these could be isochromosomes. The occurrence of isochromosomes in colorectal adenomas has been reported only in cell lines derived from familial polyposis carcinoma on chromosomes 1, 14 and translocation on chromosomes 17 and X [59]. In patient P5, trisomy of chromosomes 7, 13, and X occurred in adenoma tissue. A similar result was obtained by Longy et al., who analyzed 25 colonic adenomatous polyps from patients with average age of 66 years, and found the most frequent trisomy on chromosomes 7 (in eight cases) and 13 (in seven cases) using a direct method of chromosome visualization as in prenatal analysis [60]. Another study involving 20 colorectal non-progressed adenomas described the most common gains on chromosomes 3, 7, 13, and 20 and the most common structural rearrangements on chromosomes 1, 13, 17, and 18 using chromosome banding analysis for direct chromosome visualization [61]. Chromosome trisomy in adenoma tissue was a common feature in patient P16 (chr: 3, 5, 6, 7, 8, 12, 13, 15, 19, 20, 21, X, and Y), suggesting that the cells of pathological clone in adenoma tissue had a triploid set of chromosomes, and one of them was gradually lost. Despite the scarcity of studies published on CIN in colorectal adenomas, we hypothesize that chromosome trisomy in this aberrant kind of tissue is a common feature. The only link between all three patients with disrupted karyotypes was their young age at the time of adenoma resection compared to the rest of the study set. We assume that young age is one of the factors contributing to such extensive karyotype diversity in cells of adenoma tissue. Another hypothesis could be that the occurrence of adenomas at such a young age is precisely due to a large-scale change in the genome. Due to the finding of adenomas at such a young age, patients will continue to be monitored and have been advised to have a more frequent colonoscopy examination. Over the last 40 years, the incidence and mortality of CRC have increased in individuals under the age of 50, especially those aged 40–44. In the United States, 11% of colon cancer cases and 18% of rectal cancer cases affect patients under the age of 50. In addition to hereditary CRC syndromes, which appear at an early age (20–30 years), the main causes are sedentary lifestyle, obesity, and associated diabetes mellitus [62,63]. Therefore, it would be appropriate to monitor chromosomal rearrangements in adenoma samples of such young people (<50 years) and to validate the aberrations found in a wider group of patients. In this study, our candidates would be mainly the gain of *MALAT1* and *TALAM1* and the loss of *TSC2*.

From another point of view, we must consider the loss of *TSC2*, which occurred in five patients (P1, P4, P6, P7 and P16) out of 16. Tuberin, a product of *TSC2*, interacts with the hamartin protein, a product of the *TSC1*, in the cell. The main function of these proteins is to activate GTPase proteins, cyclins and many other proteins participating in cell cycle regulation to control the growth and size of cells, thus playing the role of tumor suppressors [20]. Disruption of the signaling pathway moderated by TSC1/TSC2 inhibition complex is often found in cancer development; in particular, loss of TSC2 function leads to hyperactivation of the mechanistic target of rapamycin complex 1 (mTORC1), a protein complex responsible for activation of protein translation [21,64]. Increased de novo protein translation enhanced endoplasmic reticulum (ER) stress, a common sign in cancer cells [65]. *TSC1* and *TSC2* are also potent regulators of the expression of a transmembrane protein named Programmed cell death ligand 1 (PD-L1), a target of inhibitors in non–small cell lung cancer treatment. The deficiency of *TSC2* showed up-regulation of PD-L1 in human lung cancer cell lines [66]. *TSC1* and *TSC2* are growth suppressor genes, therefore, we conclude that loss of *TSC2* could contribute to adenoma development.

In seven patients from the entire study group, no change was found at the chromosomal level in adenoma tissues compared to adjacent tissues. The negative finding can be justified by the fact that CIN does not occur in all colorectal tumors, but only in 70% of cases [67], thus making the expectation of CIN occurring in colorectal adenomas even lower. Furthermore, using aCGH, we were unable to detect driver mutations in genes that may be responsible for the development of adenoma. The negative findings could also be caused by insufficient genome coverage in the applied array. We could not detect aberrations in mosaicism lower than 15% certainty. Another factor was that aCGH did not capture balanced rearrangements that can disrupt genes and gene architecture or affect position effect expression at break sites.

The hypothesis that CIN will already appear early in adenoma tissue has been confirmed. Nine samples out of 16 showed CIN, representing 56% of the total set, while in CRC, CIN occurred in 65–70% [12]. We assumed that CIN would be slightly less represented in adenomas than in CRC, which was also confirmed. The histological classification of the adenomas according to clinical experience did not correlate with the extent of CIN among all samples of adenomas. Samples from patients were intentionally selected with the closest possible histological similarity to find their common feature also at the chromosomal level. The results showed a great diversity of CIN across the whole study group. An important finding was that the disrupted karyotype in adenoma tissue cells in the third group of patients (P4, P5 and P16) was also associated with young age, which may have contributed to such variable CIN. The aberrations found in the third group should receive more attention and validation with samples from patients under 50 years of age. The aberrations found in this study (losses of *TSC2*, *COL1A1*, *NOTCH1*, *MIR4673* and *GNAS*, and gains of *MALAT1* and *TALAM1*) may serve as candidate biomarkers for early detection of colorectal cancer onset.

The main clinical significance of our findings will be shown as we continue to monitor these patients to observe how/if their health conditions develop differently (e.g., whether the group with large changes will have more adenomas or progress to cancer). Our results raise doubts as to whether the classification based on histology is sufficient and suggest it should be extended to genetic analysis (e.g., detection of chromosomal instability or detection of specific changes—the diagnosis of the *MALAT1*/*TALAM1* region is offered).

Nowadays, it is crucial to identify biomarkers with sufficient sensitivity and specificity that can predict the early transformation of intestinal adenoma into adenocarcinoma. Detection of disease at its origin would help suppress the development into cancer and thus improve the patient's prognosis. These new biomarkers could serve physicians as indicators for colonoscopy and predict the frequency of this examination.

## 4. Materials and Methods

### 4.1. Sample Collection

The study included 16 individuals (Table 4) with adenomas of tubular or tubulo-villous histology that underwent recommended colonoscopy examination due to prevention or for intestinal discomfort, such as diarrhea or flatulence. The collection of samples was carried out in cooperation with the Department of Gastroenterology at Thomayer Hospital in Prague and with the Clinic of Hepatogastroenterology at the Institute for Clinical and Experimental Medicine, (Prague, Czech Republic) between March 2017 and December 2020. The study was approved by the ethical committees of both institutions. Sample biopsies of adenomas and adjacent tissue were placed into stabilization solution RNA and then stored in a deep-freeze box at −80 °C. Adenomas were transferred to histopathological examination to confirm that the samples did not show any signs of malignancy.

All subjects included in the study provided written informed consent to participate in the study and to use their biological samples for genetic analyses, in accordance with the Helsinki Declaration. The design of the study was also approved by the Ethics Committee of the Institute of Experimental Medicine, Prague, Czech Republic.

**Table 4.** Clinical and histological characteristics. [a] Female, [b] Male, [c] Tubular, [d] Tubulo-villous, [e] Low grade, [f] High grade. [g] For small samples, only two dimensions were given by pathologists.

| Sample ID | Gender | Age | Histology Type of Adenoma | Size of Adenoma (mm) | Vienna Classification | Grade | Localization |
|---|---|---|---|---|---|---|---|
| P1 | F [a] | 61 | T [c] | 15 × 8 × 10 | 3 | LG [e] | colon |
| P2 | M [b] | 60 | TL [d] | 7 × 4 × 4 | 4.1. | HG [f] | colon |
| P3 | F | 56 | T | 24 × 12 × 8 | 3 | LG | rectum |
| P4 | F | 29 | T | 8 × 8 [g] | 3 | LG | colon |
| P5 | F | 43 | T | 7 × 5 [g] | 3 | LG | rectum |
| P6 | M | 64 | T | 4 × 2 [g] | 3 | LG | colon |
| P7 | M | 63 | T | 3 × 4 [g] | 3 | LG | colon |
| P8 | M | 67 | T | 4 × 2 × 2 | 3 | LG | colon |
| P9 | M | 61 | T | 5 × 5 [g] | 4.1. | HG | rectum |
| P10 | M | 68 | T | 12 × 8 × 10 | 3 | LG | rectum |
| P11 | F | 44 | TL | 10 × 10 × 4 | 3 | LG | rectum |
| P12 | M | 53 | T | 3 × 10 [g] | 3 | LG | colon |
| P13 | M | 54 | T | 9 × 3 × 3 | 3 | LG | colon |
| P14 | M | 49 | TL | 10 × 3 × 2 | 3 | LG | colon |
| P15 | F | 57 | T | 2 × 2 [g] | 3 | LG | colon |
| P16 | M | 43 | TL | 18 × 13 × 11 | 3 | LG | rectum |

### 4.2. DNA Extraction

The genomic DNA was extracted from disrupted adenomas and adjacent tissues using AllPrep DNA/RNA Mini Kit (Qiagen, Düsseldorf, Germany) according to the standard protocol. MagNa Lyser Green Beads (Roche, Munich, Germany) were used for tissue disruption in a homogenizer (MagNaLyser Instrument, Version 4, Roche, Mannheim, Germany). The concentration of isolated DNA was measured by Qubit™ dsDNA BR Assay Kit (Invitrogen, Waltham, MA, USA) on Qubit 3.0 fluorometer (Invitrogen, Waltham, MA, USA).

### 4.3. Comparative Genomic Hybridization Array Design

Arrays used in the present study were designed to cover genomic regions bearing the cancer-associated genes by SurePrint G3 Cancer CGH+ single nucleotide polymorphism (SNP) Microarray Kit, 4 × 180 K (Agilent, Santa Clara, CA, USA). Instead of commercial DNA included within the kit, DNA isolated from the adjacent tissue from each patient was used as a source of reference DNA. For DNA labeling with Cy5 and Cy3 labels, Sure Tag Complete DNA Labeling enzyme kit was purchased (Agilent, Santa Clara, CA, USA); Cy5 label was applied as a reference for adjacent tissues and Cy3 label for adenoma tissues. The input amount of DNA into the labeling was on average 850 ng. Hybridization was performed by using an Oligo aCGH/ChIP-on-chip Hybridization kit (Agilent, Santa Clara, CA, USA).

### 4.4. Array Processing and Bioinformatics Data Analysis

SureScan Microarray Scanner instrument (Agilent, Santa Clara, CA, USA) with program Agilent G3 CGH corresponding to the barcode of arrays was used for array scanning. Data were processed in Agilent CytoGenomics software with the default analysis method—CGH v2. Due to the use of patient tissues as reference samples, SNP probes could not be analyzed.

The percentage of mosaicism was estimated according to the formula below (1), where $\delta$ is the observed fold change of the mean log ratio ($\delta = 2^{logR}$). The formula was introduced by Cheung et al. [68] to determine the presence of two or multiple cell lines ($\hat{\alpha} \leq 15\%$ implies the absence or low frequency of mosaic, and $\hat{\alpha} > 15\%$ indicates the presence of more than one clone in the cell population of the sample). Formula (1) is designed to calculate mosaicism in autosomes; we have modified the formula for heterochromosomes for male samples (2).

$$\hat{\alpha} = \left|\frac{\delta - 1}{0.5}\right| \times 100 \quad (1)$$

$$\hat{\alpha} = \left|\frac{\delta - 1}{1}\right| \times 100 \quad (2)$$

## 5. Conclusions

Using the aCGH method, we analyzed paired samples of colorectal adenomas and adjacent tissue from a total of 16 patients with histologically similar samples. The presence of CIN in the precancerous stages was confirmed in 56% of the adenomas. The significant gain was found on 11q13.1, encoding for *MALAT1* and *TALAM1* lncRNAs in five patients. We further identified several losses on chromosomes 1p, 1q, 4p, 5q, 6q, 7q, 9p, 9q, 11q, 16p, 16q, 17q, 20q, and 22q and gains in the 1p, 6q, 7p, and 11q regions. Overall, losses outweighed gains in adenoma tissue. Losses that were identified in at least two patients included the *TSC2, COL1A1, NOTCH1, MIR4673* and *GNAS* genes. *TSC2* loss was detected in five patients; since it is a tumor suppressor gene, we assume that its absence is involved in adenoma formation. Seven patients did not display any CIN in adenoma tissue at all. The study provided novel insight into chromosomal rearrangements in colorectal adenomas and provided new candidate biomarker lncRNA MALAT1 for further investigation.

**Author Contributions:** Conceptualization A.S., J.D., V.V. and P.V.; writing—review and editing A.S., J.D., V.V., K.C., K.T., J.J., J.K., P.V. and S.S.; writing—original draft preparation A.S., J.D., J.K., V.V. and K.C.; data curation and investigation A.S., J.D., J.K., J.J. and S.S.; funding acquisition P.V., T.H., V.V., P.K., L.V. and K.C.; project administration A.S., P.V., L.V. and V.V.; supervision J.D., V.V. and P.V. All authors have read and agreed to the published version of the manuscript.

**Funding:** This project was supported by the Grant Agency of the Ministry of Health of the Czech Republic (AZV NV18-03-00199); The Grant Agency of the Czech Republic (GACR 22-05942S); This work was supported by the Cooperatio Program, research area Oncology and Haematology; A.S., K.C., J.J. and V.V. have been supported by funds from the Grant Schemes project at UK, reg. n. CZ.02.2.69/0.0/0.0/19_073 project n. START/MED/052.

**Institutional Review Board Statement:** All of the experiment protocols in the manuscript involving human data were in accordance with national/international/institutional guidelines or the Declaration of Helsinki.

**Informed Consent Statement:** Written informed consent was obtained from the patients to publish this paper.

**Data Availability Statement:** The data presented in this study are available on request from the corresponding author. The data are not publicly available due to privacy and ethical restrictions.

**Conflicts of Interest:** The authors declare no conflict of interest.

## References


1. Cernat, L.; Blaj, C.; Jackstadt, R.; Brandl, L.; Engel, J.; Hermeking, H.; Jung, A.; Kirchner, T.; Horst, D. Colorectal Cancers Mimic Structural Organization of Normal Colonic Crypts. *PLoS ONE* **2014**, *9*, e104284. [CrossRef] [PubMed]
2. Bray, F.; Ferlay, J.; Soerjomataram, I.; Siegel, R.L.; Torre, L.A.; Jemal, A. Global cancer statistics 2018: GLOBOCAN estimates of incidence and mortality worldwide for 36 cancers in 185 countries. *CA Cancer J. Clin.* **2018**, *68*, 394–424. [CrossRef] [PubMed]
3. Rawla, P.; Sunkara, T.; Barsouk, A. Epidemiology of colorectal cancer: Incidence, mortality, survival, and risk factors. *Prz. Gastroenterol.* **2019**, *14*, 89–103. [CrossRef] [PubMed]
4. Loeve, F.; Boer, R.; Zauber, A.G.; Van Ballegooijen, M.; Van Oortmarssen, G.J.; Winawer, S.J.; Habbema, J.D. National Polyp Study data: Evidence for regression of adenomas. *Int. J. Cancer* **2004**, *111*, 633–639. [CrossRef] [PubMed]
5. Carvalho, B.; Sillars-Hardebol, A.H.; Postma, C.; Mongera, S.; Terhaar Sive Droste, J.; Obulkasim, A.; van de Wiel, M.; van Criekinge, W.; Ylstra, B.; Fijneman, R.J.; et al. Colorectal adenoma to carcinoma progression is accompanied by changes in gene expression associated with ageing, chromosomal instability, and fatty acid metabolism. *Cell. Oncol.* **2012**, *35*, 53–63. [CrossRef] [PubMed]

6. Muller, M.F.; Ibrahim, A.E.; Arends, M.J. Molecular pathological classification of colorectal cancer. *Virchows Arch.* **2016**, *469*, 125–134. [CrossRef]
7. Kwong, L.N.; Dove, W.F. APC and its modifiers in colon cancer. *Adv. Exp. Med. Biol.* **2009**, *656*, 85–106. [CrossRef]
8. Cross, W.; Kovac, M.; Mustonen, V.; Temko, D.; Davis, H.; Baker, A.M.; Biswas, S.; Arnold, R.; Chegwidden, L.; Gatenbee, C.; et al. The evolutionary landscape of colorectal tumorigenesis. *Nat. Ecol. Evol.* **2018**, *2*, 1661–1672. [CrossRef]
9. Heald, B.; Mester, J.; Rybicki, L.; Orloff, M.S.; Burke, C.A.; Eng, C. Frequent Gastrointestinal Polyps and Colorectal Adenocarcinomas in a Prospective Series of PTEN Mutation Carriers. *Gastroenterology* **2010**, *139*, 1927–1933. [CrossRef]
10. Tyagi, A.; Sharma, A.K.; Damodaran, C. A Review on Notch Signaling and Colorectal Cancer. *Cells* **2020**, *9*, 1549. [CrossRef]
11. Jungwirth, J.; Urbanova, M.; Boot, A.; Hosek, P.; Bendova, P.; Siskova, A.; Svec, J.; Kment, M.; Tumova, D.; Summerova, S.; et al. Mutational analysis of driver genes defines the colorectal adenoma: In situ carcinoma transition. *Sci. Rep.* **2022**, *12*, 2570. [CrossRef] [PubMed]
12. Pino, M.S.; Chung, D.C. The chromosomal instability pathway in colon cancer. *Gastroenterology* **2010**, *138*, 2059–2072. [CrossRef] [PubMed]
13. Siskova, A.; Cervena, K.; Kral, J.; Hucl, T.; Vodicka, P.; Vymetalkova, V. Colorectal Adenomas-Genetics and Searching for New Molecular Screening Biomarkers. *Int. J. Mol. Sci.* **2020**, *21*, 3260. [CrossRef] [PubMed]
14. Fernández, L.C.; Torres, M.; Real, F.X. Somatic mosaicism: On the road to cancer. *Nat. Rev. Cancer* **2016**, *16*, 43–55. [CrossRef]
15. Statello, L.; Guo, C.-J.; Chen, L.-L.; Huarte, M. Gene regulation by long non-coding RNAs and its biological functions. *Nat. Rev. Mol. Cell Biol.* **2021**, *22*, 96–118. [CrossRef]
16. Ji, P.; Diederichs, S.; Wang, W.; Böing, S.; Metzger, R.; Schneider, P.M.; Tidow, N.; Brandt, B.; Buerger, H.; Bulk, E.; et al. MALAT-1, a novel noncoding RNA, and thymosin β4 predict metastasis and survival in early-stage non-small cell lung cancer. *Oncogene* **2003**, *22*, 8031–8041. [CrossRef]
17. Goyal, B.; Yadav, S.R.M.; Awasthee, N.; Gupta, S.; Kunnumakkara, A.B.; Gupta, S.C. Diagnostic, prognostic, and therapeutic significance of long non-coding RNA MALAT1 in cancer. *Biochim. Biophys. Acta Rev. Cancer* **2021**, *1875*, 188502. [CrossRef]
18. Shi, Z.Z.; Zhang, Y.M.; Shang, L.; Hao, J.J.; Zhang, T.T.; Wang, B.S.; Liang, J.W.; Chen, X.; Zhang, Y.; Wang, G.Q.; et al. Genomic profiling of rectal adenoma and carcinoma by array-based comparative genomic hybridization. *BMC Med. Genom.* **2012**, *5*, 52. [CrossRef]
19. Douglas, E.J.; Fiegler, H.; Rowan, A.; Halford, S.; Bicknell, D.C.; Bodmer, W.; Tomlinson, I.P.; Carter, N.P. Array comparative genomic hybridization analysis of colorectal cancer cell lines and primary carcinomas. *Cancer Res.* **2004**, *64*, 4817–4825. [CrossRef]
20. Rosner, M.; Hanneder, M.; Siegel, N.; Valli, A.; Hengstschlager, M. The tuberous sclerosis gene products hamartin and tuberin are multifunctional proteins with a wide spectrum of interacting partners. *Mutat. Res.* **2008**, *658*, 234–246. [CrossRef]
21. Slattery, M.L.; Herrick, J.S.; Lundgreen, A.; Fitzpatrick, F.A.; Curtin, K.; Wolff, R.K. Genetic variation in a metabolic signaling pathway and colon and rectal cancer risk: mTOR, PTEN, STK11, RPKAA1, PRKAG2, TSC1, TSC2, PI3K and Akt1. *Carcinogenesis* **2010**, *31*, 1604–1611. [CrossRef] [PubMed]
22. Jiang, K.; Liu, H.; Xie, D.; Xiao, Q. Differentially expressed genes ASPN, COL1A1, FN1, VCAN and MUC5AC are potential prognostic biomarkers for gastric cancer. *Oncol. Lett.* **2019**, *17*, 3191–3202. [CrossRef] [PubMed]
23. Lin, P.; Tian, P.; Pang, J.; Lai, L.; He, G.; Song, Y.; Zheng, Y. Clinical significance of COL1A1 and COL1A2 expression levels in hypopharyngeal squamous cell carcinoma. *Oncol. Lett.* **2020**, *20*, 803–809. [CrossRef]
24. Zou, X.; Feng, B.; Dong, T.; Yan, G.; Tan, B.; Shen, H.; Huang, A.; Zhang, X.; Zhang, M.; Yang, P.; et al. Up-regulation of type I collagen during tumorigenesis of colorectal cancer revealed by quantitative proteomic analysis. *J. Proteom.* **2013**, *94*, 473–485. [CrossRef] [PubMed]
25. Huang, T.; Zhou, Y.; Cheng, A.S.; Yu, J.; To, K.F.; Kang, W. NOTCH receptors in gastric and other gastrointestinal cancers: Oncogenes or tumor suppressors? *Mol. Cancer* **2016**, *15*, 80. [CrossRef]
26. Arcaroli, J.J.; Tai, W.M.; McWilliams, R.; Bagby, S.; Blatchford, P.J.; Varella-Garcia, M.; Purkey, A.; Quackenbush, K.S.; Song, E.-K.; Pitts, T.M.; et al. A NOTCH1 gene copy number gain is a prognostic indicator of worse survival and a predictive biomarker to a Notch1 targeting antibody in colorectal cancer. *Int. J. Cancer* **2016**, *138*, 195–205. [CrossRef] [PubMed]
27. Dokumcu, K.; Simonian, M.; Farahani, R.M. miR4673 improves fitness profile of neoplastic cells by induction of autophagy. *Cell Death Dis.* **2018**, *9*, 1068. [CrossRef]
28. Thomis, D.C.; Berg, L.J. The role of Jak3 in lymphoid development, activation, and signaling. *Curr. Opin. Immunol.* **1997**, *9*, 541–547. [CrossRef]
29. Bastepe, M.; Juppner, H. GNAS locus and pseudohypoparathyroidism. *Horm. Res.* **2005**, *63*, 65–74. [CrossRef]
30. Liu, C.; McKeone, D.M.; Walker, N.I.; Bettington, M.L.; Leggett, B.A.; Whitehall, V.L.J. GNAS mutations are present in colorectal traditional serrated adenomas, serrated tubulovillous adenomas and serrated adenocarcinomas with adverse prognostic features. *Histopathology* **2017**, *70*, 1079–1088. [CrossRef]
31. Zauber, P.; Marotta, S.P.; Sabbath-Solitare, M. GNAS gene mutation may be present only transiently during colorectal tumorigenesis. *Int. J. Mol. Epidemiol. Genet.* **2016**, *7*, 24–31. [PubMed]
32. Vashisht, A.A.; Wohlschlegel, J.A. Chapter 8—Role of Human Xeroderma Pigmentosum Group D (XPD) Helicase in Various Cellular Pathways. In *Helicases from All Domains of Life*; Tuteja, R., Ed.; Academic Press: Cambridge, MA, USA, 2019; pp. 125–139. [CrossRef]

33. Zhang, X.; Hamblin, M.H.; Yin, K.J. The long noncoding RNA Malat1: Its physiological and pathophysiological functions. *RNA Biol.* **2017**, *14*, 1705–1714. [CrossRef]
34. Gutschner, T.; Hammerle, M.; Diederichs, S. MALAT1—A paradigm for long noncoding RNA function in cancer. *J. Mol. Med.* **2013**, *91*, 791–801. [CrossRef]
35. Biswas, S.; Thomas, A.A.; Chen, S.; Aref-Eshghi, E.; Feng, B.; Gonder, J.; Sadikovic, B.; Chakrabarti, S. MALAT1: An Epigenetic Regulator of Inflammation in Diabetic Retinopathy. *Sci. Rep.* **2018**, *8*, 6526. [CrossRef] [PubMed]
36. Gutschner, T.; Hämmerle, M.; Eißmann, M.; Hsu, J.; Kim, Y.; Hung, G.; Revenko, A.; Arun, G.; Stentrup, M.; Groß, M.; et al. The Noncoding RNA *MALAT1* Is a Critical Regulator of the Metastasis Phenotype of Lung Cancer Cells. *Cancer Res.* **2013**, *73*, 1180–1189. [CrossRef] [PubMed]
37. Meseure, D.; Vacher, S.; Lallemand, F.; Alsibai, K.D.; Hatem, R.; Chemlali, W.; Nicolas, A.; De Koning, L.; Pasmant, E.; Callens, C.; et al. Prognostic value of a newly identified MALAT1 alternatively spliced transcript in breast cancer. *Br. J. Cancer* **2016**, *114*, 1395–1404. [CrossRef] [PubMed]
38. Li, C.; Cui, Y.; Liu, L.F.; Ren, W.B.; Li, Q.Q.; Zhou, X.; Li, Y.L.; Li, Y.; Bai, X.Y.; Zu, X.B. High Expression of Long Noncoding RNA MALAT1 Indicates a Poor Prognosis and Promotes Clinical Progression and Metastasis in Bladder Cancer. *Clin. Genitourin. Cancer* **2017**, *15*, 570–576. [CrossRef]
39. Yang, L.; Bai, H.S.; Deng, Y.; Fan, L. High MALAT1 expression predicts a poor prognosis of cervical cancer and promotes cancer cell growth and invasion. *Eur. Rev. Med. Pharmacol. Sci.* **2015**, *19*, 3187–3193.
40. Konishi, H.; Ichikawa, D.; Yamamoto, Y.; Arita, T.; Shoda, K.; Hiramoto, H.; Hamada, J.; Itoh, H.; Fujita, Y.; Komatsu, S.; et al. Plasma level of metastasis-associated lung adenocarcinoma transcript 1 is associated with liver damage and predicts development of hepatocellular carcinoma. *Cancer Sci.* **2016**, *107*, 149–154. [CrossRef]
41. Ji, Q.; Zhang, L.; Liu, X.; Zhou, L.; Wang, W.; Han, Z.; Sui, H.; Tang, Y.; Wang, Y.; Liu, N.; et al. Long non-coding RNA MALAT1 promotes tumour growth and metastasis in colorectal cancer through binding to SFPQ and releasing oncogene PTBP2 from SFPQ/PTBP2 complex. *Br. J. Cancer* **2014**, *111*, 736–748. [CrossRef]
42. Xiong, M.; Wu, M.; Dan, P.; Huang, W.; Chen, Z.; Ke, H.; Chen, Z.; Song, W.; Zhao, Y.; Xiang, A.P.; et al. LncRNA DANCR represses Doxorubicin-induced apoptosis through stabilizing MALAT1 expression in colorectal cancer cells. *Cell Death Dis.* **2021**, *12*, 24. [CrossRef] [PubMed]
43. Xu, C.; Yang, M.; Tian, J.; Wang, X.; Li, Z. MALAT-1: A long non-coding RNA and its important 3′ end functional motif in colorectal cancer metastasis. *Int. J. Oncol.* **2011**, *39*, 169–175. [CrossRef] [PubMed]
44. Li, J.; Cui, Z.; Li, H.; Lv, X.; Gao, M.; Yang, Z.; Bi, Y.; Zhang, Z.; Wang, S.; Zhou, B.; et al. Clinicopathological and prognostic significance of long noncoding RNA MALAT1 in human cancers: A review and meta-analysis. *Cancer Cell Int.* **2018**, *18*, 109. [CrossRef] [PubMed]
45. Peng, Y.; Croce, C.M. The role of MicroRNAs in human cancer. *Signal Transduct. Target. Ther.* **2016**, *1*, 15004. [CrossRef] [PubMed]
46. Feng, C.; Zhao, Y.; Li, Y.; Zhang, T.; Ma, Y.; Liu, Y. LncRNA MALAT1 Promotes Lung Cancer Proliferation and Gefitinib Resistance by Acting as a miR-200a Sponge. *Arch. De Bronconeumol.* **2019**, *55*, 627–633. [CrossRef]
47. Xu, Y.; Zhang, X.; Hu, X.; Zhou, W.; Zhang, P.; Zhang, J.; Yang, S.; Liu, Y. The effects of lncRNA MALAT1 on proliferation, invasion and migration in colorectal cancer through regulating SOX9. *Mol. Med.* **2018**, *24*, 52. [CrossRef]
48. Su, K.; Wang, N.; Shao, Q.; Liu, H.; Zhao, B.; Ma, S. The role of a ceRNA regulatory network based on lncRNA MALAT1 site in cancer progression. *Biomed. Pharmacother.* **2021**, *137*, 111389. [CrossRef]
49. Ji, Q.; Cai, G.; Liu, X.; Zhang, Y.; Wang, Y.; Zhou, L.; Sui, H.; Li, Q. MALAT1 regulates the transcriptional and translational levels of proto-oncogene RUNX2 in colorectal cancer metastasis. *Cell Death Dis.* **2019**, *10*, 378. [CrossRef]
50. Chaleshi, V.; Irani, S.; Alebouyeh, M.; Mirfakhraie, R.; Aghdaei, H.A. Association of lncRNA-p53 regulatory network (lincRNA-p21, lincRNA-ROR and MALAT1) and p53 with the clinicopathological features of colorectal primary lesions and tumors. *Oncol. Lett.* **2020**, *19*, 3937–3949. [CrossRef]
51. Aisner, D.L.; Nguyen, T.T.; Paskulin, D.D.; Le, A.T.; Haney, J.; Schulte, N.; Chionh, F.; Hardingham, J.; Mariadason, J.; Tebbutt, N.; et al. ROS1 and ALK fusions in colorectal cancer, with evidence of intratumoral heterogeneity for molecular drivers. *Mol. Cancer Res.* **2014**, *12*, 111–118. [CrossRef]
52. Oneyama, C.; Yoshikawa, Y.; Ninomiya, Y.; Iino, T.; Tsukita, S.; Okada, M. Fer tyrosine kinase oligomer mediates and amplifies Src-induced tumor progression. *Oncogene* **2016**, *35*, 501–512. [CrossRef] [PubMed]
53. Medico, E.; Russo, M.; Picco, G.; Cancelliere, C.; Valtorta, E.; Corti, G.; Buscarino, M.; Isella, C.; Lamba, S.; Martinoglio, B.; et al. The molecular landscape of colorectal cancer cell lines unveils clinically actionable kinase targets. *Nat. Commun.* **2015**, *6*, 7002. [CrossRef] [PubMed]
54. Zhang, L.; Shay, J.W. Multiple Roles of APC and its Therapeutic Implications in Colorectal Cancer. *J. Natl. Cancer Inst.* **2017**, *109*, djw332. [CrossRef] [PubMed]
55. Kohonen-Corish, M.R.J.; Sigglekow, N.D.; Susanto, J.; Chapuis, P.H.; Bokey, E.L.; Dent, O.F.; Chan, C.; Lin, B.P.C.; Seng, T.J.; Laird, P.W.; et al. Promoter methylation of the mutated in colorectal cancer gene is a frequent early event in colorectal cancer. *Oncogene* **2007**, *26*, 4435–4441. [CrossRef]
56. Liu, Y.; Wang, G.; Liang, Z.; Mei, Z.; Wu, T.; Cui, A.; Liu, C.; Cui, L. Lysyl oxidase: A colorectal cancer biomarker of lung and hepatic metastasis. *Thorac. Cancer* **2018**, *9*, 785–793. [CrossRef]

57. Hirsch, D.; Camps, J.; Varma, S.; Kemmerling, R.; Stapleton, M.; Ried, T.; Gaiser, T. A new whole genome amplification method for studying clonal evolution patterns in malignant colorectal polyps. *Genes. Chromosomes Cancer* **2012**, *51*, 490–500. [CrossRef]
58. Hermsen, M.; Postma, C.; Baak, J.; Weiss, M.; Rapallo, A.; Sciutto, A.; Roemen, G.; Arends, J.W.; Williams, R.; Giaretti, W.; et al. Colorectal adenoma to carcinoma progression follows multiple pathways of chromosomal instability. *Gastroenterology* **2002**, *123*, 1109–1119. [CrossRef]
59. Paraskeva, C.; Buckle, B.G.; Sheer, D.; Wigley, C.B. The isolation and characterization of colorectal epithelial cell lines at different stages in malignant transformation from familial polyposis coli patients. *Int. J. Cancer* **1984**, *34*, 49–56. [CrossRef]
60. Longy, M.; Saura, R.; Schouler, L.; Mauhin, C.; Goussot, J.F.; Grison, O.; Couzigou, P. Chromosomal analysis of colonic adenomatous polyps. *Cancer Genet. Cytogenet.* **1990**, *49*, 249–257. [CrossRef]
61. Bomme, L.; Bardi, G.; Pandis, N.; Fenger, C.; Kronborg, O.; Heim, S. Cytogenetic analysis of colorectal adenomas: Karyotypic comparisons of synchronous tumors. *Cancer Genet. Cytogenet.* **1998**, *106*, 66–71. [CrossRef]
62. Ahnen, D.J.; Wade, S.W.; Jones, W.F.; Sifri, R.; Mendoza Silveiras, J.; Greenamyer, J.; Guiffre, S.; Axilbund, J.; Spiegel, A.; You, Y.N. The increasing incidence of young-onset colorectal cancer: A call to action. *Mayo. Clin. Proc.* **2014**, *89*, 216–224. [CrossRef] [PubMed]
63. Loomans-Kropp, H.A.; Umar, A. Increasing Incidence of Colorectal Cancer in Young Adults. *J. Cancer Epidemiol.* **2019**, *2019*, 9841295. [CrossRef] [PubMed]
64. Mieulet, V.; Lamb, R.F. Tuberous sclerosis complex: Linking cancer to metabolism. *Trends Mol. Med.* **2010**, *16*, 329–335. [CrossRef] [PubMed]
65. Johnson, C.E.; Dunlop, E.A.; Seifan, S.; McCann, H.D.; Hay, T.; Parfitt, G.J.; Jones, A.T.; Giles, P.J.; Shen, M.H.; Sampson, J.R.; et al. Loss of tuberous sclerosis complex 2 sensitizes tumors to nelfinavir-bortezomib therapy to intensify endoplasmic reticulum stress-induced cell death. *Oncogene* **2018**, *37*, 5913–5925. [CrossRef]
66. Huang, Q.; Li, F.; Hu, H.; Fang, Z.; Gao, Z.; Xia, G.; Ng, W.L.; Khodadadi-Jamayran, A.; Chen, T.; Deng, J.; et al. Loss of TSC1/TSC2 sensitizes immune checkpoint blockade in non-small cell lung cancer. *Sci. Adv.* **2022**, *8*, eabi9533. [CrossRef]
67. Sansregret, L.; Vanhaesebroeck, B.; Swanton, C. Determinants and clinical implications of chromosomal instability in cancer. *Nat. Rev. Clin. Oncol.* **2018**, *15*, 139–150. [CrossRef]
68. Cheung, S.; Shaw, C.; Scott, D.; Patel, A.; Sahoo, T.; Bacino, C.; Pursley, A.; Li, J.; Erickson, R.; Gropman, A.; et al. Microarray-based CGH detects chromosomal mosaicism not revealed by conventional cytogenetics. *Am. J. Med. Genetics. Part A* **2007**, *143A*, 1679–1686. [CrossRef]

MDPI
St. Alban-Anlage 66
4052 Basel
Switzerland
Tel. +41 61 683 77 34
Fax +41 61 302 89 18
www.mdpi.com

*International Journal of Molecular Sciences* Editorial Office
E-mail: ijms@mdpi.com
www.mdpi.com/journal/ijms